VOLUME FIVE HUNDRED AND TWENTY ONE

METHODS IN ENZYMOLOGY

G Protein Coupled Receptors

Trafficking and Oligomerization

METHODS IN ENZYMOLOGY

VOLUME FIVE HUNDRED AND TWENTY ONE

METHODS IN ENZYMOLOGY

G Protein Coupled Receptors

Trafficking and Oligomerization

Edited by

P. MICHAEL CONN

Divisions of Reproductive Sciences and Neuroscience (ONPRC) & Departments of Pharmacology and Physiology Cell and Developmental Biology, and Obstetrics and Gynecology (OHSU)
Beaverton, OR, USA

[redacted]

ELSEVIER

AMSTERDAM • BOSTON • HEIDELBERG • LONDON
NEW YORK • OXFORD • PARIS • SAN DIEGO
SAN FRANCISCO • SINGAPORE • SYDNEY • TOKYO
Academic Press is an imprint of Elsevier

Academic Press is an imprint of Elsevier
525 B Street, Suite 1900, San Diego, CA 92101-4495, USA
225 Wyman Street, Waltham, MA 02451, USA
The Boulevard, Langford Lane, Kidlington, Oxford, OX51GB, UK
32, Jamestown Road, London NW1 7BY, UK
Radarweg 29, PO Box 211, 1000 AE Amsterdam, The Netherlands

First edition 2013

ISBN: 978-0-12-391862-8
ISSN: 0076-6879

Printed and bound in United States of America
13 14 15 16 11 10 9 8 7 6 5 4 3 2 1

CONTENTS

Contributors *xiii*
Preface *xix*
Volumes in Series *xxi*

Section 1
GPCR Trafficking

1. Therapeutic Rescue of Misfolded/Mistrafficked Mutants: Automation-Friendly High-Throughput Assays for Identification of Pharmacoperone Drugs of GPCRs **3**

David C. Smithson, Jo Ann Janovick, and P. Michael Conn

1. Introduction 4
2. Choosing Pharmacoperone Model Systems 5
3. Selection of Endpoint Measures 7
4. Assay Automation 11
5. Data Analysis 12
6. Hit Follow-Up Experiments 12
7. Conclusions 14

Acknowledgments 14
References 14

2. Trafficking of the Follitropin Receptor **17**

Alfredo Ulloa-Aguirre, James A. Dias, George Bousfield, Ilpo Huhtaniemi, and Eric Reiter

1. Introduction 18
2. Outward Trafficking Defective FSHR Mutants. Studying Plasma Membrane Expression of the FSHR 23
3. Studying Oligomerization of Intracellular and Cell Surface-Expressed FSHRs 29
4. Phosphorylation, Internalization, and Recycling of the FSHR (Downward Trafficking) 33

Acknowledgments 41
References 41

3. Single-Molecule Imaging Technique to Study the Dynamic Regulation of GPCR Function at the Plasma Membrane **47**

B. E. Snaar-Jagalska, A. Cambi, T. Schmidt, and S. de Keijzer

1. Introduction 48

2. Labeling of GPCRs and Sample Preparation for SPT 48
3. Single-Molecule Imaging 56
4. Data Analysis 61
5. Concluding Remarks 64
Acknowledgments 65
References 65

4. GPCR Oligomerization and Receptor Trafficking **69**

Richard J. Ward, Tian-Rui Xu, and Graeme Milligan

1. Introduction 70
2. GPCR Expression Using the Flp-In™ T-Rex™ System 72
3. Detecting GPCR Internalization by Fluorescent Microscopy: The Orexin OX_1 Receptor 74
4. SNAP–CLIP Tagging 75
5. Detecting GPCR Oligomerization by other Resonance Energy Transfer Techniques 82
6. Biotinylation Studies 84
7. ER Trapping, Pharmacological Chaperones, and the Use of Engineered Synthetic Ligands 87
Acknowledgments 89
References 89

5. β-Arrestins and G Protein-Coupled Receptor Trafficking **91**

Dong Soo Kang, Xufan Tian, and Jeffrey L. Benovic

1. Introduction 92
2. Arrestin Expression 92
3. Assays to Measure GPCR Trafficking 97
4. Evaluating the Role of β-Arrestins in GPCR Trafficking 102
5. Summary 103
Acknowledgments 105
References 105

6. Tracking Cell Surface Mobility of GPCRs Using α-Bungarotoxin-Linked Fluorophores **109**

Saad Hannan, Megan E. Wilkins, Philip Thomas, and Trevor G. Smart

1. Introduction 110
2. Methodology 113
3. Validating the BBS Tag 117
4. Experimental Applications for the BBS Tag 120

5. Image Analysis 123
6. Other Applications 126
7. Overview 128
References 128

7. Regulatory Mechanism of G Protein-Coupled Receptor Trafficking to the Plasma Membrane: A Role for mRNA Localization **131**

Kusumam Joseph, Eleanor K. Spicer, and Baby G. Tholanikunnel

1. Introduction 132
2. Purification and Identification of β_2-AR mRNA-Binding Proteins 134
3. Functional Characterization of β_2-AR mRNA-Binding Proteins in Receptor Expression and Function 137
4. Role of RNA-Binding Protein HuR in Receptor Trafficking to the Plasma Membrane 139
5. β_2-AR mRNA Localization in Cells 141
6. Concluding Remarks and Future Perspectives 146
Acknowledgments 147
References 147

8. Dissecting Trafficking and Signaling of Atypical Chemokine Receptors **151**

Elena Borroni, Cinzia Cancellieri, Massimo Locati, and Raffaella Bonecchi

1. Introduction 152
2. ACR Trafficking 153
3. Atypical Chemokine Receptor Signaling 158
4. Summary 166
References 167

Section 2
Trafficking Motifs

9. Systematic and Quantitative Analysis of G Protein-Coupled Receptor Trafficking Motifs **171**

Carl M. Hurt, Vincent K. Ho, and Timothy Angelotti

1. Introduction 172
2. Motif Screening by Immunofluorescent Staining 173
3. Analysis of Receptor Functionality 175
4. Biochemical Analysis of ER/Golgi Trafficking 179

5. Quantitative Analysis of GPCR Trafficking 181
6. Summary 184
Acknowledgments 186
References 186

10. Identification of Endoplasmic Reticulum Export Motifs for G Protein-Coupled Receptors 189

Guangyu Wu

1. Introduction 190
2. Experimental Approaches to Identify ER Export Motifs for GPCRs 191
3. Conclusions 199
Acknowledgment 200
References 200

11. Amino Acid Residues of G-Protein-Coupled Receptors Critical for Endoplasmic Reticulum Export and Trafficking 203

Motonao Nakamura, Daisuke Yasuda, Nobuaki Hirota, Teruyasu Yamamoto, Satoshi Yamaguchi, Takao Shimizu, and Teruyuki Nagamune

1. Introduction 204
2. Generation of Mutant GPCRs 205
3. Examination of Mutant GPCR Trafficking 205
4. ER Export of Mutant GPCRs by Specific Ligands 209
5. Functional Analysis of Surface-Trafficked Mutant GPCRs in Living Cells 210
6. Conclusion 214
Acknowledgments 215
References 215

Section 3
GPCR Oligomerization

12. G-Protein-Coupled Heteromers: Regulation in Disease 219

Ivone Gomes, Achla Gupta, and Lakshmi A. Devi

1. Introduction 220
2. Generation of Heteromer-Selective mAbs 221
3. ELISA for Detection of Receptor Heteromers 228
4. Immunofluorescence for Visualization of Receptor Heteromers 230
5. Immunoprecipitation and Western Blotting 232
6. Summary and Perspectives 235

Acknowledgment 236
References 236

13. Hetero-oligomerization and Specificity Changes of G Protein-Coupled Purinergic Receptors: Novel Insight into Diversification of Signal Transduction **239**

Tokiko Suzuki, Kazunori Namba, Natsumi Mizuno, and Hiroyasu Nakata

1. Introduction 240
2. Measurement of GPCR Dimerization 242
3. Receptor Pharmacology 250
4. Conclusion 255

References 256

14. Bimolecular Fluorescence Complementation Analysis of G Protein-Coupled Receptor Dimerization in Living Cells **259**

Karin F.K. Ejendal, Jason M. Conley, Chang-Deng Hu, and Val J. Watts

1. Introduction 260
2. Generation of GPCR–BiFC Fusion Proteins 262
3. Detection of GPCR Interactions using BiFC and Fluorescence Microscopy 266
4. Microscopic Detection of GPCR Interactions by mBiFC 270
5. Fluorometric Detection of GPCR Dimerization using BiFC and mBiFC 273
6. Summary 275

Acknowledgments 276
References 276

15. G Protein–Coupled Receptor Heterodimerization in the Brain **281**

Dasiel O. Borroto-Escuela, Wilber Romero-Fernandez, Pere Garriga, Francisco Ciruela, Manuel Narvaez, Alexander O. Tarakanov, Miklós Palkovits, Luigi F. Agnati, and Kjell Fuxe

1. Introduction 282
2. *In Situ* PLA for Demonstrating Receptor Heteromers and Their Receptor–Receptor Interactions in Brain Tissue 283
3. Brain Tissue Preparation 283
4. Proximity Probes: Conjugation of Oligonucleotides to Antibodies 287
5. PLA Reactions, Reagents, and Solutions 287
6. Quantitative PLA Image Analysis 289
7. Advantages and Disadvantages of the PLA Method 289
8. Application 291

Acknowledgments 293
References 293

16. Experimental Strategies for Studying G Protein-Coupled Receptor Homo- and Heteromerization with Radioligand Binding and Signal Transduction Methods 295

Roberto Maggio, Cristina Rocchi, and Marco Scarselli

1. Introduction 296
2. Radioligand Binding in Receptor Oligomerization 296
3. Signal Transduction in Receptor Oligomerization 302
4. Domain Swapping in Receptor Oligomerization 306
5. Concluding Remarks 307

References 308

17. Analysis of GPCR Dimerization Using Acceptor Photobleaching Resonance Energy Transfer Techniques 311

Marta Busnelli, Mario Mauri, Marco Parenti, and Bice Chini

1. Introduction 312
2. Resonance Energy Transfer 313
3. Fluorescent Resonance Energy Transfer 313
4. BRET 321
5. Conclusions 326

Acknowledgments 327
References 327

18. Techniques for the Discovery of GPCR-Associated Protein Complexes 329

Avais Daulat, Pascal Maurice, and Ralf Jockers

1. Introduction 330
2. TAP of GPCR and Its Associated Protein Complexes 331
3. Alternative Methodology to Perform TAP of GPCR-Associated Protein Complexes 337
4. Purification of GPCR-Associated Protein Complexes by Peptide Affinity Chromatography 339
5. Conclusion 343

Acknowledgments 344
References 344

19. Expression, Purification, and Analysis of G-Protein-Coupled Receptor Kinases **347**

Rachel Sterne-Marr, Alison I. Baillargeon, Kevin R. Michalski, and John J. G. Tesmer

1. Introduction 348
2. Expression and Purification of GRKs 349
3. GRK Functional Assays 357

Acknowledgments 364

References 364

20. Modern Methods to Investigate the Oligomerization of Glycoprotein Hormone Receptors (TSHR, LHR, FSHR) **367**

Marco Bonomi and Luca Persani

1. Introduction 368
2. Resonance Energy Transfer Techniques 369
3. Experimental Procedures 376

Acknowledgments 381

References 381

Author Index *385*

Subject Index *407*

CONTRIBUTORS

Luigi F. Agnati
IRCCS Lido, Venice, Italy

Timothy Angelotti
Department of Anesthesia/CCM, Stanford University Medical School, Stanford, California, USA

Alison I. Baillargeon
Department of Chemistry and Biochemistry, Siena College, Morrell Science Center, Loudonville, New York, USA

Jeffrey L. Benovic
Department of Biochemistry and Molecular Biology, Thomas Jefferson University, Philadelphia, Pennsylvania, USA

Raffaella Bonecchi
Humanitas Clinical and Research Center, and Department of Medical Biotechnologies, and Translational Medicine, Università degli Studi di Milano, Rozzano, Italy

Marco Bonomi
Lab of Experimental Endocrinology and Metabolism, and Division of Endocrine and Metabolic Diseases, Ospedale San Luca, IRCCS Istituto Auxologico Italiano, Milan, Italy

Elena Borroni
Humanitas Clinical and Research Center, and Department of Medical Biotechnologies and Translational Medicine, Università degli Studi di Milano, Rozzano, Italy

Dasiel O. Borroto-Escuela
Department of Neuroscience, Karolinska Institutet, Stockholm, Sweden

George Bousfield
Studium Consortium for Research and Training in Reproductive Sciences (sCORTS), Tours, France, and Department of Biological Sciences, Wichita State University, Wichita, Kansas, USA

Marta Busnelli
CNR Institute of Neuroscience, and Department of Medical Biotechnology and Translational Medicine, University of Milan, Milan, Italy

A. Cambi
Department of Tumor Immunology, Nijmegen Centre for Molecular Life Sciences, Radboud University Nijmegen Medical Centre, Nijmegen, and Nanobiophysics, MIRA Institute for Biomedical Technology and Technical medicine, University of Twente, Enschede, The Netherlands

Cinzia Cancellieri
Humanitas Clinical and Research Center, and Department of Medical Biotechnologies and Translational Medicine, Università degli Studi di Milano, Rozzano, Italy

Bice Chini
CNR Institute of Neuroscience, University of Milan, Milan, Italy

Francisco Ciruela
Unitat de Farmacologia, Departament Patologia i Terapèutica Experimental, Universitat de Barcelona, Barcelona, Spain

Jason M. Conley
Department of Medicinal Chemistry and Molecular Pharmacology, College of Pharmacy, Purdue University, West Lafayette, Indiana, USA

P. Michael Conn
Divisions of Reproductive Sciences and Neuroscience, Oregon National Primate Research Center, Beaverton, and Departments of Physiology and Pharmacology, Cell Biology and Human Development and Obstetrics and Gynecology, Oregon Health & Science University, Portland, Oregon, USA

Avais Daulat
INSERM, U1016, Institut Cochin; CNRS UMR 8104, and Université Paris Descartes, Paris, France

S. de Keijzer
Department of Tumor Immunology, Nijmegen Centre for Molecular Life Sciences, Radboud University Nijmegen Medical Centre, Nijmegen, The Netherlands

Lakshmi A. Devi
Department of Pharmacology and Systems Therapeutics, and The Friedman Brain Institute, Mount Sinai School of Medicine, New York, New York, USA

James A. Dias
Studium Consortium for Research and Training in Reproductive Sciences (sCORTS), Tours, France, and New York State Department of Health and Department of Biomedical Sciences, Wadsworth Center, School of Public Health, University at Albany, Albany, USA

Karin F.K. Ejendal
Department of Medicinal Chemistry and Molecular Pharmacology, College of Pharmacy, Purdue University, West Lafayette, Indiana, USA

Kjell Fuxe
Department of Neuroscience, Karolinska Institutet, Stockholm, Sweden

Pere Garriga
Departament d'Enginyeria Química, Universitat Politècnica de Catalunya, Barcelona, Spain

Ivone Gomes
Department of Pharmacology and Systems Therapeutics, Mount Sinai School of Medicine, New York, New York, USA

Achla Gupta
Department of Pharmacology and Systems Therapeutics, Mount Sinai School of Medicine, New York, New York, USA

Saad Hannan
Department of Neuroscience, Physiology and Pharmacology, University College London, London, United Kingdom

Nobuaki Hirota
Department of Biochemistry and Molecular Biology, Graduate School of Medicine, The University of Tokyo, Tokyo, Japan

Vincent K. Ho
Department of Anesthesia/CCM, Stanford University Medical School, Stanford, California, USA

Chang-Deng Hu
Department of Medicinal Chemistry and Molecular Pharmacology, College of Pharmacy, Purdue University, West Lafayette, Indiana, USA

Ilpo Huhtaniemi
Studium Consortium for Research and Training in Reproductive Sciences (sCORTS), Tours, France, and Institute of Reproductive and Developmental Biology, Imperial College London, London, United Kingdom

Carl M. Hurt
Department of Anesthesia/CCM, Stanford University Medical School, Stanford, California, USA

Jo Ann Janovick
Divisions of Reproductive Sciences and Neuroscience, Oregon National Primate Research Center, Beaverton, Oregon, USA

Ralf Jockers
INSERM, U1016, Institut Cochin; CNRS UMR 8104, and Université Paris Descartes, Paris, France

Kusumam Joseph
Department of Biochemistry and Molecular Biology, Medical University of South Carolina, Charleston, South Carolina, USA

Dong Soo Kang
Department of Biochemistry and Molecular Biology, Thomas Jefferson University, Philadelphia, Pennsylvania, USA

Massimo Locati
Humanitas Clinical and Research Center, and Department of Medical Biotechnologies and Translational Medicine, Università degli Studi di Milano, Rozzano, Italy

Roberto Maggio
Department of Biotechnological and Applied Clinical Sciences, University of L'Aquila, L'Aquila, Italy

Mario Mauri
Department of Experimental Medicine, University of Milan-Bicocca, Monza, Italy

Pascal Maurice*
INSERM, U1016, Institut Cochin; CNRS UMR 8104, and Université Paris Descartes, Paris, France

Kevin R. Michalski
Department of Chemistry and Biochemistry, Siena College, Morrell Science Center, Loudonville, New York, USA

Graeme Milligan
Molecular Pharmacology Group, Institute of Molecular, Cell and Systems Biology, College of Medical, Veterinary and Life Sciences, University of Glasgow, Glasgow, United Kingdom

Natsumi Mizuno
Department of Cellular Signaling, and Department of Pharmacotherapy of Lifestyle Related Diseases, Graduate School of Pharmaceutical Sciences, Tohoku University, Sendai, Japan

Teruyuki Nagamune
Department of Chemistry and Biotechnology, Graduate School of Engineering, and Center for NanoBio Integration, The University of Tokyo, Tokyo, Japan

Motonao Nakamura
Department of Biochemistry and Molecular Biology, Graduate School of Medicine, The University of Tokyo, Tokyo, Japan

Hiroyasu Nakata
Department of Brain Development and Neural Regeneration, Tokyo Metropolitan Institute of Medical Science, Tokyo, Japan

Kazunori Namba
Otolaryngology/Laboratory of Auditory Disorders, National Institute of Sensory Organs, National Tokyo Medical Center, Tokyo, Japan

Manuel Narvaez
Department of Physiology, School of Medicine, University of Málaga, Málaga, Spain

Miklós Palkovits
Human Brain Tissue Bank, Semmelweis University, Budapest, Hungary

Marco Parenti
Department of Experimental Medicine, University of Milan-Bicocca, Monza, Italy

Luca Persani
Lab of Experimental Endocrinology and Metabolism, and Division of Endocrine and Metabolic Diseases, Ospedale San Luca, IRCCS Istituto Auxologico Italiano, and Department of Medical Sciences, University of Milan, Milan, Italy

Eric Reiter
Studium Consortium for Research and Training in Reproductive Sciences (sCORTS), Tours; BIOS Group, INRA, Unité Physiologie de la Reproduction et des Comportements, Nouzilly; CNRS, Nouzilly, and Université François Rabelais, Tours, France

*Present address: FRE CNRS/URCA 3481, UFR Sciences Exactes et Naturelles, Reims, France

Cristina Rocchi
Department of Biotechnological and Applied Clinical Sciences, University of L'Aquila, L'Aquila, Italy

Wilber Romero-Fernandez
Department of Neuroscience, Karolinska Institutet, Stockholm, Sweden

Marco Scarselli
Department of Translational Research and of New Technologies, University of Pisa, Pisa, Italy

T. Schmidt
Physics of Life Processes, Leiden Institute of Physics, Leiden University, Leiden, The Netherlands

Takao Shimizu
Department of Biochemistry and Molecular Biology, Graduate School of Medicine, The University of Tokyo, Tokyo, Japan

Trevor G. Smart
Department of Neuroscience, Physiology and Pharmacology, University College London, London, United Kingdom

David C. Smithson
Oregon Translational Research and Drug Development Institute, Portland, Oregon, USA

B.E. Snaar-Jagalska
Cell Biology, Leiden Institute of Biology, Leiden University, Leiden, The Netherlands

Eleanor K. Spicer
Department of Biochemistry and Molecular Biology, Medical University of South Carolina, Charleston, South Carolina, USA

Rachel Sterne-Marr
Biology Department, Siena College, Morrell Science Center, Loudonville, New York, USA

Tokiko Suzuki
Department of Cellular Signaling, Graduate School of Pharmaceutical Sciences, Tohoku University, Sendai, Japan

Alexander O. Tarakanov
Russian Academy of Sciences, St. Petersburg Institute for Informatics and Automation, Saint Petersburg, Russia

John J.G. Tesmer
Life Sciences Institute and the Department of Pharmacology, University of Michigan, Ann Arbor, Michigan, USA

Baby G. Tholanikunnel
Department of Biochemistry and Molecular Biology, Medical University of South Carolina, Charleston, South Carolina, USA

Philip Thomas
Department of Neuroscience, Physiology and Pharmacology, University College London, London, United Kingdom

Xufan Tian
Department of Biochemistry and Molecular Biology, Thomas Jefferson University, Philadelphia, Pennsylvania, USA

Alfredo Ulloa-Aguirre
Studium Consortium for Research and Training in Reproductive Sciences (sCORTS), Tours, France, and Division of Reproductive Health, Research Center in Population Health, National Institute of Public Health, México D.F., Mexico

Richard J. Ward
Molecular Pharmacology Group, Institute of Molecular, Cell and Systems Biology, College of Medical, Veterinary and Life Sciences, University of Glasgow, Glasgow, United Kingdom

Val J. Watts
Department of Medicinal Chemistry and Molecular Pharmacology, College of Pharmacy, Purdue University, West Lafayette, Indiana, USA

Megan E. Wilkins
Department of Neuroscience, Physiology and Pharmacology, University College London, London, United Kingdom

Guangyu Wu
Department of Pharmacology and Toxicology, Georgia Health Sciences University, Augusta, Georgia, USA

Tian-Rui Xu
Molecular Pharmacology Group, Institute of Molecular, Cell and Systems Biology, College of Medical, Veterinary and Life Sciences, University of Glasgow, Glasgow, United Kingdom

Satoshi Yamaguchi
Department of Chemistry and Biotechnology, Graduate School of Engineering, and Center for NanoBio Integration, The University of Tokyo, Tokyo, Japan

Teruyasu Yamamoto
Department of Chemistry and Biotechnology, Graduate School of Engineering, and Center for NanoBio Integration, The University of Tokyo, Tokyo, Japan

Daisuke Yasuda
Department of Biochemistry and Molecular Biology, Graduate School of Medicine, The University of Tokyo, Tokyo, Japan

PREFACE

G protein-coupled receptors (GPCRs) constitute the largest family of validated drug targets; mutations in GPCRs are the underlying cause of more than 30 diseases. These plasma membrane proteins are utilized by cells to mediate responses to sensory stimuli, hormones, and neurotransmitters. Some estimates are that as much at 4% of the human genome may be reserved for GPCRs; this is testimony to the large number of uses to which nature has put these interesting and highly interactive molecules.

Understanding the relation between receptor structure and function frequently explains the underlying pathology of disease and presents therapeutic and prophylactic opportunities. Accordingly, this volume provides descriptions of the range of methods used to analyze these important signal transducers and the authors explain how these methods are able to provide important biological insights.

Authors were selected based on research contributions in the area about which they have written and based on their ability to describe their methodological contribution in a clear and reproducible way. They have been encouraged to make use of graphics and comparisons to other methods, and to provide tricks and approaches not revealed in prior publications that make it possible to adapt methods to other systems.

The editor wants to express appreciation to the contributors for providing their contributions in a timely fashion, to the senior editors for guidance, and to the staff at Academic Press for helpful input.

P. Michael Conn
Portland, Oregon, USA
March 2012

METHODS IN ENZYMOLOGY

VOLUME I. Preparation and Assay of Enzymes
Edited by SIDNEY P. COLOWICK AND NATHAN O. KAPLAN

VOLUME II. Preparation and Assay of Enzymes
Edited by SIDNEY P. COLOWICK AND NATHAN O. KAPLAN

VOLUME III. Preparation and Assay of Substrates
Edited by SIDNEY P. COLOWICK AND NATHAN O. KAPLAN

VOLUME IV. Special Techniques for the Enzymologist
Edited by SIDNEY P. COLOWICK AND NATHAN O. KAPLAN

VOLUME V. Preparation and Assay of Enzymes
Edited by SIDNEY P. COLOWICK AND NATHAN O. KAPLAN

VOLUME VI. Preparation and Assay of Enzymes (*Continued*)
Preparation and Assay of Substrates
Special Techniques
Edited by SIDNEY P. COLOWICK AND NATHAN O. KAPLAN

VOLUME VII. Cumulative Subject Index
Edited by SIDNEY P. COLOWICK AND NATHAN O. KAPLAN

VOLUME VIII. Complex Carbohydrates
Edited by ELIZABETH F. NEUFELD AND VICTOR GINSBURG

VOLUME IX. Carbohydrate Metabolism
Edited by WILLIS A. WOOD

VOLUME X. Oxidation and Phosphorylation
Edited by RONALD W. ESTABROOK AND MAYNARD E. PULLMAN

VOLUME XI. Enzyme Structure
Edited by C. H. W. HIRS

VOLUME XII. Nucleic Acids (Parts A and B)
Edited by LAWRENCE GROSSMAN AND KIVIE MOLDAVE

VOLUME XIII. Citric Acid Cycle
Edited by J. M. LOWENSTEIN

VOLUME XIV. Lipids
Edited by J. M. LOWENSTEIN

Volume XV. Steroids and Terpenoids
Edited by Raymond B. Clayton

Volume XVI. Fast Reactions
Edited by Kenneth Kustin

Volume XVII. Metabolism of Amino Acids and Amines (Parts A and B)
Edited by Herbert Tabor and Celia White Tabor

Volume XVIII. Vitamins and Coenzymes (Parts A, B, and C)
Edited by Donald B. McCormick and Lemuel D. Wright

Volume XIX. Proteolytic Enzymes
Edited by Gertrude E. Perlmann and Laszlo Lorand

Volume XX. Nucleic Acids and Protein Synthesis (Part C)
Edited by Kivie Moldave and Lawrence Grossman

Volume XXI. Nucleic Acids (Part D)
Edited by Lawrence Grossman and Kivie Moldave

Volume XXII. Enzyme Purification and Related Techniques
Edited by William B. Jakoby

Volume XXIII. Photosynthesis (Part A)
Edited by Anthony San Pietro

Volume XXIV. Photosynthesis and Nitrogen Fixation (Part B)
Edited by Anthony San Pietro

Volume XXV. Enzyme Structure (Part B)
Edited by C. H. W. Hirs and Serge N. Timasheff

Volume XXVI. Enzyme Structure (Part C)
Edited by C. H. W. Hirs and Serge N. Timasheff

Volume XXVII. Enzyme Structure (Part D)
Edited by C. H. W. Hirs and Serge N. Timasheff

Volume XXVIII. Complex Carbohydrates (Part B)
Edited by Victor Ginsburg

Volume XXIX. Nucleic Acids and Protein Synthesis (Part E)
Edited by Lawrence Grossman and Kivie Moldave

Volume XXX. Nucleic Acids and Protein Synthesis (Part F)
Edited by Kivie Moldave and Lawrence Grossman

Volume XXXI. Biomembranes (Part A)
Edited by Sidney Fleischer and Lester Packer

VOLUME XXXII. Biomembranes (Part B)
Edited by SIDNEY FLEISCHER AND LESTER PACKER

VOLUME XXXIII. Cumulative Subject Index Volumes I-XXX
Edited by MARTHA G. DENNIS AND EDWARD A. DENNIS

VOLUME XXXIV. Affinity Techniques (Enzyme Purification: Part B)
Edited by WILLIAM B. JAKOBY AND MEIR WILCHEK

VOLUME XXXV. Lipids (Part B)
Edited by JOHN M. LOWENSTEIN

VOLUME XXXVI. Hormone Action (Part A: Steroid Hormones)
Edited by BERT W. O'MALLEY AND JOEL G. HARDMAN

VOLUME XXXVII. Hormone Action (Part B: Peptide Hormones)
Edited by BERT W. O'MALLEY AND JOEL G. HARDMAN

VOLUME XXXVIII. Hormone Action (Part C: Cyclic Nucleotides)
Edited by JOEL G. HARDMAN AND BERT W. O'MALLEY

VOLUME XXXIX. Hormone Action (Part D: Isolated Cells, Tissues, and Organ Systems)
Edited by JOEL G. HARDMAN AND BERT W. O'MALLEY

VOLUME XL. Hormone Action (Part E: Nuclear Structure and Function)
Edited by BERT W. O'MALLEY AND JOEL G. HARDMAN

VOLUME XLI. Carbohydrate Metabolism (Part B)
Edited by W. A. WOOD

VOLUME XLII. Carbohydrate Metabolism (Part C)
Edited by W. A. WOOD

VOLUME XLIII. Antibiotics
Edited by JOHN H. HASH

VOLUME XLIV. Immobilized Enzymes
Edited by KLAUS MOSBACH

VOLUME XLV. Proteolytic Enzymes (Part B)
Edited by LASZLO LORAND

VOLUME XLVI. Affinity Labeling
Edited by WILLIAM B. JAKOBY AND MEIR WILCHEK

VOLUME XLVII. Enzyme Structure (Part E)
Edited by C. H. W. HIRS AND SERGE N. TIMASHEFF

VOLUME XLVIII. Enzyme Structure (Part F)
Edited by C. H. W. HIRS AND SERGE N. TIMASHEFF

VOLUME XLIX. Enzyme Structure (Part G)
Edited by C. H. W. HIRS AND SERGE N. TIMASHEFF

VOLUME L. Complex Carbohydrates (Part C)
Edited by VICTOR GINSBURG

VOLUME LI. Purine and Pyrimidine Nucleotide Metabolism
Edited by PATRICIA A. HOFFEE AND MARY ELLEN JONES

VOLUME LII. Biomembranes (Part C: Biological Oxidations)
Edited by SIDNEY FLEISCHER AND LESTER PACKER

VOLUME LIII. Biomembranes (Part D: Biological Oxidations)
Edited by SIDNEY FLEISCHER AND LESTER PACKER

VOLUME LIV. Biomembranes (Part E: Biological Oxidations)
Edited by SIDNEY FLEISCHER AND LESTER PACKER

VOLUME LV. Biomembranes (Part F: Bioenergetics)
Edited by SIDNEY FLEISCHER AND LESTER PACKER

VOLUME LVI. Biomembranes (Part G: Bioenergetics)
Edited by SIDNEY FLEISCHER AND LESTER PACKER

VOLUME LVII. Bioluminescence and Chemiluminescence
Edited by MARLENE A. DELUCA

VOLUME LVIII. Cell Culture
Edited by WILLIAM B. JAKOBY AND IRA PASTAN

VOLUME LIX. Nucleic Acids and Protein Synthesis (Part G)
Edited by KIVIE MOLDAVE AND LAWRENCE GROSSMAN

VOLUME LX. Nucleic Acids and Protein Synthesis (Part H)
Edited by KIVIE MOLDAVE AND LAWRENCE GROSSMAN

VOLUME 61. Enzyme Structure (Part H)
Edited by C. H. W. HIRS AND SERGE N. TIMASHEFF

VOLUME 62. Vitamins and Coenzymes (Part D)
Edited by DONALD B. MCCORMICK AND LEMUEL D. WRIGHT

VOLUME 63. Enzyme Kinetics and Mechanism (Part A: Initial Rate and Inhibitor Methods)
Edited by DANIEL L. PURICH

VOLUME 64. Enzyme Kinetics and Mechanism
(Part B: Isotopic Probes and Complex Enzyme Systems)
Edited by DANIEL L. PURICH

VOLUME 65. Nucleic Acids (Part I)
Edited by LAWRENCE GROSSMAN AND KIVIE MOLDAVE

VOLUME 66. Vitamins and Coenzymes (Part E)
Edited by DONALD B. MCCORMICK AND LEMUEL D. WRIGHT

VOLUME 67. Vitamins and Coenzymes (Part F)
Edited by DONALD B. MCCORMICK AND LEMUEL D. WRIGHT

VOLUME 68. Recombinant DNA
Edited by RAY WU

VOLUME 69. Photosynthesis and Nitrogen Fixation (Part C)
Edited by ANTHONY SAN PIETRO

VOLUME 70. Immunochemical Techniques (Part A)
Edited by HELEN VAN VUNAKIS AND JOHN J. LANGONE

VOLUME 71. Lipids (Part C)
Edited by JOHN M. LOWENSTEIN

VOLUME 72. Lipids (Part D)
Edited by JOHN M. LOWENSTEIN

VOLUME 73. Immunochemical Techniques (Part B)
Edited by JOHN J. LANGONE AND HELEN VAN VUNAKIS

VOLUME 74. Immunochemical Techniques (Part C)
Edited by JOHN J. LANGONE AND HELEN VAN VUNAKIS

VOLUME 75. Cumulative Subject Index Volumes XXXI, XXXII, XXXIV–LX
Edited by EDWARD A. DENNIS AND MARTHA G. DENNIS

VOLUME 76. Hemoglobins
Edited by ERALDO ANTONINI, LUIGI ROSSI-BERNARDI, AND EMILIA CHIANCONE

VOLUME 77. Detoxication and Drug Metabolism
Edited by WILLIAM B. JAKOBY

VOLUME 78. Interferons (Part A)
Edited by SIDNEY PESTKA

VOLUME 79. Interferons (Part B)
Edited by SIDNEY PESTKA

VOLUME 80. Proteolytic Enzymes (Part C)
Edited by LASZLO LORAND

VOLUME 81. Biomembranes (Part H: Visual Pigments and Purple Membranes, I)
Edited by LESTER PACKER

VOLUME 82. Structural and Contractile Proteins (Part A: Extracellular Matrix)
Edited by LEON W. CUNNINGHAM AND DIXIE W. FREDERIKSEN

VOLUME 83. Complex Carbohydrates (Part D)
Edited by VICTOR GINSBURG

VOLUME 84. Immunochemical Techniques (Part D: Selected Immunoassays)
Edited by JOHN J. LANGONE AND HELEN VAN VUNAKIS

VOLUME 85. Structural and Contractile Proteins (Part B: The Contractile Apparatus and the Cytoskeleton)
Edited by DIXIE W. FREDERIKSEN AND LEON W. CUNNINGHAM

VOLUME 86. Prostaglandins and Arachidonate Metabolites
Edited by WILLIAM E. M. LANDS AND WILLIAM L. SMITH

VOLUME 87. Enzyme Kinetics and Mechanism (Part C: Intermediates, Stereo-chemistry, and Rate Studies)
Edited by DANIEL L. PURICH

VOLUME 88. Biomembranes (Part I: Visual Pigments and Purple Membranes, II)
Edited by LESTER PACKER

VOLUME 89. Carbohydrate Metabolism (Part D)
Edited by WILLIS A. WOOD

VOLUME 90. Carbohydrate Metabolism (Part E)
Edited by WILLIS A. WOOD

VOLUME 91. Enzyme Structure (Part I)
Edited by C. H. W. HIRS AND SERGE N. TIMASHEFF

VOLUME 92. Immunochemical Techniques (Part E: Monoclonal Antibodies and General Immunoassay Methods)
Edited by JOHN J. LANGONE AND HELEN VAN VUNAKIS

VOLUME 93. Immunochemical Techniques (Part F: Conventional Antibodies, Fc Receptors, and Cytotoxicity)
Edited by JOHN J. LANGONE AND HELEN VAN VUNAKIS

VOLUME 94. Polyamines
Edited by HERBERT TABOR AND CELIA WHITE TABOR

VOLUME 95. Cumulative Subject Index Volumes 61–74, 76–80
Edited by EDWARD A. DENNIS AND MARTHA G. DENNIS

VOLUME 96. Biomembranes [Part J: Membrane Biogenesis: Assembly and Targeting (General Methods; Eukaryotes)]
Edited by SIDNEY FLEISCHER AND BECCA FLEISCHER

VOLUME 97. Biomembranes [Part K: Membrane Biogenesis: Assembly and Targeting (Prokaryotes, Mitochondria, and Chloroplasts)]
Edited by SIDNEY FLEISCHER AND BECCA FLEISCHER

VOLUME 98. Biomembranes (Part L: Membrane Biogenesis: Processing and Recycling)
Edited by SIDNEY FLEISCHER AND BECCA FLEISCHER

VOLUME 99. Hormone Action (Part F: Protein Kinases)
Edited by JACKIE D. CORBIN AND JOEL G. HARDMAN

VOLUME 100. Recombinant DNA (Part B)
Edited by RAY WU, LAWRENCE GROSSMAN, AND KIVIE MOLDAVE

VOLUME 101. Recombinant DNA (Part C)
Edited by RAY WU, LAWRENCE GROSSMAN, AND KIVIE MOLDAVE

VOLUME 102. Hormone Action (Part G: Calmodulin and Calcium-Binding Proteins)
Edited by ANTHONY R. MEANS AND BERT W. O'MALLEY

VOLUME 103. Hormone Action (Part H: Neuroendocrine Peptides)
Edited by P. MICHAEL CONN

VOLUME 104. Enzyme Purification and Related Techniques (Part C)
Edited by WILLIAM B. JAKOBY

VOLUME 105. Oxygen Radicals in Biological Systems
Edited by LESTER PACKER

VOLUME 106. Posttranslational Modifications (Part A)
Edited by FINN WOLD AND KIVIE MOLDAVE

Volume 122. Vitamins and Coenzymes (Part G)
Edited by Frank Chytil and Donald B. McCormick

Volume 123. Vitamins and Coenzymes (Part H)
Edited by Frank Chytil and Donald B. McCormick

Volume 124. Hormone Action (Part J: Neuroendocrine Peptides)
Edited by P. Michael Conn

Volume 125. Biomembranes (Part M: Transport in Bacteria, Mitochondria, and Chloroplasts: General Approaches and Transport Systems)
Edited by Sidney Fleischer and Becca Fleischer

Volume 126. Biomembranes (Part N: Transport in Bacteria, Mitochondria, and Chloroplasts: Protonmotive Force)
Edited by Sidney Fleischer and Becca Fleischer

Volume 127. Biomembranes (Part O: Protons and Water: Structure and Translocation)
Edited by Lester Packer

Volume 128. Plasma Lipoproteins (Part A: Preparation, Structure, and Molecular Biology)
Edited by Jere P. Segrest and John J. Albers

Volume 129. Plasma Lipoproteins (Part B: Characterization, Cell Biology, and Metabolism)
Edited by John J. Albers and Jere P. Segrest

Volume 130. Enzyme Structure (Part K)
Edited by C. H. W. Hirs and Serge N. Timasheff

Volume 131. Enzyme Structure (Part L)
Edited by C. H. W. Hirs and Serge N. Timasheff

Volume 132. Immunochemical Techniques (Part J: Phagocytosis and Cell-Mediated Cytotoxicity)
Edited by Giovanni Di Sabato and Johannes Everse

Volume 133. Bioluminescence and Chemiluminescence (Part B)
Edited by Marlene DeLuca and William D. McElroy

Volume 134. Structural and Contractile Proteins (Part C: The Contractile Apparatus and the Cytoskeleton)
Edited by Richard B. Vallee

Volume 135. Immobilized Enzymes and Cells (Part B)
Edited by Klaus Mosbach

VOLUME 136. Immobilized Enzymes and Cells (Part C)
Edited by KLAUS MOSBACH

VOLUME 137. Immobilized Enzymes and Cells (Part D)
Edited by KLAUS MOSBACH

VOLUME 138. Complex Carbohydrates (Part E)
Edited by VICTOR GINSBURG

VOLUME 139. Cellular Regulators (Part A: Calcium- and Calmodulin-Binding Proteins)
Edited by ANTHONY R. MEANS AND P. MICHAEL CONN

VOLUME 140. Cumulative Subject Index Volumes 102–119, 121–134

VOLUME 141. Cellular Regulators (Part B: Calcium and Lipids)
Edited by P. MICHAEL CONN AND ANTHONY R. MEANS

VOLUME 142. Metabolism of Aromatic Amino Acids and Amines
Edited by SEYMOUR KAUFMAN

VOLUME 143. Sulfur and Sulfur Amino Acids
Edited by WILLIAM B. JAKOBY AND OWEN GRIFFITH

VOLUME 144. Structural and Contractile Proteins (Part D: Extracellular Matrix)
Edited by LEON W. CUNNINGHAM

VOLUME 145. Structural and Contractile Proteins (Part E: Extracellular Matrix)
Edited by LEON W. CUNNINGHAM

VOLUME 146. Peptide Growth Factors (Part A)
Edited by DAVID BARNES AND DAVID A. SIRBASKU

VOLUME 147. Peptide Growth Factors (Part B)
Edited by DAVID BARNES AND DAVID A. SIRBASKU

VOLUME 148. Plant Cell Membranes
Edited by LESTER PACKER AND ROLAND DOUCE

VOLUME 149. Drug and Enzyme Targeting (Part B)
Edited by RALPH GREEN AND KENNETH J. WIDDER

VOLUME 150. Immunochemical Techniques (Part K: *In Vitro* Models of B and T Cell Functions and Lymphoid Cell Receptors)
Edited by GIOVANNI DI SABATO

VOLUME 151. Molecular Genetics of Mammalian Cells
Edited by MICHAEL M. GOTTESMAN

VOLUME 152. Guide to Molecular Cloning Techniques
Edited by SHELBY L. BERGER AND ALAN R. KIMMEL

VOLUME 153. Recombinant DNA (Part D)
Edited by RAY WU AND LAWRENCE GROSSMAN

VOLUME 154. Recombinant DNA (Part E)
Edited by RAY WU AND LAWRENCE GROSSMAN

VOLUME 155. Recombinant DNA (Part F)
Edited by RAY WU

VOLUME 156. Biomembranes (Part P: ATP-Driven Pumps and Related Transport: The Na, K-Pump)
Edited by SIDNEY FLEISCHER AND BECCA FLEISCHER

VOLUME 157. Biomembranes (Part Q: ATP-Driven Pumps and Related Transport: Calcium, Proton, and Potassium Pumps)
Edited by SIDNEY FLEISCHER AND BECCA FLEISCHER

VOLUME 158. Metalloproteins (Part A)
Edited by JAMES F. RIORDAN AND BERT L. VALLEE

VOLUME 159. Initiation and Termination of Cyclic Nucleotide Action
Edited by JACKIE D. CORBIN AND ROGER A. JOHNSON

VOLUME 160. Biomass (Part A: Cellulose and Hemicellulose)
Edited by WILLIS A. WOOD AND SCOTT T. KELLOGG

VOLUME 161. Biomass (Part B: Lignin, Pectin, and Chitin)
Edited by WILLIS A. WOOD AND SCOTT T. KELLOGG

VOLUME 162. Immunochemical Techniques (Part L: Chemotaxis and Inflammation)
Edited by GIOVANNI DI SABATO

VOLUME 163. Immunochemical Techniques (Part M: Chemotaxis and Inflammation)
Edited by GIOVANNI DI SABATO

VOLUME 164. Ribosomes
Edited by HARRY F. NOLLER, JR., AND KIVIE MOLDAVE

VOLUME 165. Microbial Toxins: Tools for Enzymology
Edited by SIDNEY HARSHMAN

VOLUME 166. Branched-Chain Amino Acids
Edited by ROBERT HARRIS AND JOHN R. SOKATCH

VOLUME 167. Cyanobacteria
Edited by LESTER PACKER AND ALEXANDER N. GLAZER

VOLUME 168. Hormone Action (Part K: Neuroendocrine Peptides)
Edited by P. MICHAEL CONN

VOLUME 169. Platelets: Receptors, Adhesion, Secretion (Part A)
Edited by JACEK HAWIGER

VOLUME 170. Nucleosomes
Edited by PAUL M. WASSARMAN AND ROGER D. KORNBERG

VOLUME 171. Biomembranes (Part R: Transport Theory: Cells and Model Membranes)
Edited by SIDNEY FLEISCHER AND BECCA FLEISCHER

VOLUME 172. Biomembranes (Part S: Transport: Membrane Isolation and Characterization)
Edited by SIDNEY FLEISCHER AND BECCA FLEISCHER

VOLUME 173. Biomembranes [Part T: Cellular and Subcellular Transport: Eukaryotic (Nonepithelial) Cells]
Edited by SIDNEY FLEISCHER AND BECCA FLEISCHER

VOLUME 174. Biomembranes [Part U: Cellular and Subcellular Transport: Eukaryotic (Nonepithelial) Cells]
Edited by SIDNEY FLEISCHER AND BECCA FLEISCHER

VOLUME 175. Cumulative Subject Index Volumes 135–139, 141–167

VOLUME 176. Nuclear Magnetic Resonance (Part A: Spectral Techniques and Dynamics)
Edited by NORMAN J. OPPENHEIMER AND THOMAS L. JAMES

VOLUME 177. Nuclear Magnetic Resonance (Part B: Structure and Mechanism)
Edited by NORMAN J. OPPENHEIMER AND THOMAS L. JAMES

VOLUME 178. Antibodies, Antigens, and Molecular Mimicry
Edited by JOHN J. LANGONE

VOLUME 179. Complex Carbohydrates (Part F)
Edited by VICTOR GINSBURG

VOLUME 180. RNA Processing (Part A: General Methods)
Edited by JAMES E. DAHLBERG AND JOHN N. ABELSON

VOLUME 181. RNA Processing (Part B: Specific Methods)
Edited by JAMES E. DAHLBERG AND JOHN N. ABELSON

VOLUME 182. Guide to Protein Purification
Edited by MURRAY P. DEUTSCHER

VOLUME 183. Molecular Evolution: Computer Analysis of Protein and Nucleic Acid Sequences
Edited by RUSSELL F. DOOLITTLE

VOLUME 184. Avidin-Biotin Technology
Edited by MEIR WILCHEK AND EDWARD A. BAYER

VOLUME 185. Gene Expression Technology
Edited by DAVID V. GOEDDEL

VOLUME 186. Oxygen Radicals in Biological Systems (Part B: Oxygen Radicals and Antioxidants)
Edited by LESTER PACKER AND ALEXANDER N. GLAZER

VOLUME 187. Arachidonate Related Lipid Mediators
Edited by ROBERT C. MURPHY AND FRANK A. FITZPATRICK

VOLUME 188. Hydrocarbons and Methylotrophy
Edited by MARY E. LIDSTROM

VOLUME 189. Retinoids (Part A: Molecular and Metabolic Aspects)
Edited by LESTER PACKER

VOLUME 190. Retinoids (Part B: Cell Differentiation and Clinical Applications)
Edited by LESTER PACKER

VOLUME 191. Biomembranes (Part V: Cellular and Subcellular Transport: Epithelial Cells)
Edited by SIDNEY FLEISCHER AND BECCA FLEISCHER

VOLUME 192. Biomembranes (Part W: Cellular and Subcellular Transport: Epithelial Cells)
Edited by SIDNEY FLEISCHER AND BECCA FLEISCHER

VOLUME 193. Mass Spectrometry
Edited by JAMES A. MCCLOSKEY

Volume 194. Guide to Yeast Genetics and Molecular Biology
Edited by Christine Guthrie and Gerald R. Fink

Volume 195. Adenylyl Cyclase, G Proteins, and Guanylyl Cyclase
Edited by Roger A. Johnson and Jackie D. Corbin

Volume 196. Molecular Motors and the Cytoskeleton
Edited by Richard B. Vallee

Volume 197. Phospholipases
Edited by Edward A. Dennis

Volume 198. Peptide Growth Factors (Part C)
Edited by David Barnes, J. P. Mather, and Gordon H. Sato

Volume 199. Cumulative Subject Index Volumes 168–174, 176–194

Volume 200. Protein Phosphorylation (Part A: Protein Kinases: Assays, Purification, Antibodies, Functional Analysis, Cloning, and Expression)
Edited by Tony Hunter and Bartholomew M. Sefton

Volume 201. Protein Phosphorylation (Part B: Analysis of Protein Phosphorylation, Protein Kinase Inhibitors, and Protein Phosphatases)
Edited by Tony Hunter and Bartholomew M. Sefton

Volume 202. Molecular Design and Modeling: Concepts and Applications (Part A: Proteins, Peptides, and Enzymes)
Edited by John J. Langone

Volume 203. Molecular Design and Modeling: Concepts and Applications (Part B: Antibodies and Antigens, Nucleic Acids, Polysaccharides, and Drugs)
Edited by John J. Langone

Volume 204. Bacterial Genetic Systems
Edited by Jeffrey H. Miller

Volume 205. Metallobiochemistry (Part B: Metallothionein and Related Molecules)
Edited by James F. Riordan and Bert L. Vallee

Volume 206. Cytochrome P450
Edited by Michael R. Waterman and Eric F. Johnson

Volume 207. Ion Channels
Edited by Bernardo Rudy and Linda E. Iverson

Volume 208. Protein–DNA Interactions
Edited by Robert T. Sauer

Volume 209. Phospholipid Biosynthesis
Edited by Edward A. Dennis and Dennis E. Vance

Volume 210. Numerical Computer Methods
Edited by Ludwig Brand and Michael L. Johnson

Volume 211. DNA Structures (Part A: Synthesis and Physical Analysis of DNA)
Edited by David M. J. Lilley and James E. Dahlberg

Volume 212. DNA Structures (Part B: Chemical and Electrophoretic Analysis of DNA)
Edited by David M. J. Lilley and James E. Dahlberg

Volume 213. Carotenoids (Part A: Chemistry, Separation, Quantitation, and Antioxidation)
Edited by Lester Packer

Volume 214. Carotenoids (Part B: Metabolism, Genetics, and Biosynthesis)
Edited by Lester Packer

Volume 215. Platelets: Receptors, Adhesion, Secretion (Part B)
Edited by Jacek J. Hawiger

Volume 216. Recombinant DNA (Part G)
Edited by Ray Wu

Volume 217. Recombinant DNA (Part H)
Edited by Ray Wu

Volume 218. Recombinant DNA (Part I)
Edited by Ray Wu

Volume 219. Reconstitution of Intracellular Transport
Edited by James E. Rothman

Volume 220. Membrane Fusion Techniques (Part A)
Edited by Nejat Düzgüneş

Volume 221. Membrane Fusion Techniques (Part B)
Edited by Nejat Düzgüneş

Volume 222. Proteolytic Enzymes in Coagulation, Fibrinolysis, and Complement Activation (Part A: Mammalian Blood Coagulation

Factors and Inhibitors)
Edited by Laszlo Lorand and Kenneth G. Mann

Volume 223. Proteolytic Enzymes in Coagulation, Fibrinolysis, and Complement Activation (Part B: Complement Activation, Fibrinolysis, and Nonmammalian Blood Coagulation Factors)
Edited by Laszlo Lorand and Kenneth G. Mann

Volume 224. Molecular Evolution: Producing the Biochemical Data
Edited by Elizabeth Anne Zimmer, Thomas J. White, Rebecca L. Cann, and Allan C. Wilson

Volume 225. Guide to Techniques in Mouse Development
Edited by Paul M. Wassarman and Melvin L. DePamphilis

Volume 226. Metallobiochemistry (Part C: Spectroscopic and Physical Methods for Probing Metal Ion Environments in Metalloenzymes and Metalloproteins)
Edited by James F. Riordan and Bert L. Vallee

Volume 227. Metallobiochemistry (Part D: Physical and Spectroscopic Methods for Probing Metal Ion Environments in Metalloproteins)
Edited by James F. Riordan and Bert L. Vallee

Volume 228. Aqueous Two-Phase Systems
Edited by Harry Walter and Göte Johansson

Volume 229. Cumulative Subject Index Volumes 195–198, 200–227

Volume 230. Guide to Techniques in Glycobiology
Edited by William J. Lennarz and Gerald W. Hart

Volume 231. Hemoglobins (Part B: Biochemical and Analytical Methods)
Edited by Johannes Everse, Kim D. Vandegriff, and Robert M. Winslow

Volume 232. Hemoglobins (Part C: Biophysical Methods)
Edited by Johannes Everse, Kim D. Vandegriff, and Robert M. Winslow

Volume 233. Oxygen Radicals in Biological Systems (Part C)
Edited by Lester Packer

Volume 234. Oxygen Radicals in Biological Systems (Part D)
Edited by Lester Packer

Volume 235. Bacterial Pathogenesis (Part A: Identification and Regulation of Virulence Factors)
Edited by Virginia L. Clark and Patrik M. Bavoil

VOLUME 236. Bacterial Pathogenesis (Part B: Integration of Pathogenic Bacteria with Host Cells)
Edited by VIRGINIA L. CLARK AND PATRIK M. BAVOIL

VOLUME 237. Heterotrimeric G Proteins
Edited by RAVI IYENGAR

VOLUME 238. Heterotrimeric G-Protein Effectors
Edited by RAVI IYENGAR

VOLUME 239. Nuclear Magnetic Resonance (Part C)
Edited by THOMAS L. JAMES AND NORMAN J. OPPENHEIMER

VOLUME 240. Numerical Computer Methods (Part B)
Edited by MICHAEL L. JOHNSON AND LUDWIG BRAND

VOLUME 241. Retroviral Proteases
Edited by LAWRENCE C. KUO AND JULES A. SHAFER

VOLUME 242. Neoglycoconjugates (Part A)
Edited by Y. C. LEE AND REIKO T. LEE

VOLUME 243. Inorganic Microbial Sulfur Metabolism
Edited by HARRY D. PECK, JR., AND JEAN LEGALL

VOLUME 244. Proteolytic Enzymes: Serine and Cysteine Peptidases
Edited by ALAN J. BARRETT

VOLUME 245. Extracellular Matrix Components
Edited by E. RUOSLAHTI AND E. ENGVALL

VOLUME 246. Biochemical Spectroscopy
Edited by KENNETH SAUER

VOLUME 247. Neoglycoconjugates (Part B: Biomedical Applications)
Edited by Y. C. LEE AND REIKO T. LEE

VOLUME 248. Proteolytic Enzymes: Aspartic and Metallo Peptidases
Edited by ALAN J. BARRETT

VOLUME 249. Enzyme Kinetics and Mechanism (Part D: Developments in Enzyme Dynamics)
Edited by DANIEL L. PURICH

VOLUME 250. Lipid Modifications of Proteins
Edited by PATRICK J. CASEY AND JANICE E. BUSS

VOLUME 251. Biothiols (Part A: Monothiols and Dithiols, Protein Thiols, and Thiyl Radicals)
Edited by LESTER PACKER

VOLUME 252. Biothiols (Part B: Glutathione and Thioredoxin; Thiols in Signal Transduction and Gene Regulation)
Edited by LESTER PACKER

VOLUME 253. Adhesion of Microbial Pathogens
Edited by RON J. DOYLE AND ITZHAK OFEK

VOLUME 254. Oncogene Techniques
Edited by PETER K. VOGT AND INDER M. VERMA

VOLUME 255. Small GTPases and Their Regulators (Part A: Ras Family)
Edited by W. E. BALCH, CHANNING J. DER, AND ALAN HALL

VOLUME 256. Small GTPases and Their Regulators (Part B: Rho Family)
Edited by W. E. BALCH, CHANNING J. DER, AND ALAN HALL

VOLUME 257. Small GTPases and Their Regulators (Part C: Proteins Involved in Transport)
Edited by W. E. BALCH, CHANNING J. DER, AND ALAN HALL

VOLUME 258. Redox-Active Amino Acids in Biology
Edited by JUDITH P. KLINMAN

VOLUME 259. Energetics of Biological Macromolecules
Edited by MICHAEL L. JOHNSON AND GARY K. ACKERS

VOLUME 260. Mitochondrial Biogenesis and Genetics (Part A)
Edited by GIUSEPPE M. ATTARDI AND ANNE CHOMYN

VOLUME 261. Nuclear Magnetic Resonance and Nucleic Acids
Edited by THOMAS L. JAMES

VOLUME 262. DNA Replication
Edited by JUDITH L. CAMPBELL

VOLUME 263. Plasma Lipoproteins (Part C: Quantitation)
Edited by WILLIAM A. BRADLEY, SANDRA H. GIANTURCO, AND JERE P. SEGREST

VOLUME 264. Mitochondrial Biogenesis and Genetics (Part B)
Edited by GIUSEPPE M. ATTARDI AND ANNE CHOMYN

VOLUME 265. Cumulative Subject Index Volumes 228, 230–262

VOLUME 266. Computer Methods for Macromolecular Sequence Analysis
Edited by RUSSELL F. DOOLITTLE

VOLUME 267. Combinatorial Chemistry
Edited by JOHN N. ABELSON

VOLUME 268. Nitric Oxide (Part A: Sources and Detection of NO; NO Synthase)
Edited by LESTER PACKER

VOLUME 269. Nitric Oxide (Part B: Physiological and Pathological Processes)
Edited by LESTER PACKER

VOLUME 270. High Resolution Separation and Analysis of Biological Macromolecules (Part A: Fundamentals)
Edited by BARRY L. KARGER AND WILLIAM S. HANCOCK

VOLUME 271. High Resolution Separation and Analysis of Biological Macromolecules (Part B: Applications)
Edited by BARRY L. KARGER AND WILLIAM S. HANCOCK

VOLUME 272. Cytochrome P450 (Part B)
Edited by ERIC F. JOHNSON AND MICHAEL R. WATERMAN

VOLUME 273. RNA Polymerase and Associated Factors (Part A)
Edited by SANKAR ADHYA

VOLUME 274. RNA Polymerase and Associated Factors (Part B)
Edited by SANKAR ADHYA

VOLUME 275. Viral Polymerases and Related Proteins
Edited by LAWRENCE C. KUO, DAVID B. OLSEN, AND STEVEN S. CARROLL

VOLUME 276. Macromolecular Crystallography (Part A)
Edited by CHARLES W. CARTER, JR., AND ROBERT M. SWEET

VOLUME 277. Macromolecular Crystallography (Part B)
Edited by CHARLES W. CARTER, JR., AND ROBERT M. SWEET

VOLUME 278. Fluorescence Spectroscopy
Edited by LUDWIG BRAND AND MICHAEL L. JOHNSON

VOLUME 279. Vitamins and Coenzymes (Part I)
Edited by DONALD B. MCCORMICK, JOHN W. SUTTIE, AND CONRAD WAGNER

VOLUME 280. Vitamins and Coenzymes (Part J)
Edited by DONALD B. MCCORMICK, JOHN W. SUTTIE, AND CONRAD WAGNER

VOLUME 297. Photosynthesis: Molecular Biology of Energy Capture
Edited by LEE MCINTOSH

VOLUME 298. Molecular Motors and the Cytoskeleton (Part B)
Edited by RICHARD B. VALLEE

VOLUME 299. Oxidants and Antioxidants (Part A)
Edited by LESTER PACKER

VOLUME 300. Oxidants and Antioxidants (Part B)
Edited by LESTER PACKER

VOLUME 301. Nitric Oxide: Biological and Antioxidant Activities (Part C)
Edited by LESTER PACKER

VOLUME 302. Green Fluorescent Protein
Edited by P. MICHAEL CONN

VOLUME 303. cDNA Preparation and Display
Edited by SHERMAN M. WEISSMAN

VOLUME 304. Chromatin
Edited by PAUL M. WASSARMAN AND ALAN P. WOLFFE

VOLUME 305. Bioluminescence and Chemiluminescence (Part C)
Edited by THOMAS O. BALDWIN AND MIRIAM M. ZIEGLER

VOLUME 306. Expression of Recombinant Genes in Eukaryotic Systems
Edited by JOSEPH C. GLORIOSO AND MARTIN C. SCHMIDT

VOLUME 307. Confocal Microscopy
Edited by P. MICHAEL CONN

VOLUME 308. Enzyme Kinetics and Mechanism (Part E: Energetics of Enzyme Catalysis)
Edited by DANIEL L. PURICH AND VERN L. SCHRAMM

VOLUME 309. Amyloid, Prions, and Other Protein Aggregates
Edited by RONALD WETZEL

VOLUME 310. Biofilms
Edited by RON J. DOYLE

VOLUME 311. Sphingolipid Metabolism and Cell Signaling (Part A)
Edited by ALFRED H. MERRILL, JR., AND YUSUF A. HANNUN

VOLUME 312. Sphingolipid Metabolism and Cell Signaling (Part B)
Edited by ALFRED H. MERRILL, JR., AND YUSUF A. HANNUN

Volume 313. Antisense Technology
(Part A: General Methods, Methods of Delivery, and RNA Studies)
Edited by M. Ian Phillips

Volume 314. Antisense Technology (Part B: Applications)
Edited by M. Ian Phillips

Volume 315. Vertebrate Phototransduction and the Visual Cycle
(Part A)
Edited by Krzysztof Palczewski

Volume 316. Vertebrate Phototransduction and the Visual Cycle (Part B)
Edited by Krzysztof Palczewski

Volume 317. RNA–Ligand Interactions (Part A: Structural Biology Methods)
Edited by Daniel W. Celander and John N. Abelson

Volume 318. RNA–Ligand Interactions (Part B: Molecular Biology Methods)
Edited by Daniel W. Celander and John N. Abelson

Volume 319. Singlet Oxygen, UV-A, and Ozone
Edited by Lester Packer and Helmut Sies

Volume 320. Cumulative Subject Index Volumes 290–319

Volume 321. Numerical Computer Methods (Part C)
Edited by Michael L. Johnson and Ludwig Brand

Volume 322. Apoptosis
Edited by John C. Reed

Volume 323. Energetics of Biological Macromolecules (Part C)
Edited by Michael L. Johnson and Gary K. Ackers

Volume 324. Branched-Chain Amino Acids (Part B)
Edited by Robert A. Harris and John R. Sokatch

Volume 325. Regulators and Effectors of Small GTPases
(Part D: Rho Family)
Edited by W. E. Balch, Channing J. Der, and Alan Hall

Volume 326. Applications of Chimeric Genes and Hybrid Proteins
(Part A: Gene Expression and Protein Purification)
Edited by Jeremy Thorner, Scott D. Emr, and John N. Abelson

VOLUME 327. Applications of Chimeric Genes and Hybrid Proteins (Part B: Cell Biology and Physiology)
Edited by JEREMY THORNER, SCOTT D. EMR, AND JOHN N. ABELSON

VOLUME 328. Applications of Chimeric Genes and Hybrid Proteins (Part C: Protein–Protein Interactions and Genomics)
Edited by JEREMY THORNER, SCOTT D. EMR, AND JOHN N. ABELSON

VOLUME 329. Regulators and Effectors of Small GTPases (Part E: GTPases Involved in Vesicular Traffic)
Edited by W. E. BALCH, CHANNING J. DER, AND ALAN HALL

VOLUME 330. Hyperthermophilic Enzymes (Part A)
Edited by MICHAEL W. W. ADAMS AND ROBERT M. KELLY

VOLUME 331. Hyperthermophilic Enzymes (Part B)
Edited by MICHAEL W. W. ADAMS AND ROBERT M. KELLY

VOLUME 332. Regulators and Effectors of Small GTPases (Part F: Ras Family I)
Edited by W. E. BALCH, CHANNING J. DER, AND ALAN HALL

VOLUME 333. Regulators and Effectors of Small GTPases (Part G: Ras Family II)
Edited by W. E. BALCH, CHANNING J. DER, AND ALAN HALL

VOLUME 334. Hyperthermophilic Enzymes (Part C)
Edited by MICHAEL W. W. ADAMS AND ROBERT M. KELLY

VOLUME 335. Flavonoids and Other Polyphenols
Edited by LESTER PACKER

VOLUME 336. Microbial Growth in Biofilms (Part A: Developmental and Molecular Biological Aspects)
Edited by RON J. DOYLE

VOLUME 337. Microbial Growth in Biofilms (Part B: Special Environments and Physicochemical Aspects)
Edited by RON J. DOYLE

VOLUME 338. Nuclear Magnetic Resonance of Biological Macromolecules (Part A)
Edited by THOMAS L. JAMES, VOLKER DÖTSCH, AND ULI SCHMITZ

VOLUME 339. Nuclear Magnetic Resonance of Biological Macromolecules (Part B)
Edited by THOMAS L. JAMES, VOLKER DÖTSCH, AND ULI SCHMITZ

Volume 340. Drug–Nucleic Acid Interactions
Edited by Jonathan B. Chaires and Michael J. Waring

Volume 341. Ribonucleases (Part A)
Edited by Allen W. Nicholson

Volume 342. Ribonucleases (Part B)
Edited by Allen W. Nicholson

Volume 343. G Protein Pathways (Part A: Receptors)
Edited by Ravi Iyengar and John D. Hildebrandt

Volume 344. G Protein Pathways (Part B: G Proteins and Their Regulators)
Edited by Ravi Iyengar and John D. Hildebrandt

Volume 345. G Protein Pathways (Part C: Effector Mechanisms)
Edited by Ravi Iyengar and John D. Hildebrandt

Volume 346. Gene Therapy Methods
Edited by M. Ian Phillips

Volume 347. Protein Sensors and Reactive Oxygen Species (Part A: Selenoproteins and Thioredoxin)
Edited by Helmut Sies and Lester Packer

Volume 348. Protein Sensors and Reactive Oxygen Species (Part B: Thiol Enzymes and Proteins)
Edited by Helmut Sies and Lester Packer

Volume 349. Superoxide Dismutase
Edited by Lester Packer

Volume 350. Guide to Yeast Genetics and Molecular and Cell Biology (Part B)
Edited by Christine Guthrie and Gerald R. Fink

Volume 351. Guide to Yeast Genetics and Molecular and Cell Biology (Part C)
Edited by Christine Guthrie and Gerald R. Fink

Volume 352. Redox Cell Biology and Genetics (Part A)
Edited by Chandan K. Sen and Lester Packer

Volume 353. Redox Cell Biology and Genetics (Part B)
Edited by Chandan K. Sen and Lester Packer

VOLUME 354. Enzyme Kinetics and Mechanisms (Part F: Detection and Characterization of Enzyme Reaction Intermediates)
Edited by DANIEL L. PURICH

VOLUME 355. Cumulative Subject Index Volumes 321–354

VOLUME 356. Laser Capture Microscopy and Microdissection
Edited by P. MICHAEL CONN

VOLUME 357. Cytochrome P450, Part C
Edited by ERIC F. JOHNSON AND MICHAEL R. WATERMAN

VOLUME 358. Bacterial Pathogenesis (Part C: Identification, Regulation, and Function of Virulence Factors)
Edited by VIRGINIA L. CLARK AND PATRIK M. BAVOIL

VOLUME 359. Nitric Oxide (Part D)
Edited by ENRIQUE CADENAS AND LESTER PACKER

VOLUME 360. Biophotonics (Part A)
Edited by GERARD MARRIOTT AND IAN PARKER

VOLUME 361. Biophotonics (Part B)
Edited by GERARD MARRIOTT AND IAN PARKER

VOLUME 362. Recognition of Carbohydrates in Biological Systems (Part A)
Edited by YUAN C. LEE AND REIKO T. LEE

VOLUME 363. Recognition of Carbohydrates in Biological Systems (Part B)
Edited by YUAN C. LEE AND REIKO T. LEE

VOLUME 364. Nuclear Receptors
Edited by DAVID W. RUSSELL AND DAVID J. MANGELSDORF

VOLUME 365. Differentiation of Embryonic Stem Cells
Edited by PAUL M. WASSAUMAN AND GORDON M. KELLER

VOLUME 366. Protein Phosphatases
Edited by SUSANNE KLUMPP AND JOSEF KRIEGLSTEIN

VOLUME 367. Liposomes (Part A)
Edited by NEJAT DÜZGÜNEŞ

VOLUME 368. Macromolecular Crystallography (Part C)
Edited by CHARLES W. CARTER, JR., AND ROBERT M. SWEET

VOLUME 369. Combinational Chemistry (Part B)
Edited by GUILLERMO A. MORALES AND BARRY A. BUNIN

Volume 370. RNA Polymerases and Associated Factors (Part C)
Edited by Sankar L. Adhya and Susan Garges

Volume 371. RNA Polymerases and Associated Factors (Part D)
Edited by Sankar L. Adhya and Susan Garges

Volume 372. Liposomes (Part B)
Edited by Nejat Düzgüneş

Volume 373. Liposomes (Part C)
Edited by Nejat Düzgüneş

Volume 374. Macromolecular Crystallography (Part D)
Edited by Charles W. Carter, Jr., and Robert W. Sweet

Volume 375. Chromatin and Chromatin Remodeling Enzymes (Part A)
Edited by C. David Allis and Carl Wu

Volume 376. Chromatin and Chromatin Remodeling Enzymes (Part B)
Edited by C. David Allis and Carl Wu

Volume 377. Chromatin and Chromatin Remodeling Enzymes (Part C)
Edited by C. David Allis and Carl Wu

Volume 378. Quinones and Quinone Enzymes (Part A)
Edited by Helmut Sies and Lester Packer

Volume 379. Energetics of Biological Macromolecules (Part D)
Edited by Jo M. Holt, Michael L. Johnson, and Gary K. Ackers

Volume 380. Energetics of Biological Macromolecules (Part E)
Edited by Jo M. Holt, Michael L. Johnson, and Gary K. Ackers

Volume 381. Oxygen Sensing
Edited by Chandan K. Sen and Gregg L. Semenza

Volume 382. Quinones and Quinone Enzymes (Part B)
Edited by Helmut Sies and Lester Packer

Volume 383. Numerical Computer Methods (Part D)
Edited by Ludwig Brand and Michael L. Johnson

Volume 384. Numerical Computer Methods (Part E)
Edited by Ludwig Brand and Michael L. Johnson

Volume 385. Imaging in Biological Research (Part A)
Edited by P. Michael Conn

Volume 386. Imaging in Biological Research (Part B)
Edited by P. Michael Conn

VOLUME 387. Liposomes (Part D)
Edited by NEJAT DÜZGÜNEŞ

VOLUME 388. Protein Engineering
Edited by DAN E. ROBERTSON AND JOSEPH P. NOEL

VOLUME 389. Regulators of G-Protein Signaling (Part A)
Edited by DAVID P. SIDEROVSKI

VOLUME 390. Regulators of G-Protein Signaling (Part B)
Edited by DAVID P. SIDEROVSKI

VOLUME 391. Liposomes (Part E)
Edited by NEJAT DÜZGÜNEŞ

VOLUME 392. RNA Interference
Edited by ENGELKE ROSSI

VOLUME 393. Circadian Rhythms
Edited by MICHAEL W. YOUNG

VOLUME 394. Nuclear Magnetic Resonance of Biological Macromolecules (Part C)
Edited by THOMAS L. JAMES

VOLUME 395. Producing the Biochemical Data (Part B)
Edited by ELIZABETH A. ZIMMER AND ERIC H. ROALSON

VOLUME 396. Nitric Oxide (Part E)
Edited by LESTER PACKER AND ENRIQUE CADENAS

VOLUME 397. Environmental Microbiology
Edited by JARED R. LEADBETTER

VOLUME 398. Ubiquitin and Protein Degradation (Part A)
Edited by RAYMOND J. DESHAIES

VOLUME 399. Ubiquitin and Protein Degradation (Part B)
Edited by RAYMOND J. DESHAIES

VOLUME 400. Phase II Conjugation Enzymes and Transport Systems
Edited by HELMUT SIES AND LESTER PACKER

VOLUME 401. Glutathione Transferases and Gamma Glutamyl Transpeptidases
Edited by HELMUT SIES AND LESTER PACKER

VOLUME 402. Biological Mass Spectrometry
Edited by A. L. BURLINGAME

VOLUME 403. GTPases Regulating Membrane Targeting and Fusion
Edited by WILLIAM E. BALCH, CHANNING J. DER, AND ALAN HALL

VOLUME 404. GTPases Regulating Membrane Dynamics
Edited by WILLIAM E. BALCH, CHANNING J. DER, AND ALAN HALL

VOLUME 405. Mass Spectrometry: Modified Proteins and Glycoconjugates
Edited by A. L. BURLINGAME

VOLUME 406. Regulators and Effectors of Small GTPases: Rho Family
Edited by WILLIAM E. BALCH, CHANNING J. DER, AND ALAN HALL

VOLUME 407. Regulators and Effectors of Small GTPases: Ras Family
Edited by WILLIAM E. BALCH, CHANNING J. DER, AND ALAN HALL

VOLUME 408. DNA Repair (Part A)
Edited by JUDITH L. CAMPBELL AND PAUL MODRICH

VOLUME 409. DNA Repair (Part B)
Edited by JUDITH L. CAMPBELL AND PAUL MODRICH

VOLUME 410. DNA Microarrays (Part A: Array Platforms and Web-Bench Protocols)
Edited by ALAN KIMMEL AND BRIAN OLIVER

VOLUME 411. DNA Microarrays (Part B: Databases and Statistics)
Edited by ALAN KIMMEL AND BRIAN OLIVER

VOLUME 412. Amyloid, Prions, and Other Protein Aggregates (Part B)
Edited by INDU KHETERPAL AND RONALD WETZEL

VOLUME 413. Amyloid, Prions, and Other Protein Aggregates (Part C)
Edited by INDU KHETERPAL AND RONALD WETZEL

VOLUME 414. Measuring Biological Responses with Automated Microscopy
Edited by JAMES INGLESE

VOLUME 415. Glycobiology
Edited by MINORU FUKUDA

VOLUME 416. Glycomics
Edited by MINORU FUKUDA

VOLUME 417. Functional Glycomics
Edited by MINORU FUKUDA

VOLUME 418. Embryonic Stem Cells
Edited by IRINA KLIMANSKAYA AND ROBERT LANZA

VOLUME 419. Adult Stem Cells
Edited by IRINA KLIMANSKAYA AND ROBERT LANZA

VOLUME 420. Stem Cell Tools and Other Experimental Protocols
Edited by IRINA KLIMANSKAYA AND ROBERT LANZA

VOLUME 421. Advanced Bacterial Genetics: Use of Transposons and Phage for Genomic Engineering
Edited by KELLY T. HUGHES

VOLUME 422. Two-Component Signaling Systems, Part A
Edited by MELVIN I. SIMON, BRIAN R. CRANE, AND ALEXANDRINE CRANE

VOLUME 423. Two-Component Signaling Systems, Part B
Edited by MELVIN I. SIMON, BRIAN R. CRANE, AND ALEXANDRINE CRANE

VOLUME 424. RNA Editing
Edited by JONATHA M. GOTT

VOLUME 425. RNA Modification
Edited by JONATHA M. GOTT

VOLUME 426. Integrins
Edited by DAVID CHERESH

VOLUME 427. MicroRNA Methods
Edited by JOHN J. ROSSI

VOLUME 428. Osmosensing and Osmosignaling
Edited by HELMUT SIES AND DIETER HAUSSINGER

VOLUME 429. Translation Initiation: Extract Systems and Molecular Genetics
Edited by JON LORSCH

VOLUME 430. Translation Initiation: Reconstituted Systems and Biophysical Methods
Edited by JON LORSCH

VOLUME 431. Translation Initiation: Cell Biology, High-Throughput and Chemical-Based Approaches
Edited by JON LORSCH

VOLUME 432. Lipidomics and Bioactive Lipids: Mass-Spectrometry–Based Lipid Analysis
Edited by H. ALEX BROWN

VOLUME 433. Lipidomics and Bioactive Lipids: Specialized Analytical Methods and Lipids in Disease
Edited by H. ALEX BROWN

VOLUME 434. Lipidomics and Bioactive Lipids: Lipids and Cell Signaling
Edited by H. ALEX BROWN

VOLUME 435. Oxygen Biology and Hypoxia
Edited by HELMUT SIES AND BERNHARD BRÜNE

VOLUME 436. Globins and Other Nitric Oxide-Reactive Protiens (Part A)
Edited by ROBERT K. POOLE

VOLUME 437. Globins and Other Nitric Oxide-Reactive Protiens (Part B)
Edited by ROBERT K. POOLE

VOLUME 438. Small GTPases in Disease (Part A)
Edited by WILLIAM E. BALCH, CHANNING J. DER, AND ALAN HALL

VOLUME 439. Small GTPases in Disease (Part B)
Edited by WILLIAM E. BALCH, CHANNING J. DER, AND ALAN HALL

VOLUME 440. Nitric Oxide, Part F Oxidative and Nitrosative Stress in Redox Regulation of Cell Signaling
Edited by ENRIQUE CADENAS AND LESTER PACKER

VOLUME 441. Nitric Oxide, Part G Oxidative and Nitrosative Stress in Redox Regulation of Cell Signaling
Edited by ENRIQUE CADENAS AND LESTER PACKER

VOLUME 442. Programmed Cell Death, General Principles for Studying Cell Death (Part A)
Edited by ROYA KHOSRAVI-FAR, ZAHRA ZAKERI, RICHARD A. LOCKSHIN, AND MAURO PIACENTINI

VOLUME 443. Angiogenesis: *In Vitro* Systems
Edited by DAVID A. CHERESH

VOLUME 444. Angiogenesis: *In Vivo* Systems (Part A)
Edited by DAVID A. CHERESH

VOLUME 445. Angiogenesis: *In Vivo* Systems (Part B)
Edited by DAVID A. CHERESH

VOLUME 446. Programmed Cell Death, The Biology and Therapeutic Implications of Cell Death (Part B)
Edited by ROYA KHOSRAVI-FAR, ZAHRA ZAKERI, RICHARD A. LOCKSHIN, AND MAURO PIACENTINI

Volume 447. RNA Turnover in Bacteria, Archaea and Organelles
Edited by Lynne E. Maquat and Cecilia M. Arraiano

Volume 448. RNA Turnover in Eukaryotes: Nucleases, Pathways and Analysis of mRNA Decay
Edited by Lynne E. Maquat and Megerditch Kiledjian

Volume 449. RNA Turnover in Eukaryotes: Analysis of Specialized and Quality Control RNA Decay Pathways
Edited by Lynne E. Maquat and Megerditch Kiledjian

Volume 450. Fluorescence Spectroscopy
Edited by Ludwig Brand and Michael L. Johnson

Volume 451. Autophagy: Lower Eukaryotes and Non-Mammalian Systems (Part A)
Edited by Daniel J. Klionsky

Volume 452. Autophagy in Mammalian Systems (Part B)
Edited by Daniel J. Klionsky

Volume 453. Autophagy in Disease and Clinical Applications (Part C)
Edited by Daniel J. Klionsky

Volume 454. Computer Methods (Part A)
Edited by Michael L. Johnson and Ludwig Brand

Volume 455. Biothermodynamics (Part A)
Edited by Michael L. Johnson, Jo M. Holt, and Gary K. Ackers (Retired)

Volume 456. Mitochondrial Function, Part A: Mitochondrial Electron Transport Complexes and Reactive Oxygen Species
Edited by William S. Allison and Immo E. Scheffler

Volume 457. Mitochondrial Function, Part B: Mitochondrial Protein Kinases, Protein Phosphatases and Mitochondrial Diseases
Edited by William S. Allison and Anne N. Murphy

Volume 458. Complex Enzymes in Microbial Natural Product Biosynthesis, Part A: Overview Articles and Peptides
Edited by David A. Hopwood

Volume 459. Complex Enzymes in Microbial Natural Product Biosynthesis, Part B: Polyketides, Aminocoumarins and Carbohydrates
Edited by David A. Hopwood

Volume 474. Thiol Redox Transitions in Cell Signaling, Part B Cellular Localization and Signaling
Edited by Enrique Cadenas and Lester Packer

Volume 475. Single Molecule Tools, Part B: Super-Resolution, Particle Tracking, Multiparameter, and Force Based Methods
Edited by Nils G. Walter

Volume 476. Guide to Techniques in Mouse Development, Part A Mice, Embryos, and Cells, 2nd Edition
Edited by Paul M. Wassarman and Philippe M. Soriano

Volume 477. Guide to Techniques in Mouse Development, Part B Mouse Molecular Genetics, 2nd Edition
Edited by Paul M. Wassarman and Philippe M. Soriano

Volume 478. Glycomics
Edited by Minoru Fukuda

Volume 479. Functional Glycomics
Edited by Minoru Fukuda

Volume 480. Glycobiology
Edited by Minoru Fukuda

Volume 481. Cryo-EM, Part A: Sample Preparation and Data Collection
Edited by Grant J. Jensen

Volume 482. Cryo-EM, Part B: 3-D Reconstruction
Edited by Grant J. Jensen

Volume 483. Cryo-EM, Part C: Analyses, Interpretation, and Case Studies
Edited by Grant J. Jensen

Volume 484. Constitutive Activity in Receptors and Other Proteins, Part A
Edited by P. Michael Conn

Volume 485. Constitutive Activity in Receptors and Other Proteins, Part B
Edited by P. Michael Conn

Volume 486. Research on Nitrification and Related Processes, Part A
Edited by Martin G. Klotz

Volume 487. Computer Methods, Part C
Edited by Michael L. Johnson and Ludwig Brand

Volume 488. Biothermodynamics, Part C
Edited by Michael L. Johnson, Jo M. Holt, and Gary K. Ackers

Volume 489. The Unfolded Protein Response and Cellular Stress, Part A
Edited by P. Michael Conn

Volume 490. The Unfolded Protein Response and Cellular Stress, Part B
Edited by P. Michael Conn

Volume 491. The Unfolded Protein Response and Cellular Stress, Part C
Edited by P. Michael Conn

Volume 492. Biothermodynamics, Part D
Edited by Michael L. Johnson, Jo M. Holt, and Gary K. Ackers

Volume 493. Fragment-Based Drug Design Tools,
Practical Approaches, and Examples
Edited by Lawrence C. Kuo

Volume 494. Methods in Methane Metabolism, Part A
Methanogenesis
Edited by Amy C. Rosenzweig and Stephen W. Ragsdale

Volume 495. Methods in Methane Metabolism, Part B
Methanotrophy
Edited by Amy C. Rosenzweig and Stephen W. Ragsdale

Volume 496. Research on Nitrification and Related Processes, Part B
Edited by Martin G. Klotz and Lisa Y. Stein

Volume 497. Synthetic Biology, Part A
Methods for Part/Device Characterization and Chassis Engineering
Edited by Christopher Voigt

Volume 498. Synthetic Biology, Part B
Computer Aided Design and DNA Assembly
Edited by Christopher Voigt

Volume 499. Biology of Serpins
Edited by James C. Whisstock and Phillip I. Bird

Volume 500. Methods in Systems Biology
Edited by Daniel Jameson, Malkhey Verma, and Hans V. Westerhoff

Volume 501. Serpin Structure and Evolution
Edited by James C. Whisstock and Phillip I. Bird

Volume 502. Protein Engineering for Therapeutics, Part A
Edited by K. Dane Wittrup and Gregory L. Verdine

Volume 503. Protein Engineering for Therapeutics, Part B
Edited by K. Dane Wittrup and Gregory L. Verdine

Volume 504. Imaging and Spectroscopic Analysis of Living Cells
Optical and Spectroscopic Techniques
Edited by P. Michael Conn

Volume 505. Imaging and Spectroscopic Analysis of Living Cells
Live Cell Imaging of Cellular Elements and Functions
Edited by P. Michael Conn

Volume 506. Imaging and Spectroscopic Analysis of Living Cells
Imaging Live Cells in Health and Disease
Edited by P. Michael Conn

Volume 507. Gene Transfer Vectors for Clinical Application
Edited by Theodore Friedmann

Volume 508. Nanomedicine
Cancer, Diabetes, and Cardiovascular, Central Nervous System, Pulmonary and Inflammatory Diseases
Edited by Nejat Düzgüneş

Volume 509. Nanomedicine
Infectious Diseases, Immunotherapy, Diagnostics, Antifibrotics, Toxicology and Gene Medicine
Edited by Nejat Düzgüneş

Volume 510. Cellulases
Edited by Harry J. Gilbert

Volume 511. RNA Helicases
Edited by Eckhard Jankowsky

Volume 512. Nucleosomes, Histones & Chromatin, Part A
Edited by Carl Wu and C. David Allis

Volume 513. Nucleosomes, Histones & Chromatin, Part B
Edited by Carl Wu and C. David Allis

Volume 514. Ghrelin
Edited by Masayasu Kojima and Kenji Kangawa

Volume 515. Natural Product Biosynthesis by Microorganisms and Plants, Part A
Edited by David A. Hopwood

SECTION 1

GPCR Trafficking

CHAPTER ONE

Therapeutic Rescue of Misfolded/Mistrafficked Mutants: Automation-Friendly High-Throughput Assays for Identification of Pharmacoperone Drugs of GPCRs

David C. Smithson[*], **Jo Ann Janovick**[†], **P. Michael Conn**[†,‡,1]
[*]Oregon Translational Research and Drug Development Institute, Portland, Oregon, USA
[†]Divisions of Reproductive Sciences and Neuroscience, Oregon National Primate Research Center, Beaverton, Oregon, USA
[‡]Departments of Physiology and Pharmacology, Cell Biology and Human Development and Obstetrics and Gynecology, Oregon Health & Science University, Portland, Oregon, USA
[1]Corresponding author: e-mail address: connm@ohsu.edu

Contents

1. Introduction 4
2. Choosing Pharmacoperone Model Systems 5
3. Selection of Endpoint Measures 7
   3.1 V2R assay endpoint measure 8
   3.2 GnRHR assay endpoint measure 9
4. Assay Automation 11
5. Data Analysis 12
6. Hit Follow-Up Experiments 12
7. Conclusions 14
Acknowledgments 14
References 14

Abstract

Mutations cause protein folding defects that result in cellular misrouting of otherwise functional proteins. Such mutations are responsible for a wide range of disease states, especially among G-protein coupled receptors. Drugs which serve as chemical templates and promote the proper folding of these proteins are valuable therapeutic molecules since they return functional proteins to the proper site of action. Small molecules have been identified that are able to function as pharmacological chaperones or "pharmacoperones" and stabilize the correct conformations of their target proteins with high specificity. Most of these are also agonists or antagonists of the proteins of interest,

Methods in Enzymology, Volume 521
ISSN 0076-6879
http://dx.doi.org/10.1016/B978-0-12-391862-8.00001-6

complicating potential therapeutic use. This is due, in part, to the fact that the majority of these were discovered during high-throughput screening campaigns using assays designed to detect agonists and antagonists, rather than compounds which improve the trafficking of misrouted mutants. The assays described in this report are designed specifically to identify compounds which result in the reactivation and correct trafficking of misfolded gonadotropin releasing hormone receptor and vasopressin type 2 receptor mutants, rather than those which act as agonists directly. The system reported is a generalizable approach amenable to use in automated (robotic) high-throughput screening efforts and can be used to identify compounds which affect protein conformation without necessarily acting as direct agonists or antagonists.

1. INTRODUCTION

Pharmacoperone drugs (from "pharmacological chaperone") are small molecules that enter cells and serve as molecular scaffolding in order to cause otherwise misfolded mutant proteins to fold and route correctly within the cell. Many known pharmacoperones are also agonists or antagonists because they have come from high-throughput screens that were originally designed with a view toward identification of such congeners as lead drug candidates, not pharmacoperones, as such. New pharmacoperones identified from screens may be found that are not also agonists or antagonists.

In principle, the pharmacoperone-rescue approach applies to a diverse array of human diseases that result from protein misfolding—among these are cystic fibrosis (Amaral, 2006; Dormer et al., 2001; Galietta et al., 2001; Zhang et al., 2003) hypogonadotropic hypogonadism (HH; Ulloa-Aguirre, Janovick, Leanos-Miranda, & Conn, 2003), nephrogenic diabetes insipidus (Bernier, Lagace, Bichet, & Bouvier, 2004; Bichet, 2006; Morello & Bichet, 2001), retinitis pigmentosa (Noorwez et al., 2004), hypercholesterolemia, cataracts (Benedek, Pande, Thurston, & Clark, 1999), neurodegenerative diseases Huntington's, Alzheimer's, Parkinson's (Heiser et al., 2000; Forloni et al., 2002; Permanne et al., 2002; Soto et al., 2000; Muchowski & Wacker, 2005), particular cancers (Peng, Li, Chen, Sebti, & Chen, 2003), α1 trypsin deficiency and lysosomal storage disease (Bottomley, 2011; Fan, 2003), mucopolysaccharidosis type IIIC (Feldhammer, Durand, & Pshezhetsky, 2009).

In the case of certain proteins (e.g., the gonadotropin releasing hormone receptor (GnRHR), vasopressin type 2 receptor (V2R), and rhodopsin), this approach has succeeded with a striking number of different mutants of individual proteins (Conn, Ulloa-Aguirre, Ito, & Janovick, 2007), supporting the view that pharmacoperones will become powerful weapons in our therapeutic arsenal (Conn et al., 2007). For this reason, we have created a generalizable screening technique that allows identification of specific pharmacoperones from chemical libraries. In the present report, we describe steps that have been taken to make this assay automation-friendly.

Functional rescue of misfolded mutant receptors by small nonpeptide molecules has been demonstrated. These small, target-specific molecules (pharmacological chaperones or "pharmacoperones") serve as molecular templates, promote correct folding, and allow otherwise misfolded mutants to pass the scrutiny of the cellular quality control system (QCS) and be expressed at the plasma membrane (PM) where they function similar to wild-type (WT) proteins. It has also become apparent that the pharmacoperone approach is able to rescue (Janovick et al., 2007), in addition to nascent proteins (Conn, Leanos-Miranda, & Janovick, 2002; Janovick et al., 2003), proteins that have previously been retained and may evoke ER-stress responses (Marciniak & Ron, 2006; Ron, 2002; Ron & Hubbard, 2008).

In the case of the GnRHR, drugs that rescue one mutant, typically rescue many mutants, even if the mutations are located at distant sites (extracellular loops, intracellular loops, transmembrane helices). This increases the value of these drugs. These drugs are typically identified, *post hoc*, from "hits" in screens designed to detect antagonists or agonists. The therapeutic utility of pharmacoperones has been limited due to the absence of assays which enable identification of pharmacoperones *per se* (and may not be antagonists) and which are amenable to use in automated high-throughput screening campaigns.

2. CHOOSING PHARMACOPERONE MODEL SYSTEMS

G-protein-coupled receptors (GPCRs), which include the GnRHR and V2R, comprise the largest family of validated drug targets; 30–50% of approved drugs derive their benefits by selective targeting of GPCRs (Gruber, Muttenthaler, & Freissmuth, 2010). Mutations in GPCRs are known to be responsible for over 30 disorders, including cancers, heritable obesity, and endocrine diseases. Normally, GPCRs are subjected to a stringent QCS in the endoplasmic reticulum (ER). The QCS insures that only correctly folded proteins enter the pathway leading to the PM. This system consists of both protein chaperones that retain misfolded proteins and enzyme-like proteins that participate in catalysis of the folding process. It has become apparent that point mutations may result in the production of misfolded and disease-causing proteins that are unable to reach their functional destinations in the cell because they are retained by the QCS, even though they may retain function. Pharmacoperone activity has been identified in these targets *post hoc* by us (Conn & Ulloa-Aguirre, 2010; Janovick et al., 2003) and others (Bernier, Bichet, & Bouvier, 2004) for the GnRHR and V2R systems, respectively.

We previously described a generalizable screening approach for pharmacoperone drugs based on measurement of gain of activity in HeLa cells stably expressing the mutants of two different model GPCRs (hGnRHR[E^{90}K] or hV2R[L^{83}Q]) (Conn & Janovick, 2011; Janovick, Park, & Conn, 2011). These cells turn off expression of the receptor mutant gene of interest in the presence of tetracycline and its analogs, which provides a convenient means to identify false positives.

There are several advantages to using stable cell lines when developing assays for use in high-throughput screening campaigns.

1. Stable cell lines produce a reproducible response upon receptor stimulation and give rise to a high signal window. In transiently transfected cells, a high proportion of nontransfected cells may be present. Untransfected cells reduce the maximum signal since they do not contribute to production of stimulated endpoint.
2. Stable cells are convenient since they do not require separate transfection for each experiment.
3. A special feature of the cell lines developed is that the GPCRs or GPCR mutants are expressed under the control of the tetracycline-controlled transactivator (tTA). The tetracycline-regulated expression system is based on two components: a Tet-dependent transcription activator (tTA), which is a fusion between the Tet repressor of transposon TN10 and transcription factor binding domains of the herpes simplex protein VP16, and secondly, a tTA-responsive promoter, composed of seven Tet repressor binding sites (TetO7) immediately upstream of an RNA polymerase II transcriptional start site of the cytomegalovirus IE promoter (CMVm). When both elements are present in the cell, tTA binds to TetO7 and activates transcription at its neighboring initiation site. In the absence of tetracycline, the GPCR is expressed, but in the presence of tetracycline, the GPCR is not measurably expressed. This model allows use of the GPCR to measure signal in the HTS (i.e., no tetracycline) and the identical background cell, lacking the expressed GPCR, to serve as a negative control (i.e., with tetracycline), thereby isolating false positives that may activate cellular functions other than the GPCR target.

Mutant models of the human gonadotropin releasing hormone and vasopressin 2 receptors were selected for use in identifying novel pharmacoperone compounds. These mutants are known to be misrouted, misfolded proteins (Conn et al., 2007), hGnRHR[E^{90}K] (Conn et al., 2002), and hV2R[L^{83}Q] (Morello et al., 2000) and are naturally occurring in patients with HH and nephrogenic diabetes insipidus, respectively. Stable HeLa cells were created expressing these proteins under control of a tetracycline-off promoter.

Stable cell lines were generated as follows:

1. The stable HeLa (tTA) cell line was obtained from Peter Seeburg (Max-Planck-Institut für Medizinische Forschung, Molekulare Neurobiologie Jahnstraße 29, 69120 Heidelberg, Germany).
2. Cells were maintained in DFG growth medium (DMEM/10%FCS/ 20 μg/ml Gentamicin) and grown at 37 °C, 5% CO_2 in a humidified atmosphere until about 90% confluent.
3. The cells were washed with Dulbecco's PBS, and then trypsinized to detach the cells. Growth medium will be added to the cells to dilute out the trypsin which will be centrifuged to pellet the cells.
4. The hGnRHR[E^{90}K] or hV2R[L^{83}Q] mutants (in pTRE2-Hygromycin vector) were transfected into the stable HeLa (tTA) cell line.
5. Selection antibiotics were used at 400 μg/ml G418 plus 200 μg/ml Hygromycin. Single colonies were selected and screened for expression of the mutant GPCRs.
6. The dual stable cell lines were maintained using 200 μg/ml G418 plus 100 μg/ml Hygromycin in growth medium. Subcloning was used to select the best-expressing lines.

3. SELECTION OF ENDPOINT MEASURES

Assays which are to be used in high-throughput settings must satisfy several important criteria:

1. Assays should be amenable to use in high-density microplates (384 or 1536 well).
2. Endpoint reads should be fluorescent-, luminescent-, or absorbance-based and should not involve radioactive labels if at all possible.
3. The signal window (high signal to background) should be large enough and the signal-to-noise ratio high enough that compounds with the desired activity can be identified. This is often quantitated using the Z' statistic as described by Zhang, Chung, and Oldenburg, (1999).

The use of HeLa cells stably expressing the model mutant receptors addresses the first criteria since this cell type has been widely used in both high-throughput and ultra-high-throughput applications (Madoux et al., 2010). However, the endpoint measures used in the original publication of this assay either used radioactive markers (in the case of GnRHR) or were chromatographic in nature (in the case of V2R), neither of which are generally tenable methods for high-throughput screening. In order to overcome these limitations, alternative endpoint measures were selected and optimized as described below.

3.1. V2R assay endpoint measure

Since pharmacoperone compounds act by restoring both proper folding and trafficking of their target receptors, an endpoint measure which captures restoration of an active signaling pathway is needed. For this reason, cAMP, a downstream effector molecule upregulated by V2R activation was chosen as an appropriate method to quantitate the effects of potential pharmacoperones. Another advantage of this approach is the wide range of methods available to quantitate cellular cAMP levels many of which are amenable to high-throughput applications (Martikkala, Rozwandowicz-Jansen, Hanninen, Petaja-Repo, & Harma, 2011; Pantel et al., 2011; Prystay, Gagne, Kasila, Yeh, & Banks, 2001). For this particular protocol, we chose the cAMP-Glo™ reagent available from Promega which uses an enzyme-linked method to quantitate cAMP levels (Kumar, Hsiao, Vidugiriene, & Goueli, 2007). This approach resulted in a robust signal and a Z' value of 0.7, more than sufficient for high-throughput screening (Fig. 1.1A). The positive control used in this experiment was SR121463, a selective nonpeptidyl V2R agonist known to act as a pharmacoperone (Ranadive et al., 2009; Serradeil-Le Gal, 2001).

1. Plate 30 μl HeLa cells stably expressing V2R $L^{83}Q$ at 1800 cells per well in a 384-well tissue culture treated microplate. Cells are cultured and plated in DMEM, 10% FBS, and 20 μg/ml Gentamicin.
2. Grow cells at 37 °C, 5% CO_2 for 48 h.
3. Add test compounds and positive/negative control compounds to appropriate wells. Compounds are generally prepared as DMSO stock solutions. In general, final DMSO percentage should be kept constant and should never be greater than 1% by volume.
   a. A minimum of 8 high signal (DMSO alone) and 8 low signal (1 μM SR121463B) wells should be on each plate. We generally use 16 high and 16 low wells to monitor assay performance and select active compounds.

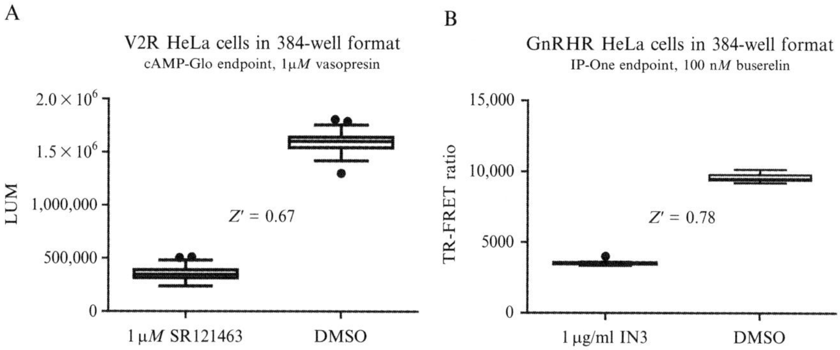

Figure 1.1 (A) Performance of the cAMP-Glo endpoint measure used with HeLa cells expressing V2R $L^{83}Q$. The assay was performed using a partially automated approach as described in Section 4. The data presented are of raw endpoints from a representative 384-well microplate. $n=16$ for both DMSO and positive control populations. (B) Performance of the IP-One endpoint measure used with HeLa cells expressing GnRHR $E^{90}K$. The assay was performed using a partially automated approach as described in Section 4. The data presented are of a representative 384-well microplate. $n=16$ for both DMSO and positive control populations.

   b. It is also useful to include a series of SR121463B dosages on the plate as well to monitor assay performance. We generally start at 1 μ*M* and go down to 0.0005 μ*M* in a 1:3 dilution series.

4. Incubate cells with test compounds at 37 °C, 5% CO_2 for 16 h.

5. Wash cells 3 × 60 μl with DMEM, 0.1% (w/v) BSA and 20 μg/ml Gentamicin plus 1% DMSO by volume.
   a. Wash 1–10 min at 37 °C
   b. Wash 2–10 min at 37 °C
   c. Wash 3–20 min at 37 °C

6. Remove growth media from wells and add 15 μl stimulation buffer (wash media plus 1 μ*M* vasopressin and 500 μ*M* 3-Isobutyl-1-methylxanthine (IBMX).

   Note: IBMX is necessary to stabilize cAMP present in the wells.

7. Incubate microplate at 37 °C, 5% CO_2 for 2 h.

8. Proceed with cAMP-Glo™ assay per the manufacturer's directions.

9. Measure luminescence in a suitable plate reader.

3.2. GnRHR assay endpoint measure

To quantitate GnRHR activation, we chose to measure inositol phosphate one (IP1), a downstream effector molecule which accumulates following receptor activation and which is stable in the presence of LiCl. The commercially available homogenous time-resolved fluorescence energy transfer

(TR-FRET) reagent IP-One™ from Cisbio was selected to measure IP1 levels in test wells following activation of the receptor with Buserelin (a metabolically stable agonist of the hGnRHR). The TR-FRET signal generated using this approach is inversely proportional to the amount of IP1 present in the test well. This reagent has been shown to be usable in high-throughput applications for other GPCRs and in this application resulted in a Z' value of 0.8 (Fig. 1.1B.) (Zhang et al., 2010). The positive control for this experiment was the indole IN3 ([(2*S*)-2-[5-[2-(2-azabicyclo[2.2.2]oct-2-yl)-1,1-dimethyl-2-oxoethyl]-2-(3,5-dimethylphenyl)-1*H*-indol-3-yl]-*N*-(2-pyridin-4-ylethyl)propan-1-amine]), an antagonist of GnRHR known to function as a pharmacoperone (Ulloa-Aguirre et al., 2003). IN3 was obtained from Merck and Company.

1. Plate 30 μl HeLa cells stably expressing hGnRHR[$E^{90}K$] at 3600 cells per well in a 384-well tissue culture treated microplate. Cells are cultured and plated in DMEM, 10% FBS, and 20 μg/ml Gentamicin.
2. Incubate cells for 48 h at 37 °C and 5% CO_2.
3. Add test compounds and positive/negative control compounds to appropriate wells. Compounds are generally prepared as DMSO stock solutions. In general, final DMSO percentage should be kept constant and should never be greater than 1% by volume.
   - a. A minimum of eight high signal (DMSO alone) and eight low signal (1 μg/ml IN3) wells should be on each plate.
   - b. It is also useful to include a series of IN3 dosages on the plate as well to monitor assay performance. We generally start at 1 μg/ml and go down to 0.0005 μg/ml in a 1:3 dilution series.
4. Incubate cells for 4 h at 37 °C and 5% CO_2.
5. Wash cells 3 × 60 μl with DMEM, 0.1% (w/v) BSA, and 20 μg/ml Gentamicin plus 1% DMSO by volume.
   - a. Wash 1–10 min at 37 °C
   - b. Wash 2–10 min at 37 °C
   - c. Wash 3–10 min at 37 °C
6. Remove growth media from wells and add 20 μl stimulation buffer (100 n*M* Buserelin, 10 m*M* Hepes, 1 m*M* $CaCl_2$, 0.5 m*M* $MgCl_2$, 4.2 m*M* KCl, 146 m*M* NaCl, 5.5 m*M* glucose, 50 m*M* LiCl, pH 7.4)
   - a. Other buffers may be substituted, but should include 50 m*M* LiCl to stabilize IP1 in the wells.
   - b. Do not use phosphate-based buffers for this.
7. Incubate microplate at 37 °C, 5% CO_2 for 2 h.
8. Proceed with IP-One™ protocol per the manufacturer's directions.
9. Read TR-FRET signal on appropriate plate reader.

4. ASSAY AUTOMATION

This protocols described in Section 3 are amenable to automation in a reasonably equipped high-throughput screening laboratory (Fig. 1.2). The required instruments to accomplish this are

1. A bulk liquid dispenser (e.g., Thermo-Fisher Wellmate™ or Multidrop™).
   Note: Used to plate cells and dispense all bulk reagents.
2. An automated multichannel liquid handling instrument (e.g., Caliper Sciclone, Beckman Coulter Biomek, Tecan Freedom Evo or equivalent).
   Note: Used to add test and control compounds.
3. An automated plate washer capable of aspirating media from a 384-well microplate (e.g., Biotek ELx405).
   Note: Used for all washes.
4. A microplate reader with luminescence and time resolved fluorescence measurements.

The general automated protocol is as follows:

1. Plate 30 μl cells using a Thermo-fisher Welllmate™ into white polystyrene Corning Costar tissue culture treated plates.

Figure 1.2 A view of the automated liquid handling station used to perform the assays described in this report. (See Color Insert.)

2. Incubate plates in an offline incubator for 48 h at 37 °C and 5% CO_2.
   Note: Incubators should be left undisturbed during this time to minimize evaporation and edge effects.
3. Add 5 μl appropriately diluted test and control compounds using the Caliper Sciclone™ automated liquid handling instrument.
4. Incubate plates in offline incubator at 37 °C and 5% CO_2 for either 4 h (GnRHR assay) or 16 h (V2R assay).
5. Remove media from all wells using a Biotek ELx405 automated plate washer.
6. Add 60 μl wash media to plate using Thermo-Fisher Wellmate™.
   Note: It is helpful to set the dispensing speed lower for this to avoid disturbing the cell monolayer.
7. Incubate plates in offline incubator for the times indicated in Section 3.
8. Repeat steps 5–7 two more times.
9. Remove wash media using Biotek ELx405 automated plate washer.
10. Add appropriate volume of stimulation buffer using the Thermo-Fisher Wellmate™.
11. Incubate plates in offline incubator at 37 °C and 5% CO_2 for 2 h.
12. Perform endpoint read (cAMP-Glo™ or IP-One™) per kit manufactures protocol.
    Note: All reagents in both of kits used can be dispensed using the Thermo-Fisher Wellmate™.

5. DATA ANALYSIS

Raw endpoint values should be calculated as directed by the kit manufactures. In order to identify compounds with the desired activities, we used a robust statistical approach to pick those which displayed a significantly different signal from the negative control population. Robust statistical approaches to defining outliers are more resistant to nonnormal populations as well as extreme outliers, both of which are often found in high-throughput screening data sets (Emerson & Strenio, 1983). By using these criteria, which are defined on a plate-by-plate basis, it is possible to identify active compounds without applying artificial percent activity cutoff values.

6. HIT FOLLOW-UP EXPERIMENTS

In order to rule out the possibility that compounds showing GPCR activation are not acting at other cellular receptors, it is important to repeat the described experiments in the presence of 1 μg/ml doxycycline, which prevents the mutant GPCRs from being expressed. If the activation signal is still seen in the presence of doxycycline, it can be inferred that it is due to an off-target effect rather than pharmacoperone activity. Note that on-target activity does not indicate that compounds are directly interacting with the mutant GPCRs. A doxycycline-dependent signal simply shows that the compound is causing recruitment of the mutant GPCR to the cellular surface. Further studies are necessary to establish a direct interaction between the small molecule and the target of interest (Conn et al., 2002; Janovick et al., 2003; Morello et al., 2000).

It is also important to note that pharmacoperone compounds often do not follow classical dose–response behavior. In the case of SR121463B and the V2R assay, a linear nonsaturating response is observed (Fig. 1.3A). However, for IN3 and GnRHR, a classical saturating response is observed (Fig. 1.3B). For this reason, using standard dose dependency experiments for classifying hit compounds should be done only with caution as normal curve fitting protocols using a sigmoidal model may fail.

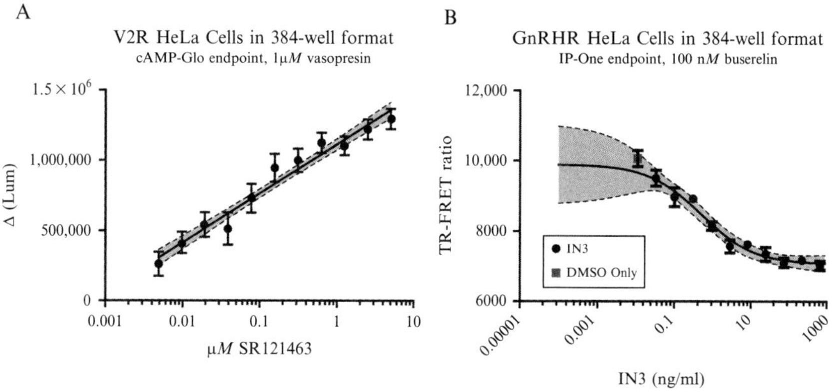

Figure 1.3 (A) A dose dependence curve showing the nonsaturating behavior of the V2R $L^{83}Q$ pharmacoperone SR121463B. $n=8$ for each tested concentration. The data were fit to a semilogarithmic straight line ($y=m(\log([SR121463])+b)$ in Graphpad PRISM 5.02 using a least-squares method. The Δ luminescence endpoint used for these calculations was obtained by subtracting the luminescence value of the treated well from the trimmed mean of DMSO only treated wells (negative control wells) as recommended by the cAMP-Glo manufacturer. The data presented here were obtained using the semi-automated protocol described in Section 4. (B) A dose dependence curve showing the saturating behavior of IN3 with GnRHR $E^{90}K$. The data were fit to a standard four-parameter sigmoidal dose–response curve in Graphpad PRISM 5.02 using a least-squares method. $n=8$ for each tested concentration. The data presented here were obtained using the semi-automated protocol described in Section 4. (For color version of this figure, the reader is referred to the online version of this chapter.)

7. CONCLUSIONS

The methods described and characterized here provide the basis of automated, robotic primary screens for pharmacoperones that detect drugs which rescue GPCR mutants of specific receptors. This approach will identify structures that would have been missed in screens that were designed to select only agonists or antagonists. Nonantagonistic pharmacoperones have a therapeutic advantage since they will not compete for endogenous agonists and may not have to be washed out once rescue has occurred and before activation by endogenous or exogenous agonists.

ACKNOWLEDGMENTS

This work was supported by National Institutes of Health Grants DK85040, OD012220, OD011092, and the Oregon Translational Research & Drug Development Institute (OTRADI), Innovation and Commercialization Fund (OICF) supported by awards from the Oregon Innovation Council and U.S. Department of Commerce Economic Development Administration i6 Challenge Award (EDA077906632).

REFERENCES

Amaral, M. D. (2006). Therapy through chaperones: Sense or antisense? Cystic fibrosis as a model disease. *Journal of Inherited Metabolic Disease, 29*, 477–487.

Benedek, G. B., Pande, J., Thurston, G. M., & Clark, J. I. (1999). Theoretical and experimental basis for the inhibition of cataract. *Progress in Retinal and Eye Research, 18*, 391–402.

Bernier, V., Bichet, D. G., & Bouvier, M. (2004). Pharmacological chaperone action on G-protein-coupled receptors. *Current Opinion in Pharmacology, 4*, 528–533.

Bernier, V., Lagace, M., Bichet, D. G., & Bouvier, M. (2004). Pharmacological chaperones: Potential treatment for conformational diseases. *Trends in Endocrinology and Metabolism, 15*, 222–228.

Bichet, D. G. (2006). Nephrogenic diabetes insipidus. *Advances in Chronic Kidney Disease, 13*, 96–104.

Bottomley, S. P. (2011). The structural diversity in alpha1-antitrypsin misfolding. *EMBO Reports, 12*, 983–984 PMC3185355.

Conn, P. M., & Janovick, J. A. (2011). Pharmacoperone identification for therapeutic rescue of misfolded mutant proteins. *Frontiers in Endocrinology (Lausanne), 2*.

Conn, P. M., Leanos-Miranda, A., & Janovick, J. A. (2002). Protein origami: Therapeutic rescue of misfolded gene products. *Molecular Interventions, 2*, 308–316.

Conn, P. M., & Ulloa-Aguirre, A. (2010). Trafficking of G-protein-coupled receptors to the plasma membrane: Insights for pharmacoperone drugs. *Trends in Endocrinology and Metabolism, 21*, 190–197 PMC2831145.

Conn, P. M., Ulloa-Aguirre, A., Ito, J., & Janovick, J. A. (2007). G protein-coupled receptor trafficking in health and disease: Lessons learned to prepare for therapeutic mutant rescue in vivo. *Pharmacological Reviews, 59*, 225–250.

Dormer, R. L., Derand, R., McNeilly, C. M., Mettey, Y., Bulteau-Pignoux, L., Metaye, T., et al. (2001). Correction of delF508-CFTR activity with benzo(c)quinolizinium compounds through facilitation of its processing in cystic fibrosis airway cells. *Journal of Cell Science, 114*, 4073–4081.

Emerson, J. D., & Strenio, J. (1983). Boxplots and batch comparison. In D. C. Hoaglin, F. Mosteller & J. Tukey (Eds.), *Understanding robust and exploratory data analysis* (pp. 59–96). New York: John Wiley & Sons.

Fan, J. Q. (2003). A contradictory treatment for lysosomal storage disorders: Inhibitors enhance mutant enzyme activity. *Trends in Pharmacological Sciences, 24*, 355–360.

Feldhammer, M., Durand, S., & Pshezhetsky, A. V. (2009). Protein misfolding as an underlying molecular defect in mucopolysaccharidosis III type C. *PloS One, 4*, e7434.

Forloni, G., Terreni, L., Bertani, I., Fogliarino, S., Invernizzi, R., Assini, A., et al. (2002). Protein misfolding in Alzheimer's and Parkinson's disease: Genetics and molecular mechanisms. *Neurobiology of Aging, 23*, 957–976.

Galietta, L. J., Springsteel, M. F., Eda, M., Niedzinski, E. J., By, K., Haddadin, M. J., et al. (2001). Novel CFTR chloride channel activators identified by screening of combinatorial libraries based on flavone and benzoquinolizinium lead compounds. *The Journal of Biological Chemistry, 276*, 19723–19728.

Gruber, C. W., Muttenthaler, M., & Freissmuth, M. (2010). Ligand-based peptide design and combinatorial peptide libraries to target G protein-coupled receptors. *Current Pharmaceutical Design, 16*, 3071–3088.

Heiser, V., Scherzinger, E., Boeddrich, A., Nordhoff, E., Lurz, R., Schugardt, N., et al. (2000). Inhibition of huntingtin fibrillogenesis by specific antibodies and small molecules: Implications for Huntington's disease therapy. *Proceedings of the National Academy of Sciences of the United States of America, 97*, 6739–6744.

Janovick, J. A., Brothers, S. P., Cornea, A., Bush, E., Goulet, M. T., Ashton, W. T., et al. (2007). Refolding of misfolded mutant GPCR: Post-translational pharmacoperone action in vitro. *Molecular and Cellular Endocrinology, 272*, 77–85 PMC2169380.

Janovick, J. A., Goulet, M., Bush, E., Greer, J., Wettlaufer, D. G., & Conn, P. M. (2003). Structure-activity relations of successful pharmacologic chaperones for rescue of naturally occurring and manufactured mutants of the gonadotropin-releasing hormone receptor. *The Journal of Pharmacology and Experimental Therapeutics, 305*, 608–614.

Janovick, J. A., Park, B. S., & Conn, P. M. (2011). Therapeutic rescue of misfolded mutants: Validation of primary high throughput screens for identification of pharmacoperone drugs. *PloS One, 6*, e22784 PMC3144936.

Kumar, M., Hsiao, K., Vidugiriene, J., & Goueli, S. A. (2007). A bioluminescent-based, HTS-compatible assay to monitor G-protein-coupled receptor modulation of cellular cyclic AMP. *Assay and Drug Development Technologies, 5*, 237–245.

Madoux, F., Simanski, S., Chase, P., Mishra, J. K., Roush, W. R., Ayad, N. G., et al. (2010). An ultra-high throughput cell-based screen for wee1 degradation inhibitors. *Journal of Biomolecular Screening, 15*, 907–917 3082437.

Marciniak, S. J., & Ron, D. (2006). Endoplasmic reticulum stress signaling in disease. *Physiological Reviews, 86*, 1133–1149.

Martikkala, E., Rozwandowicz-Jansen, A., Hanninen, P., Petaja-Repo, U., & Harma, H. (2011). A homogeneous single-label time-resolved fluorescence cAMP assay. *Journal of Biomolecular Screening, 16*, 356–362.

Morello, J. P., & Bichet, D. G. (2001). Nephrogenic diabetes insipidus. *Annual Review of Physiology, 63*, 607–630.

Morello, J. P., Salahpour, A., Laperriere, A., Bernier, V., Arthus, M. F., Lonergan, M., et al. (2000). Pharmacological chaperones rescue cell-surface expression and function of misfolded V2 vasopressin receptor mutants. *The Journal of Clinical Investigation, 105*, 887–895.

Muchowski, P. J., & Wacker, J. L. (2005). Modulation of neurodegeneration by molecular chaperones. *Nature Reviews. Neuroscience, 6*, 11–22.

Noorwez, S. M., Malhotra, R., McDowell, J. H., Smith, K. A., Krebs, M. P., & Kaushal, S. (2004). Retinoids assist the cellular folding of the autosomal dominant retinitis pigmentosa opsin mutant P23H. *The Journal of Biological Chemistry, 279*, 16278–16284.

Pantel, J., Williams, S. Y., Mi, D., Sebag, J., Corbin, J. D., Weaver, C. D., et al. (2011). Development of a high throughput screen for allosteric modulators of melanocortin-4 receptor signaling using a real time cAMP assay. *European Journal of Pharmacology, 660*, 139–147 3175485.

Peng, Y., Li, C., Chen, L., Sebti, S., & Chen, J. (2003). Rescue of mutant p53 transcription function by ellipticine. *Oncogene, 22*, 4478–4487.

Permanne, B., Adessi, C., Saborio, G. P., Fraga, S., Frossard, M. J., Van Dorpe, J., et al. (2002). Reduction of amyloid load and cerebral damage in a transgenic mouse model of Alzheimer's disease by treatment with a beta-sheet breaker peptide. *The FASEB Journal, 16*, 860–862.

Prystay, L., Gagne, A., Kasila, P., Yeh, L. A., & Banks, P. (2001). Homogeneous cell-based fluorescence polarization assay for the direct detection of cAMP. *Journal of Biomolecular Screening, 6*, 75–82.

Ranadive, S. A., Ersoy, B., Favre, H., Cheung, C. C., Rosenthal, S. M., Miller, W. L., et al. (2009). Identification, characterization and rescue of a novel vasopressin-2 receptor mutation causing nephrogenic diabetes insipidus. *Clinical Endocrinology, 71*, 388–393.

Ron, D. (2002). Translational control in the endoplasmic reticulum stress response. *The Journal of Clinical Investigation, 110*, 1383–1388.

Ron, D., & Hubbard, S. R. (2008). How IRE1 reacts to ER stress. *Cell, 132*, 24–26.

Serradeil-Le Gal, C. (2001). An overview of SR121463, a selective non-peptide vasopressin V(2) receptor antagonist. *Cardiovascular Drug Reviews, 19*, 201–214.

Soto, C., Kascsak, R. J., Saborio, G. P., Aucouturier, P., Wisniewski, T., Prelli, F., et al. (2000). Reversion of prion protein conformational changes by synthetic beta-sheet breaker peptides. *Lancet, 355*, 192–197.

Ulloa-Aguirre, A., Janovick, J. A., Leanos-Miranda, A., & Conn, P. M. (2003). Misrouted cell surface receptors as a novel disease aetiology and potential therapeutic target: The case of hypogonadotropic hypogonadism due to gonadotropin-releasing hormone resistance. *Expert Opinion on Therapeutic Targets, 7*, 175–185.

Zhang, J. H., Chung, T. D., & Oldenburg, K. R. (1999). A simple statistical parameter for use in evaluation and validation of high throughput screening assays. *Journal of Biomolecular Screening, 4*, 67–73.

Zhang, J. Y., Kowal, D. M., Nawoschik, S. P., Dunlop, J., Pausch, M. H., & Peri, R. (2010). Development of an improved IP(1) assay for the characterization of 5-HT(2C) receptor ligands. *Assay and Drug Development Technologies, 8*, 106–113.

Zhang, X. M., Wang, X. T., Yue, H., Leung, S. W., Thibodeau, P. H., Thomas, P. J., et al. (2003). Organic solutes rescue the functional defect in delta F508 cystic fibrosis transmembrane conductance regulator. *The Journal of Biological Chemistry, 278*, 51232–51242.

CHAPTER TWO

Trafficking of the Follitropin Receptor

Alfredo Ulloa-Aguirre[*,†,1], **James A. Dias**[*,‡,1], **George Bousfield**[*,§], **Ilpo Huhtaniemi**[*,¶], **Eric Reiter**[*,||,#,**]

[*]Studium Consortium for Research and Training in Reproductive Sciences (sCORTS), Tours, France
[†]Division of Reproductive Health, Research Center in Population Health, National Institute of Public Health, México D.F., Mexico
[‡]New York State Department of Health and Department of Biomedical Sciences, Wadsworth Center, School of Public Health, University at Albany, Albany, USA
[§]Department of Biological Sciences, Wichita State University, Wichita, Kansas, USA
[¶]Institute of Reproductive and Developmental Biology, Imperial College London, London, United Kingdom
[||]BIOS Group, INRA, Unité Physiologie de la Reproduction et des Comportements, Nouzilly, France
[#]CNRS, Nouzilly, France
[**]Université François Rabelais, Tours, France
[1]Corresponding authors: e-mail address: aulloaa@unam.mx; jdias@uamail.albany.edu

Contents

1. Introduction 18
2. Outward Trafficking Defective FSHR Mutants. Studying Plasma Membrane Expression of the FSHR 23
   2.1 Methods to study FSH binding in FSHR-expressing human embryonic kidney-293 cells 24
   2.2 Methods to detect total FSHR expression by gel electrophoresis (SDS-PAGE) and Western immunoblotting 25
3. Studying Oligomerization of Intracellular and Cell Surface-Expressed FSHRs 29
   3.1 Detection of FSHR self-association at the plasma membrane by cell surface fluorescence resonance energy transfer 30
   3.2 Coimmunoprecipitation of c-myc-tagged and FLAG-tagged FSHR to detect intracellular association of FSHRs 32
4. Phosphorylation, Internalization, and Recycling of the FSHR (Downward Trafficking) 33
   4.1 Phosphorylation of the FSHR and β-arrestin recruitment 34
   4.2 Internalization in equilibrium and nonequilibrium conditions 36
   4.3 Recycling and degradation of internalized FSHRs 39

Acknowledgments 41
References 41

Abstract

The follitropin or follicle-stimulating hormone receptor (FSHR) belongs to a highly conserved subfamily of the G protein-coupled receptor (GPCR) superfamily and is mainly expressed in specific cells in the gonads. As any other GPCR, the newly synthesized FSHR

Methods in Enzymology, Volume 521
ISSN 0076-6879
http://dx.doi.org/10.1016/B978-0-12-391862-8.00002-8

has to be correctly folded and processed in order to traffic to the cell surface plasma membrane and interact with its cognate ligand. In this chapter, we describe in detail the conditions and procedures used to study outward trafficking of the FSHR from the endoplasmic reticulum to the plasma membrane. We also describe some methods to analyze phosphorylation, β-arrestin recruitment, internalization, and recycling of this particular receptor, which have proved useful in our hands for dissecting its downward trafficking and fate following agonist stimulation.

1. INTRODUCTION

The pituitary gonadotropic hormones (GPH), follitropin (FSH) and lutropin (LH), as well as placental choriongonadotropin (hCG), are glycoprotein hormones that play an essential role in gonadal function. Their cognate receptors (FSHR and LHCGR) belong, together with the thyroid-stimulating hormone receptor, to a highly conserved subfamily of the G protein-coupled receptor (GPCR) superfamily. FSHR and LHCGR are expressed by specific cells in the gonads (Vassart, Pardo, & Costagliola, 2004). The FSHR is expressed in ovarian granulosa cells where its action is required for growth and maturation of ovarian follicles and for granulosa cell estrogen production. In the testis, FSH supports the metabolism of Sertoli cells, thereby indirectly maintaining spermatogenesis. GPH receptors are characterized by a large NH_2-terminal extracellular domain (ECD), where recognition and binding of their cognate ligands occur. This ectodomain comprises of a central structural motif of nine imperfect leucine-rich repeats (LRRs), a motif that is shared with a number of other membrane receptors that are involved in ligand selectivity and specific protein–protein interactions (Bogerd, 2007). The carboxyl-terminal end of the large ECD displays the so-called hinge region, which has recently been structurally characterized (Jiang et al., 2012). The hinge region structurally links the leucine-rich ECD with the serpentine transmembrane domain of GPH receptors and depending on the GPH receptor, this region may be involved in high-affinity hormone binding, receptor activation, intramolecular signal transduction, and/or silencing the basal activity of the receptor in the absence of ligand (Mueller, Jaeschke, Gunther, & Paschke, 2009). It is interesting to note that in contrast to the LHCGR, which appears insensitive to effects on hormone binding when trafficking is impaired by mutagenesis, the FSHR is particularly sensitive to mutations of the primary sequence. For example, when mutations impair trafficking of this receptor from the endoplasmic reticulum (ER) to the cell surface plasma membrane (PM) [at the ECD, exoloop 3 (Rozell, Wang, Liu, & Segaloff, 1995), and the carboxyl-terminal domain (C-tail) (Song, Ji, Beauchamp, Isaacs, & Ji, 2001)], it is not possible to recover hormone-binding activity (Nechamen & Dias, 2000; Rozell et al., 1995; Song et al., 2001). This may be due to an inherent flexibility of the human (h) FSHR that results in metastability and, conversely, a greater stability of the LHCGR. Along these lines, it is worth noting that, with one exception (Peltoketo et al., 2010; Tao, Mizrachi, & Segaloff, 2002), activating mutations in the *LHCGR* do not appear to translate into activating mutations when analogous mutations are made in the FSHR. In humans, the *FSHR* gene is located on chromosome 2p21–p16 (Gromoll, Ried, Holtgreve-Grez, Nieschlag, & Gudermann, 1994). The coding region of this gene consists of 10 exons, ranging in size between 69 and 1234 bp, and nine introns, ranging in size between 108 and 15 kb (Gromoll, Pekel, & Nieschlag, 1996). The receptor protein consists of 695 amino acid residues; the first 17 amino acids encode a signal sequence, resulting in a mature FSHR of 678 amino acid residues long, with a molecular weight of ~75 kDa as predicted from its cDNA (Dias et al., 2002). However, glycosylation of its three or four possible glycosylation sites may give rise to receptor forms with approximate weights of 80–87 kDa for the mature receptor (see below).

Upon agonist binding, the activated FSHR stimulates a number of intracellular signaling pathways. The canonical Gαs/cAMP/PKA signaling pathway has been recognized as a key effector mechanism of LH and FSH biological action for more than 20 years. However, gonadotropin receptors have also been reported to couple to other G protein subtypes and activate a number of distinct effector enzymes (reviewed in Ulloa-Aguirre, Crepieux, Poupon, Maurel, & Reiter, 2011) depending on the particular developmental stage of the host cells (Musnier et al., 2009).

One interesting signaling cascade involved in the so-called *functional selectivity* of the FSHR is that mediated by β-arrestins (Ulloa-Aguirre et al., 2011). β-Arrestins are versatile adapter proteins that form complexes with GPCRs following receptor activation. These particular scaffold molecules play a major role in desensitization and downward trafficking (internalization and recycling) of many GPCRs (see below). In addition, they play an instrumental role as regulators of intracellular signaling events via activation of several pathways, including the MAPK pathway (Reiter & Lefkowitz, 2006). In the case of the FSHR, biased extracellular signal-regulated kinase signaling has been shown to occur at the FSHR bound to a modified agonist (Wehbi, Decourtye, et al., 2010; Wehbi, Tranchant, et al., 2010) or when its PM expression is severely reduced (Tranchant et al., 2011). In this case, β-arrestins recruited to the agonist-bound receptor assemble a MAPK module, whereas G protein-dependent signaling is impaired.

Similar to other proteins, GPCRs have to be correctly folded in order to pass through the ER quality control system (Ulloa-Aguirre & Conn, 2009). In fact, mutations that affect the folding process of these receptors result in intracellular retention or increased degradation of the folding intermediates. Several GPCR-interacting proteins that support trafficking to the PM have been identified. Among these are calnexin and calreticulin, which bind a broad range of glycoproteins facilitating proper folding of intermediate molecules (Helenius, Trombetta, Hebert, & Simons, 1997). In the case of the FSHR, coimmunoprecipitation experiments have shown that the folding process of the precursor involves interactions with these two chaperones (Mizrachi & Segaloff, 2004; Rozell, Davis, Chai, & Segaloff, 1998) as well as with the protein disulfide isomerase (PDI; Mizrachi & Segaloff, 2004). PDI is an ER-resident enzyme involved in disulfide bond formation of folding intermediates and that probably acts as a cochaperone with calnexin and calreticulin during their association with the glycoprotein hormone receptors (Mizrachi & Segaloff, 2004).

Properly folded and assembled secretory proteins are segregated from ER-resident proteins into COPII-coated vesicles for exporting to the Golgi to be further processed before being sent to their final destination. Several recently identified motifs have been shown to be involved in GPCR exit from the ER and the Golgi. Among these motifs is the $F(x)_6LL$ sequence identified in the C-tail of several GPCRs, including the glycoprotein hormone receptors (Duvernay, Zhou, & Wu, 2004). In the hFSHR, this export motif is located between amino acid residues 616 (residues are numbered according to the mature receptor, minus the 17 amino acid residues leader sequence) and 624 (Ulloa-Aguirre & Conn, 2009). The C-tail of the hFSHR contains the minimal BBXXB motif reversed in its juxtamembrane region (in the rat this motif is rather BBXB), which in other GPCRs is involved in G protein coupling (Timossi et al., 2004). The last two residues of the BXXBB motif (R617 and R618) and the preceding F616 constitute the NH_2 end of the highly conserved $F(x)_6LL$ motif. Consequently, mutations in these residues impair receptor trafficking and cell membrane localization of the receptor (Timossi et al., 2004; Zarinan et al., 2010). The third intracellular loop (iL) of the hFSHR also contains this motif and replacement of all its basic residues with alanine impairs trafficking of the receptor to the PM (Timossi et al., 2004). Receptor mutants with this sequence (e.g., R556A) are retained intracellularly and, when present in the heterozygous state, exert dominant negative effects on their wild-type (WT) counterpart, probably through forming misfolded mutant:WT receptor complexes (Zarinan et al., 2010) (see below).

Posttranslational modifications are important for GPCR export to the cell surface (Ulloa-Aguirre & Conn, 2009). Three modifications are particularly important: glycosylation, palmitoylation, and phosphorylation. Glycosylation facilitates folding of protein precursors by increasing their solubility and stabilizing protein conformation (Helenius & Aebi, 2004). The hFSHR bears four potential glycosylation sites (as defined by the consensus sequence N-X-Ser/Thr, where X is any amino acid except proline) at positions 174, 182, 276, and 301 of the extracellular NH_2-terminal domain (Dias et al., 2002). The only direct biochemical evidence that exists as to which sites are glycosylated in the hFSHR is derived from the crystal structure of the ECD residues 25–250 (Fan et al., 2005). The structure shows that carbohydrate is attached at residue N174 which protrudes into solvent, whereas no carbohydrate is attached at residue N182, which protrudes from the flat beta sheet into the hormone–receptor binding interface. No structural information is available for residues 276 and 301 at this time. The naturally occurring mutations Ala172Val and Asn174Ile cause a profound defect in targeting the receptor protein to the cell membrane (Huhtaniemi & Themmen, 2005), confirming the importance of glycosylation of FSHR at position 174 and that glycosylation at this particular location is essential for trafficking of the receptor to the PM. In contrast, it has been shown that the rat FSHR is glycosylated at two of three glycosylation consensus sequences (174, 182, and 276) and that the presence of carbohydrates at either one of these residues (N174 or N276) is sufficient for receptor folding and trafficking to the PM (Davis, Liu, & Segaloff, 1995).

Another posttranslational modification important for GPCR trafficking to the PM is palmitoylation. Cysteine residues in the COOH-terminus of several GPCR, belonging mainly to family A, have been shown to be the target for S-acylation with palmitic acid, and this posttranslational modification is often required for efficient delivery of the protein to the cell membrane. In this vein, the FSHR is not an exception. In fact, the hFSHR exhibits in its COOH-terminus (C-tail) two conserved cysteine residues (at positions 629 and 655) and one nonconserved Cys residue at position 627 (Uribe et al., 2008). Employing site-directed mutagenesis, it has been recently shown that the hFSHR is palmitoylated at all cysteine residues, regardless of their location in the C-tail of the receptor (Uribe et al., 2008). Apparently, S-acylation at C627 and C655 is not essential for efficient

FSHR cell surface membrane expression, whereas S-acylation at C629 is, as replacement of this residue with glycine or alanine reduced detection of the mature form of the receptor by ~40–70%. Further, when all palmitoylation sites are removed from the FSHR, cell surface PM expression is reduced to ~10–30%.

Agonist-induced phosphorylation is a general regulatory mechanism affecting downward trafficking of most GPCRs (Reiter & Lefkowitz, 2006). The FSHR has been reported to be phosphorylated by second messenger-dependent kinases PKA and PKC but also by G protein-coupled receptor kinases (GRK) 2, 3, 5, and 6 in various models (Kara et al., 2006; Krishnamurthy, Galet, & Ascoli, 2003; Lazari, Liu, Nakamura, Benovic, & Ascoli, 1999; Marion et al., 2002; Nakamura, Hipkin, & Ascoli, 1998; Troispoux et al., 1999). PKA and PKC contribute to both agonist-dependent (homologous) and agonist-independent (heterologous) desensitization of the receptor. GRK-mediated phoshorylation leads to more complex effects as they are centrally involved in homologous desensitization while simultaneously regulating β-arrestin recruitment and subsequent β-arrestin effects on receptor internalization through clathrin-coated pits and G protein-independent signaling. A cluster of five serines and threonines located in the C-tail of the FSHR has been shown to account for the bulk of FSH-induced phosphorylation as a result of GRK2 action (Kara et al., 2006). It is also well documented that β-arrestins are recruited to the GRK-phosphorylated and agonist-occupied FSHR (Kara et al., 2006; Krishnamurthy, Galet, et al., 2003; Lazari et al., 1999; Marion et al., 2002; Nakamura et al., 1998; Troispoux et al., 1999). In addition to GRK2, GRK5 and 6 have also been found to contribute to the same processes in HEK293 cells, though to a lesser extent (Kara et al., 2006). Interestingly, β-arrestins recruited to GRK2- or GRK5/6-phosphorylated FSHR have been suggested to exert distinct intracellular functions (Kara et al., 2006; Reiter & Lefkowitz, 2006). It is well established that β-arrestin 1 and 2 binding to GRK-phosphorylated FSHR leads to the internalization and recycling of the receptor (Kara et al., 2006; Kishi, Krishnamurthy, Galet, Bhaskaran, & Ascoli, 2002; Lazari et al., 1999; Nakamura et al., 1998; Piketty, Kara, Guillou, Reiter, & Crepieux, 2006). As also reported for other GPCR, GRK2-phosphorylated FSHR has been reported to predominate in the β-arrestin-mediated desensitization process whereas GRK5 and 6-induced phosphorylation of the activated FSHR is required for β-arrestin-dependent signaling pathway in HEK293 cells (Kara et al., 2006; Reiter & Lefkowitz, 2006).

Given the importance of cellular trafficking and correct placement of the receptor in the PM for normal function, in the following sections, we will briefly describe the basic methods used to analyze: (1) Human FSHR protein expression by radioligand binding assays and immunoblotting, (2) cell surface PM and intracellular association of FSHRs by acceptor photobleaching fluorescence resonance energy transfer (FRET) and coimmunoprecipitation, (3) phosphorylation and β-arrestins recruitment to the FSHR, and (4) internalization and recycling of the FSHR. Sections start with brief background information related to the application of the method(s) described.

2. OUTWARD TRAFFICKING DEFECTIVE FSHR MUTANTS. STUDYING PLASMA MEMBRANE EXPRESSION OF THE FSHR

As mentioned above, the hFSHR is unusually sensitive to mutations. In fact, all naturally occurring mutations in the ECD of this receptor (I143T, A172V, N174I, D207V, and P329R) drastically impair targeting of the receptor to the PM and thereby limit hormone binding and intracellular signaling. Mutations at or near the transmembrane region have minimal effects on FSH binding but impair to variable extent signal transduction. Therefore, it seems that the location, rather than nature of the amino acid alteration, is a stronger determinant of the functional response. As described above, A172 is the first amino acid in the perfectly conserved stretch of five amino acids in gonadotropin receptors (AFNQT), which is also the locus of loss-of-function mutations in the LHCGR (Gromoll et al., 2002). Valine in this position as well as isoleucine in position 174 may interfere with structural integrity of the LRR which hosts the glycosylation site, and perturbation of this structure likely impairs proper receptor LRR formation, especially the alpha helical portion of the LRR. The putative loss of glycosylation may affect trafficking to the PM. In fact, it has been shown that when the A172V mutant is overexpressed *in vitro*, only a very small proportion of the mutated receptor is present on the PM but is functional, though it has yet to be determined whether this form of the receptor is glycosylated or not at the N174 site. Instead, most of the mutated receptor is sequestered inside the cell, explaining the inactivation mechanism (Huhtaniemi & Alevizaki, 2007; Rannikko et al., 2002). Mutations at the NH_2-terminal end of the hFSHR ectodomain also affect expression of the receptor. Alanine scanning mutagenesis of this region has identified two regions encompassing amino acids 12–14 and 22–30 whose primary sequence is important for receptor trafficking (Nechamen & Dias, 2000, 2003). In particular, mutations at F13, I23, D26, L27, R29, and N30 considerably reduced cell surface expression due to impaired intracellular trafficking (Nechamen & Dias, 2003). Remarkably, mutations at these sites impair proper glycosylation of the receptor but this is likely due to inappropriate amino terminal folding and trapping of these intermediates by surveillance proteins which then block appropriate glycosylation processing of endoglycosidase H (an enzyme that cleaves asparagine-linked mannose-rich oligosaccharides, but not highly processed complex oligosaccharides from glycoproteins)-sensitive molecules in the ER–Golgi (Nechamen & Dias, 2003).

Radioligand receptor-binding assays, which detect only functional (i.e., ligand bindable) receptor molecules and Western immunoblotting, which detects both immature and mature forms of receptor have been used to detect PM expression of the FSHR. For the latter, the anti-human FSHR antibody mAb 106.105 has been employed (Lindau-Shepard, Brumberg, Peterson, & Dias, 2001). This antibody was generated by immunization of mice with the ECD of the hFSHR (residues 1–350) expressed in insect cells and purified. The monoclonal antibody maps to the peptide epitope comprised by amino acid residues 300–315 (317–332 when leader sequence is included), without showing reactivity for the rat FSHR. For mature, binding-competent FSHRs both methods appear to correlate well with each other (Nechamen & Dias, 2003; Zarinan et al., 2010). Additional FSHR antibodies prepared against bacterial-expressed FSHR-ECD are available from the American Tissue Type Collection, which appear suitable for this method (Vannier, Loosfelt, Meduri, Pichon, & Milgrom, 1996) (see also Section 4.1 for NH_2-terminally tagged FSHR constructs).

2.1. Methods to study FSH binding in FSHR-expressing human embryonic kidney-293 cells

Cultured human embryonic kidney-293 (HEK-293) cells stably or transiently expressing the WT hFSHR or mutants are used. In the case of transiently transfected cells, binding assays may be performed 48 h after transfection.

2.1.1 Radioreceptor assay to determine affinity constant and number of binding sites

This assay is an equilibrium displacement binding isotherm assay necessary to calculate the K_D and the number of cell surface receptors, and is performed under conditions where no internalization can occur. The radioreceptor assay (RRA) method we have used is the following:

Prepare low specific activity (25–30 μCi/μg protein) ^{125}I-FSH by diluting ^{125}I-FSH in RRA buffer (50 mM Tris, pH 7.5; 25 mM $MgCl_2$; 0.3% bovine serum albumin) so that final concentration of tracer is 20 ng/ml ($\sim 5 \times 10^5$cpm/50 μl). In our experience, iodination of 20–50 μg highly purified pituitary FSH using lactoperoxidase yields a FSH tracer with the desired specific activity (Weiner & Dias, 1992). Highly purified natural or recombinant hFSH is available from Prospec (Ness-Ziona, Israel). Prepare unlabeled pituitary FSH samples in 100 μl RRA buffer. Make eight dilutions and assay each in triplicate (1000, 300, 100, 30, 10, 3, 1, and 0 ng); the 1000 ng dose will be employed to calculate the nonspecific binding (NSB). Alternatively, untransfected HEK-293 may be used to determine NSB. Thereafter, wash transfected HEK-293 cells (cultured in 100 × 20 mm dishes) twice with unsupplemented medium and detach cells with 1 ml trypsin solution. This brief treatment of cells with trypsin does not decrease radiolabeled FSH binding, but this condition should be verified in each laboratory because the quality of trypsin preparations may differ. Add 9 ml supplemented 10% bovine serum albumin-containing medium and count cells. Remove medium by centrifugation and resuspend cells in RRA buffer at a density of 150,000 cells/150 μl. Based on the results of total (specific) ^{125}I-FSH binding (see above), adjust the number of cells transfected with expression-deficient FSHR plasmids to yield a comparable cpm of ^{125}I-FSH binding in the RRA in order to accurately compare K_D values between WT receptor and mutants. Add 150 μl of the cell suspension to culture tubes, then each dose of unlabeled FSH (in 100 μl), and finally 50 μl of the ^{125}I-FSH solution. Allow equilibrium by incubating the mixture for 18 h at room temperature (RT) with constant shaking. At the end of the incubation period, add 3-ml ice-cold RRA buffer and centrifuge at 1500 × g for 30 min at 4 °C. Transfer the tubes to sponge racks for holding and decanting assay tubes into a proper container, and dry the pellet by inverting the tubes over a paper towel, allowing the sample to air dry for 2–3 min. Count pellets in a gamma counter, plot data (Uribe et al., 2008), and analyze using any available computer software (Cohen, Bariteau, Magenis, & Dias, 2003).

2.2. Methods to detect total FSHR expression by gel electrophoresis (SDS-PAGE) and Western immunoblotting

The anti-human FSHR antibody mAb 106.105 detects both mature (m.w. ~80 kDa), membrane-anchored as well as immature, underglycosylated, high mannose ER-localized forms (m.w. ~75 kDa) of the receptor (Lindau-Shepard et al., 2001; Nechamen & Dias, 2003; Fig. 2.1A). Expression-defective hFSHR mutants are retained intracellularly and endoglycosidase

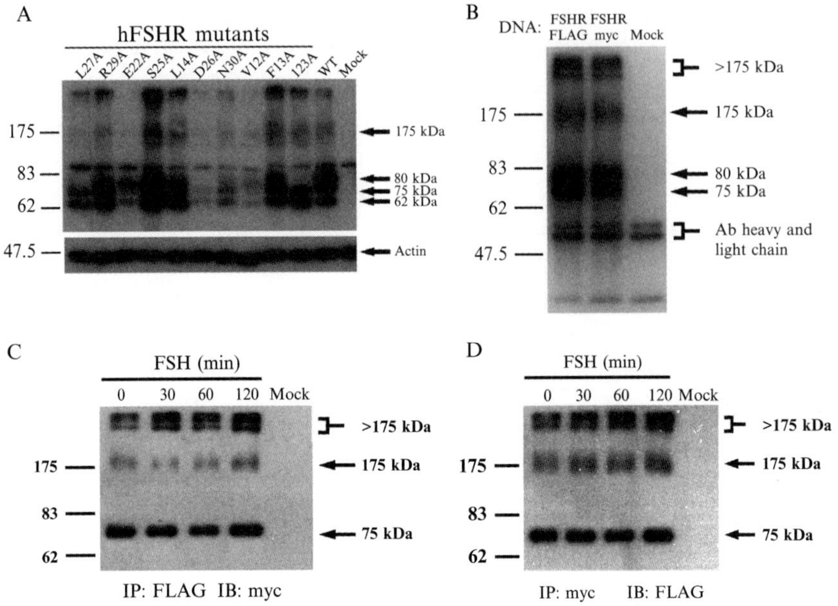

Figure 2.1 (A) and (B) Visualization of the hFSHR by immunoblotting. HEK 293 cells were transiently transfected with the hFSHR constructs indicated using Lipofectamine reagent according to the manufacturer's recommendations. Forty-eight hours later, cells were solubilized in Igepal/DOC lysis buffer, protein content of lysates was quantified by BCA protein assay, equal amounts of protein were analyzed by SDS-PAGE and Western blot (A) or immunoprecipitated with 5 μg of mAb 106.105, and then analyzed by SDS-PAGE and Western blot (B). Both blots were probed with 5 μg of mAb 106.105 followed by goat anti-mouse HRP conjugate secondary antibody at 1:10,000. In (A), the blot was reprobed with anti-actin Ab (1:5000) followed by goat anti-rabbit HRP conjugate secondary Ab (1:10,000). For reasons which are unclear at present, the 62 kDa form of receptor is not coimmunoprecipitated. The 75 and 175 kDa bands represent immature forms of the receptor which have not yet had the C-tail clipped. The 175 kDa form is either irreversibly aggregated or irreversibly bound to, for example, a chaperone protein. (C) and (D) Detection of intracellular oligomerization by coimmunoprecipitation and immunoblotting. HEK 293 cells were cotransfected with FLAG- and myc-tagged FSHR or mock transfected and were then treated with 1.2 n*M* FSH for the indicated times. Cell lysates were immunoprecipitated with anti-FLAG (C) or anti-myc (D) mAbs and then analyzed by SDS and Western blots. Immunoblots were probed with biotinylated myc mAb and HRP-conjugated streptavidin to detect myc-tagged FSHR in FLAG immunoprecipitates (C) or with HRP-conjugated FLAG mAb to detect FLAG-tagged FSHR in myc immunoprecipitates (D). *Figures B–D are reproduced from Thomas et al. (2007) with permission from the authors and The Endocrine Society. Copyright 2007, The Endocrine Society.*

H treatment causes a gel shift of the 75 kDa band representing immature, incompletely processed FSHR down to the deglycosylated 55 kDa band, while the mature form band is unaffected (Nechamen & Dias, 2003). HEK-293 cells stably or transiently expressing the FSHR and cultured in 60 × 20 mm culture plates may be used to detect the FSHR by SDS-PAGE and Western immunoblotting. If transiently transfected cells are used, this procedure may be performed 48 h after transfection.

2.2.1 SDS-PAGE

Prepare resolving and stacking gel solutions for 7.5% discontinuous buffer system SDS-PAGE (Laemmli, 1970) and polymerize in an assembled casting frame and stand for a 8.3 × 7.3 cm gel electrophoresis apparatus (e.g., Mini-PROTEAN II for 1-D vertical gel electrophoresis; Bio-Rad, Hercules, CA); also prepare the sample buffer (1 × SDS sample buffer: 0.0625 *M* Tris, pH 6.8; 2% SDS; 10% glycerol; 0.1 *M* DTT; 0.01% bromophenol blue) and running electrophoresis buffer (1 × electrophoresis buffer: 0.025 *M* Tris, 0.192 *M* glycine, 0.1% SDS). To prepare protein extracts, remove medium from FSHR-expressing cultured cells, wash twice with 1 ml ice-cold 1 × PBS, and harvest cells in Igepal–DOC lysis buffer (10 m*M* Tris, pH 7.5; 1% Igepal; 0.4% deoxycholate; 140 m*M* NaCl; 5 m*M* EDTA) supplemented with protease inhibitor cocktail tablets (Roche Diagnostics, Indianapolis, IN). Using a Dounce homogenizer, extracts are prepared with 10 strokes using a tight fitting pestle. Clean extracts by centrifugation at 13,000 rpm for 10 min at 4 °C and transfer supernatants to 1.5 ml microfuge vials. Determine protein concentration to be loaded onto gels as well as the volume to load per well, depending on different comb configurations (e.g., 10–15 μg protein/gel lane in a volume of 15–20 μl for 0.75 mm combs). If samples are too concentrated, dilute with 1 × PBS before adding sample buffer (2 μl sample + 18 μl 1 × PBS + 3.3 μl 6 × sample buffer). Warm samples for 5 min at 60 °C; it is critically important to *not* boil the samples as is usually done with SDS gels because this will cause aggregation of the receptor/sample. Spin tubes at 13,000 rpm for ~1 min at RT in microfuge and load samples using appropriate gel-loading tips. Place electrode assembly with loaded samples into the electrophoresis chamber, fill the electrode assembly carefully and then the electrophoresis chamber both with 1 × electrophoresis buffer. Run gels at 85 V for stacking gel and 100–150 V for resolving gel (can increase to 200 V for resolving gel if running on ice).

2.2.2 Western immunoblotting

After SDS-PAGE, transfer mini-gels to PVDF (polyvinylidene difluoride) membranes (0.45-μm pore size) by electroblotting (wet transfer or semidry transfer); make sure that prestained molecular weight markers transfer to the membrane matrix. After protein electrotransfer, wash membrane with Ponceau-S red staining solution (0.1% (w/v) Ponceau-S in 1% (v/v) acetic acid) to rapidly determine efficiency and evenness of transfer of electrophoresed proteins, then destain with 1 × TBST wash buffer (0.01 *M* Tris–HCl, 0.15 *M* NaCl, 0.05% Tween 20). Incubate membranes in 10-ml blocking buffer (5% (w/v) nonfat milk/1 × TBST) for 1 h at RT or overnight at 4 °C on platform rocker. After blocking, wash blots with 50 ml 1 × TBST three times, 5–10 min each on an orbital shaker. Incubate blots with blocking buffer containing 5 μg purified anti-human FSHR antibody mAb 106.105 overnight at 4 °C or 1 h at RT on platform rocker. Wash blots again 3 × with 50 ml 1 × TBST on orbital shaker, and then incubate blots with goat anti-mouse antibody (secondary antibody) conjugated to horseradish peroxidase (HRP) (at a 1:10,000 dilution in blocking buffer) for 1 h at RT or overnight at 4 °C on rocker. After incubation with the secondary antibody, wash blots with 50 ml 1 × TBST on orbital shaker three times, 5–10 min each, and then add 5 ml enhanced chemiluminescent substrate reagent and incubate for 5 min. Place blot in between transparency film, gently remove air bubbles and expose to autoradiography film (e.g., Kodak Biomax light autoradiography film, 13 × 18 cm; Perkin Elmer Life Sciences, Santa Clara, CA) in a darkroom. It is recommended to expose for 1 min to check signal intensity and then continue with longer or shorter exposures as needed (maximal. exposure length = 60 min, minimal exposure length = 1 s). For high backgrounds, wash blot in 1 × TBST then redevelop. Figure 2.1A shows a typical autoradiogram from an immunoblot of cells expressing WT hFSHR and traffic-defective hFSHR mutants. In the above described SDS-PAGE system, the 80-kDa band represents the mature receptor with complex oligosaccharides; this form is the only species of FSHR biotinylated by cell-impermeant biotin (Thomas et al., 2007). The 62-kDa band corresponds to the high mannose precursor glycoprotein localized in the ER and the 75-kDa band represents immature form of FSHR not yet expressed on the cell surface.

2.2.3 Immunoprecipitation of the FSHR

In order to detect the interaction of FSHR with other proteins or to study the posttranslational modifications of FSHR, it is necessary to isolate the receptor from cellular constituents. Since the mAb 106.105 antibody recognizes both mature and immature forms of the hFSHR, SDS-PAGE and Western blotting also can be applied to visualize this receptor and its interacting proteins from cell extracts immunoprecipitated with this antibody (Nechamen & Dias, 2003). To this end, confluent hFSHR-expressing cells cultured in two 60 mm dishes (which provide enough extract for at least two immunoprecipitations) are lysed as described above with Igepal/DOC lysis buffer. Five hundred micrograms of protein from each lysate are precleared by incubating with 100 μl protein A-agarose beads (Millipore, Billerica, MA) for 30 min at 4 °C with end-over-end rotation. Agarose beads are pelleted in a microfuge (5 min) and the supernatants are carefully poured off into a clean tube. Immunoprecipitations are performed with 5 μg mAb 106.105 (or mouse IgG2b isotype control) overnight at 4 °C (rocking is not needed). Antibody–protein complexes are then pulled-down by adding 100 μl protein A-agarose beads followed by incubation at 4 °C for 2 h with end-over-end rotation. After incubation, extracts are underlayered with a 0.5 ml cushion of 0.5 × lysis buffer/30% sucrose and spun in a microfuge at 13,000 rpm for 20 s. Beads are washed twice more with lysis buffer (0.5 ×) and finally resuspended in 60 μl 2 × Laemmli buffer and stored at − 20 °C until used for SDS-PAGE and Western immunoblotting. Before SDS-PAGE, the suspension is heated to 60 °C for 5 min, spun for 1 min, and the supernatants are carefully poured off into clean tubes. Usually 40–60 μl/tube are recovered.

When immunoblotting is performed using monoclonal anti-hFSHR antibody, both the mature and immature forms of the FSHR are detected. Importantly, this procedure also allows a better identification of high-molecular-weight FSHR bands with molecular weights ≥175 kDa, which may represent immature FSHRs complexed with a chaperone or a cargo protein that *are not* dissociable by SDS (Thomas et al., 2007; Fig. 2.1B).

3. STUDYING OLIGOMERIZATION OF INTRACELLULAR AND CELL SURFACE-EXPRESSED FSHRs

As with many other GPCRs, the FSHR forms dimers or oligomers early during receptor biosynthesis (Thomas et al., 2007), which may be important to allow correct intracellular trafficking of the complex to the PM as well as for signal diversification through coupling to multiple G proteins (Nechamen, Thomas, & Dias, 2007). The mechanism and extent of FSHR self-association is not known. The ligand-bound ectodomain of the FSHR has been crystallized in two forms. (Fan et al., 2005; Jiang et al., 2012). The structures show how follicle-stimulating hormone binds to the curved inner surface of the receptor ectodomain in a hand-clasp fashion and both α- and β-subunits interact with the β-strand residues of the receptor. In one crystal structure of the ectodomain, two monomers were weakly associated, suggesting that the FSHR could form a dimer of two occupied receptors, if the interacting surface was via the alpha helical portion of the LRR ectodomain (Fan & Hendrickson, 2005, 2007). However, the FSHR-ECD/FSHR-ECD-interacting interface is weak and is not likely to be the major stabilizing force for dimerization. Further, the area of contact is small and may not represent a physiologically relevant protein interaction site for dimerization. The more recent crystal structure, which included the entire 350 amino acid ectodomain demonstrated an additional mode of association of hormone with the extracellular domain that included the hinge regions of the receptor, and a trimeric receptor structure (Jiang et al, 2012). Also chimeric, mutational, and interfering peptide fragment studies have suggested that the association between hFSHR monomers might occur through the TM domains (Guan et al., 2010; Osuga et al., 1997; Zarinan et al., 2010) or the C-tail (Zarinan et al., 2010), as it has been demonstrated for the β_2-adrenergic receptor (Hebert et al., 1996) and the dopamine D_2-receptor homodimerization (Lee, O'Dowd, Rajaram, Nguyen, & George, 2003; Ng et al., 1996) and μ- and δ-opiod receptors hetero-oligomerization (Fan et al., 2005).

3.1. Detection of FSHR self-association at the plasma membrane by cell surface fluorescence resonance energy transfer

Membrane receptor oligomerization has to date been primarily studied using FRET which occurs over intermolecular distances between 1.0 and 10.0 nm and is thus a way to assess protein–protein interaction even under conditions of very small clusters of proteins in the PM (Kenworthy, 2001; Kenworthy & Edidin, 1998; Siegel et al., 2000). The FSHR poses a special problem because, unlike other GPCRs or even the LHCGR which has been studied extensively with FRET assays (Lei et al., 2007), the FSHR is clipped at the C-tail and, therefore, it is not possible to employ fluorescent protein fusions at the carboxyl-terminus of the receptor, as is the standard of practice with other GPCRs. In an alternative approach, it has been possible to employ the mAb 106.105 conjugated to Alexa 568 and to Alexa 647 to successfully detect the presence of FSHR oligomers at the PM (Thomas et al., 2007). The use of differentially labeled mAb 106.105 or mAb 106.105-Fab′ fragments for hFSHR FRET obviates the need for C-tail fusion proteins, which may be useful to detect FSHR–FSHR interactions that occur within the cell *but not* at the PM level as the C-tail-terminal end of the FSHR (but not that of the LHCGR) is subjected to proteolytic processing during biosynthesis (Thomas et al., 2007). In fact, C-tail epitope tags in mature, cell surface-expressed FSHR usually are undetectable.

The Alexa 568 and Alexa 647 chromofluor pair meets the requisite characteristic to produce FRET, in which the emission spectrum of the donor (Alexa 568) overlaps the absorption spectrum of the acceptor (Alexa 647). The efficiency of energy transfer (E) from donor to acceptor is highly dependent on the distance between the chromofluors, and is often reported as *E*%. The easiest way to quantify the absolute efficiency of transfer is to measure donor emission before and after photobleaching of the acceptor. The increase in emission (dequenching) of the donor is a direct measure of the FRET efficiency (E), and is calculated from: $E\% = [1 - ($donor emission prior to acceptor photobleach)/(donor emission after acceptor photobleach)$] \times 100$. Analyses are performed directly on captured confocal images, using a cell or its subregion as its own internal standard after photobleaching (Siegel et al., 2000). This procedure is called acceptor photobleaching FRET. The acceptor photobleaching cell surface fluorescence resonance energy transfer (CSFRET) method employed to detect dimerization of the hFSHR is the following:

Human embryonic kidney-293 cells stably expressing the hFSHR and plated in 35-mm Petri dishes with glass coverslip bottoms (MatTek Corp, Ashland, MA) are placed on ice following exchange of tissue culture medium with ice-cold PBS. Cells are then incubated with a 1:1 (μg/μg) mixture of mAb 106.105-Alexa 568 or mAb 106.105-Alexa 647 (directly labeled according to the manufacturer's instructions with Alexa 568 or Alexa 647; Molecular Probes, Inc, Eugene, OR), for 30 min at 4 °C. The cells are washed twice with ice-cold PBS for 5 min, and then fixed with 4% formaldehyde in PBS freshly prepared from a 16% formaldehyde solution. Following fixation, the cells are washed in PBS at RT two times for 5 min. The cells are then covered with PBS and imaged on a Zeiss LSM 510 META confocal microscope system on an Axiomat 200 *M* inverted microscope equipped with a $63 \times$ 1.4 NA, oil immersion differential interference contrast lens. Alexa 568 images are collected using the 540-nm laser line from a HeNe laser and 545 dichroic mirror and a band pass filter of 565–615 nm. Alexa 647 images are collected using the 633-nm laser line from a HeNe laser and 545 dichroic mirror and a band pass filter of 650–710 nm. The pinhole is set at 1.32 Airy units and a Z resolution of $\sim$2.0 μm. Images may be collected at 12-bit intensity resolution over 512×512 pixels at a pixel dwell time of 6.4 μs. FRET is recorded by examining the dequenching of the

Alexa 568 in a region of interest (ROI) following photobleaching of the Alexa 647 by the 633 nm HeNe laser line for 30–90 s at maximum power. This irradiation results in greater than 95% photodestruction of the acceptor fluor.

Images may be analyzed using, for example, a LSM FRET tool that is integrated with the LSM 510 collection software (Zeiss, Inc. Thornwood, NY). FRET analyses are performed on ROI drawn directly on captured confocal images. The analysis uses a cell or its subregion as its own internal standard after photobleaching (Siegel et al., 2000).

3.2. Coimmunoprecipitation of c-myc-tagged and FLAG-tagged FSHR to detect intracellular association of FSHRs

Dimerization of intracellular FSHRs may be detected by coimmunoprecipitating FSHR–FSHR complexes in cell extracts from HEK-293 cells (grown to 90–100% confluency on 60 × 20-mm dishes) previously cotransfected with 1 μg each of plasmids (pRK5 vector) encoding either FLAG (DYKDDDDK) or c-myc (EQKLISEEDL) carboxyl-terminally tagged FSHR, and incubated for an additional 24–48 h. Epitope tags, which are inserted at the carboxyl-terminal end of the FSHR C-tail, are subject to proteolytic processing *before* trafficking to the PM (Thomas et al., 2007). Coimmunoprecipitation with either anti-FLAG or anti-c-myc antibodies should only detect associated, immature forms of the FSHR (Fig. 2.1C and D). The immunoprecipitation procedure is performed as described above employing mAb 106.105, anti-FLAG M2 mAb (Sigma Chemical Co., St. Louis, MO), or anti-c-myc clone 9E10 mAb (American Type Culture Collection, Manassas, VA) at a dilution of 5 μg Ab/500 μg protein extract. Immunoprecipitates are then analyzed by 7.5% SDS-PAGE and Western immunoblotting.

Blots from FLAG immunoprecipitates are probed with 10 μg c-myc mAb labeled with biotin [following the manufacturer's recommendations for EZ-Link Sulfo-NHS-LC-biotin reagent (Pierce, Rockford, IL)]. Blots are then washed and incubated with 10 μg HRP-conjugated-streptavidin (Pierce) to reveal c-myc-tagged FSHR in FLAG immunoprecipitates. Meanwhile, extracts immunoprecipitated with anti-c-myc mAb are probed with HRP-conjugated FLAG M2 mAb (Sigma) (1:5000) to detect FLAG-tagged FSHR in c-myc immunoprecipitates. Signal is developed using enhanced chemiluminescence substrate reagent as described above. In both cases (as well as when mAb 106.105 immunoprecipitates are probed with either biotinylated anti-c-myc or with HRP-conjugated FLAG M2 mAb), blots will mainly show immature and high molecular weight forms of the FSHR. Note that the membrane expressed form of FSHR represented as the 80-kDa band in Fig. 2.1B is absent in Fig. 2.1C and D when staining with the tag-specific antibodies, indicating that the tag and some portion of the C-tail is clipped before insertion into the PM.

4. PHOSPHORYLATION, INTERNALIZATION, AND RECYCLING OF THE FSHR (DOWNWARD TRAFFICKING)

As with other GPCRs, the FSHR undergoes homologous desensitization. The receptor becomes rapidly phosphorylated on serine and threonine residues located at iLs-1 and -3 and the C-tail in response to agonist stimulation, and this phosphorylation facilitates agonist-induced functional uncoupling and internalization through β-arrestins-mediated mechanisms (Kara et al., 2006). The FSHR bears a conserved Ser/Thr cluster in its C-tail and, accordingly, may be recognized as a "class B" seven-transmembrane receptor, in which high-affinity recruitment of both β-arrestin-1 and -2 equally occurs (Reiter & Lefkowitz, 2006). In addition to promoting receptor/G protein uncoupling, β-arrestins target such desensitized receptors to clathrin-coated pits for endocytosis by functioning as adaptor proteins that link the receptors to components (e.g., clathrin molecules) of the endocytic machinery. The majority of the internalized FSH/FSHR complexes then accumulates in endosomes and subsequently recycles back to the cell surface PM, where bound hormone dissociates, allowing preservation of potential responsiveness of the receptor to further hormonal stimulation (Krishnamurthy, Kishi, et al., 2003). Nevertheless, a fraction of FSH/FSHR complexes may be routed to the degradation pathway, which takes place in the lysosomes (Dias, 1986; Krishnamurthy, Kishi, et al., 2003). Cell surface residence of the FSHR at the cell membrane is additionally governed by ubiquitination of the receptor at residues located in the iL-3. Residues that regulate downward trafficking (internalization and postendocytic fate) of the FSHR from the PM are mainly located in the intracellular domains of the receptor, and include T371, S373, T378 (iL-1), S546–S549 (iL-3), and T639–T644 (C-tail) (phosphorylation and β-arrestins interaction); D550 (iL-3), P671, L672, H674, Q677, and N678 (C-tail) (post-endocytic fate); and K555 (plus other still unidentified residues at the iL-3) (ubiquitination) (Cohen et al., 2003; Krishnamurthy, Galet, et al., 2003; Krishnamurthy, Kishi, et al., 2003; Ulloa-Aguirre, Zarinan, Pasapera, Casas-Gonzalez, & Dias, 2007).

In this section, we will briefly describe the methods to study FSHR phosphorylation and coupling to β-arrestins as well as some methods to analyze FSHR internalization and recycling.

4.1. Phosphorylation of the FSHR and β-arrestin recruitment

Detection of either phosphorylation or β-arrestin recruitment necessitates immunoprecipitation of the FSHR. In the two following sections, we describe an immunoprecipitation procedure that relies on a FLAG-tagged version of the hFSHR which is different from the one described in Section 3.2. In this construct, FLAG epitope was inserted in the NH_2-terminal of the FSHR and the native signal sequence was replaced by the one from a hemagglutinin of influenza virus (Guan, Kobilka, & Kobilka, 1992; Reiter et al., 2001). Alternatively, FSHR could be immunoprecipitated with mAb 106.105 antibody as described in Section 2.2.3. For both protocols, HEK-293 cells are grown in DMEM supplemented with 10% heat-inactivated FBS, 10 U/ml penicillin, and 10 μg/ml streptomycin. When 70–80% confluent, cells are transiently transfected with 66 ng/cm^2 of the plasmid encoding the FLAG-tagged FSHR.

4.1.1 Method to detect phosphorylation of the FSHR

All experiments should be performed with cells expressing equivalent amounts of FSHR at the PM. One 10-cm Petri dish is prepared for each experimental condition. Forty to 72 h after transfection, cells are incubated for 1 h at 37 °C in phosphate-free DMEM (Invitrogen, Life Technologies, Inc.) containing 0.1 mCi ^{32}P/ml (Kara et al., 2006). Cells are then stimulated for 5 min with 100 ng/ml of FSH. After stimulation, cells are placed on ice, washed thrice with ice-cold PBS and harvested in 1 ml glycerol buffer (50 m*M* HEPES; 0.5% Nonidet P-40; 250 m*M* NaCl; 2 m*M* EDTA; and 10% glycerol, pH 8.0) containing protease and phosphatase inhibitors (0.2 m*M* phenylmethylsulfonyl fluoride, 1 m*M* Na_3VO_4, and 10 μg/ml aprotinin). The lysates are tumbled for 3 h at 4 °C and then spun at 15,000 rpm for 20 min at 4 °C. At this stage, aliquots of total lysates are saved for each condition and stored at −20 °C until further analysis. The remaining supernatants are then immunoprecipitated by tumbling overnight at 4 °C with M2-conjugated agarose beads (Sigma). After immunoprecipitation, the beads are washed five times with Tris-buffered saline and eluted with 50 μl of Laemmli buffer. Eluates are then incubated at 42 °C for 30 min and loaded on a 10% SDS-PAGE. After migration, the gel is dried on a paper and autoradiographed. The phosphorylation signal can be revealed and analyzed using either a PhosphorImager (Molecular Dynamics, Amersham Biosciences, Pittsburgh, PA) or X-ray film. Autoradiographic analysis should reveal an agonist-induced band corresponding to the FSHR molecular weight. Little to no signal should be observed with the WT FSHR in the absence of FSH stimulation.

4.1.2 Method to analyze β-arrestin recruitment to the FSHR

Endogenous β-arrestins are coimmunoprecipitated with the activated receptor after sulfo-dithio-bis[succinimidyl propionate] (DSP, Pierce, Rockford, IL) cross-linking using a method adapted from Luttrell et al. (2001). All experiments should be performed with cells expressing equivalent amounts of FSHR at the PM. Prepare one 10-cm culture dish per condition. Forty to 72 h after transfection, cells are serum-starved for 4 h in DMEM supplemented with antibiotics, 10 m*M* HEPES, and 0.1% BSA. Cells are then stimulated with 100 ng/ml FSH in 4 ml of PBS/10 m*M* HEPES per culture dish. Approximately 10–15 min before the end of stimulation, prepare fresh DSP reagent in a 50-ml conical tube. Dilute 8 mg of DSP in 4 ml DMSO and then add 4 ml PBS, 10 m*M* HEPES drop-wise. Incubate the DSP mixture for 5–10 min at RT. Slowly add 1 ml of the DSP mixture to the cells placed on an orbital shaker (minimum speed). Incubate cell dishes at RT for 30 min with gentle agitation. The cross-linking reaction is quenched with Tris–HCl (pH 7.3) to a 20 m*M* final concentration. Culture dishes are then placed on ice and washed three times with ice-cold PBS/10 m*M* HEPES. After carefully removing the final wash, ~600 μl of RIPA buffer (150 m*M* NaCl; 50 m*M* Tris, pH 8.0; 5 m*M* EDTA; 1% Nonidet P-40) with freshly added proteases inhibitor cocktail (4 μg/ml aprotinin, 20 μg/ml leupeptin, 2 m*M* phenylmethylsulfonyl fluoride, 1 μg/ml pepstatin A, 2 m*M* benzamidin, 2 m*M* $NaVO_3$, 10 μg/ml trypsin inhibitor) are added to each dish and cells are scraped. The lysates are tumbled for at least 4 h at 4 °C. Thereafter, the pellets are centrifuged at 15,000 rpm for 20 min at 4 °C. Aliquots of total lysates are collected and stored at −20 °C until further analysis.

The remaining lysates are incubated with M2-agarose beads (25 μl of bead suspension) overnight at 4 °C. The beads are washed three times with RIPA buffer and eluted with 50 μl of Laemmli buffer. The samples are incubated at RT for 1 h before loading on 10% SDS-PAGE. Proteins are transferred to nitrocellulose membrane (Whatman, Inc., Piscataway, NJ) and incubated overnight with an anti-β-arrestin antibody. The polyclonal A1CT antibody (from Dr. R. J. Lefkowitz, Duke University Medical Center, Durham, NC) cross-reacting with β-arrestins 1 and 2, can be used at a 1:3000 dilution in Tris-buffered saline, 0.1% Tween 20, and 5% nonfat dry milk. Alternatively, the commercially available anti-pan arrestin antibody (ab2914) from Abcam (Cambridge, MA) can also be used (at a 1:1000 dilution). FLAG M2 monoclonal antibody (1:20,000) in Tris-buffered saline, 0.1% Tween 20, and 5% nonfat dry milk is used to monitor the levels of immunoprecipitated FSHR across the different conditions. The blots are revealed using an enhanced chemiluminescence reaction ECL (Amersham Pharmacia Biotech, Pittsburgh, PA). Two bands corresponding to β-arrestin 1 (~51 kDa) and β-arrestin 2 (~49 kDa) are detected in FSH-stimulated cells.

4.2. Internalization in equilibrium and nonequilibrium conditions

It is not clear if internalization of FSHR occurs in the absence of ligand occupancy, in a constitutive manner. Measurement of FSHR internalization is done indirectly using radiolabeled FSH. Upon occupancy, FSHR internalization proceeds concomitant with activation of adenylate cyclase, association with arrestins, endocytosis, degradation, and recycling (Fig. 2.2). Consideration of whether to use equilibrium or nonequilibrium assay conditions is based on the outcome measurements desired (Cohen et al., 2003; Wiley & Cunningham, 1982). In the case on nonequilibrium conditions, initial rates of internalization are calculated, uninfluenced by recycling, and degradation of the radiolabeled tracer. It is important to remember that the tracer is the beacon for internalization, and that it is degraded and recycled under equilibrium conditions. Internalization rate of the receptor has never been measured directly by measuring the receptor itself. In addition, the reversibility of FSH binding decreases with increasing incubation time (Andersen, Curatolo, & Reichert, 1983). Measuring internalization under equilibrium conditions provides an estimate of internalization rate that is undoubtedly affected by recycling and degradation, and could in essence be considered a beta phase internalization rate. Concentration of receptor is also an issue to consider, as well as the genetic background of the cells used to study internalization. Thus, internalization rates of FSH studied in native primary cultures of Sertoli cells (Dias, 1986) are different from internalization rates determined in HEK293 cells overexpressing FSHR (Cohen et al., 2003).

In internalization assays, cells remain attached to plates so it is possible to distinguish between surface binding and internalized radiolabeled ligand.

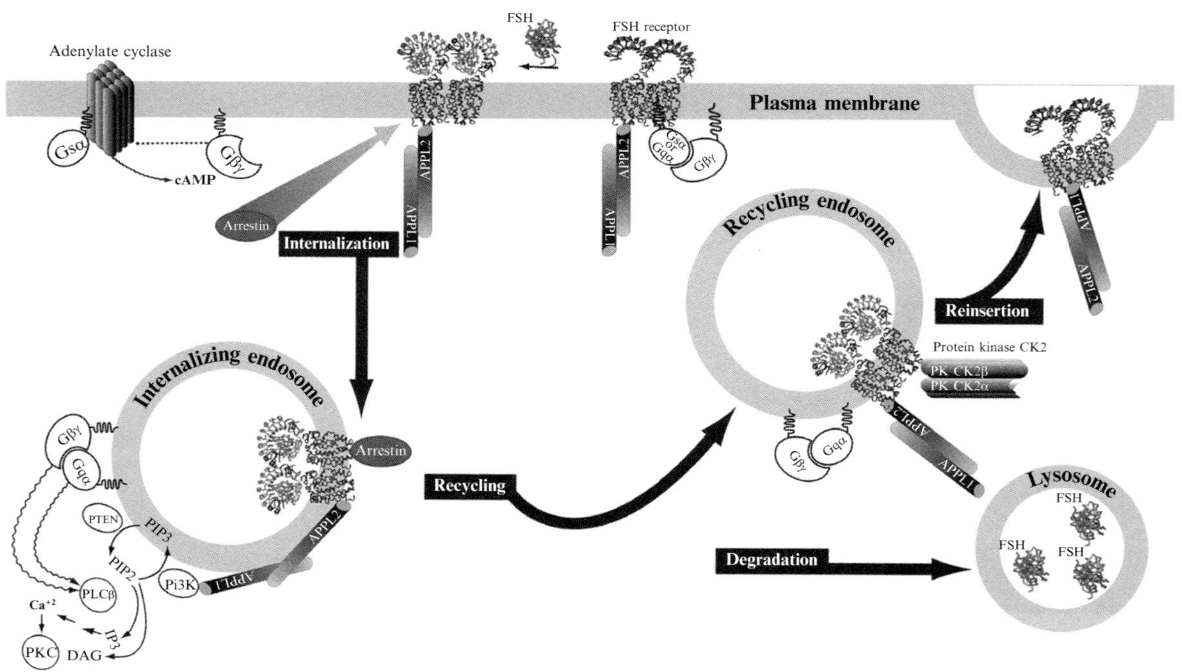

Figure 2.2 Downward traffic of the FSHR is linked to intracellular signaling pathways. FSH binding to FSHR induces the exchange of GDP from the stimulatory subunit ($G\alpha_s$) of the G protein heterotrimer. Upon binding of GTP, the G protein heterotrimer dissociates and $G\alpha_s$ activates adenylate cyclase, while the $\beta\gamma$ dimer activates phospholipase Cβ. These are the canonical pathways for FSH activation. The receptor iL1 and iL3 are subject to phosphorylation by GRKs and subsequent binding of arrestins. Internalization via arrestins likely follows the same early endosome pathway as that of APPL proteins; the latter interacts constitutively with the iL1 of the FSHR. Mutation of K376A in FSHR prevents association of APPL1 with FSHR and loss of FSH-induced intracellular calcium release and inositol triphosphate (IP3) production, with no effect on internalization. Knockdown of arrestin results in a loss of FSH-stimulated ERK phosphorylation. FSHR is ubiquitinated, likely enabling its sorting into recycling vesicles with bound FSH or fused with lysosomes where unliganded FSH is degraded. Recycled unoccupied receptor in theory can engage FSH for another round of FSH stimulation.

For studying internalization in equilibrium conditions, seed cells into 24-well plate(s) at a density of 1×10^5 cells/well in 1 ml DMEM supplemented with antibiotics and 10% FBS (seed extra wells if possible to normalize binding data to cell number or protein concentration). Incubate at 37 °C, 5% CO_2 in a humidified incubator until ~70–80% confluent. Transfect cells with the FSHR or control cDNA and incubate cells at 37 °C, 5% CO_2 for 48 h. The day of the assay, remove medium and replace with 250 μl serum-free medium only or serum-free medium containing 1 μg purified unlabeled FSH (to NSB wells) to account for nonspecific binding. Preincubate for 30 min at 37 °C, 5% CO_2. At the end of the preincubation time, add 50 μl of a ^{125}I-FSH solution (~1×10^6cpm/50 μl) to all wells to achieve a final concentration of ~40 ng/ml; final volume should be 300 μl/well. Incubate at 37 °C, 5% CO_2 for one additional hour and thereafter aspirate media gently, put plates on ice and wash twice with 1 ml ice-cold 1 × PBS. Add 300 μl ice-cold elution buffer (50 m*M* glycine; 100 m*M* NaCl, pH 3.0). Incubate on ice for 20 min and then remove and transfer elution buffer to glass tubes. Count each tube's radioactivity content in a gamma counter to measure "cell surface"-bound ^{125}I-FSH. Add 500 μl of 2 N NaOH to wells at RT and place on orbital shaker for 1 h (at RT) to solubilize cells. Remove and transfer NaOH samples to glass tubes and count in a gamma counter to measure "cell-associated" ^{125}I-FSH (Kluetzman, Thomas, Nechamen, & Dias, 2011). A protein or DNA assay or cell count on extra wells can be done to normalize binding data with respect to protein concentration or cell number.

For nonequilibrium conditions, it is essential to use a concentration of radiolabeled FSH that will achieve rapid saturation of all receptors. However, this is counterbalanced by the high level of radioactivity that must be handled. Considering these two issues, typically ^{125}I-FSH is diluted in serum-free medium so that final concentration is 40 ng/ml (~1×10^6cpm/50 μl), which is high enough to assure almost immediate binding (Cohen et al., 2003; Kluetzman et al., 2011). In the nonequilibrium internalization assay, ^{125}I-FSH tracer is allowed to bind for short periods of time (0–50 min) and cells are harvested in order to study *initial* phases of internalization. Transfected cells are seeded into 24-well plates at 1.5×10^5 cells/ml/well and incubated until subconfluency. On the day of the experiment, aspirate a row of four wells (time point = 50 min), add 250-μl serum-free medium with (one well) or without (three wells) 1 μg purified FSH and then add 50 μl of ^{125}I-FSH to all wells; stagger time points (t = 0, 5, 10, 20, and 30) for adding tracer so that all wells are harvested at same time. Incubate at 37 °C, 5% CO_2 and continue until time point *zero*. Immediately after adding trace to time *zero* wells, aspirate all incubation wells, put plate on ice, wash two times with ice-cold PBS, and then add 300 μl ice-cold elution buffer to recover cell surface-bound ^{125}I-FSH. Thereafter, wash cells once with cold PBS, and add 500 μl 2 N NaOH to measure cell-associated ^{125}I-FSH after transferring the NaOH samples to glass tubes. Do cell count on a separate 24-well plate to normalize counts with respect to cell number.

4.3. Recycling and degradation of internalized FSHRs

Note that in Section 4.2, the procedures result in a readout that represents a "snapshot" of radioligand binding, which does not account for radiolabeled tracer that was bound and internalized prior to acid elution of surface-bound radioligand. In addition, the acid-soluble fraction likely includes recycled radioligand depending on when the sample is taken (Kluetzman et al., 2011). The base solubilized fraction of radiolabeled ligand represents internalized hormone but also includes hormone, which is in the process of recycling to the cell membrane. This fraction of recycled hormone can be assessed by incubating cells with excess unlabeled ligand following removal of radioligand from the cell surface and washing. Subsequent acid elution will yield recycled radioligand (Kluetzman et al., 2011).

4.3.1 Assessment of recycled FSHR

Following internalization, the radiolabeled FSH/FSHR complex can be targeted to the lysosomes or it may recycle to the cell surface. Upon recycling to the cell surface, the radiolabeled FSH can be quantified by acid elution, provided that sufficient unlabeled FSH is present to chase the labeled FSH. The following procedure was described recently and reflects a way to assess this fraction of internalized FSHR (Kluetzman et al., 2011).

Forty-eight hours following transfection (see above), transfected cells are incubated in the presence of excess (40 ng/ml) ^{125}I-FSH for 1 h at 37 °C, 5% CO_2 in SFM (to allow for internalization). After the incubation, ^{125}I-FSH is removed, monolayers are washed twice with cold PBS, and surface-bound hormone is eluted by incubating in cold isotonic pH 3.0 buffer on ice for 2 min. Cells are then washed twice again with cold PBS and placed back into the incubator in warm assay medium containing 1 μg/ml unlabeled FSH (to prevent the reassociation of ^{125}I-FSH released from the cells back into the medium. If this is not done, it will not be possible to determine recycled FSH because the capacity of FSH binding is such that radiolabeled

FSH does not get chased to the cell surface), and a second incubation at 37 °C for different times (0–120 min) is conducted to allow the cells to process the hormone that had been internalized during the first incubation. At each time point, medium is removed for trichloroacetic acid (TCA) precipitation and then cell surface-bound ^{125}I-hFSH is eluted as above, followed by lysis with 2 N NaOH to determine internalized radiolabeled hormone. At time point *zero*, unlabeled hormone is immediately removed, surface-bound ^{125}I-hFSH is eluted, and cell-associated ^{125}I-hFSH is released by 2 N NaOH lysis. Cell-associated radioactivity at this time point represents "initial ^{125}I-hFSH internalized" that is used for relative comparisons at later time points. The total recycled ^{125}I-hFSH represents the *summation* of surface-eluted ^{125}I-hFSH and TCA-precipitable material at each time point analyzed (see below).

4.3.2 Assessment of degraded FSH

Radiolabeled FSH, which is internalized and then trafficked to lysosomes, can be released into the acid compartment of the lysosome and then degraded. Degraded radiolabeled FSH is an indication of the pool of FSHR which has trafficked to the lysosome (Kluetzman et al., 2011).

Following binding and internalization, it is possible to assess the TCA-soluble radioactivity in the media, which represents degraded ^{125}I-FSH. In contrast, TCA-insoluble radioactivity released from the cells subsequent to acid elution of surface-bound radiolabeled FSH, represents ^{125}I-FSH that has been internalized and then recycled to the cell surface. These two parameters are determined by addition of an equal volume of 20% TCA to medium collected during the second incubation (during which an excess of unlabeled FSH was present). Insoluble material is removed by centrifugation at 3000 × *g* for 30 min. The pellets are washed with an additional aliquot of 10% TCA, and then tubes are spun again, supernatant is removed, and the radioactivity in pellets, supernatants, and washes is determined in a gamma counter. The radioactivity determined in the wash supernatant is added to that determined in the supernatant (TCA-soluble) fractions.

In some experiments, it may be desired to inhibit internalization of radiolabeled FSH as a control to be used in assessing degradation. Two approaches can be used. In one approach, cells are incubated with 0.45 *M* sucrose in DMEM, 10% FBS for 1 h prior to and during ^{125}I-hFSH incubation in serum-free medium. An additional approach uses dominant negative dynamin expression to inhibit internalization (Damke, Baba, Warnock, & Schmid, 1994). Both approaches prevent the internalization of receptor and radiolabeled FSH. However, the sucrose method may affect binding, whereas the dynamin method does not have this effect (Kluetzman et al., 2011).

ACKNOWLEDGMENTS

The authors wish to thank supports from CONACyT (Mexico) (grant 86881 to A. U.-A.), The National Institutes of Health (Bethesda, MD) (grants HD18407 to J. A. D., and AG029531 to G. R. B.), The Wellcome Trust and Academy of Finland (to I. H.), and the Région Centre, Institut National de la Recherche Agronomique and Agence Nationale de la Recherche (E. R.). We also acknowledge LE STUDIUM (Orleans, France) for supporting sCORTS. R. Thomas and T. Zariñán assisted in the preparation of this chapter.

REFERENCES

Andersen, T. T., Curatolo, L. M., & Reichert, L. E., Jr. (1983). Follitropin binding to receptors in testis: Studies on the reversibility and thermodynamics of the reaction. *Molecular and Cellular Endocrinology*, *33*, 37–52.

Bogerd, J. (2007). Ligand-selective determinants in gonadotropin receptors. *Molecular and Cellular Endocrinology*, *260–262*, 144–152.

Cohen, B. D., Bariteau, J. T., Magenis, L. M., & Dias, J. A. (2003). Regulation of follitropin receptor cell surface residency by the ubiquitin-proteasome pathway. *Endocrinology*, *144*, 4393–4402.

Damke, H., Baba, T., Warnock, D. E., & Schmid, S. L. (1994). Induction of mutant dynamin specifically blocks endocytic coated vesicle formation. *The Journal of Cell Biology*, *127*, 915–934.

Davis, D., Liu, X., & Segaloff, D. L. (1995). Identification of the sites of N-linked glycosylation on the follicle-stimulating hormone (FSH) receptor and assessment of their role in FSH receptor function. *Molecular Endocrinology*, *9*, 159–170.

Dias, J. A. (1986). Effect of transglutaminase substrates and polyamines on the cellular sequestration and processing of follicle-stimulating hormone by rat Sertoli cells. *Biology of Reproduction*, *35*, 49–58.

Dias, J. A., Cohen, B. D., Lindau-Shepard, B., Nechamen, C. A., Peterson, A. J., & Schmidt, A. (2002). Molecular, structural, and cellular biology of follitropin and follitropin receptor. *Vitamins and Hormones*, *64*, 249–322.

Duvernay, M. T., Zhou, F., & Wu, G. (2004). A conserved motif for the transport of G protein-coupled receptors from the endoplasmic reticulum to the cell surface. *The Journal of Biological Chemistry*, *279*, 30741–30750.

Fan, Q. R., & Hendrickson, W. A. (2005). Structure of human follicle-stimulating hormone in complex with its receptor. *Nature*, *433*, 269–277.

Fan, Q. R., & Hendrickson, W. A. (2007). Assembly and structural characterization of an authentic complex between human follicle stimulating hormone and a hormone-binding ectodomain of its receptor. *Molecular and Cellular Endocrinology*, *260–262*, 73–82.

Fan, T., Varghese, G., Nguyen, T., Tse, R., O'Dowd, B. F., & George, S. R. (2005). A role for the distal carboxyl tails in generating the novel pharmacology and G protein activation profile of mu and delta opioid receptor hetero-oligomers. *The Journal of Biological Chemistry*, *280*, 38478–38488.

Gromoll, J., Pekel, E., & Nieschlag, E. (1996). The structure and organization of the human follicle-stimulating hormone receptor (FSHR) gene. *Genomics*, *35*, 308–311.

Gromoll, J., Ried, T., Holtgreve-Grez, H., Nieschlag, E., & Gudermann, T. (1994). Localization of the human FSH receptor to chromosome 2 p21 using a genomic probe comprising exon 10. *Journal of Molecular Endocrinology*, *12*, 265–271.

Gromoll, J., Schulz, A., Borta, H., Gudermann, T., Teerds, K. J., Greschniok, A., et al. (2002). Homozygous mutation within the conserved Ala-Phe-Asn-Glu-Thr motif of exon 7 of the LH receptor causes male pseudohermaphroditism. *European Journal of Endocrinology*, *147*, 597–608.

Guan, X. M., Kobilka, T. S., & Kobilka, B. K. (1992). Enhancement of membrane insertion and function in a type IIIb membrane protein following introduction of a cleavable signal peptide. *The Journal of Biological Chemistry*, *267*, 21995–21998.

Guan, R., Wu, X., Feng, X., Zhang, M., Hebert, T. E., & Segaloff, D. L. (2010). Structural determinants underlying constitutive dimerization of unoccupied human follitropin receptors. *Cellular Signalling*, *22*, 247–256.

Hebert, T. E., Moffett, S., Morello, J. P., Loisel, T. P., Bichet, D. G., Barret, C., et al. (1996). A peptide derived from a beta2-adrenergic receptor transmembrane domain inhibits both receptor dimerization and activation. *The Journal of Biological Chemistry*, *271*, 16384–16392.

Helenius, A., & Aebi, M. (2004). Roles of N-linked glycans in the endoplasmic reticulum. *Annual Review of Biochemistry*, *73*, 1019–1049.

Helenius, A., Trombetta, E., Hebert, D., & Simons, J. F. (1997). Calnexin, calreticulin and the folding glycoproteins. *Trends in Biochemical Sciences*, *7*, 193–200.

Huhtaniemi, I., & Alevizaki, M. (2007). Mutations along the hypothalamic-pituitary-gonadal axis affecting male reproduction. *Reproductive Biomedicine Online*, *15*, 622–632.

Huhtaniemi, I. T., & Themmen, A. P. (2005). Mutations in human gonadotropin and gonadotropin-receptor genes. *Endocrine*, *26*, 207–217.

Jiang, X., Liu, H., Chen, X., Chen, P.-H., Fischer, D., Sriraman, V., et al. (2012). Structure of follicle-stimulating hormone in comlex with the entire ectodoman of its receptor. *Proceedngs of the National Academy of Sciences of the United States of America*, *109*, 12491–12496.

Kara, E., Crepieux, P., Gauthier, C., Martinat, N., Piketty, V., Guillou, F., et al. (2006). A phosphorylation cluster of five serine and threonine residues in the C-terminus of the follicle-stimulating hormone receptor is important for desensitization but not for beta-arrestin-mediated ERK activation. *Molecular Endocrinology*, *20*, 3014–3026.

Kenworthy, A. K. (2001). Imaging protein-protein interactions using fluorescence resonance energy transfer microscopy. *Methods*, *24*, 289–296.

Kenworthy, A. K., & Edidin, M. (1998). Distribution of a glycosylphosphatidylinositol-anchored protein at the apical surface of MDCK cells examined at a resolution of <100 A using imaging fluorescence resonance energy transfer. *The Journal of Cell Biology*, *142*, 69–84.

Kishi, H., Krishnamurthy, H., Galet, C., Bhaskaran, R. S., & Ascoli, M. (2002). Identification of a short linear sequence present in the C-terminal tail of the rat follitropin receptor that modulates arrestin-3 binding in a phosphorylation-independent fashion. *The Journal of Biological Chemistry*, *277*, 21939–21946.

Kluetzman, K. S., Thomas, R. M., Nechamen, C. A., & Dias, J. A. (2011). Decreased degradation of internalized follicle-stimulating hormone caused by mutation of aspartic acid 6.30(550) in a protein kinase-CK2 consensus sequence in the third intracellular loop of human follicle-stimulating hormone receptor. *Biology of Reproduction*, *84*, 1154–1163.

Krishnamurthy, H., Galet, C., & Ascoli, M. (2003). The association of arrestin-3 with the follitropin receptor depends on receptor activation and phosphorylation. *Molecular and Cellular Endocrinology, 204*, 127–140.

Krishnamurthy, H., Kishi, H., Shi, M., Galet, C., Bhaskaran, R. S., Hirakawa, T., et al. (2003). Postendocytotic trafficking of the follicle-stimulating hormone (FSH)-FSH receptor complex. *Molecular Endocrinology, 17*, 2162–2176.

Laemmli, U. K. (1970). Cleavage of structural proteins during the assembly of the head of bacteriophage T4. *Nature, 227*, 680–685.

Lazari, M. F., Liu, X., Nakamura, K., Benovic, J. L., & Ascoli, M. (1999). Role of G protein-coupled receptor kinases on the agonist-induced phosphorylation and internalization of the follitropin receptor. *Molecular Endocrinology, 13*, 866–878.

Lee, S. P., O'Dowd, B. F., Rajaram, R. D., Nguyen, T., & George, S. R. (2003). D2 dopamine receptor homodimerization is mediated by multiple sites of interaction, including an intermolecular interaction involving transmembrane domain 4. *Biochemistry, 42*, 11023–11031.

Lei, Y., Hagen, G. M., Smith, S. M., Liu, J., Barisas, G., & Roess, D. A. (2007). Constitutively-active human LH receptors are self-associated and located in rafts. *Molecular and Cellular Endocrinology, 260–262*, 65–72.

Lindau-Shepard, B., Brumberg, H. A., Peterson, A. J., & Dias, J. A. (2001). Reversible immunoneutralization of human follitropin receptor. *Journal of Reproductive Immunology, 49*, 1–19.

Luttrell, L. M., Roudabush, F. L., Choy, E. W., Miller, W. E., Field, M. E., Pierce, K. L., et al. (2001). Activation and targeting of extracellular signal-regulated kinases by beta-arrestin scaffolds. *Proceedings of the National Academy of Sciences of the United States of America, 98*, 2449–2454.

Marion, S., Robert, F., Crepieux, P., Martinat, N., Troispoux, C., Guillou, F., et al. (2002). G protein-coupled receptor kinases and beta arrestins are relocalized and attenuate cyclic 3',5'-adenosine monophosphate response to follicle-stimulating hormone in rat primary Sertoli cells. *Biology of Reproduction, 66*, 70–76.

Mizrachi, D., & Segaloff, D. L. (2004). Intracellularly located misfolded glycoprotein hormone receptors associate with different chaperone proteins than their cognate wild-type receptors. *Molecular Endocrinology, 18*, 1768–1777.

Mueller, S., Jaeschke, H., Gunther, R., & Paschke, R. (2009). The hinge region: An important receptor component for GPHR function. *Trends in Endocrinology and Metabolism, 21*, 111–122.

Musnier, A., Heitzler, D., Boulo, T., Tesseraud, S., Durand, G., Lecureuil, C., et al. (2009). Developmental regulation of p70 S6 kinase by a G protein-coupled receptor dynamically modelized in primary cells. *Cellular and Molecular Life Sciences, 66*, 3487–3503.

Nakamura, K., Hipkin, R. W., & Ascoli, M. (1998). The agonist-induced phosphorylation of the rat follitropin receptor maps to the first and third intracellular loops. *Molecular Endocrinology, 12*, 580–591.

Nechamen, C. A., & Dias, J. A. (2000). Human follicle stimulating hormone receptor trafficking and hormone binding sites in the amino terminus. *Molecular and Cellular Endocrinology, 166*, 101–110.

Nechamen, C. A., & Dias, J. A. (2003). Point mutations in follitropin receptor result in ER retention. *Molecular and Cellular Endocrinology, 201*, 123–131.

Nechamen, C. A., Thomas, R. M., & Dias, J. A. (2007). APPL1, APPL2, Akt2 and FOXO1a interact with FSHR in a potential signaling complex. *Molecular and Cellular Endocrinology, 260–262*, 93–99.

Ng, G. Y., O'Dowd, B. F., Lee, S. P., Chung, H. T., Brann, M. R., Seeman, P., et al. (1996). Dopamine D2 receptor dimers and receptor-blocking peptides. *Biochemical and Biophysical Research Communications, 227*, 200–204.

Osuga, Y., Hayashi, M., Kudo, M., Conti, M., Kobilka, B., & Hsueh, A. J. (1997). Co-expression of defective luteinizing hormone receptor fragments partially reconstitutes ligand-induced signal generation. *The Journal of Biological Chemistry, 272,* 25006–25012.

Peltoketo, H., Strauss, L., Karjalainen, R., Zhang, M., Stamp, G. W., Segaloff, D. L., et al. (2010). Female mice expressing constitutively active mutants of FSH receptor present with a phenotype of premature follicle depletion and estrogen excess. *Endocrinology, 151,* 1872–1883.

Piketty, V., Kara, E., Guillou, F., Reiter, E., & Crepieux, P. (2006). Follicle-stimulating hormone (FSH) activates extracellular signal-regulated kinase phosphorylation independently of beta-arrestin- and dynamin-mediated FSH receptor internalization. *Reproductive Biology and Endocrinology, 4,* 33.

Rannikko, A., Pakarinen, P., Manna, P. R., Beau, I., Misrahi, M., Aittomaki, K., et al. (2002). Functional characterization of the human FSH receptor with an inactivating Ala189Val mutation. *Molecular Human Reproduction, 8,* 311–317.

Reiter, E., & Lefkowitz, R. J. (2006). GRKs and beta-arrestins: Roles in receptor silencing, trafficking and signaling. *Trends in Endocrinology and Metabolism, 17,* 159–165.

Reiter, E., Marion, S., Robert, F., Troispoux, C., Boulay, F., Guillou, F., et al. (2001). Kinase-inactive G-protein-coupled receptor kinases are able to attenuate follicle-stimulating hormone-induced signaling. *Biochemical and Biophysical Research Communications, 282,* 71–78.

Rozell, T. G., Davis, D. P., Chai, Y., & Segaloff, D. L. (1998). Association of gonadotropin receptor precursors with the protein folding chaperone calnexin. *Endocrinology, 139,* 1588–1593.

Rozell, T. G., Wang, H., Liu, X., & Segaloff, D. L. (1995). Intracellular retention of mutant gonadotropin receptors results in loss of hormone binding activity of the follitropin receptor but not of the lutropin/choriogonadotropin receptor. *Molecular Endocrinology, 9,* 1727–1736.

Siegel, R. M., Chan, F. K., Zacharias, D. A., Swofford, R., Holmes, K. L., Tsien, R. Y., et al. (2000). Measurement of molecular interactions in living cells by fluorescence resonance energy transfer between variants of the green fluorescent protein. *Science's STKE, 2000,* l1.

Song, Y. S., Ji, I., Beauchamp, J., Isaacs, N. W., & Ji, T. H. (2001). Hormone interactions to Leu-rich repeats in the gonadotropin receptors. I. Analysis of Leu-rich repeats of human luteinizing hormone/chorionic gonadotropin receptor and follicle-stimulating hormone receptor. *The Journal of Biological Chemistry, 276,* 3426–3435.

Tao, Y. X., Mizrachi, D., & Segaloff, D. L. (2002). Chimeras of the rat and human FSH receptors (FSHRs) identify residues that permit or suppress transmembrane 6 mutation-induced constitutive activation of the FSHR via rearrangements of hydrophobic interactions between helices 6 and 7. *Molecular Endocrinology, 16,* 1881–1892.

Thomas, R. M., Nechamen, C. A., Mazurkiewicz, J. E., Muda, M., Palmer, S., & Dias, J. A. (2007). Follice-stimulating hormone receptor forms oligomers and shows evidence of carboxyl-terminal proteolytic processing. *Endocrinology, 148,* 1987–1995.

Timossi, C., Ortiz-Elizondo, C., Pineda, D. B., Dias, J. A., Conn, P. M., & Ulloa-Aguirre, A. (2004). Functional significance of the BBXXB motif reversed present in the cytoplasmic domains of the human follicle-stimulating hormone receptor. *Molecular and Cellular Endocrinology, 223,* 17–26.

Tranchant, T., Durand, G., Gauthier, C., Crepieux, P., Ulloa-Aguirre, A., Royere, D., et al. (2011). Preferential beta-arrestin signalling at low receptor density revealed by functional characterization of the human FSH receptor A189 V mutation. *Molecular and Cellular Endocrinology, 331,* 109–118.

Troispoux, C., Guillou, F., Elalouf, J. M., Firsov, D., Iacovelli, L., De Blasi, A., et al. (1999). Involvement of G protein-coupled receptor kinases and arrestins in desensitization to follicle-stimulating hormone action. *Molecular Endocrinology, 13*, 1599–1614.

Ulloa-Aguirre, A., & Conn, P. M. (2009). Targeting of G protein-coupled receptors to the plasma membrane in health and disease. *Frontiers in Bioscience, 14*, 973–994.

Ulloa-Aguirre, A., Crepieux, P., Poupon, A., Maurel, M. C., & Reiter, E. (2011). Novel pathways in gonadotropin receptor signaling and biased agonism. *Reviews in Endocrine and Metabolic Disorders, 12*, 259–274.

Ulloa-Aguirre, A., Zarinan, T., Pasapera, A. M., Casas-Gonzalez, P., & Dias, J. A. (2007). Multiple facets of follicle-stimulating hormone receptor function. *Endocrine, 32*, 251–263.

Uribe, A., Zarinan, T., Perez-Solis, M. A., Gutierrez-Sagal, R., Jardon-Valadez, E., Pineiro, A., et al. (2008). Functional and structural roles of conserved cysteine residues in the carboxyl-terminal domain of the follicle-stimulating hormone receptor in human embryonic kidney 293 cells. *Biology of Reproduction, 78*, 869–882.

Vannier, B., Loosfelt, H., Meduri, G., Pichon, C., & Milgrom, E. (1996). Anti-human FSH receptor monoclonal antibodies: Immunochemical and immunocytochemical characterization of the receptor. *Biochemistry, 35*, 1358–1366.

Vassart, G., Pardo, L., & Costagliola, S. (2004). A molecular dissection of the glycoprotein hormone receptors. *Trends in Biochemical Sciences, 29*, 119–126.

Wehbi, V., Decourtye, J., Piketty, V., Durand, G., Reiter, E., & Maurel, M. C. (2010). Selective modulation of follicle-stimulating hormone signaling pathways with enhancing equine chorionic gonadotropin/antibody immune complexes. *Endocrinology, 151*, 2788–2799.

Wehbi, V., Tranchant, T., Durand, G., Musnier, A., Decourtye, J., Piketty, V., et al. (2010). Partially deglycosylated equine LH preferentially activates beta-arrestin-dependent signaling at the follicle-stimulating hormone receptor. *Molecular Endocrinology, 24*, 561–573.

Weiner, R. S., & Dias, J. A. (1992). Biochemical analyses of proteolytic nicking of the human glycoprotein hormone alpha-subunit and its effect on conformational epitopes. *Endocrinology, 131*, 1026–1036.

Wiley, H. S., & Cunningham, D. D. (1982). The endocytotic rate constant. A cellular parameter for quantitating receptor-mediated endocytosis. *The Journal of Biological Chemistry, 257*, 4222–4229.

Zarinan, T., Perez-Solis, M. A., Maya-Nunez, G., Casas-Gonzalez, P., Conn, P. M., Dias, J. A., et al. (2010). Dominant negative effects of human follicle-stimulating hormone receptor expression-deficient mutants on wild-type receptor cell surface expression. Rescue of oligomerization-dependent defective receptor expression by using cognate decoys. *Molecular and Cellular Endocrinology, 321*, 112–122.

CHAPTER THREE

Single-Molecule Imaging Technique to Study the Dynamic Regulation of GPCR Function at the Plasma Membrane

B.E. Snaar-Jagalska[*], A. Cambi[†,‡], T. Schmidt[§], S. de Keijzer[†,1]

[*]Cell Biology, Leiden Institute of Biology, Leiden University, Leiden, The Netherlands
[†]Department of Tumor Immunology, Nijmegen Centre for Molecular Life Sciences, Radboud University Nijmegen Medical Centre, Nijmegen, The Netherlands
[‡]Nanobiophysics, MIRA Institute for Biomedical Technology and Technical medicine, University of Twente, Enschede, The Netherlands
[§]Physics of Life Processes, Leiden Institute of Physics, Leiden University, Leiden, The Netherlands
[1]Corresponding author: e-mail address: s.dekeijzer@ncmls.ru.nl

Contents

1. Introduction 48
2. Labeling of GPCRs and Sample Preparation for SPT 48
   2.1 Glass cleaning for SPT 49
   2.2 cAR1 labeling with eYFP 49
   2.3 cAR1 labeling with Halo-TMR 52
   2.4 EP2 receptor labeling with quantum dots 53
3. Single-Molecule Imaging 56
   3.1 SPT setup 57
   3.2 Image acquisition 58
4. Data Analysis 61
   4.1 Localization of single molecules 61
   4.2 Analyzing the mobility of single GPCRs 62
5. Concluding Remarks 64
Acknowledgments 65
References 65

Abstract

The lateral diffusion of a G-protein-coupled receptor (GPCR) in the plasma membrane determines its interaction capabilities with downstream signaling molecules and critically modulates its function. Mechanisms that control GPCR mobility, like compartmentalization, enable a cell to fine-tune its response through local changes in the rate, duration, and extent of signaling. These processes are known to be highly dynamic and tightly regulated in time and space, usually not completely synchronized in time. Therefore, bulk studies such as protein biochemistry or conventional confocal

Methods in Enzymology, Volume 521
ISSN 0076-6879
http://dx.doi.org/10.1016/B978-0-12-391862-8.00003-X

microscopy will only yield information on the average properties of the interactions and are compromised by poor time resolution. Single-particle tracking (SPT) in living cells is a key approach to directly monitor the function of a GPCR within its natural environment and to obtain unprecedented detailed information about receptor mobility, binding kinetics, aggregation states, and domain formation. This review provides a detailed description on how to perform single GPCR tracking experiments.

1. INTRODUCTION

G-protein-coupled receptors (GPCRs), localized in the plasma membrane, transmit the extracellular cues to intracellular processes via protein–protein interactions. Therefore, in addition to acting as a selective barrier, the plasma membrane serves as a platform for initiating and regulating signaling pathways. It is proposed that the effectiveness of the signaling cascade depends on the organization and lateral mobility of the signaling components within the plasma membrane (Allen, Halverson-Tamboli, & Rasenick, 2007; De Keijzer, Galloway, Harms, Devreotes, & Iglesias, 2011; De Keijzer, Serge, et al., 2008; Head et al., 2006; Ostrom & Insel, 2004; Pucadyil & Chattopadhyay, 2007; Zuo, Ushio-Fukai, Hilenski, & Alexander, 2004). The signaling efficiency depends on the probability of interaction between, for example, the GPCR and its G proteins, and this is controlled by the localization and mobility of GPCRs and G proteins in the plasma membrane. Mechanisms exist that can control the localization and mobility, such as compartmentalization caused by cytoskeletal contacts, lipid environment, or protein–protein interactions. Such a machinery allows for the spatiotemporal fine-tuning of a cell's response to extracellular signals. Because GPCR dynamics are transient, fast, and usually involve few, if not individual, proteins, single-molecule microscopy is mandatory to monitor these very early signaling events. This review describes methods to visualize the dynamics of single GPCRs on the plasma membrane of living cells using single-particle tracking (SPT). We take the *Dictyostelium* cAMP receptor cAR1 and the human prostaglandin receptor EP2 as examples of target GPCRs for single-molecule imaging.

2. LABELING OF GPCRs AND SAMPLE PREPARATION FOR SPT

We will discuss the sample preparation of three different types of GPCR labeling: (1) C-terminal fusion of the yellow fluorescent protein to cAR1, (2) C-terminal fusion of the HaloTag® protein to cAR1 subsequently labeled

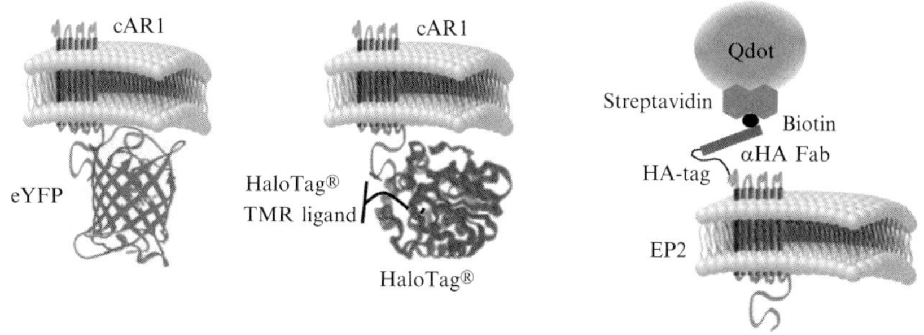

Figure 3.1 Sketch of different labeling techniques to visualize single GPCRs. Discussed in this review is C-terminal fusion of the *Dictyostelium* cAR1 receptor with eYFP and HaloTag® protein, with the latter labeled with the HaloTag® tetramethylrhodamine (TMR) ligand prior to imaging. Furthermore, the N-terminal labeling of HA epitope-tagged EP2 receptor with Qdot® 655 streptavidin conjugated with biotinylated anti-HA Fab fragment is discussed.

with HaloTag® ligand carrying a fluorophore, and (3) N-terminal fusion of the HA (human influenza hemagglutinin) epitope tag to EP2 subsequently labeled with quantum dots via anti-HA Fab fragments (Fig. 3.1).

2.1. Glass cleaning for SPT

Dust on glass or culture treatment of glass can induce noise (background signal) during SPT experiments and therefore the imaging glass needs thorough cleaning. The following protocol was established by the group of A. Kusumi (personal communication, Kyoto University, Japan).

1. Clean two-well Lab-Tek™ II chambered coverglass (Nunc Brand, Sanbio BV, The Netherlands) by subsequent ultrasonification (15 min) in soap, H_2O, 0.1 *M* KOH, and H_2O.
2. Rinse the Lab-Tek for ~15 min under running water.
3. Dry upside down.
4. This can be done for multiple Lab-Tek at a time and store them in ethanol. It is possible to reuse Lab-Tek II chambered coverglasses with this cleaning method for up to two to three times (until the glue is affected and the Lab-Tek starts to leak).

2.2. cAR1 labeling with eYFP

The cAMP receptor, cAR1 is a GPCR that is functionally related to chemokine receptors, which mediate chemotactic responses of leukocytes. cAR1 regulates *Dictyostelium discoideum* chemotaxis toward the ligand cAMP, which is a highly conserved mechanism and *D. discoideum* has proven indispensable as a model organism to investigate the processes underlying chemotaxis. The ability to chemotax in a very shallow gradient is believed to depend on the spatial restriction of the responses along the anterior–posterior axis as displayed by the localization of certain proteins to either the leading or the trailing edge of the cells (Garcia & Parent, 2008). By applying SPT, we showed that the mobility of the cAR1 receptor molecule exhibits an asymmetry between the leading and trailing edge, which, in turn, accounts for the primary decision in the spatially restricted response of the downstream signaling pathway during cell migration (De Keijzer, Serge, et al., 2008; Van Hemert, Lazova, Snaar-Jagaska, & Schmidt, 2010). Therefore, high spatial and temporal information on the dynamics of the GPCR obtained with SPT indicated that directional sensing is directly related to the acceleration of the activation step of the G-protein-linked signaling pathways at the leading edge. This biophysical approach has significantly improved our understanding of the cell's compass.

2.2.1 *Culture and transformation of* Dictyostelium *cells*

1. Culture cells at 22 °C in HL5 medium containing 5 g proteose peptone, 5 g thiotone E peptone, 10 g glucose, 5 g yeast extract, 0.35 g Na_2H PO_4,$\cdot 7H_2O$, 0.35 g KH_2PO_4, 0.05 g dihydrostreptomycin sulfate dissolved in 1 L water with pH set to 6.5 (Autoclave for 20 min to sterilize). Supplement with 100 μg/ml ampicillin and with 100 μg/ml mixture of penicillin and streptomycin (1:1).
2. Before transformation, pellet 5×10^6 cells (wt or *car1* null mutant) and wash them twice with ice-cold H-50 buffer containing 20 m*M* HEPES, 50 m*M* KCl, 10 m*M* NaCl, 1 m*M* $MgSO_4$, 5 m*M* $NaHCO_3$, and 1 m*M* NAH_2PO_4 (pH 7 and filter sterilized with 0.22-μm pore size filter).
3. For transformation, resuspend the cells in 100 ml of H-50 buffer and add 1–10 μg of the *Dictyostelium* cAR1 expression plasmid (Parent, Blacklock, Froehlich, Murphy, & Devreotes, 1998) containing cAR1-eYFP and geneticin selection cassette.
4. Transfer the cells to a cold 0.1-cm gap electroporation cuvette and electroporate at 0.85 kV and 25 μF two times with a 5-s break in between with a gene pulser electroporation system (Bio-Rad).
5. Incubate the cuvette on ice for 5 min before transferring the cells to a 100-mm tissue culture dish containing 10 ml HL5 supplemented with 100 μg/ml ampicillin and with 100 μg/ml mixture of penicillin and streptomycin (1:1). Add selection, 5–20 μg/ml of geneticin, after 1 day.
6. Grow the cells until single colonies can be observed and pick cells from a single colony to start a culture. For populations, this step is not necessary.

7. Test the functionality of the cAR1-YPF construct by expressing the fluorescent receptor in cells. This should rescue the aggregation-deficient phenotype of the *car1* null mutant (De Keijzer, Serge, et al., 2008).

2.2.2 Development of Dictyostelium cells and sample preparation

The endogenous cAR1 receptor is expressed during the developmental program of *Dictyostelium* cells and is important for the aggregation of single cells into a multicellular body (Konijn & Van Haastert, 1987). To ensure that the dynamics of the cAR1 receptor were studied at the appropriate developmental stage of the cells and that all cAR1 signaling partners are expressed, the developmental program of *Dictyostelium* cells is initiated before measurements.

1. Replace the HL5 medium with low fluorescence medium (Liu et al., 2002) 16 h before measurement. Low fluorescence medium contains 1 ml of 500 m*M* NH_4Cl, 200 m*M* $MgCl_2$ and 10 m*M* $CaCl_2$, 1 ml of 50 m*M* $FeCl_3$. Add 0.1 ml of 4.84 g Na_2-EDTA·$2H_2O$, 2.3 g Zn $SO_4{\cdot}H_2O$, 1.11 g H_3BO_4, 0.51 g $MnCl_2{\cdot}4H_2O$, 0.17 g $CoCl_2{\cdot}6H_2O$, 0.15 g $CuSO_4{\cdot}5H_2O$, and 0.10 g $(NH_4)_6MO7O_{24}{\cdot}4H_2O$ dissolved in 100 ml water with pH set to 6.5. Finally, add 11 g glucose·H_2O, 5 ml of 1 *M* K_2HPO_4, and 5 g casein peptone (do not use vitamin-free casein peptone) and add water to a final volume of 1 L, adjust the pH to 6.5, and filter sterilize the low fluorescence medium (0.22-μm pore size, Stericup® from Millipore).
2. Harvest the cells and develop the cells by shaking the cells at a density of 2×10^7 cells/ml for 6 h in development buffer containing 5 m*M* Na_2H PO_4, 5 m*M* KH_2PO_4, 1 m*M* $CaCl_2$, 2 m*M* $MgCl_2$, pH set to 6.5.
3. Start cAMP pulsing 1 h after the initiation of starvation by adding cAMP (100 n*M* final concentration) every 6 min for 5 h.
4. Wash the cells with phosphate buffer (PB) containing 0.534 g Na_2H PO_4, 0.952 g KH_2PO_4 in 1 L of H_2O with pH set to 6.5. Resuspend to a concentration of 1×10^5 cells/ml in PB and add 1 ml of suspension in a two-well Lab-Tek II chambered coverglass.
5. Allow the cells to adhere for 15 min.

2.2.3 Flattening of Dictyostelium cells to increase basal membrane surface for TIRF

In order to facilitate an optimal cell-to-glass contact necessary for total internal reflection fluorescence (TIRF) microscopy, overlay the *Dictyostelium* cells with an agarose layer (De Keijzer et al., 2011; Fukui, Yumura, & Yumura, 1987). A larger contact area of the basal plasma membrane allows for the measurement of more single molecules in one cell, as the density of molecules should not be too high in order to separate the signals as different molecules (<1 μm^2).

1. Dissolve 1 g of agarose-M (LKB Pharmacia: cat. #2206-101) in 50 ml of PB buffer, 10 m*M* Na_2HPO_4/KH_2PO_4, pH 6.5 to make a 2% (w/v) solution. Autoclave for 15 min and store at 4 °C.
2. Redissolve the agarose in a microwave slowly (to prevent agarose from boiling). Using a fine glass cutter with a diamond tip, cut a Corning 22 × 22 mm, no. 1½ cover slip (cat. #2870) into strips of about 3 mm wide. These strips are 0.19 mm thick and serve as spacers. Place a piece of ethanol-cleaned slide glass on a heated surface (heating block for example) and align the two spacers on both edges of the slide using forceps.
3. Drop about 1 ml of hot agarose on the slide and overlay with the second ethanol-cleaned slide glass on top of the agarose drop. Gently but quickly hold down the slide glass to spread the agarose before it solidifies.
4. Remove the slide sandwich from the heated surface and wait until the agarose has formed a gel. Using a pair of razor blades, remove excess agarose from edges of the slide sandwich. Store the sandwich in a Petri dish containing about 20 ml of PB buffer and keep it in a refrigerator.
5. Carefully remove the upper slide glass from the sandwich with a pair of forceps to handle the slide. Remove excess buffer from the cells and leave approximately 100–200 μl buffer on top of the cells. Cut the agarose gel with a razor in a rectangle fitting the Lab-Tek chamber and pick up the agarose gel with a flat forceps and gently cover the cells with it. Make sure that the sides of the gel do not touch the sides of the Lab-Tek chamber; otherwise, the cells will dry out very rapidly.

2.3. cAR1 labeling with Halo-TMR

In order to study the type of diffusion displayed by cAR1 in the plasma membrane, long-term (~seconds) tracking of the receptors in the plasma membrane is required. Green fluorescent protein mutants, like eYFP, bleach rapidly resulting in short trajectories even with TIRFM. The HaloTag® technology (Promega, USA) allows for longer imaging at the single-molecule level. The HaloTag technology is based on the formation of a covalent bond between the HaloTag protein (34 kDa) that is fused to the GPCR and synthetic ligands that carry fluorophores. These fluorophores, for example, tetramethylrhodamine (TMR), are more photostable than the green fluorescent protein mutants and thus can be longer imaged. Another advantage is that, by controlling the concentration of the Halo-TMR ligand, the number of fluorescently tagged receptors can be regulated. Other similar technologies are available like SNAP- or CLIP-tag (New England Biolabs) and the ACP- or MCP-tag (New England Biolabs). Using this labeling technique, we showed that microtubules are very important in organizing the movement of the cAR1 receptor in the plasma membrane (De Keijzer et al., 2011).

1. Culture and transform the Dictyostelium cells as described in Section 2.2.1 with pJK1, a *Dictyostelium* extrachromosomal expression vector (Kim et al., 1997), containing the protein target of interest (cAR1) fused to full-length HaloTag® protein (Promega, USA).
2. Initiate the developmental program of cells containing the construct as described in Section 2.2.2.
3. After resuspension of the cells in PB, add 0.01 μM HaloTag® TMR ligand (Promega, USA) for 15 min that covalently binds the HaloTag® protein. This concentration allows for the observation of discrete signals ($<1\ \mu m^2$) from cAR1-Halo-TMR necessary for SPT (this concentration needs to be optimized for every GPCR type). Wash away unbound ligand by centrifugation and resuspend the cells to a concentration of 1×10^5 cells/ml in PB. Before starting experiments, check for non-specific labeling by performing the same labeling procedure on non-transfected cells.
4. Add 1 ml of cell suspension in a two-well Lab-Tek chambered coverglass and allow the cells to adhere for 15 min before you overlay the cells with an agar sheet (as described in Section 2.2.3).

2.4. EP2 receptor labeling with quantum dots

Prostaglandin E2 (PGE2) is a lipid mediator derived from cyclooxygenase-catalyzed metabolism of arachidonic acid and exhibits the most versatile actions in a wide variety of tissues (Coleman, Smith, & Narumiya, 1994; Smith, 1992). In immune cells, including macrophages, T cells, and dendritic cells, PGE2 potently and highly selectively modulates the development, cytokine production, and migration via the GPCRs EP2 and EP4 (Coleman et al., 1994; Narumiya, 2003). The distinguishing feature of the EP2 and EP4 prostanoid receptors is that their signaling is predominantly transduced by a $G\alpha_s$ protein, through which receptor activation is associated with an increase in adenylate cyclase activity and elevated intracellular cAMP levels (Honda et al., 1993; Regan et al., 1994). Interestingly, EP4 appears to be able to activate additional signaling pathways via coupling to cAMP-inhibitory Gα_i protein, responsible for the activation of phosphatidylinositol 3-kinase (Fujino & Regan, 2006) suggesting cross talk between the receptors. However, how this cross talk is regulated at the molecular level is unknown. Therefore, single PGE2 receptor tracking will provide more insight in PGE2 signaling regulation. Here, we describe the strategy to monitor single EP2 receptors during signaling. As intracellular fusion proteins of EP2 (GFP, SNAP-tag) were not correctly expressed at the plasma membrane, a different labeling technique was followed using the small HA Epitope Tag (YPYDVPDYA) fused to the extracellular N-terminus of EP2 (Desai, April, Nwaneshiudu, & Ashby, 2000). The receptor molecules are subsequently labeled with streptavidin-coated quantum dots conjugated to biotinylated anti-HA Fab fragments. Of course, other epitope tags, for example, AU1, c-Myc with antibodies directed against them, as well as antibodies directed against the GPCR of interest can be used. To prevent unwanted aggregation of GPCRs that may influence the lateral diffusion profile, use biotinylated Fabs or single chain (scFv) antibodies of choice conjugated to streptavidin-coated quantum dots.

2.4.1 Transform human embryonic kidney (HEK293) cells with HA-EP2

1. Culture HEK293 cells at 37 °C and 5% CO_2 in culture flasks (standard T75, Greiner Bio-one) containing DMEM (Gibco, Life Technologies, The Netherlands) supplied with 10% fetal bovine serum (FBS, Greiner Bio-one, Austria), 1% MEM nonessential amino acids (NEAA, Gibco), and 0.5% Antibiotic–Antimytotic (AA, Gibco).
2. Harvest the cultured cells 48 h prior to measurements by washing 1 × with phosphate-buffered saline (PBS) and incubating the cells for 2–3 min at 37 °C with 2 ml (T75 culture flask) of 0.05% Trypsin-EDTA buffer. Centrifuge the cells for 5 min, 450 RCF and resuspend in phenol red-free DMEM supplied with 10% FBS, 1% NEAA, and 0.5% AA. Using phenol red-free DMEM will decrease autofluorescence and when this is still very high, culturing the HEK293 cells in phenol red-free DMEM is also an option. Plate 2×10^5 cells/ml in a four-well chamber cleaned Lab-Tek.
3. Let the cells adhere overnight and transfect the cells with the plasmid vector carrying the HA-tagged EP2 gene using Metafectene (Biontex, Germany) according to the protocol supplied by the company.

2.4.2 Preparation of aHA-Qdot conjugate

1. On the day of the SPT experiment, make a 40 n*M* solution of the Qdot® streptavidin (Invitrogen, Life Technologies, The Netherlands) conjugate of desired wavelength in PBS containing 8 g NaCl, 0.2 g KCl, 1.44 g

Na_2HPO_4, and 0.24 g KH_2PO_4 in 1 L of H_2O with pH set to 7.4 and supplemented with 1% BSA. We use the 655 nm Qdots (Qdot® 655 streptavidin conjugate) because intrinsic autofluorescence of the cells is decreasing at higher wavelengths and the intensity of the 655 nm Qdots is much higher when excited at 491 nm compared to the lower wavelength Qdots (e.g., Qdot® 565 from Invitrogen). Incubate the Qdot solution on ice for 5 min.

2. Make a 40 n*M* solution of the anti-HA Biotin, high-affinity (3F10) Fab fragments (Roche, Germany) in PBS/w 1% BSA and incubate on ice for 5 min.
3. Mix the two solutions and agitate for 1–2 h rotating at 4 °C. Store the conjugate on ice. The conjugate should not be stored longer than a day. It is highly recommended to make a fresh solution for every experiment.

The following steps are optional when high background labeling is observed on untransfected cells.

4. Preabsorb 200 μl of the 20 n*M* aHA-QD conjugate with 800 μl phenol red-free RPMI 1640 supplied with 1% ultraglutamine-1 (Lonza, Lonza Benelux, The Netherlands), 0.5% AA, and 0.1% BSA with untransfected HEK293 cells at 37 °C for 15 min.
5. Centrifuge the conjugated mix at 3400 RCF for 5 min to remove cell debris.
6. Recover the supernatant containing ~4 n*M* aHA-Qdot conjugate and store on ice.

2.4.3 Labeling of HA-EP2-expressing HEK293 cells with the aHA-Qdot conjugate

The following steps are done with prewarmed (37 °C) Tyrode's buffer containing 135 m*M* NaCl, 10 m*M* KCl, 0.4 m*M* $MgCl_2$, 1 m*M* $CaCl_2$, and 10 m*M* HEPES, pH 7.2 (only when no CO_2 is available on the imaging setup). This buffer can be stored at 4 °C. Add 20 m*M* glucose and 0.1% BSA fresh before use.

1. Remove the medium from the transfected HEK293 cells and wash the cells 1 × with Tyrode's buffer.
2. Add 400 μl Tyrode's buffer and 100 μ*M* aHA-Qdot conjugate and incubate for 5 min at 37 °C. This amount depends on the number of target GPCR expressed in the cell and should be optimized for each GPCR type to obtain discrete signals necessary for SPT (<1 μm^2).
3. Wash the sample for three to five times with Tyrode's buffer.
4. Add 400 μl Tyrode's buffer for imaging.

3. SINGLE-MOLECULE IMAGING

To monitor GPCR mobility, which requires high spatial and temporal resolution, wide-field microscopy is the method of choice because it allows simultaneous imaging of the whole area of view. Consequently, the time resolution of this imaging technique is only limited by the acquisition speed of the camera system and the fluorescence properties of the labeled species. TIRF microscopy, a wide-field technique, is advantageous over conventional wide-field microscopy, as it limits the excitation volume to a region close to the cover slip surface thereby reducing out-of-focus noise (autofluorescence or fluorescence from other cellular compartments) and thus improving spatial resolution without compromising acquisition speed. TIRF microscopy uses an evanescent wave (a very thin electromagnetic field) generated by reflected light at the glass–cell surface to selectively excite fluorophores in a restricted region of less than a 100-nm vertical distance to the cover slip–specimen interface. This section thickness is approximately 1/10th of a confocal section, which dramatically improves the signal-to-background noise ratio and consequently the spatial resolution by avoiding excitation of fluorophores in deeper regions of the cell. Objective-type TIRF microscopy is easily implemented on a wide-field microscopic setup by changing the illumination of the excitation light from standard wide field to TIRF by shifting the position of the beam focus at the back focal plane of the objective from center to edge (Fig. 3.2). The choice of using wide field versus TIRF depends on the biological questions one wants to address. Although TIRF is optimal for reducing the autofluorescence, it can only image GPCR dynamics at the basal plasma membrane, where the GPCR mobility might find hindrance from mechanism controlling cell attachment (cytoskeletal network, integrins, etc.). Although wide-field imaging can be used to image GPCRs in the apical membrane, the use of eYFP in wide field is very difficult because of the spectral overlap with autofluorescence of the cells. Moreover, at high eYFP-labeled receptor expression, it is not possible to directly image at single-molecule level in wide field because all signals in the cells are acquired. Imaging eYFP in wide field is possible, but photobleaching of most of the autofluorescence and signal might be necessary, a procedure described in Section 3.2 (De Keijzer, Serge, et al., 2008). With respect to eYFP, TMR is certainly better, but quantum dots are most likely the best tools for single-molecule imaging in wide field. Another point to consider is that many cell types have a round shape, making the 2D surface to image very small at the apical side. Therefore, in order to get

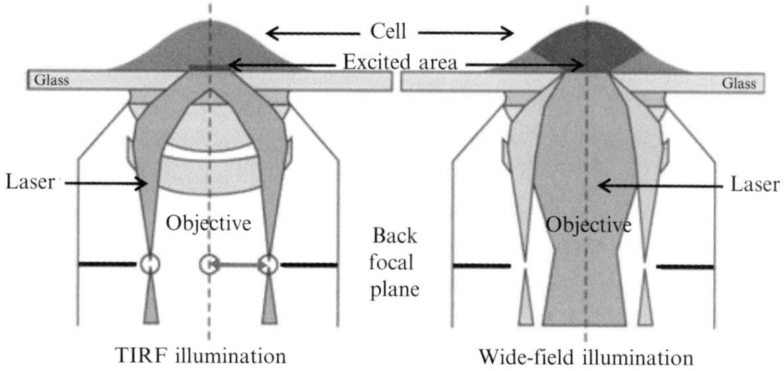

Figure 3.2 Switching from wide-field imaging to TIRF. The fluorophores in the cell are excited by laser-mediated excitation. In wide field, where the light comes from the center of the objective, most of the cells are excited. In TIRF, the position of the beam focus at the back focal plane of the objective is shifted from the center (wide field) to the edge of the objective changing the incident angle of the light beam. The evanescent wave (a very thin electromagnetic field) generated by the reflected light at the glass–cell surface selectively excites fluorophores in a restricted region of less than a 100 nm vertical distance to the cover slip–specimen interface. (See Color Insert.)

quantitative data (enough signals to analyze), a high number of experiments need to be performed in wide field compared to TIRF.

3.1. SPT setup

Our optical configuration consists of a wide-field microscopy setup with an objective-type TIRFM unit (both from Olympus) that allows for the control of the incident angle of the light through the objective. Details about the components are shown in Table 3.1. To reach the required light intensity for single-molecule detection, the fluorophores are excited by laser-mediated excitation. In order to obtain total internal reflection, the path of the light beam needs to be shifted from the center (wide field) to the edge of the objective changing the incident angle of the light beam (Fig. 3.2). Refractive index differences between the glass and water phases determine how light is refracted or reflected at the interface as a function of the incident angle. Beyond the critical angle, the beam of light is totally internally reflected from the glass/water interface. High spatial accuracy critically depends on the achieved signal-to-noise ratio, which is further determined by the sensitivity and time resolution of the fluorescence detector. Electron-multiplying CCD cameras (EMCCD, ImageEM by Hamamatsu, Japan) make use of on-chip electron amplification before read-out and presently deliver the fastest image acquisition via high

Table 3.1 Wide-field setup with TIRFM unit

TIRFM component	Examples of equipment	Manufacturer
Inverted microscope	IX-71	Olympus
Vibration isolation table		
Objective	UAPON 150 × (100 ×) OTIRF, 1.45 NA	Olympus
Laser excitation	LAS-491-100 (491 nm, 100 mW) for eYFP, Qdot 655 LAS-561-100 (561 nm, 100 mW) for TMR	Olympus
Angle control of the incident light	MITICO 4-line TIRFM unit with motorized incident angle control	Olympus
Ultrasensitive camera	ImagEM CCD (C9100-13)	Hamamatsu
Objective heater	MATS-ULH Lens Heater + Thermo Controller	Tokai Hit
Heatable insert for microscope stage	Heating Insert P Lab-Tek™ (for Lab-Tek chambered coverglasses)	Pecon
Temperature regulator	Tempcontrol 37-2 digital 2-channel	Pecon
Hardware/software to control setup (lasers, shutters, CCD camera, etc.)	PC-CTR XCellence (PC with real-time controller)	Olympus
	XCellence software	
Dichroic mirror and filters	CMR-U-M4TIR-SBX quadruple (405/491/561/640) filter set	Olympus

quantum yield and optimum resolution. This detector type further supports a sufficiently fast read-out of 32 frames/s in full-frame mode (512 × 512 pixels) and up to 170 frames/s when operating in subregions of the chip (64 × 64 pixels). Optimal control of the TIRFM units such as high speed 1 ms TTL laser shutters (Olympus) and dichroic mirrors (Olympus) with appropriate hardware (PC-CTR XCellence PC with real-time controller, Olympus) and software (XCellence, Olympus) provides exact timing of the illumination time, intensity, and wavelengths.

3.2. Image acquisition

1. Turn on the setup (turn on the laser, CCD camera, water cooling, PC, etc.). The manufacturer usually provides a manual containing this information.

2. Mount the heating insert on the stage and the objective heater on the objective and turn the temperature controller on at the desirable temperature (37 °C for HEK293 cells and room temperature for *Dictyostelium* cells).
3. Shut the laser by closing the shutter and set the sample on the stage of the setup. Wait a few minutes until the system is stable (no observable drift in focus due to temperature change).
4. Bring the cells in focus by observation under transmitted light illumination.
5. Turn off the transmitted light and take a background image to determine the EMCCD shot noise. Adjust the power of the laser (1–5% for 100 mW laser), open the shutter, and search for a cell expressing the labeled GPCR while exciting the sample in wide field.
6. For TIRFM, adjust the incident angle of the laser beam to the calculated critical angle based on glass thickness, wavelength, and objective type (calculation is implemented in XCellence software). This will get you close to the actual critical angle. Increase the laser power when necessary. Now manually adjust the angle in very small steps until you lose the signal and go back a little until you obtain signal again. This position should be optimal TIRF. Practice finding optimal TIRF angle using fluorescent dyes or quantum dots adherent to the glass surface. Adjust the incident angle while focusing until clear spots on the glass appear. Remember that in TIRF, you should not be able to focus inside the cell.
7. Set the image frequency. The image frequency is a tradeoff between the camera acquisition speed and the imaging region of interest. The camera acquisition speed is faster when operating in subregions of the chip (up to 170 frames/s when using 64 × 64 pixels). However, with a 150 × objective, a subregion of 256 × 256 pixels is necessary to have a complete cell in the field of view. Therefore, we used a time lag of 30–40 ms between images (De Keijzer et al., 2011; De Keijzer, Serge, et al., 2008).
8. Set the illumination time and laser intensity so that individual spots ($<1\ \mu m^{-2}$) are observed. In order to reach a density of eYFP-labeled receptors at which individual molecules can be observed, it might be necessary to photobleach cells with high laser power for a few seconds prior to imaging. The unbleached population of receptors is a fully representative subpopulation of all receptors as photobleaching occurs at random. Using the HaloTag technology or quantum dots, the optimal concentration of HaloTag TMR ligand or quantum dots to label cells with in order to observe individual spots should be determined.
9. Acquire images through the CCD (Fig. 3.3A).

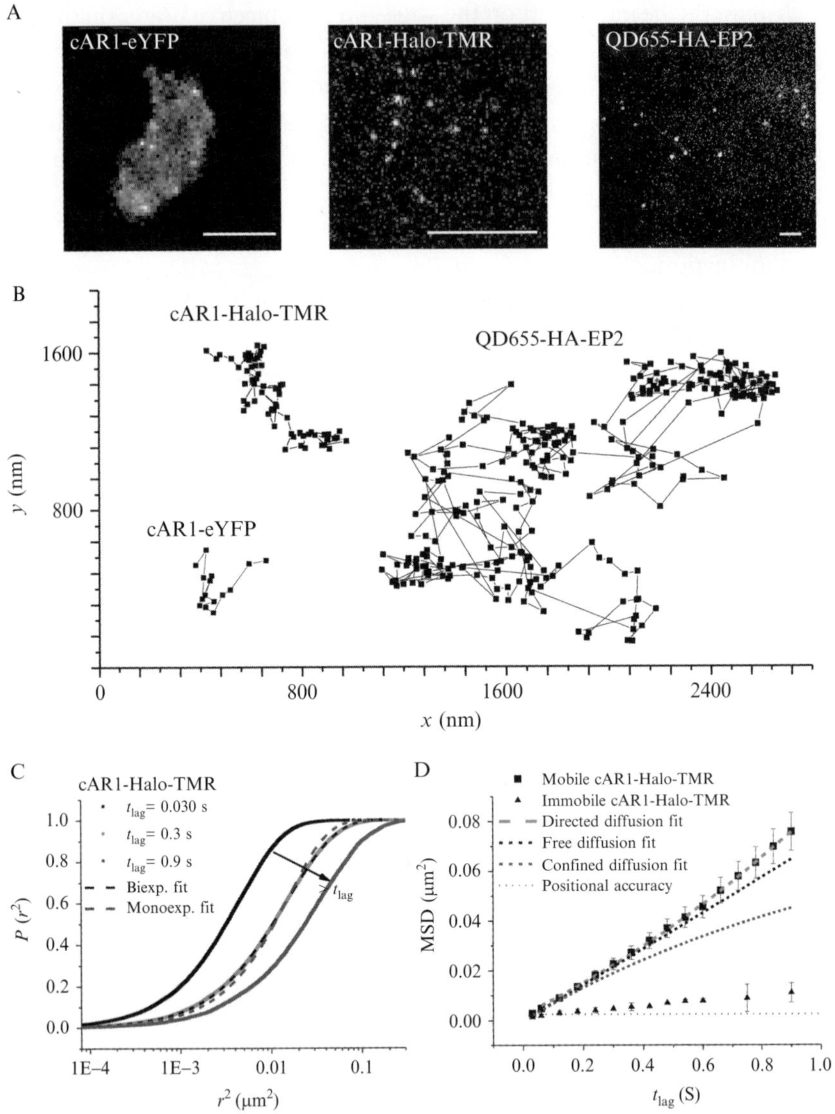

Figure 3.3 Data analysis. (A) Images from a stack showing fluorescence signals from cAR1-eYFP, cAR1-Halo-TMR in live *Dictyostelium* cells, and QD655-HA-EP2 in live HEK293 cells, which are characterized as diffraction-limited spots. Scale bar is 5 μm. (B) Representative trajectories of the three different molecules. (C) Cumulative probability distributions ($P(r^2)$) of the square displacements (r^2) of cAR1-Halo-TMR in *Dictyostelium* cells at three different t_{lag} ($t_{lag}=30$, 300, and 900 ms). The CPD plot at $t_{lag}=300$ ms is described with a monoexponential fit (gray dashed curve) and a biexponential fit (black dashed curve). The biexponential fit correctly described the cAR1-Halo-TMR CPD plot

4. DATA ANALYSIS

4.1. Localization of single molecules

1. Localize the single molecules using dedicated software. We use software packages written in Matlab (Mathworks) (De Keijzer, Snaar-Jagalska, Spaink, & Schmidt, 2008) that filters the images after background subtraction using a Gaussian correlation filter. Subsequently, a threshold criterion is applied taking into account the noise in each image. This initial analysis provided the starting values for a nonlinear fitting procedure of two-dimensional Gaussian profiles to the original images (Marquardt, 1963). The high image contrast characterized by the signal-to-noise ratio allows for determination of the position of each molecule with an accuracy of 20–40 nm. As signals from quantum dots are much stronger, that is, more photons, than from eYFP, quantum dots will have a better positional accuracy than eYFP.
2. Validate the signals as individual emitters by constructing cut-off criteria regarding the intensity, the width, and the relative errors in all fitting parameters of single-molecule signals. These criteria can be obtained by imaging single molecules on glass (TMR, Qdots) or on artificial lipid bilayers (eYFP) (Harms, Cognet, Lommerse, Blab, & Schmidt, 2001). Use these cut-off criteria to validate the signal obtained in cells as individual emitting molecules. A second validation that can be used for eYFP and TMR is to observe the intensity of the signal in consecutive images and determine single-step photobleaching events (De Keijzer et al., 2011; De Keijzer, Serge, et al., 2008).

giving two mean square displacements ($r_1^2(t)$,$r_2^2(t)$) and the fraction size (α). (D) Mean square displacement (MSD) data of the two single cAR1-Halo-TMR molecule fractions are plotted versus t_{lag}. Seventy percent (α) of the molecules were mobile, while the remaining 30% ($1-\alpha$) of molecules were considered immobile on the timescales measured (30 ms) because the MSD values stayed really close to the positional accuracy for all time lags. The MSD versus t_{lag} plot of mobile cAR1-Halo-TMR is fitted by three different models describing free diffusion (black dashed line), diffusion with confinement (dark gray dashed line) and with flow (light gray dashed line). The fitting results showed that cAR1-Halo-TMR moved directionally with $D=0.015\pm0.002$ μm^2/s and velocity of $v=0.16\pm0.02$ μm/s in the basal plasma membrane of live *Dictyostelium* cells. These data were published in De Keijzer et al. (2011).

4.2. Analyzing the mobility of single GPCRs

4.2.1 Tracing single molecules

1. Use the determined single-molecule positions (x,y coordinates) to construct the trajectories from the positional shifts in consecutive images (Fig. 3.3B). We allow for continuous tracking of single molecules if the signal is not picked up in a few consecutive images with a maximum of 10 images for quantum dots and 5 images for TMR and eYFP. This way we prevent truncated trajectories due to blinking of quantum dots or small plasma membrane movements resulting in molecules moving in and out of focus. Despite skipping images, the time lag between the positions of the trajectory is always correct because the position of the molecule is linked to the number of the image in the stack.
2. Analyze the trajectories by plotting the cumulative probability distribution (CPD) of the square displacements (distance between molecules in consecutive images). The CPDs $P(r_i^2, t_{\mathrm{lag}})$ are constructed from the single-molecule trajectories by counting the number of square displacements with values $\leq r^2$ and subsequent normalization by the total number of data points (Schutz, Schindler, & Schmidt, 1997) (Fig. 3.3C). Fit the plot with

$$\widetilde{P}(r^2, t) = 1 - \exp\left(-\frac{r^2}{r_0^2}\right). \qquad [3.1]$$

Equation (3.1) is a model which describes the cumulative probability that a particle is found within a circle of radius r at time t if it was at the origin at time $t=0$ (Fig. 3.3C). Simulations revealed that at least 200 data points of displacement values per time lag are needed to get an accurate fit (De Keijzer, Snaar-Jagalska, et al., 2008), which yields the mean square displacement (MSD $= r_0^2$). When the CPD of the square displacements exhibits a biphasic behavior that cannot be described by a monoexponential (Eq. 3.1), Eq. (3.1) can be extrapolated to a biexponential. Provided that the system under study segregates into two components, Eq. (3.1) becomes (Schutz et al., 1997)

$$\widetilde{P}(r^2, t) = 1 - \left[\alpha \cdot \exp\left(-\frac{r^2}{r_1^2(t)}\right) + (1-\alpha) \cdot \exp\left(-\frac{r^2}{r_2^2(t)}\right)\right]. \qquad [3.2]$$

Describing the data with this biexponential fit results in two receptor populations with each of their MSDs (MSD$_i = r_1^2$ and r_2^2) and the size of the relative fractions, α and $(1-\alpha)$, respectively (Fig. 3.3C).

3. Calculate in a similar fashion the MSD for increasing time lags. The data for increasing time lags are constructed from the same data set taken with $t_{\text{lag}} = 30$ ms by taking the 1-step ($t_{\text{lag}} = 30$ ms), 2-step ($t_{\text{lag}} = 60$ ms) up to the n-step ($t_{\text{lag}} = n \times 30$ ms) displacements. In order to construct the CPD data sets for increasing time lags from the same data set taken with one time lag, long trajectories are necessary. When long trajectories are not available (e.g., with eYFP), the MSD versus time lag can be constructed by making 1-step displacements data sets for different imaging frequencies.
4. Plot the MSD values against the time lags (Fig. 3.3D) so that the mobility of the GPCR molecules can be further analyzed.

4.2.2 Determine type of diffusion and diffusion coefficient

The heterogenic microscopic organization of the plasma membrane imparts different types of motion onto GPCRs, like immobility, free Brownian diffusion, transient confinement in dynamic microdomains, hop diffusion across membrane picket fences, and directed flow (Owen, Williamson, Rentero, & Gaus, 2009). Quantifying the mode of diffusion and the diffusion coefficient of GPCR molecules enables insight into the influence of membrane nanoenvironment on GPCR function. Quantification is established by fitting the MSD (r_i^2) versus time lag (t_{lag}) plots with models describing different modes of diffusion (Table 3.2; Fig. 3.3D).

1. In case of free or Brownian motion, the MSD varies linearly with time lag and the fit returns the diffusion constant D.
2. When the diffusion is hindered by obstructions or trapping, the MSD will grow slower with time lag and is characterized by a power-law

Table 3.2 Different models to describe the lateral mobility of a protein

Diffusion mode	Model	Parameters
Free or Brownian	$r_i^2 = 4Dt_{\text{lag}}$	D = diffusion constant
Anomalous	$r_i^2 = \Gamma t_{\text{lag}}^{\alpha}$	Γ = diffusion parameter with unit $\mu m^2/s^{\alpha}$
Confined	$r_i^2 = \frac{L^2}{3}\left(1 - \exp\left(\frac{-12D_{\text{init}}t_{\text{lag}}}{L^2}\right)\right)$	D_{init} = initial diffusion constant, L = confinement size
Directed	$r_i^2 = 4Dt_{\text{lag}} + v^2 t_{\text{lag}^2}$	D = diffusion constant, v = constant velocity

dependence with exponent $\alpha < 1$. Such behavior is called anomalous subdiffusion and the fit returns the diffusion parameter Γ of unit $\mu m^2/s^\alpha$. For free diffusion $\alpha = 1$, the diffusion parameter is given by the regular diffusion constant $\Gamma = 4D$.

3. With confined diffusion, the MSD initially increases with time but levels off to a constant value for longer time lags. Assuming that diffusion is free within a square of side length L surrounded by an impermeable reflecting barrier, the MSD depends on L and the initial diffusion coefficient D_{init} (Kusumi et al., 1993).
4. When the GPCR displays a directed motion in the plasma membrane, the MSD versus time lag plot will show a quadratic shape and can be fitted with a model describing diffusion with diffusion constant D and an additional flow of constant velocity v (Qian, Sheetz, & Elson, 1991).
5. To improve the fitting procedure of the MSD versus time lag plots with the models, one should take into account the constant offset in r_i^2 due to the error in positional accuracy. Determination of molecule positions by a fitting algorithm is a random process and leads to diffusion-like mobility of an immobile object leading to the error in positional accuracy. The positional accuracy (σ) can be calculated from the mean error in fitted x and y coordinates by $\Delta x^2 + \Delta y^2 = \sigma^2$.
6. Obtaining a good fit with the different diffusion modes requires the MSD up to sufficiently large time lags. For example, if a GPCR is diffusing with 0.5 mm^2/s within a confinement zone of 1 mm, then the plateau in the MSD versus time lag plot, indicating confined diffusion will only be reached on the order of a few seconds. When the data set is not sufficiently long to get an accurate fit, one might choose to analyze the first MSD versus time lag data points with the model for free diffusion to obtain the initial diffusion coefficient.

5. CONCLUDING REMARKS

During the past decade, it has become clear that the plasma membrane is compartmentalized in domains that are diverse in size and composition. Dynamic assembly of spatially separated signaling platforms enables a cell to tune cellular outputs in response to different input stimuli. The activity of GPCRs is subject to precise regulation of localization in space and time. However, the mechanisms orchestrating such a degree of plasma membrane organization remain enigmatic. Single-molecule microscopy permits to study localization and lateral mobility of GPCRs in the plasma membrane.

Analysis of single GPCR trajectories and fitting these to different models for diffusion will determine whether molecules are hindered in their diffusion (i.e., confined) or actively transported (i.e., directed motion). The cause of a certain type of diffusion can be directly tested using cytoskeleton-disrupting compounds (e.g., latrunculin, nocodazole), chemical or enzymatic treatments that inhibit cholesterol (e.g., nystatin, MCD, cholesterol oxidase) or sphingolipid (e.g., zaragozic acid, sphingomyelinase) content and metabolism that perturb plasma membrane electrostatic potential and asymmetry (e.g., synaptojanin2, wortmannin, sphingosine) or that affect size and distribution of lipid rafts (e.g., n-3 PUFA). Furthermore, the investigation of GPCR dynamics upon engagement with extracellular ligands, specific agonists, and antagonists, as well as introduction of GPCR domain mutants will unravel the processing of extracellular signal across the plasma membrane into cellular responses.

Binding of an individual ligand to a GPCR occurs within milliseconds and is followed by amplification steps, finally leading to modulation of gene expression and other cellular functions, such as shape change and migration. Single-molecule microscopy allows to study the function of one or very few GPCR molecules in the full panoply of regulatory mechanisms and to show how a particular outcome is being achieved by decisions made during early steps in receptor complex assembly and signal transduction. Hence, single-molecule microscopy will lead to a better understanding of the primary processes in signal transduction and eventually improve our knowledge of fundamental aspects of human diseases, including those caused by alterations in GPCRs or G protein signaling, autoimmunity, and cancer progression.

ACKNOWLEDGMENTS

The research leading to these results has received funding from the Dutch CW/NWO project 700-50-032 (B. E. S.-J.), the European Commission's Seventh Framework Programme (FP7-ICT-2011-7) under grant agreement n° 288263 (NanoVista) (S. d. K. and A. C.), and the International Human Frontier Science Program Organization (RGY 0072/2008) (S. d. K. and A. C.). A. C. is the recipient of a Meervoud grant from the Netherlands organization for scientific research (NWO).

REFERENCES

Allen, J. A., Halverson-Tamboli, R. A., & Rasenick, M. M. (2007). Lipid raft microdomains and neurotransmitter signalling. *Nature Reviews. Neuroscience*, *8*, 128–140.

Coleman, R. A., Smith, W. L., & Narumiya, S. (1994). International Union of Pharmacology classification of prostanoid receptors: Properties, distribution, and structure of the receptors and their subtypes. *Pharmacological Reviews*, *46*, 205–229.

De Keijzer, S., Galloway, J., Harms, G. S., Devreotes, P. N., & Iglesias, P. A. (2011). Disrupting microtubule network immobilizes amoeboid chemotactic receptor in the plasma membrane. *Biochimica et Biophysica Acta*, *1808*, 1701–1708.

De Keijzer, S., Serge, A., van Hemert, F., Lommerse, P. H. M., Lamers, G. E. M., Spaink, H. P., et al. (2008). A spatially restricted increase in receptor mobility is involved in directional sensing during Dictyostelium discoideum chemotaxis. *Journal of Cell Science*, *121*, 1750–1757.

De Keijzer, S., Snaar-Jagalska, B. E., Spaink, H. P., & Schmidt, T. (2008). Single-molecule imaging of cellular reactions. In R. Rigler & H. Vogel (Eds.), *Single molecules and nanotechnology*, Vol. 12, (pp. 107–129). *Springer Series in Biophysics 12*, Berlin Heidelberg: Springer-Verlag.

Desai, S., April, H., Nwaneshiudu, C., & Ashby, B. (2000). Comparison of agonist-induced internalization of the human EP2 and EP4 prostaglandin receptors: Role of the carboxyl terminus in EP4 receptor sequestration. *Molecular Pharmacology*, *58*, 1279–1286.

Fujino, H., & Regan, J. W. (2006). EP(4) prostanoid receptor coupling to a pertussis toxin-sensitive inhibitory G protein. *Molecular Pharmacology*, *69*, 5–10.

Fukui, Y., Yumura, S., & Yumura, T. K. (1987). Agar-overlay immunofluorescence: High-resolution studies of cytoskeletal components and their changes during chemotaxis. *Methods in Cell Biology*, *28*, 347–356.

Garcia, G. L., & Parent, C. A. (2008). Signal relay during chemotaxis. *Journal of Microscopy*, *231*, 529–534.

Harms, G. S., Cognet, L., Lommerse, P. H., Blab, G. A., & Schmidt, T. (2001). Autofluorescent proteins in single-molecule research: Applications to live cell imaging microscopy. *Biophysical Journal*, *80*, 2396–2408.

Head, B. P., Patel, H. H., Roth, D. M., Murray, F., Swaney, J. S., Niesman, I. R., et al. (2006). Microtubules and actin microfilaments regulate lipid raft/caveolae localization of adenylyl cyclase signaling components. *The Journal of Biological Chemistry*, *281*, 26391–26399.

Honda, A., Sugimoto, Y., Namba, T., Watabe, A., Irie, A., Negishi, M., et al. (1993). Cloning and expression of a cDNA for mouse prostaglandin E receptor EP2 subtype. *The Journal of Biological Chemistry*, *268*, 7759–7762.

Kim, J. Y., Soede, R. D., Schaap, P., Valkema, R., Borleis, J. A., Van Haastert, P. J., et al. (1997). Phosphorylation of chemoattractant receptors is not essential for chemotaxis or termination of G-protein-mediated responses. *The Journal of Biological Chemistry*, *272*, 27313–27318.

Konijn, T. M., & Van Haastert, P. J. M. (1987). Measurement of chemotaxis in dictyostelium. *Methods in Cell Biology*, *28*, 283–298.

Kusumi, A., Sako, Y., & Yamamoto, M. (1993). Confined lateral diffusion of membrane receptors as studied by single particle tracking (nanovid microscopy). Effects of calcium-induced differentiation in cultured epithelial cells. *Biophysical Journal*, *65*, 2021–2040.

Liu, T., Mirschberger, C., Chooback, L., Arana, Q., Dal Sacco, Z., MacWilliams, H., et al. (2002). Altered expression of the 100 kDa subunit of the Dictyostelium vacuolar proton pump impairs enzyme assembly, endocytic function and cytosolic pH regulation. *Journal of Cell Science*, *115*, 1907–1918.

Marquardt, D. W. (1963). An algorithm for least-squares estimation of nonlinear parameters. *Journal of the Society for Industrial and Applied Mathematics*, *11*, 431–441.

Narumiya, S. (2003). Prostanoids in immunity: Roles revealed by mice deficient in their receptors. *Life Sciences*, *74*, 391–395.

Ostrom, R. S., & Insel, P. A. (2004). The evolving role of lipid rafts and caveolae in G protein-coupled receptor signaling: Implications for molecular pharmacology. *British Journal of Pharmacology*, *143*, 235–245.

Owen, D. M., Williamson, D., Rentero, C., & Gaus, K. (2009). Quantitative microscopy: Protein dynamics and membrane organisation. *Traffic, 10*, 962–971.

Parent, C. A., Blacklock, B. J., Froehlich, W. M., Murphy, D. B., & Devreotes, P. N. (1998). G protein signaling events are activated at the leading edge of chemotactic cells. *Cell, 95*, 81–91.

Pucadyil, T. J., & Chattopadhyay, A. (2007). The human serotonin1A receptor exhibits G-protein-dependent cell surface dynamics. *Glycoconjugate Journal, 24*, 25–31.

Qian, H., Sheetz, M. P., & Elson, E. L. (1991). Single particle tracking. Analysis of diffusion and flow in two-dimensional systems. *Biophysical Journal, 60*, 910–921.

Regan, J. W., Bailey, T. J., Pepperl, D. J., Pierce, K. L., Bogardus, A. M., Donello, J. E., et al. (1994). Cloning of a novel human prostaglandin receptor with characteristics of the pharmacologically defined EP2 subtype. *Molecular Pharmacology, 46*, 213–220.

Schutz, G. J., Schindler, H., & Schmidt, T. (1997). Single-molecule microscopy on model membranes reveals anomalous diffusion. *Biophysical Journal, 73*, 1073–1080.

Smith, W. L. (1992). Prostanoid biosynthesis and mechanisms of action. *The American Journal of Physiology, 263*, F181–F191.

Van Hemert, F., Lazova, M. D., Snaar-Jagaska, B. E., & Schmidt, T. (2010). Mobility of G proteins is heterogeneous and polarized during chemotaxis. *Journal of Cell Science, 123*, 2922–2930.

Zuo, L., Ushio-Fukai, M., Hilenski, L. L., & Alexander, R. W. (2004). Microtubules regulate angiotensin II type 1 receptor and Rac1 localization in caveolae/lipid rafts: Role in redox signaling. *Arteriosclerosis, Thrombosis, and Vascular Biology, 24*, 1223–1228.

CHAPTER FOUR

GPCR Oligomerization and Receptor Trafficking

Richard J. Ward, Tian-Rui Xu, Graeme Milligan[1]
Molecular Pharmacology Group, Institute of Molecular, Cell and Systems Biology, College of Medical, Veterinary and Life Sciences, University of Glasgow, Glasgow, United Kingdom
[1]Corresponding author: e-mail address: graeme.milligan@glasgow.ac.uk

Contents

1. Introduction 70
2. GPCR Expression Using the Flp-In™ T-Rex™ System 72
   2.1 Inducible expression of a single GPCR: Studies of homomers 72
   2.2 Constitutive expression of a second GPCR: Studies of heteromers 73
3. Detecting GPCR Internalization by Fluorescent Microscopy: The Orexin OX_1 Receptor 74
4. SNAP–CLIP Tagging 75
   4.1 Imaging of SNAP/CLIP-tagged GPCRs as an alternative to the use of autofluorescent proteins 76
   4.2 Using SNAP and CLIP tags to label GPCRs for microscopy 77
   4.3 Using SNAP and CLIP tags to detect oligomerization by htrFRET 78
   4.4 Using SNAP and CLIP tags to measure internalization of coexpressed GPCRs 80
5. Detecting GPCR Oligomerization by other Resonance Energy Transfer Techniques 82
   5.1 Detection of orexin OX_1 receptor oligomerization by saturation BRET assays 82
   5.2 Oligomerization of OX_1 receptor detected by FRET imaging in living cells 83
6. Biotinylation Studies 84
   6.1 Biotinylation protection assay 86
7. ER Trapping, Pharmacological Chaperones, and the Use of Engineered Synthetic Ligands 87
Acknowledgments 89
References 89

Abstract

The effects of oligomerization of G protein-coupled receptors (GPCRs) upon their trafficking around the cell are considerable, and this raises the potential of significant impact upon the use of existing pharmacological agents and the development of new ones. Herein, we describe a number of different techniques that can be used to study receptor dimerization/oligomerization and trafficking, beginning with a cellular system which allows the expression of two GPCRs simultaneously, one under inducible control. Subsequently, we describe means to visualize and monitor the movement of GPCRs

Methods in Enzymology, Volume 521
ISSN 0076-6879
http://dx.doi.org/10.1016/B978-0-12-391862-8.00004-1

within the cell, detect oligomerization by both resonance energy transfer and more traditional biochemical approaches, and to measure the internalization of GPCRs as part of the process of receptor regulation.

1. INTRODUCTION

The concept that GPCRs exist and function as homo- or heterodimers/oligomers is now well established and supported by a great deal of experimental evidence (Milligan, 2007, 2008, 2009). While some reported examples may reflect data overinterpretation or artifact due to the system in use, it seems likely that, despite the ability of GPCRs to function in a strictly monomeric state (Kuszak et al., 2009; Whorton et al., 2007), the oligomerization concept is one which is well founded. Receptor oligomerization may go some way toward answering the question, fundamental to GPCR biology, of how a single ligand, acting at a specific receptor, can induce a set of signaling responses, which in turn lead to a range of physiological outcomes.

Receptor trafficking begins with the translocation of GPCRs to the plasma membrane from their site of synthesis in the endoplasmic reticulum (ER) and subsequent processing in the Golgi apparatus, a process that involves interactions with chaperone proteins which assist in folding and may retain misfolded proteins within the ER (Conn & Ulloa-Aguirre, 2010; Tan, Brady, Highfield Nichols, Wang, & Limbird, 2004). Once at the plasma membrane, GPCRs are available to interact with ligands present in the external environment, which may result in receptor activation, signal transduction and subsequent receptor trafficking and desensitization through the processes of endocytosis. This later event involves phosphorylation by G protein receptor and potentially other kinases, interaction with arrestins, and internalization from the cell surface in vesicles formed from clathrin-coated pits (Gainetdinov, Premont, Bohn, Lefkowitz, & Caron, 2004; Kelly, Bailey, & Henderson, 2008).

A link between GPCR quaternary structure and trafficking between the ER and the cell plasma membrane was originally exemplified by the $GABA_B$ receptor, a member of the metabotropic glutamate-like, type C receptor family. In its functional, cell membrane located form this is a constitutive heterodimer/oligomer composed of the products of two related genes, *$GABA_BR1$* and *$GABA_BR2$*. The $GABA_BR1$ polypeptide contains the binding site for γ-amino butyric acid but, when expressed alone, is retained within the ER due to the presence of an ER retention sequence in the C-terminal tail. If the R1 and

R2 polypeptides are coexpressed, then they interact and the retention sequence is masked, resulting in the export of the heteromer to the cell surface. Despite a number of efforts to adapt this idea, either by introduction of the $GABA_BR1$ sequence or other ER retention sequences into the intracellular tail of a range of GPCRs, although often resulting in ER retention of the modified receptor, these have had limited success in producing cotrapping of a coexpressed potential interaction partner (Wilson, Wilkinson, & Milligan, 2005) or other evidence to support protein–protein interactions. Furthermore, although the ER-trapped nature of oligomerization-defective mutants of class A GPCRs has also been used to define production of GPCR oligomers prior to cell surface delivery (Canals, Lopez-Gimenez, & Milligan, 2009; Kobayashi, Ogawa, Yao, Lichtarge, & Bouvier, 2009), a limited number of studies have used this approach, perhaps because the basis of GPCR oligomerization remains difficult to define.

Studies of a large number of potential heteromers of rhodopsin-like, type A receptors have been based, at least in part, on the ability of pharmacologically selective agonist ligands to promote cointernalization of the presumed heteromeric partner of the GPCR without binding directly to this polypeptide (Milligan, 2010), and this has been extended to examples in which constitutive trafficking of a GPCR results in cotrafficking of a second, coexpressed receptor (Ellis, Pediani, Canals, Milasta, & Milligan, 2006; Ward, Pediani, & Milligan, 2011b). This is clearly more challenging to assess for GPCR homomers because there is not a convenient and generally applicable means to selectively activate an individual protomer of the presumed dimer/oligomer. However, useful studies have been performed after coexpression of a wild-type GPCR along with a variant in which chemical engineering has been used to alter receptor pharmacology (Sartania, Appelbe, Pediani, & Milligan, 2007).

Although it is beyond the scope of the present chapter to describe details of such approaches, in each case, detailed exploration of how cell trafficking may be controlled by the oligomeric state of a GPCR has required methods that allow effective and regulated coexpression of both distinctly modified forms of the same GPCR (homomers) or of two different GPCRs (heteromers). They have also required effective means to visualize and quantify the location of such variants in both space and time. These topics will provide the methodological focus of the chapter.

The consequences of receptor dimerization/oligomerization upon trafficking of receptors are likely to be extensive (Milligan, 2010) and with wide ranging consequences for pharmacology and function.

2. GPCR EXPRESSION USING THE FLP-IN™ T-REX™ SYSTEM

The Flp-In™ T-Rex™ expression system (Invitrogen Life Technologies) allows the introduction of a gene or cDNA of interest (such as a GPCR) into the genome at a single defined, specific, (FRT) site (theoretically, this should be assessed via Southern analysis) to create a stable isogenic cell line without it being necessary to isolate individual clones of cells (the Flp-In™ system). Expression of the gene of interest (GOI) is controlled positively by the addition of the antibiotic tetracycline or its derivative doxycycline (T-Rex™).

2.1. Inducible expression of a single GPCR: Studies of homomers

There are a range of parental Flp-In™ or Flp-In™ T-Rex™ cell lines commercially available (Invitrogen Life Technologies), of which perhaps the most commonly used, Flp-In™ T-Rex™ 293 cells, are based upon HEK293 cells. HEK293 cells were modified by integration of two plasmids into the genome, pFRT/lacZeo (which forms the FRT integration site, maintained by the presence of zeocin in the medium) and pcDNA6/TR (which expresses the tet repressor and is maintained by the presence of blasticidin in the medium; Ward, Alvarez-Curto, & Milligan, 2010).

1. The GOI must be subcloned using standard molecular biological techniques into the plasmid pcDNA5–FRT–TO (Invitrogen Life Technologies). Amino or C-terminal epitope tags or fluorescent labels are often included to allow the GPCR to be detected later.
2. Flp-In™ T-Rex™ 293 cells are maintained in medium consisting of Dulbecco's Modified Eagle's Medium (+4.5 g l^{-1} glucose, +L-glutamine, −pyruvate), 10% fetal bovine serum (FBS), 100 μg ml^{-1} streptomycin, 100 units ml^{-1} penicillin, 100 μg ml^{-1} zeocin, and 5 μg ml^{-1} blasticidin.
3. Seed the Flp-In™ T-Rex™ 293 cells into 10-cm dishes in the medium described above and grow until 65% confluent.
4. Cotransfect the Flp-In™ T-Rex™ 293 cells with the GOI subcloned into pcDNA5/FRT/TO and the plasmid pOG44, which encodes the Flp recombinase (which then catalyzes integration). We use a plasmid ratio of 9:1, 7.2 μg pOG44 and 0.8 μg GOI containing pcDNA5/FRT/TO. Transfection may be carried out using standard protocols and reagents, such as Lipofectamine (Invitrogen Life Technologies).

Alternatively, a more economical approach is to use polyethyleneimine (25 kDa linear) as described by van Rijn et al. (2008).

5. Following transfection, the cells should be incubated overnight in medium without zeocin, as the zeocin resistance is destroyed by the integration process.
6. 24 h after transfection, replace the medium.
7. 48 h after transfection, split the cells into 10-cm dishes at densities lower than 25%. The medium should be as described above, but without zeocin.
8. 72 h after transfection, begin selection by changing the medium for that described above, but with 0.2 mg ml^{-1} hygromycin b (no zeocin). Cells will begin to die off.
9. Change the medium every 2–3 days until visible foci appear (approximately 2 weeks after transfection). The cells should then be pooled, expanded, and tested for tetracycline-regulated expression of the GOI. Alternatively, the more stable derivative doxycycline may be used, at a concentration range of 1 ng ml^{-1} to 1 μg ml^{-1}; with the most suitable concentration being assessed empirically.

The inducible nature of expression of the GOI in this system allows studies of GPCR trafficking and homomerization to be conducted at a range of expression levels to consider whether protein–protein interactions might be an artifact driven solely by mass-action and observed only at high expression levels (Pou, Mannoury la Cour, Stoddart, Millan, & Milligan, 2012; Xu, Ward, Pediani, & Milligan, 2011).

2.2. Constitutive expression of a second GPCR: Studies of heteromers

Once a stable Flp-In™ T-Rex™ GPCR cell line has been established, a second GPCR may be introduced which will express constitutively. This second GPCR is not integrated at a defined site, and as such, individual clones will need to be selected and analyzed for appropriate levels of expression. The second GPCR should be subcloned into an appropriate mammalian expression vector, such as pcDNA3.1 (Invitrogen Life Technologies), incorporating a selectable resistance gene (for instance, G418 or geneticin).

1. Seed Flp-In™ T-Rex™ 293 cells harboring the first GPCR at the inducible locus into 10-cm dishes with the medium described in the previous section (with hygromycin b, without zeocin). Maintain the cells until they are 65% confluent.

2. Transfect with 5–10 μg of the second cDNA using standard reagents, as in the previous section.
3. Change the medium for fresh 24 h after transfection.
4. 48 h after transfection, split the cells into 10-cm dishes at various densities below 20%.
5. Change medium for that described above, but supplemented with 1 mg ml^{-1} G418. Cells will begin to die off.
6. Change the medium every 2–3 days; continue until visible foci appear.
7. Select individual colonies using "cloning rings" (Ward et al., 2010), into 12-well plates using the G418-supplemented medium.
8. Determine expression levels for individual clones by ligand-binding studies, microscopy, or Western blotting.

Following isolation and characterization of the expression levels, pharmacology, cellular location, and trafficking of the constitutively expressed GPCR, induction of varying levels of the GPCR harbored at the inducible locus can be used to assess direct interactions of the two proteins as a heteromer (Alvarez-Curto, Ward, Pediani, & Milligan, 2010) and the effect of production of the heteromer on pharmacology and trafficking of the constitutively expressed receptor. Recent studies have also used this approach to identify concurrently homomers and heteromers of coexpressed dopamine receptor subtypes at the surface of such cells (Pou et al., 2012).

3. DETECTING GPCR INTERNALIZATION BY FLUORESCENT MICROSCOPY: THE OREXIN OX_1 RECEPTOR

The identification of the cellular location of a GPCR by fluorescent microscopy and its cellular trafficking is a routinely used approach. Although other techniques are used, most studies are based upon the fusion of the GPCR to an autofluorescent protein, most usually via the C-terminal. The fluorescent proteins used are often modified versions of the 26.9 kDa green fluorescent protein (GFP), originally isolated from the jellyfish *Aequorea victoria*. The modifications are designed to produce more stable proteins, which fluoresce at different wavelengths and have reduced tendency to aggregate (these are usually described as monomeric variants); the later feature is obviously of great importance when employed in the study of GPCR oligomerization. Such fusions are also used routinely to study the trafficking of GPCRs, such as the internalization of a cell membrane-localized GPCR after activation by an agonist (Lopez-Gimenez, Vilaró, & Milligan, 2008; Ward et al., 2011b; Ward, Pediani, & Milligan, 2011a). The distinct fluorescence characteristics of GFP variants allows both macro-level assessment of the general cellular location of two distinct but coexpressed GPCRs, and, if the GFP variants have excitation and emission spectra consistent with resonance energy transfer (Alvarez-Curto, Pediani, & Milligan, 2010), concurrent analysis of whether direct protein–protein interactions within a GPCR dimer/oligomer is consistent with their "colocalization" at the level of light microscopy.

In recent times, we have studied aspects of the trafficking of the orexin OX_1 receptor within both homomers and heteromers.

1. Seed Flp-In™ T-Rex™ HEK293 cells harboring an inducible VSV-G-OX_1-eYFP (Xu et al., 2011) construct into 6-well plates containing poly-D-lysine-treated glass coverslips and incubate overnight. In order to be sufficiently well separated for imaging after 2–3 days growth, the cells must be seeded at comparatively low density.
2. Induce the cells with a suitable concentration of doxycycline and incubate for 24–48 h.
3. Place the coverslips with attached cells into a microscope chamber containing physiological HEPES-buffered saline solution (130 m*M* NaCl, 0.5 m*M* KCl, 1 m*M* $CaCl_2$, 1 m*M* $MgCl_2$, 20 m*M* HEPES, and 10 m*M* D-glucose, pH 7.4) and observe and record images using appropriate excitation wavelengths and filters.
4. Treat with varying concentrations of orexin A (one of the endogenously generated peptides that activate the OX_1 receptor) and record sequential images at appropriate intervals for up to 60 min. The effect of orexin A on the distribution of receptors between the cell surface and intracellular locations can then be assessed and quantified via a range of algorithms (Xu et al., 2011).

4. SNAP–CLIP TAGGING

The recent development of SNAP- and CLIP-tagging technology (New England BioLabs, Inc.) has been of great value to the study of GPCR oligomerization and trafficking (Maurel et al., 2008). The SNAP and CLIP tags are 20-kDa polypeptides based upon the enzyme O^6-alkylguanine-DNA-alkyltransferase.

4.1. Imaging of SNAP/CLIP-tagged GPCRs as an alternative to the use of autofluorescent proteins

These sequences may be fused to the GPCR in question (Fig. 4.1A). In the majority of cases to date, this has been to the extracellular N-terminal domain to allow specific detection and imaging of GPCRs that are, or

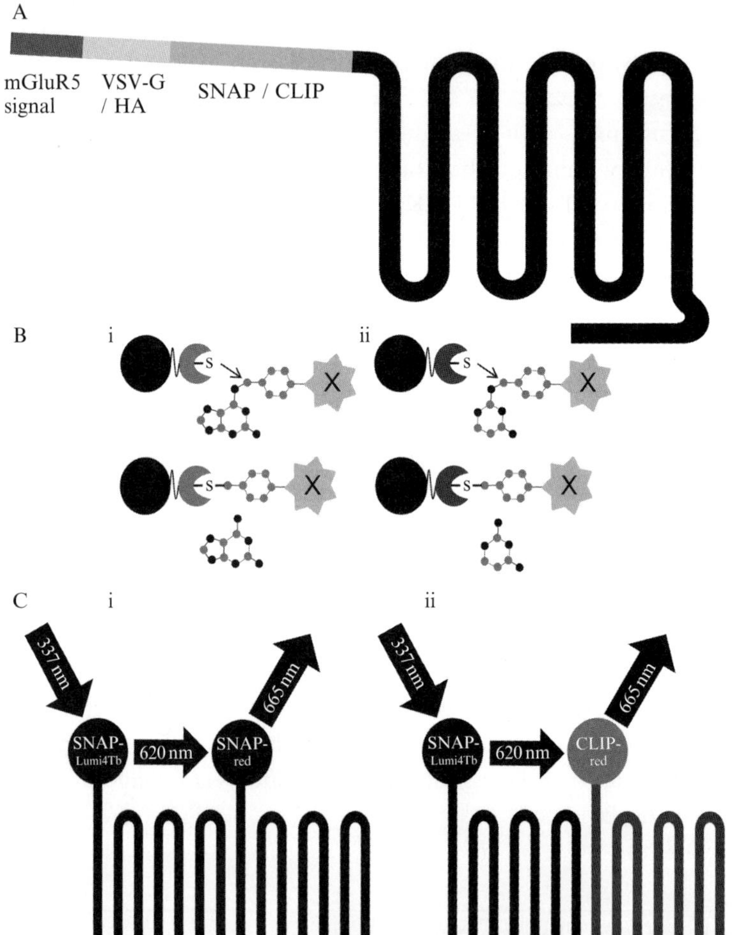

Figure 4.1 Construction of SNAP/CLIP-tagged receptors and subsequent labeling. (A) Cartoon showing the design of SNAP- and CLIP-tagged receptors. (B) Covalent labeling of the SNAP tag with a benzyl guanine derivative (i) and CLIP with a benzyl cytosine derivative (ii). (C) The Tag-lite® htrFRET system showing the detection of (i) homo- and (ii) hetero-oligomerization, using the SNAP lumi4Tb reagent as a FRET donor and either SNAP-red (i) or CLIP-red (ii) as acceptors.

previously were, at the cell surface. The SNAP and CLIP tags act as "suicide" enzymes by reacting with benzyl guanine- and benzyl cytosine-linked labels, respectively, to covalently bond the label to the SNAP or CLIP tag via a thio-ester linkage (Fig. 4.1B). A wide variety of SNAP and CLIP labels are available including SNAP/CLIP-surface (cell impermeant) and SNAP/CLIP-cell (cell permeant) fluorescent labels, biotin, blocking agents (cell permeant and impermeant), SNAP-capture magnetic or pulldown resins, vista stains for protein extracts, and benzyl guanine/cytosine building blocks for custom synthesis of labels. Indeed, the flexibility of ways to label these tags is a major advantage in their use. Probes that label the SNAP and CLIP tags do not cross-react, although in general labeling of the CLIP sequence is substantially less effective than the SNAP sequence (Pou et al., 2012; Ward et al., 2011a), and this allows the simultaneous labeling of two GPCRs with, for instance, two differently colored fluorescent labels, the labeling patterns of which can then be analyzed concurrently.

The utility of the SNAP and CLIP tags has recently been extended by the development of Tag-lite® technology by Cisbio Bioassays. This system consists of htrFRET (homogenous time-resolved fluorescence resonance energy transfer) competent parings of energy donor and acceptor species which are linked to benzyl guanine or benzyl cytosine, thus allowing them to be attached to SNAP- and CLIP-tagged proteins, (Fig. 4.1C). It is then possible to use htrFRET to monitor GPCR homomers and heteromers (and, indeed, both concurrently; Pou et al., 2012) depending upon the combination of donor and acceptor species used.

We have used these tools to monitor both trafficking and oligomerization of a number of GPCRs.

4.2. Using SNAP and CLIP tags to label GPCRs for microscopy

SNAP or CLIP tags together with VSV-G (vesicular stomatitis virus) or HA (hemagglutinin) epitope tags and a membrane targeting signal sequence from the metabotropic glutamate receptor 5 were assembled with the human orexin OX_1 and cannabinoid CB_1 receptors as shown in Fig. 4.1A (Ward et al., 2011b). These constructs were then used to build cell lines, as described in Section 2, such that one GPCR (tagged with, for instance, VSV-G and SNAP) would be under inducible control and the other (tagged HA and CLIP) expressed constitutively. The cell lines were then colabeled with appropriate fluorescent species (which may be cell permeant or impermeant) as follows.

1. Cells are seeded into 6-well plates containing ethanol washed and then 0.1 mg ml^{-1} poly-D-lysine coated circular coverslips, followed by overnight incubation and doxycycline induction as required.
2. Transfer the coverslips (with cells attached) to new 6-well plates and rinse with fresh medium.
3. Add medium with the required fluorescent SNAP and CLIP labels at appropriate concentration and incubate for 30 min at 37 °C. (The actual concentration of the label will need to be determined for a particular application, but we have found 2.5–5 μ*M* to be a good starting point.)
4. Remove the labeling medium and wash the cells three times with fresh medium.
5. The cells are then washed once in Hank's balanced salt solution (HBSS), prior to fluorescent imaging.

4.3. Using SNAP and CLIP tags to detect oligomerization by htrFRET

A similar GPCR expression cell line to that described in the previous section can be used to detect homomers or heteromers of GPCRs at the cell surface via the Tag-lite® system. This uses terbium cryptate as an htrFRET donor due to its long emission half-life, which allows biological fluorescence due to excitation from the light source to decay to background levels before FRET measurements are made. The donors are termed SNAP (or CLIP)-Lumi4Tb, while the acceptors are SNAP (or CLIP)-red or -green.

(*Note*: The CLIP tag has recently been replaced in this system with an alternative, the Halotag (Cisbio Bioassays), which is used in a similar way.)

4.3.1 htrFRET to detect GPCR heteromers

In order to study GPCR heteromers using this system, the optimal ratio of FRET donor (e.g., SNAP-Lumi4Tb) to acceptor (e.g., CLIP-red) must first be established for the cell line and GPCR expression levels. This is done as follows:

1. Seed appropriate cells (at 75,000 cells/well) into black, 96-well plates which have been treated with 0.1 mg ml^{-1} poly-D-lysine. Induce GPCR expression as required and grow overnight.
2. Remove medium and replace with 1× labeling medium containing 10 n*M* SNAP-Lumi4Tb and a range of concentrations of CLIP-red from 2000 to 15 n*M*, halving the concentration at each step. (It should be noted that the concentration of donor described here is lower than that recommended by the manufacturer, but we have found that these

concentrations of donor and acceptor result in an acceptable FRET signal, while ensuring that a minimum amount of costly reagent is used.)

3. Incubate the plates at 37 °C/5% CO_2 for 1 h and then remove the unbound donor, acceptor, and labeling medium.
4. Wash the plates with 4 × 200 μl/well labeling medium, and then add 100 μl labeling medium/well.
5. Read the plates using a PHERAstar FS htrFRET compatible reader (BMG Labtechnologies) or other appropriate instrument. Plot the 665 n*M* FRET signal from the acceptor against the concentration of acceptor as shown in Fig. 4.2A. The optimum ratio is that which gives the maximum FRET signal. In the example shown in Fig. 4.2A this is 200 n*M*.

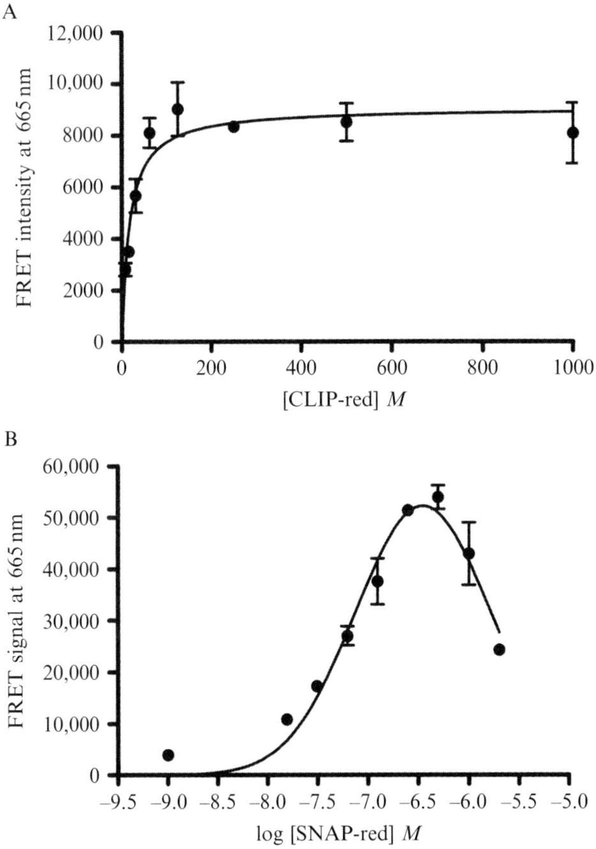

Figure 4.2 Optimization of conditions for htrFRET. (A) Optimization of a htrFRET hetero-oligomerization experiment, indicating an optimal CLIP-red concentration of 200 n*M*. (B) Optimization of a htrFRET homo-oligomerization experiment indicating an optimal SNAP-red concentration of 80 n*M*.

6. This optimized ratio should be used in any subsequent experiments to establish, for example, the effect of ligands upon the extent of GPCR heteromerization.

4.3.2 htfrFRET to detect GPCR homomers

In a similar manner to the study of GPCR heteromers, an optimum ratio of energy donor to acceptor must be established. However, in order to monitor the presence of homomers, the donor and acceptor must label the same receptor species, that is, the donor and acceptor must both be either SNAP specific or CLIP specific. In detail, the experiment is carried out in the same way as described in Section 4.3.1, but the FRET signal at 665 nm should be plotted against a logarithmic scale of acceptor concentration. Such data can then be fitted to a Gaussian equation to generate a bell-shaped distribution, as shown in Fig. 4.2B. This reflects that as both donor and acceptor are specific for the same tagged receptor species, they compete for binding and as the concentration of acceptor is increased, it will out-compete a single concentration of donor. The largest FRET signal and, therefore, the optimum ratio of donor to acceptor can then be defined from the apex of curve. This optimum ratio can then be used to label cells, such as those expressing VSV-G-SNAP-OX_1 (Xu et al., 2011).

4.4. Using SNAP and CLIP tags to measure internalization of coexpressed GPCRs

The Tag-lite® htrFRET donor (SNAP-Lumi4Tb or CLIP-Lumi4Tb) can also be used to measure N-terminally SNAP- or CLIP-tagged GPCRs at the cell surface, as an alternative to an antibody-based ELISA (enzyme-linked immunoabsorbent assay), and frequently, this produces a much superior signal to background ratio (Ward et al., 2011a). Cells are prepared as for detection of GPCR oligomers, but labeled with (only) the htrFRET donor for 30 min. After washing, the plates are read as for htrFRET, but only the emission from the donor at 620 nm is noted as this provides a direct measure of binding of the htrFRET donor to the appropriate tag and therefore, of receptor at the cell surface (Fig. 4.3). This approach was used by Ward et al. (2011b) to demonstrate that when constructs based upon SNAP- and CLIP-tagged versions of the cannabinoid CB_1 and orexin OX_1 receptors are coexpressed, treatment with the OX_1 agonist orexin A causes internalization of both receptors, despite orexin A having no direct affinity for the cannabinoid CB_1 receptor. Furthermore, it was shown that orexin A was more potent at internalizing the CB_1 than the OX_1 receptor and that

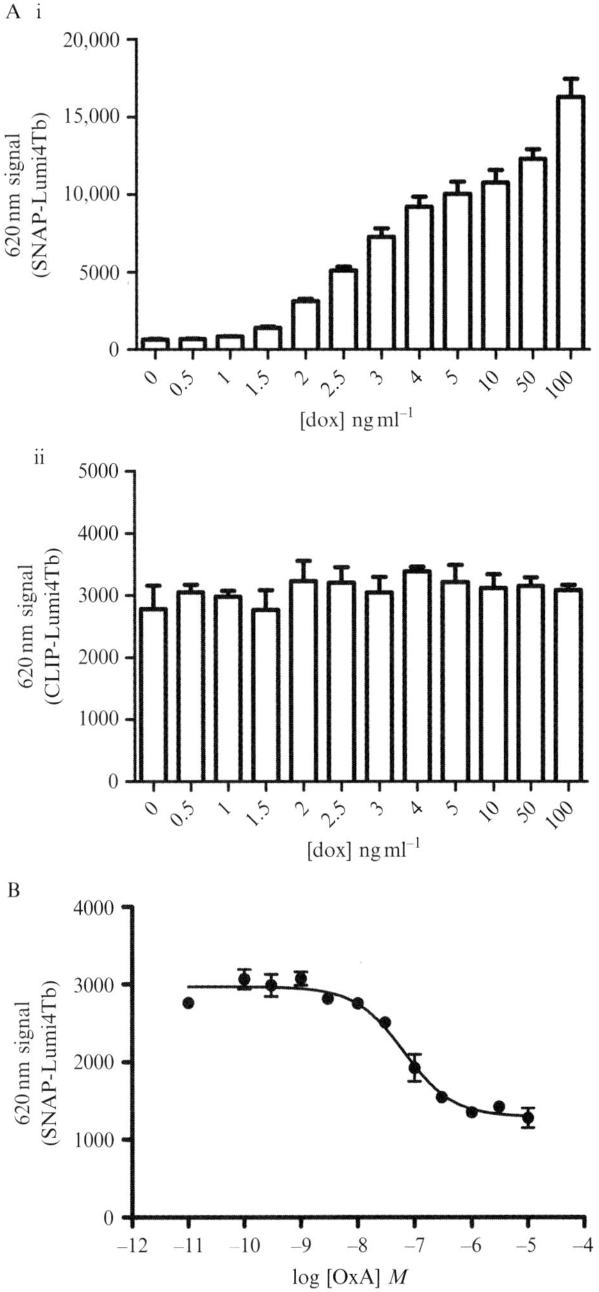

Figure 4.3 Use of the htrFRET donor to measure N-terminally tagged GPCRs at the cell surface. (A) Cells expressing the cannabinoid CB_1 receptor (tagged with SNAP) inducibly and the orexin OX_1 receptor (tagged with CLIP) constitutively were induced with a variety of concentrations of doxycycline and then labeled with SNAP-specific htrFRET donor (i) or CLIP-specific htrFRET donor (ii). Emission at 620 nm was measured using an appropriate reader. (B) Cells expressing the orexin OX_1 receptor (tagged with SNAP) were treated with varying concentrations of the agonist orexin A for 40 min and then labeled with SNAP-specific htrFRET donor and measured as described above.

selective antagonists for both the OX_1 and the CB_1 receptor were each able to prevent internalization of these receptors produced by orexin A but only when the two GPCRs were coexpressed (Ward et al., 2011b). Such studies once more provided important links between GPCR oligomerization state and cellular trafficking.

5. DETECTING GPCR OLIGOMERIZATION BY OTHER RESONANCE ENERGY TRANSFER TECHNIQUES

The determination of protein–protein proximity by any resonance energy transfer technique exploits the nonradiative transfer of energy between an energy donor and an acceptor based upon the Forster mechanism (Alvarez-Curto et al., 2010). This is highly dependent on the distance between the donor and acceptor species. As such, although there are distinct advantages to the htrFRET methods described earlier, other resonance energy transfer approaches are also used widely to investigate GPCR oligomerization in living cells when the potentially interacting GPCRs are appropriately labeled with donor and acceptor. Both bioluminescence resonance energy transfer (BRET) and fluorescence resonance energy transfer (FRET) have been use in many key studies (Alvarez-Curto et al., 2010; Milligan & Bouvier, 2005). In BRET, the energy donor is bioluminescent (generally a luciferase (Luc) which can oxidize a form of coelenterazine and in so doing generates light) (e.g., El-Asmar et al., 2005; Xu et al., 2011), while in the case of FRET, the donor is excited to emit light that is transferred to the acceptor by an external light source (e.g., Parenty, Appelbe, & Milligan, 2008; Xu et al., 2008).

5.1. Detection of orexin OX_1 receptor oligomerization by saturation BRET assays

1. Transiently cotransfect HEK-293T cells with orexin OX_1 receptor-*Renilla reniformis* Luc8 and orexin OX_1 receptor-enhanced yellow fluorescent protein (eYFP) fusion constructs (Xu et al., 2011) in 6-well plates, using the Lipofectamine 2000 reagent according the manufacturer's instructions (Invitrogen). For BRET saturation assays in which the extent of resonance energy transfer is determined at a range of energy acceptor to energy donor expression ratios (Milligan & Bouvier, 2005), cells were cotransfected with a fixed amount (0.5 μg) of the *Renilla* Luc8-tagged receptor construct and varying amounts of the eYFP-tagged receptor construct (0.5–5 μg).

2. 24 h after transfection, release the cells from the plate by addition of trypsin and transfer them to poly-D-lysine-treated white 96-well plates (for BRET and luminescence reading) and black 96-well plates (for eYFP reading).
3. At 48 h post-transfection, wash the cells in black 96-well plates with HBSS (Invitrogen) and measure the emission of eYFP on a PHERAstar FS microplate reader (BMG Labtech) or similar using the wavelengths 485 nm excitation and 520 nm emission.
4. Also wash the cells in the white 96-well plates and incubate with 100 μl/well HBSS at 37 °C for 30 min. Add coelenterazine-h (to 5 μ*M* final concentration), incubate for another 10 min in the dark at 37 °C, and then measure BRET using a PHERAstar FS reader, or similar.
5. Calculate the Net BRET signal as a function of eYFP values (energy acceptor) divided by Rluc8 values (energy donor). In the case of such saturation BRET assays, plot the Net BRET and fit curves using a nonlinear regression equation assuming a single binding site with Prism v. 5.0 software or similar.
6. To control for the effects of potential nonspecific protein–protein interactions driven by high level expression and mass-action effects, parallel experiments should be performed using GPCRs which, based on independent information, are not expected to display a significant propensity to interact.

5.2. Oligomerization of OX_1 receptor detected by FRET imaging in living cells

1. Seed HEK-293T cells into 6-well plates containing poly-D-lysine-treated glass coverslips and maintain overnight in growth medium.
2. Cotransfect the cells with orexin OX_1 constructs C-terminally tagged with either eYFP or enhanced cyan fluorescent protein (eCFP) (VSV-G-OX_1-eYFP and HA-OX_1-eCFP) using Lipofectamine 2000 reagent according the manufacturer's instructions. The VSV-G-OX_1-eYFP and HA-OX_1-eCFP constructs should also be expressed separately to permit the calculation of the extent of bleedthrough (overlap of excitation/emission spectra between the energy donor and acceptor).
3. 48 h after transfection, transfer the cells (attached to the coverslip) to a microscope chamber containing Hepes-buffered saline solution (see Section 3).
4. In the case of VSV-G-OX_1-eYFP- and HA-OX_1-eCFP-expressing cells, excitation was achieved using an Optoscan monochromator (Cairn

Research) which was set to 500/5 and 430/12 nm for the sequential excitation of eYFP and eCFP, respectively. A dual dichroic mirror (86002v2bs; Chroma) was used to reflect the excitation light through a Nikon ×40 (NA = 1.3) oil-immersion Fluor lens.

5. The resultant eCFP and eYFP fluorescence emission signals are detected using a high-speed filterwheel device (Prior Instruments) containing the following band pass emitters: HQ470/30 nm for eCFP and HQ535/30 nm for eYFP.
6. Bleedthrough coefficients should be measured using cells expressing either the eCFP- or eYFP-tagged constructs alone and defined by dividing the amount of fluorescence detected in the FRET channel (e.g., $\mathrm{FRET}_{\mathrm{eCFP-eYFP}}$) by the fluorescence detected from each fluorescent protein, in its own filter channel. Calculate corrected FRET (FRET_c) using a pixel-by-pixel methodology with the equation

$$\mathrm{FRET}_c = \mathrm{FRET} - (\text{coefficient } B \times \mathrm{CFP}) - (\text{coefficient } A \times \mathrm{YFP}).$$

7. eCFP, eYFP, and FRET values correspond to background corrected images obtained through the eCFP, eYFP, and FRET channels. *B* and *A* correspond to the values obtained for the eCFP (donor) and eYFP (acceptor) bleedthrough coefficients, respectively. To correct the FRET levels for the varying amounts of donor (eCFP) and acceptor (eYFP), normalized FRET (FRET_N) was calculated using the equation

$$\mathrm{FRET}_N = \mathrm{FRET}_c/\mathrm{eCFP},$$

where FRET_c and eCFP are equal to the fluorescence values obtained from single cells.

Further details are available in Xu et al. (2008).

6. BIOTINYLATION STUDIES

Cell surface proteins and their subsequent trafficking can be assessed in cell populations via a variety of protocols that involve the incorporation of a biotin label. The biotinylation protocols described here are based upon the use of EZ-Link® Sulfo-NHS-SS-Biotin, which is a thiol-cleavable, amine-reactive biotinylation reagent (Pierce, Rockford, IL). When cells are treated with this reagent, the *N*-hydrosulfosuccinimide (NHS) group can potentially react with the ε-amine of any lysine residues found in cell surface proteins, such as GPCRs, hence labeling them with biotin. Cell lysis, followed by pulldown with immobilized streptavidin (Pierce) allows the isolation of labeled cell surface proteins, which may then be released by treatment with a reducing agent such as 2-mercaptoethanol and analyzed by SDS-PAGE and Western blotting with an appropriate antiserum (Canals et al., 2009). Thus, the relative intensity of a band corresponding to the GPCR in question can be used to monitor, for instance, the extent of removal from the cell surface of the GPCR in response to agonist treatment. A variation on this is the biotinylation protection assay (Fig. 4.4). Herein, cell surface proteins are biotinylated, then treated with agonist to promote trafficking of the GPCR

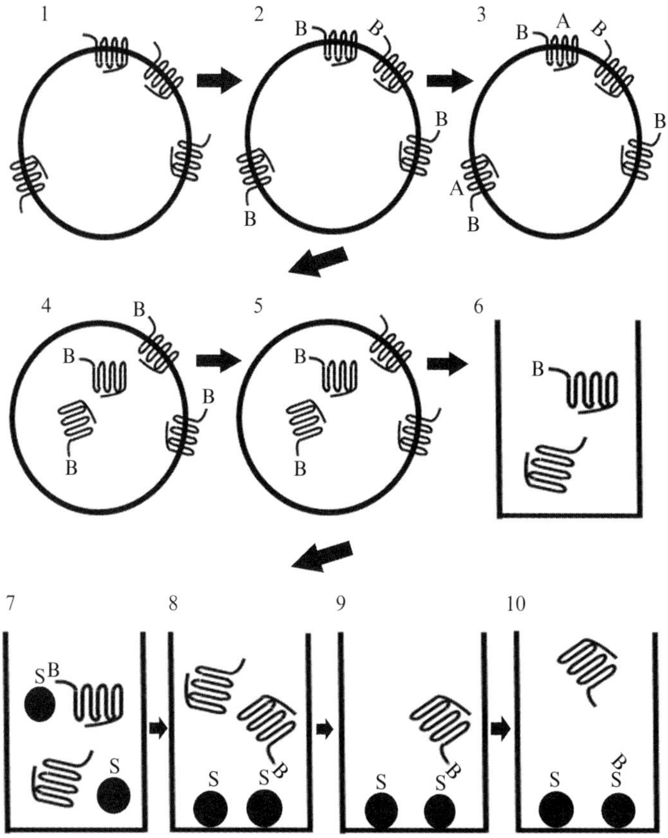

Figure 4.4 Steps involved in the biotinylation protection assay. (1) Cells expressing a GPCR of interest. (2) Cell surface expressed GPCRs are biotinylated (B). (3) Agonist (A) treatment is applied, which results in internalization of the biotinylated GPCRs (4). (5) Biotin is removed from the remaining cell surface GPCRs. (6) The cells are lysed and immobilized streptavidin (S) beads are added (7). (8) The beads are separated by centrifugation and washed (9). (10) Finally, the biotinylated, internalized GPCRs are released by β-mercaptoethanol treatment and analyzed by SDS-PAGE and Western blotting.

in question into intracellular vesicles. If remaining cell surface biotin is removed and the cells are lysed and analyzed as described above, then a GPCR that retains the initial biotin label must have moved from the cell surface into an internal location. This "protection" assay has been used to monitor internalization of heteromeric cannabinoid CB_1–orexin OX_1 receptor complexes (Ward et al., 2011b). It was shown that the CB_1 receptor was internalized both by agonists specific for the CB_1 receptor and the OX_1 peptide agonist orexin A.

6.1. Biotinylation protection assay

1. Seed cells onto 0.1 mg ml^{-1} poly-D-lysine coated 6-well plates and maintain overnight.
2. Remove the medium and place the plates on ice. Wash the cells with 1 × 2 ml/well ice-cold phosphate buffered saline (PBS).
3. Add 1 ml/well 0.3 mg ml^{-1} EZ-Link® Sulfo-NHS-SS-Biotin and incubate on ice, in the dark for 30 min.
4. Remove the biotinylation label and wash with 1 × 2 ml/well ice-cold PBS.
5. Add 1 ml/well prewarmed medium and incubate at 37 °C/5% CO_2 for varying periods of time.
6. Carry out any ligand treatments, as required within this time.
7. Remove the medium, put the plates on ice, and wash the cells with 1 × 2 ml/well ice-cold PBS.
8. Remove (usually described as "stripping") the biotin from cell surface-modified proteins by adding 2 ml/well 50 m*M* glutathione, 300 m*M* NaCl, 75 m*M* NaOH, 1% FBS. Incubate on ice at 4 °C for 30 min.
9. Remove "stripping" solution and wash the cells with 2 × 2 ml/well ice-cold PBS.
10. Lyse the cells with 500 μl 1 × RIPA buffer (radioimmunoprecipitation assay buffer, 50 m*M* HEPES, 150 m*M* NaCl, 1% Triton X-100, and 0.5% sodium deoxycholate, 10 m*M* NaF, 5 m*M* EDTA, 10 m*M* NaH_2PO_4, 5% ethylene glycol, pH 7.4) supplemented with complete protease inhibitors cocktail (Roche Diagnostics, Mannheim, Germany). Transfer to 1.5-ml tubes and incubate on a rotating wheel at 4 °C for 15 min.
11. Centrifuge at 14,000 × *g* for 30 min at 4 °C and transfer the supernatant to fresh tubes.
12. Determine the protein concentrations and equalize the samples.

13. Add 100 μl Immunopure immobilized streptavidin and incubate overnight at 4 °C on a rotating wheel.
14. Centrifuge at 2000 × *g* for 2 min at 4 °C and remove the supernatant.
15. Wash the immobilized streptavidin beads with 3 × 500 μl 1 × RIPA.
16. Add 100 μl SDS-PAGE sample buffer (containing 5% 2-mercaptoethanol), mix and incubate at 37 °C for 1 h.
17. Analyze the samples by SDS-PAGE and Western blotting with appropriate antisera.

7. ER TRAPPING, PHARMACOLOGICAL CHAPERONES, AND THE USE OF ENGINEERED SYNTHETIC LIGANDS

The necessity of heteromer formation in the ER prior to successful trafficking to the cell surface for the two gene products of the $GABA_B$ receptor provided both detailed understanding of this specific GPCR and a potential strategy to assess whether equivalent processes were required for other GPCRs, including GPCR homomers and well as other heteromers. In addition to the type C $GABA_B$ receptor, both well characterized and potential ER retention motifs can also be found in family A GPCRs, such as the arginine-based motif found in the C-terminal of the α_{2C}-adrenoceptor (Wilson et al., 2005). This motif, in a similar manner to that of the $GABA_BR1$ polypeptide, ensures that the α_{2C}-adrenoceptor is located predominantly in the ER when transiently expressed. Wilson et al. (2005) added this motif to the CXCR1 chemokine receptor C-terminal tail and noted that not only was the modified receptor unable to reach the cell membrane but also that coexpressed wild-type CXCR1 or CXCR2 chemokine receptors also became "trapped" within the ER, which, in concert with other evidence based on FRET and BRET studies, provided evidence for both homo- and heteromerization of this GPCR. An important control in these studies was that the modified CXCR1 did not trap a coexpressed α_{1A}-adrenoceptor, showing that the effect was selective (Wilson et al., 2005).

Canals et al. (2009) used a number of transmembrane domain (TMD) I and TMDIV point mutations to investigate the oligomeric structure of the α_{1B}-adrenoceptor. While some of these mutants were ER retained, and possessed close to normal structure, as they were able to bind α_1-adrenoceptor-selective ligands with near normal affinity, detailed FRET analysis suggested that quaternary organization was disrupted (Lopez-Gimenez, Canals, Pediani, & Milligan, 2007). Sustained treatment of the cells with α_1-adrenoceptor ligands, such as the antagonist prazosin, resulted in trafficking of a specific TMDI–IV mutant receptor to the cell surface, where (in the presence of the ligand) it was found to have similar quaternary structure to the wild-type receptor. Thus, the ligand was able to act as a pharmacological chaperone (Janovick et al., 2009) and either ensure correct folding and, hence, trafficking of the receptor and/or alter the conformation to enhance oligomerization and then trafficking to the cell membrane. An α_{1B}-adrenoceptor mutant designed to be unable to bind prazosin, was delivered normally to the cell surface, but it was retained in the ER when coexpressed with the TMDI–IV mutant. However, when cells coexpressing the TMDI–IV mutant and the prazosin-binding deficient but otherwise wild-type variant were treated with prazosin, both mutants were delivered to the cell surface, consistent with a requirement for oligomerization prior to cell surface delivery (Canals et al., 2009). No cell surface expression was seen when both mutations were introduced into the same receptor construct, indicating that the orthosteric binding site was required for the pharmacological chaperone function (Canals et al., 2009). Similar results have been reported for β1-adrenoceptor mutants (Kobayashi et al., 2009) that were impaired in their ability to form homomers. Treatment with lipophilic β-adrenergic ligands was able to restore effective homomerization and surface delivery.

Synthetic ligands which only have affinity for a mutationally modified GPCR that has lost affinity for its endogenous ligand have also been used to examine the interrelationship between homo-oligomerization and receptor trafficking. In particular, this approach has been used to access if a GPCR may be internalized from the cell surface as an oligomeric complex (Sartania et al., 2007). In these studies, the β2-adrenoceptor was modified to lose affinity for the agonist drug isoprenaline, which binds to and promotes internalization of the wild-type β2-adrenoceptor. However, the ligand L-158,870 was designed to bind and cause internalization only of the mutant. After coexpression of the wild-type and mutant β2-adrenoceptor forms, L-158,870 was able to internalize both, demonstrating both the presence of homomers containing both forms and the internalization of such homomers as a complex. Such studies require a high level of design to produce a pairing of ligand and modified GPCR but the development of receptor activated solely by synthetic ligand forms of muscarinic receptor subtypes, for example (Alvarez-Curto et al., 2010, 2011), may provide a means to examine whether effects seen for the β2-adrenoceptor can be applied more widely.

ACKNOWLEDGMENTS

These studies were supported by the Medical Research Council UK (grant G0900050).

REFERENCES

Alvarez-Curto, E., Pediani, J. D., & Milligan, G. (2010). Applications of fluorescence- and bioluminescence-resonance energy transfer to drug discovery at G protein-coupled receptors. *Analytical and Bioanalytical Chemistry, 398*, 167–180.

Alvarez-Curto, E., Prihandoko, R., Tautermann, C. S., Zwier, J. M., Pediani, J. D., Lohse, M. J., et al. (2011). Developing chemical genetic approaches to explore G protein-coupled receptor function: Validation of the use a Receptor Activated Solely by Synthetic Ligand (RASSL). *Molecular Pharmacology, 80*, 1033–1046.

Alvarez-Curto, E., Ward, R. J., Pediani, J. D., & Milligan, G. (2010). Ligand regulation of the quaternary organization of cell surface M_3 muscarinic acetylcholine receptors analyzed by fluorescence resonance energy transfer (FRET) imaging and homogenous time-resolved FRET. *The Journal of Biological Chemistry, 285*, 23318–23330.

Canals, M., Lopez-Gimenez, J. F., & Milligan, G. (2009). Cell surface delivery and structural re-organisation by pharmacological chaperones of an oligomerisation-defective alpha (1b)-adrenoceptor mutants demonstrates membrane targeting of GPCR oligomers. *The Biochemical Journal, 417*, 161–172.

Conn, P. M., & Ulloa-Aguirre, A. (2010). Trafficking of GPCRs to the plasma membrane: Insights for pharmacoperone drugs. *Trends in Endocrinology and Metabolism, 21*, 190–197.

El-Asmar, L., Springael, J. Y., Ballet, S., Andrieu, E. U., Vassart, G., & Parmentier, M. (2005). Evidence for negative binding cooperativity within CCR5-CCR2b heterodimers. *Molecular Pharmaceutics, 67*, 460–469.

Ellis, J., Pediani, J. D., Canals, M., Milasta, S., & Milligan, G. (2006). Orexin-1 receptor-cannabinoid CB_1 receptor hetero-dimerization results in both ligand-dependent and -independent co-ordinated alterations of receptor localization and function. *The Journal of Biological Chemistry, 281*, 38812–38824.

Gainetdinov, R. R., Premont, R. T., Bohn, L. M., Lefkowitz, R. J., & Caron, M. G. (2004). Desensitisation of G protein-coupled receptors and neuronal functions. *Annual Review of Neuroscience, 27*, 107–144.

Janovick, J. A., Patny, A., Mosley, R., Goulet, M. T., Altman, M. D., Rush, T. S., et al. (2009). Molecular mechanism of action of pharmacoperone rescue of mis-routed GPCR mutants: The GnRH receptor. *Molecular Endocrinology, 23*, 157–168.

Kelly, E., Bailey, C. P., & Henderson, G. (2008). Agonist selective mechanisms of GPCR desensitisation. *British Journal of Pharmacology, 153*, S379–S388.

Kobayashi, H., Ogawa, K., Yao, R., Lichtarge, O., & Bouvier, M. (2009). Functional rescue of beta-adrenoceptor dimerization and trafficking by pharmacological chaperones. *Traffic, 10*, 1019–1033.

Kuszak, A. J., Pitchiaya, S., Anand, J. P., Mosberg, H. I., Walter, N. G., & Sunahara, R. K. (2009). Purification and functional reconstitution of monomeric mu-opioid receptors: Allosteric modulation of agonist binding by Gi2. *The Journal of Biological Chemistry, 284*, 26732–26741.

Lopez-Gimenez, J. F., Canals, M., Pediani, J. D., & Milligan, G. (2007). The alpha1b-adrenoceptor exists as a higher order oligomer: Effective oligomerisation is required for receptor maturation, cell surface delivery and function. *Molecular Pharmacology, 71*, 1015–1029.

Lopez-Gimenez, J. F., Vilaró, M. T., & Milligan, G. (2008). Morphine Desensitization, Internalisation, and Down-Regulation of the μ Opioid Receptor is Facilitated by Serotonin 5-hydroxytryptamine$_{2A}$ Receptor Coactivation. *Molecular Pharmacology, 74*, 1278–1291.

Maurel, D., Comps-Agrar, L., Brock, C., Rives, M. L., Bourrier, E., Ayoub, M. A., et al. (2008). Cell-surface protein-protein interaction analysis with time-resolved FRET and SNAP-tag technologies: Application to GPCR oligomerisation. *Nature Methods, 5*, 561–567.

Milligan, G. (2007). G protein coupled receptor dimerisation: Molecular basis and relevance to function. *Biochimica et Biophysica Acta, 1768*, 825–835.

Milligan, G. (2008). A day in the life of a G-protein coupled receptor: The contribution to function of G-protein coupled receptor dimerisation. *British Journal of Pharmacology, 153* (Suppl. 1), S216–S229.

Milligan, G. (2009). G protein coupled receptor hetero-dimerisation: Contribution to pharmacology and function. *British Journal of Pharmacology, 158*, 5–14.

Milligan, G. (2010). The role of dimerisation in the cellular trafficking of G-protein-coupled receptors. *Current Opinion in Pharmacology, 10*, 23–29.

Milligan, G., & Bouvier, M. (2005). Methods to monitor the quaternary structure of G protein-coupled receptors. *The FEBS Journal, 272*, 2914–2925.

Parenty, G., Appelbe, S., & Milligan, G. (2008). CXCR2 chemokine receptor antagonism enhances DOP opioid receptor function via allosteric regulation of the CXCR2-DOP receptor heterodimer. *The Biochemical Journal, 412*, 245–256.

Pou, C., Mannoury la Cour, C., Stoddart, L. A., Millan, M. J., & Milligan, G. (2012). Functional homomers and heteromers of dopamine D_{2L} and D_3 receptors co-exist at the cell surface. *The Journal of Biological Chemistry, 287*, 8864–8878.

Sartania, N., Appelbe, S., Pediani, J. D., & Milligan, G. (2007). Agonist occupancy of a single monomeric element is sufficient to cause internalization of the dimeric β2-adrenoceptor. *Cell Signalling, 19*, 1928–1938.

Tan, C. M., Brady, A. E., Highfield Nichols, H., Wang, Q., & Limbird, L. E. (2004). Membrane trafficking of G protein coupled receptors. *Annual Review of Pharmacology and Toxicology, 44*, 559–609.

van Rijn, R. M., van Marle, A., Chazot, P. L., Langemeijer, E., Qin, Y., Shenton, F. C., et al. (2008). Cloning and characterisation of dominant negative splice variants of the human histamine H4 receptor. *The Biochemical Journal, 414*, 121–131.

Ward, R. J., Alvarez-Curto, E., & Milligan, G. (2010). Using the Flp-In^{TM} Trex^{TM} system to regulate GPCR expression. In G. B. Willars & R. A. J. Challis (Eds.), *Receptor signal transduction protocols*. (3rd ed). *Methods in molecular biology, 746*. New York: Springer Science and Business Media.

Ward, R. J., Pediani, J. D., & Milligan, G. (2011a). Ligand-induced internalisation of the orexin OX_1 and cannabinoid CB_1 receptors assessed via N-terminal SNAP and CLIP-tagging. *British Journal of Pharmacology, 162*, 1439–1452.

Ward, R. J., Pediani, J. D., & Milligan, G. (2011b). Hetero-multimerization of the cannabinoid CB_1 receptor and the orexin OX_1 receptor generates a unique complex in which both protomers are regulated by orexin A. *The Journal of Biological Chemistry, 286*, 37414–37428.

Whorton, M. R., Bokoch, M. P., Rasmussen, S. G., Huang, B., Zare, R. N., Kobilka, B., et al. (2007). A monomeric G-protein coupled receptor isolated in a high-density lipoprotein particle efficiently activates its G protein. *Proceedings of the National Academy of Sciences of the United States of America, 104*, 7682–7687.

Wilson, S., Wilkinson, G., & Milligan, G. (2005). The CXCR1 and CXCR2 receptors form constitutive homo and heterodimers selectively and with apparent equal affinities. *The Journal of Biological Chemistry, 280*, 28663–28674.

Xu, T. R., Baillie, G. S., Bhari, N., Houslay, T. M., Pitt, A. M., Adams, D. R., et al. (2008). Mutations of beta-arrestin 2 that limit self-association also interfere with interactions with the beta2-adrenoceptor and the ERK1/2 MAPKs: Implications for beta2-adrenoceptor signalling via the ERK1/2 MAPKs. *The Biochemical Journal, 413*, 51–60.

Xu, T. R., Ward, R. J., Pediani, J. D., & Milligan, G. (2011). The orexin OX_1 receptor exists predominantly as a homodimer in the basal state: Potential regulation of receptor organization by both agonist and antagonist ligands. *The Biochemical Journal, 439*, 171–183.

CHAPTER FIVE

β-Arrestins and G Protein-Coupled Receptor Trafficking

Dong Soo Kang, Xufan Tian, Jeffrey L. Benovic[1]
Department of Biochemistry and Molecular Biology, Thomas Jefferson University, Philadelphia, Pennsylvania, USA
[1]Corresponding author: e-mail address: benovic@mail.jci.tju.edu

Contents

1. Introduction 92
2. Arrestin Expression 92
   2.1 Using antibodies to evaluate β-arrestin expression 92
   2.2 Overexpression of arrestins 95
   2.3 Arrestin knockdown 96
3. Assays to Measure GPCR Trafficking 97
   3.1 ELISA 97
   3.2 Fluorescence activated cell sorting 99
   3.3 Ligand binding 100
   3.4 Fluorescence 100
4. Evaluating the Role of β-Arrestins in GPCR Trafficking 102
   4.1 Arrestin knockdown/knockout strategies 102
   4.2 Arrestin mutants 102
5. Summary 103
Acknowledgments 105
References 105

Abstract

Arrestins are adaptor proteins that function to regulate G protein-coupled receptor (GPCR) signaling and trafficking. There are four mammalian members of the arrestin family, two visual and two nonvisual. The visual arrestins (arrestin-1 and arrestin-4) are localized in rod and cone cells, respectively, and function to quench phototransduction by inhibiting receptor/G protein coupling. The nonvisual arrestins (β-arrestin1 and β-arrestin2, a.k.a. arrestin-2 and arrestin-3) are ubiquitously expressed and function to inhibit GPCR/G protein coupling and promote GPCR trafficking and arrestin-mediated signaling. Arrestin-mediated endocytosis of GPCRs requires the coordinated interaction of β-arrestins with clathrin, adaptor protein 2, and phosphoinositides such as PIP_2/PIP_3. These interactions are facilitated by a conformational change in β-arrestin that is thought to occur upon binding to a phosphorylated activated GPCR. In this chapter, we provide an overview of the reagents and techniques used to study β-arrestin-mediated receptor trafficking.

Methods in Enzymology, Volume 521
ISSN 0076-6879
http://dx.doi.org/10.1016/B978-0-12-391862-8.00005-3

1. INTRODUCTION

It has now been over 15 years since the discovery that β-arrestins stimulate agonist-promoted internalization of the β_2-adrenergic receptor (Ferguson et al., 1996; Goodman et al., 1996). Many subsequent studies have revealed that β-arrestins promote trafficking of many G protein-coupled receptors (GPCRs) as well as additional classes of receptors (Moore, Milano, & Benovic, 2007; Shenoy & Lefkowitz, 2011). However, there remain many receptors where this has not been investigated. Moreover, many new strategies have been developed over the past 5–10 years that have significantly advanced our ability to understand these processes. These include the use of RNA interference to knockdown β-arrestin expression, β-arrestin1/2 knockout (KO) mouse embryonic fibroblasts (MEFs) (Kohout, Lin, Perry, Conner, & Lefkowitz, 2001), and a better understanding of the mechanisms that mediate β-arrestin-promoted trafficking. This process involves the coordinated interaction of β-arrestins with the GPCR (Vishnivetskiy, Gimenez et al., 2011; Vishnivetskiy, Hosey, Benovic, & Gurevich, 2004), clathrin (Kang et al., 2009; Krupnick, Goodman, Keen & Benovic, 1997; Krupnick, Santini, Gagnon, Keen & Benovic, 1997), adaptor protein 2 (AP2) (Burtey et al., 2007; Kim & Benovic, 2002; Laporte et al., 1999; Schmid et al., 2006), and phosphoinositides (Gaidarov, Krupnick, Falck, Benovic, & Keen, 1999; Milano, Kim, Stefano, Benovic, & Brenner, 2006). Moreover, GPCR binding appears to promote β-arrestin interaction with the endocytic machinery thereby linking the binding and trafficking events (Kim & Benovic, 2002; Nobles, Guan, Xiao, Oas, & Lefkowitz, 2007; Xiao, Shenoy, Nobles, & Lefkowitz, 2004). The X-ray crystal structure of β-arrestin1 along with the regions involved in mediating GPCR binding and endocytic trafficking are shown in Fig. 5.1.

While previous publications have provided detailed methodology for evaluating arrestin-mediated trafficking of GPCRs (Gurevich, Orsini, & Benovic, 2000; Mundell, Orsini, & Benovic, 2002), here we will provide an overview of these methods highlighting the use of β-arrestin mutants, RNA interference, and β-arrestin KO MEFs as tools to study GPCR trafficking.

2. ARRESTIN EXPRESSION

2.1. Using antibodies to evaluate β-arrestin expression

The β-arrestins are ubiquitously expressed, although they are found at relatively low concentrations in most tissues and cells (typically 5–100 ng/mg protein). Nevertheless, endogenous β-arrestins in most mammalian cells are

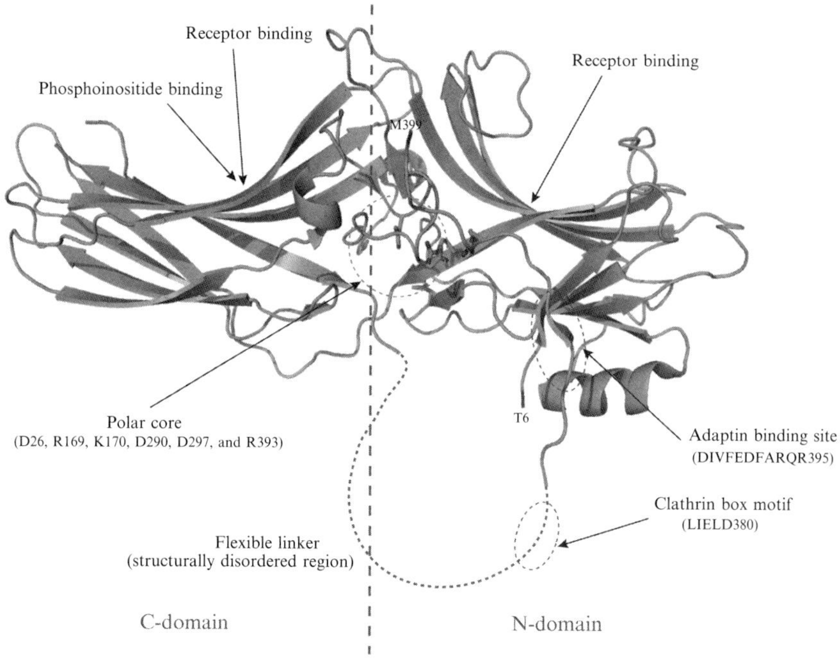

Figure 5.1 Secondary structure of β-arrestin1. Ribbon diagram of β-arrestin1 indicates the N and C domains, the polar core and binding sites for receptor, phosphoinositides, clathrin, and β2-adaptin. (For color version of this figure, the reader is referred to the online version of this chapter.)

readily detectable by immunoblotting. For example, we have used a mouse monoclonal β-arrestin1 antibody to detect endogenous β-arrestin1 and 2 (Luo, Busillo, & Benovic, 2008) as well as subtype selective polyclonal antibodies that can detect endogenous β-arrestins by immunoblotting and immunohistochemistry (Shankar et al., 2010). Here, we go through procedures that can be used to evaluate expression of endogenous β-arrestins as well as assess the effectiveness of knockdown and overexpression strategies.

2.1.1 Preparation of cell lysate

1. Prepare 10 ml of lysis buffer containing 20 m*M* Hepes, pH 7.5, 1% Triton X-100, 150 m*M* NaCl, 10 m*M* EDTA, and one tablet of complete protease inhibitor mix (Roche). The volume of lysis buffer can be scaled up or down as needed but should be made fresh and chilled on ice before use.
2. Cells grown in complete media to 75–90% confluence are rinsed several times with prechilled phosphate buffered saline (PBS). Ice-cold lysis buffer is then added (~0.5–1 ml of lysis buffer per 10 cm culture dish),

and the cells are scraped off using a rubber policeman and transferred to a centrifuge tube.

3. Cells are lysed by freezing in liquid nitrogen or a dry ice/water bath, rapid thawing in a room temperature water bath, and vortexing several times. The cell lysate is centrifuged (40,000 × *g* for 20 min) to remove particulate matter, and the supernatant is transferred to a new tube and assayed for protein concentration. The cell supernatant can be used immediately or frozen and stored at −80 °C for later analysis.

2.1.2 Electrophoresis and immunoblotting

1. Incubate 20–40 μg of cell supernatant protein (prepared as described in Section 2.1.1) with SDS sample buffer for 10 min at room temperature. It is helpful to also include a prestained standard on the gel to make sure that the transfer is effective and that the appropriate proteins are being analyzed (β-arrestin1 and β-arrestin2 migrate as distinct proteins at ~55 and 50 kDa, respectively).
2. Electrophorese the samples on a 10% SDS-polyacrylamide gel at a constant voltage of 120–150 V until the dye front is within a few millimeters of the bottom.
3. Carefully, remove and discard the stacking gel and then set up the gel for transfer to a nitrocellulose membrane. This is accomplished by layering a transfer sponge, one piece of Whatman 3 MM paper, the gel, one piece of nitrocellulose membrane, one piece of Whatman 3 MM paper, and another sponge. Everything should initially be wetted in transfer buffer (25 m*M* Tris base, 192 m*M* glycine, 20% methanol, do not adjust pH) before layering. It is also important to make sure that all air bubbles are removed as each layer is added. Proteins are transferred to the nitrocellulose membrane for 1 h at 100 V using cold transfer buffer.
4. After transfer, sample loading and transfer efficiency can be checked by staining the nitrocellulose membrane with 0.2% Ponceau S for ~1 min. Transferred proteins can then be visualized by rinsing the nitrocellulose membrane with deionized water to remove excess stain.
5. Rinse the nitrocellulose membrane several times in TBS-T (20 m*M* Tris–HCl, pH 7.5, 150 m*M* NaCl, 0.05% Tween 20). The membrane is then blocked for 1 h in 10 ml of TBS-T containing 5% (w/v) nonfat dry milk (TBS-T/milk).
6. The nitrocellulose membrane is incubated for ≥1 h at room temperature with 5 ml of a β-arrestin-specific antibody diluted in TBS-T/milk.

A wide variety of antibodies are available commercially, and we have developed some that are subtype specific (Shankar et al., 2010).

7. Wash the membrane three to five times for 10 min each in TBS-T.
8. Incubate the nitrocellulose membrane for 1 h with an affinity purified goat antimouse IgG conjugated to horseradish peroxidase (1:3000) in TBS-T/milk.
9. Wash the membrane three to five times for 10 min each in TBS-T.
10. Overlay the nitrocellulose membrane with 1–2 ml of enhanced chemiluminescence (Fisher) reagent for ~1 min, allow the blot to drip dry, wrap in Saran Wrap, and visualize on X-ray film.

2.2. Overexpression of arrestins

The overexpression of wild-type or mutant β-arrestins in mammalian cells can provide information regarding the role of arrestins in the desensitization, internalization, and signaling of a given GPCR. In general, HEK293 cells have been used extensively for such studies because they are widely available, easy to grow, and permit the expression of GPCRs and arrestins at high levels using either transient or stable transfection. The mammalian expression vector pcDNA3 (Invitrogen) is suitable for such studies. This vector features a convenient multiple cloning site, the strong cytomegalovirus promoter, the neomycin resistance gene, and a bovine growth hormone polyadenylation signal. Using 1–5 μg of plasmid DNA, we routinely obtain 25- to 50-fold overexpression of transfected β-arrestins. Both X-tremeGENE (Roche) and Lipofectamine (Invitrogen) cationic lipid-based transfection reagents yield comparable results in our hands although we routinely use X-tremeGENE due to its ease of use.

2.2.1 Cell culture

Split HEK293 cells a day prior to transfection such that the cells are ~60–75% confluent the following day (generally a 1:3 to 1:4 split of a confluent dish). The degree of confluency required depends on the toxicity of the transfection reagent used (e.g., in our experience, X-tremeGENE works effectively over a broad confluence range, whereas Lipofectamine works optimally when cells are at a higher confluence).

1. HEK293 cells are maintained in Dulbecco's Modified Eagles Medium (DMEM) supplemented with 10% fetal calf serum, 5 m*M* Hepes, pH 7.2, and 100 μg/ml penicillin and streptomycin (complete media). In general, cells are split 1:10 twice per week for maintenance. Cells are

maintained using ~10 ml of complete medium per 10 cm dish and grown at 37 °C in a humidified incubator with 5% CO_2.

2. To split HEK293 cells, aspirate the medium and carefully rinse the monolayer twice with sterile PBS. Incubate with 1 ml of a 0.05% trypsin solution containing 0.5 m*M* EDTA for 1–2 min at room temperature. Aspirate the trypsin solution, tap the plate gently to dislodge cells, and then add ~9 ml of complete medium to quench the remaining trypsin. Occasionally, cells may need a short incubation at 37 °C for complete trypsinization.

2.2.2 Expression of wild-type, mutant, or GFP-tagged β-arrestins

We generally use either X-tremeGENE (Roche) or Lipofectamine (Invitrogen) for transient expression of β-arrestins. The protocol for X-tremeGENE is described below.

1. Seed appropriate amount of HEK293 cells in 10 cm culture dish such that they reach ~60–75% confluence on the following day.
2. Combine 1 μg of wild-type, mutant, or GFP-tagged β-arrestin and 5 μg of pcDNA3 (total 6 μg of DNA) with 300 μl of serum-free DMEM and incubate for 5 min at room temperature.
3. Combine 18 μl of X-tremeGENE (at room temperature) with 300 μl of serum-free DMEM and incubate for 5 min at room temperature.
4. Combine the DNA and X-tremeGENE solutions from steps 2 and 3 and incubate at room temperature for 20–30 min.
5. Add the DNA/X-tremeGENE solution to the dish drop-by-drop and gently rock a few times to mix.
6. Incubate the dish at 37 °C for 48–72 h.

2.3. Arrestin knockdown

We have successfully knocked down β-arrestin expression using the SMARTpool siRNAs from Dharmacon RNAi Technologies (Thermo Scientific) (access numbers NP_004032 for human β-arrestin1 and NP_004304 for human β-arrestin2). An example of the effectiveness and specificity of the knockdown is shown in Fig. 5.2 (Luo et al., 2008).

1. HEK293 cells are maintained in DMEM supplemented with 10% fetal bovine serum, 25 m*M* Hepes, pH 7.2, and 0.1 m*M* nonessential amino acids in a 5% CO_2 incubator at 37 °C.
2. HEK293 cells are grown to ~90% confluence in 10 cm dishes and then transfected with 600 pmol of either β-arrestin1 or β-arrestin2 siRNA using Lipofectamine 2000 (Invitrogen) in Opti-MEM. Control siRNA from Dharmacon is transfected in a similar fashion.
3. Cells are incubated for 48 h and then split for assay the following day.

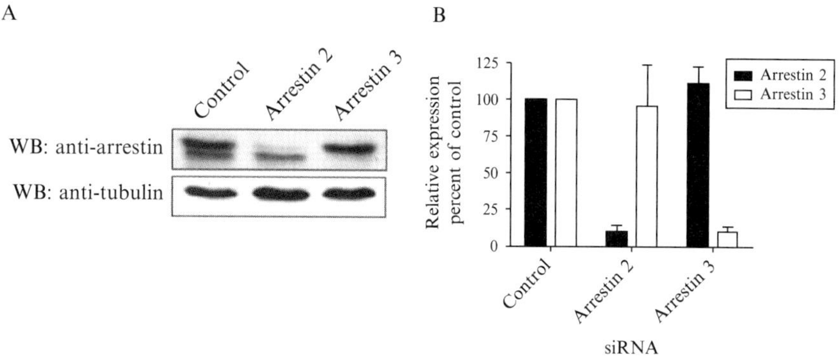

Figure 5.2 Knockdown of endogenous β-arrestin1 (arrestin-2) and β-arrestin2 (arrestin-3) in HEK293 cells using siRNAs. (A) Effective knockdown was confirmed by immunoblotting using a β-arrestin1-specific monoclonal antibody that cross-reacts with β-arrestin2 (Luo et al., 2008). (B) The level of β-arrestin expression was quantified by densitometry. Figure is from Luo et al., 2008.

3. ASSAYS TO MEASURE GPCR TRAFFICKING

For many GPCRs, β-arrestins promote internalization by targeting the receptor to clathrin-coated pits (CCPs) (Moore et al., 2007). β-Arrestin-mediated trafficking can be evaluated for any receptor in any cell type if appropriate reagents are available. However, here, we describe methods that involve transfection into heterologous cells for analysis. COS cells, which contain relatively low endogenous β-arrestin levels, are good cells to assay the promotion of internalization by β-arrestins (Krupnick, Goodman et al., 1997; Krupnick, Santini et al., 1997). Conversely, HEK293 cells, which have higher endogenous β-arrestin levels, are good cells to observe inhibition of internalization by dominant-negative arrestins or knockdown approaches (Kim & Benovic, 2002; Luo et al., 2008). Similarly, wild-type and β-arrestin KO MEFs are ideal for evaluating the involvement of β-arrestins in trafficking of a given receptor (Kohout et al., 2001).

3.1. ELISA

This section describes methods to assay receptor internalization by ELISA using the β_2-adrenergic receptor (β_2AR) as an example. Analysis by ELISA is based on the presence of an N-terminal epitope tag on the receptor, such as the hemagglutinin (HA) or FLAG, which is no longer recognized by the cognate antibody once the receptor is internalized. An epitope-tagged receptor is overexpressed in cells with or without wild-type or mutant β-arrestins as described in Section 2.2.2.

1. Prior to splitting the cells, treat wells of a 24-well plate with 0.1 mg/ml poly-L-lysine (PLL) (Sigma). Briefly, dissolve the PLL in sterile water, add ~0.25 ml to cover the well, wait 1 min, remove the PLL, and then let the well dry for ~10 min. Split the cells into the 24-well plate and incubate for 24 h in a 37 °C incubator. It is convenient to divide the 24-well plate in half vertically, using one side for untreated and the other side for agonist-treated conditions. Each 24-well plate can then accommodate four separate transfections horizontally.
2. Carefully aspirate medium from the wells, wash once with DMEM, and treat triplicate wells for the desired time with DMEM containing 0.1 m*M* ascorbic acid with or without isoproterenol (typically 10–20 μ*M*) at 37 °C. Isoproterenol is freshly prepared as a 10 m*M* stock solution in 0.1 m*M* ascorbic acid.
3. Aspirate medium and fix the cells with 0.25 ml/well of 3.7% formaldehyde (Sigma) in TBS (20 m*M* Tris, pH 7.5, 150 m*M* NaCl) for 5 min. The remainder of the assay can be performed at room temperature.
4. Wash the plate three times with TBS (0.5 ml/well) and then block each well with 0.5 ml of 1% BSA in TBS for 45 min.
5. Aspirate the blocking solution, add the primary antibody diluted 1:1000 in 1% BSA/TBS, and incubate for 1 h at room temperature. We have successfully used the 101R antiHA as raw ascites (Covance), as well as the anti-FLAG antibody M1 (Sigma). Because the binding of the M1 antibody is dependent on the presence of calcium, 1 m*M* $CaCl_2$ must be present for this step and all subsequent steps prior to color development.
6. Wash the plate three times with TBS (0.5 ml/well) and reblock with 0.5 ml of 1% BSA/TBS for 15 min.
7. Aspirate the blocking solution and add 0.25 ml/well of secondary antibody (alkaline phosphatase-conjugated goat antimouse antibody, Bio-Rad) diluted 1:1000 in 1% BSA/TBS and incubate for 1 h at room temperature.
8. Wash the plate three times with 0.5 ml/well of TBS and then develop with alkaline phosphatase substrate kit (Bio-Rad). Dilute the diethanolamine solution 1:5 in water and add one substrate tablet per 5 ml (dissolve the tablet completely by vigorous vortexing). The color of the solution should be colorless to a slight pale yellow. Add 0.25 ml of developing solution per well and develop until a bright yellow color

appears, generally 15–30 min, depending on the efficiency of transfection and receptor expression. If the signal is weak, development is aided by incubation at 37 °C. It is important not to allow development to proceed beyond the linear range of the ELISA reader. Ideally, OD_{405} readings should fall between 0.5 and 1.2 after stopping the reaction (step 9). Alternatively, one can develop using 1-Step™ ABTS (Thermo Scientific). Equilibrate 1-Step™ ABTS to room temperature, add 0.3 ml of the developing solution per well, and incubate the plate at room temperature for 10–30 min or until desired color change occurs (dark green).

9. Aliquot 0.1 ml/well of 0.4 *M* NaOH in a 96-well plate. Remove 0.1 ml from the developed wells and add to the NaOH to stop the reaction. If 1-Step™ ABTS is used, the reaction can be quenched by adding an equal amount of 1% SDS to the wells.
10. Read plate in ELISA reader at 405 nm. Subtract background from mock-transfected or untransfected control cells and calculate percentage of internalization.

3.2. Fluorescence activated cell sorting

Fluorescence activated cell sorting (FACS) can be used to quantify removal of a GPCR (e.g., the β_2AR) from the plasma membrane and determine the rate of internalization upon agonist stimulation.

1. Forty-eight hours after transfection of an epitope-tagged β_2AR, treat the cells with 10 μ*M* isoproterenol and 0.1 m*M* ascorbate for various times (usually 0–60 min).
2. Wash the cells with cold PBS, detach the cells with cell stripper solution by incubating for ~5 min at 37 °C and transfer the detached cells to a new conical tube.
3. Incubate with anti-FLAG or anti-HA antibody in cold PBS containing 1% BSA (1:500 dilution) for 2 h on ice.
4. Wash the cells three to five times with cold PBS by centrifugation at low speed (500 × *g*).
5. Incubate with FITC or desired fluorescence-conjugated secondary antibody in cold PBS containing 1% BSA (1:1000 dilution) for 1 h on ice in the dark.
6. Wash the cells three to five times with cold PBS, resuspend in 0.25 ml PBS, and then fix the cells by addition of 0.25 ml of 2% formaldehyde in PBS. Determine mean cell surface fluorescence and number of fluorescence-positive cells by flow cytometry.

3.3. Ligand binding

Radiolabeled ligands can also be used to evaluate cell surface versus total GPCR density. For example, hydrophilic radioligands that are impermeable to the plasma membrane can be used to determine receptor density on the cell surface (Kang & Leeb-Lundberg, 2002; Krupnick, Goodman et al., 1997; Krupnick, Santini et al., 1997). Alternatively, hydrophilic ligands can be used to compete with hydrophobic radiolabeled ligands for cell surface binding. In some cases, GPCR internalization can be measured by determining the amount of radiolabeled ligand that is internalized with the receptor over time. For example, internalized radiolabeled peptides can be measured after removal of cell surface radioligand by treating cells with a cold washing buffer (0.05 *M* glycine, pH 3.0 or 0.2 *M* acetate, pH 2.5, 0.5 *M* NaCl) (Lamb, De Weerd, & Leeb-Lundberg, 2001).

Cells are plated on 6-, 12-, or 24-well plates for the whole-cell binding assay. In general, 2–3 wells of the plate are used for "total binding" and 1–2 wells are used for nonspecific binding. The specific binding of the radioligand (total minus nonspecific) represents the density of cell surface receptors. Changes in cell surface receptor levels can be measured in a time-dependent manner after incubating cells with an agonist at 37 °C for various times, washing the cells with cold buffer, and then evaluating cell surface receptor density by radioligand binding as described below.

1. Incubate HEK293 cells in a binding buffer (e.g., Leibovitz's L-15 medium, pH 7.4, 0.1% BSA, complete protease inhibitor mixture) with a saturating concentration (~10-fold K_d) of a hydrophilic radioligand for a given receptor at 4 °C for 60–90 min. Nonspecific binding is determined in the presence of a cold ligand (100-fold K_d).
2. Wash three times with cold PBS containing 0.3% BSA.
3. Add 0.5 ml/well of a solubilization buffer (1% SDS, 0.1 *M* NaOH, 0.1 *M* Na_2CO_3) and incubate at 4 °C for 10 min.
4. Transfer the solubilized cell lysate to a scintillation vial, add 5 ml of scintillation cocktail and count the radioligand.

3.4. Fluorescence

3.4.1 Use of fluorophore-conjugated antibodies

To study the role of β-arrestins in GPCR trafficking, fluorescence-labeled antibodies that specifically recognize a given receptor are often used to evaluate subcellular receptor localization. Fluorophores with broad wavelength ranges are beneficial to identify colocalization of a target receptor with β-arrestins (Goodman et al., 1996), other endocytic proteins (Burtey et al., 2007; Kang et al., 2009; Laporte et al., 1999), and markers for specific intracellular compartments (Temkin et al., 2011). Preparation of microscopic slides is described below, although more detailed experimental designs and protocols can be found elsewhere (Hislop & von Zastrow, 2011; Santini & Keen, 2000).

1. Seed cells in a 6-well plate containing PLL coated coverslips.
2. Treat cells with agonist for a given time and then put the plate on ice to stop further trafficking.
3. Wash cells three times with cold PBS.
4. Fix cells with 4% formaldehyde/PBS (1 ml/well) on ice for 5–10 min.
5. Quench the formaldehyde by incubation with TBS (1 ml/well) for 20 min.
6. Wash cells three times with TBS.
7. Permeabilize cells with PBS containing 1% BSA and 0.05% Triton X-100 (1 ml/well) for 10 min.
8. Incubate cells with a blocking solution (1% BSA in PBS) for 20–30 min at room temperature.
9. Incubate cells with anti-HA/FLAG antibody for 1 h.
10. Wash three times with TBS.
11. Incubate cells with fluorophore-conjugated secondary antibody for 1 h.
12. Wash three to five times with TBS.
13. Prepare a microscopic slide with few drops of mounting solution (Vector Laboratories).
14. Place the coverslip (cell side down) onto drops of mounting solution on the microscope slide.
15. Seal the coverslip with a clear nail polish and store in a light-protected container at 4 °C until use.

3.4.2 Green fluorescence protein tagged proteins

Green fluorescence protein (GFP) is a 28 kDa β-barrel-shaped protein that has been widely used in biological sciences. In GPCR trafficking, GFP-fused receptors (Kallal & Benovic, 2000, 2002; Scherrer et al., 2006), β-arrestins (Barak, Ferguson, Zhang, Martenson, & Caron, 1997; Oakley, Laporte, Holt, Caron, & Barak, 2000), and other endocytic proteins (Zhao & Keen, 2008) have been used to visualize real time localization of target proteins. GFP-tagged proteins have also been used to study GPCR interactions that occur during trafficking using techniques such as fluorescence resonance energy transfer (FRET) and bioluminescence resonance energy transfer (BRET). In FRET, excitation of GFP-tagged protein (donor) results in energy transfer to another fluorophore-labeled protein (acceptor) if the two proteins are in a close proximity (1–10 nm range). In BRET, a luciferase-tagged protein is used as the energy donor and a GFP-tagged protein as the acceptor. FRET and BRET have been used to study intermolecular interactions between a donor and acceptor protein (Charest, Terrillon, & Bouvier, 2005; Hamdan et al., 2007; Shukla et al., 2008) as well as intramolecular interactions within a given protein (Audet, Lagace, Silversides, & Bouvier, 2010; Hamdan et al., 2007; Masri et al., 2008). Thus, these techniques enable one to monitor protein–protein interactions as well as conformational changes within a given protein that occur during the GPCR trafficking process.

4. EVALUATING THE ROLE OF β-ARRESTINS IN GPCR TRAFFICKING

4.1. Arrestin knockdown/knockout strategies

The use of siRNAs to knockdown β-arrestin expression was described in Section 2.3. This strategy has been used to effectively knockdown the expression of β-arrestin1 and β-arrestin2 and evaluate the role of these proteins in GPCR desensitization, trafficking, and signaling (Luo et al., 2008).

Another strategy that has proved very useful in understanding the role of β-arrestins in GPCR trafficking has been the use of MEFs prepared from wild-type as well as β-arrestin1 KO, β-arrestin2 KO, and β-arrestin1/2 KO mice (Kohout et al., 2001). These cells are readily transfectable and can be used to express and monitor agonist-promoted changes in GPCR localization. For example, this strategy demonstrated a role for β-arrestin1 and β-arrestin2 in endocytosis of the β_2AR and, moreover, showed that β-arrestin2 was more effective in mediating this process (Kohout et al., 2001). An example of the use of β-arrestin KO MEFs to study β_2AR internalization is shown in Fig. 5.3.

4.2. Arrestin mutants

A wide variety of β-arrestin mutants have been generated that function as effective dominant negatives to inhibit GPCR internalization. These include mutants that constitutively localize in CCPs and inhibit GPCR trafficking such as a V53D point mutant in β-arrestin1 (V54D in β-arrestin2) (Ferguson et al., 1996; Krupnick, Santini et al., 1997), C-terminal minigenes

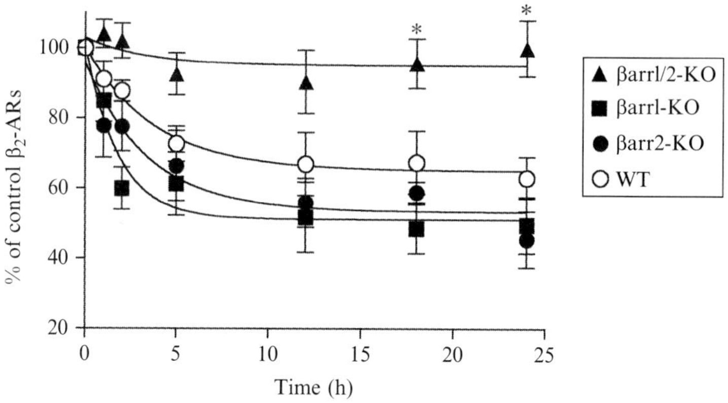

Figure 5.3 Agonist-promoted β_2AR sequestration on β-arrestin knockout MEFs. Wild-type, β-arrestin1 KO, β-arrestin2 KO, and β-arrestin1/2 KO MEFs were stimulated with 10 μ*M* isoproterenol and cell surface receptor levels were determined by ligand binding. *, $p<0.01$ between WT and βarr1/2 KO. Figure is from Kohout et al., 2001.

of β-arrestin1 (319–418) (Krupnick, Santini et al., 1997) and β-arrestin2 (310–410 and 284–409) (Laporte et al., 1999; Orsini & Benovic, 1998), and mutants that disrupt β-arrestin binding to $PIP_{2/3}$, clathrin and/or AP2 (Gaidarov et al., 1999; Kim & Benovic, 2002). These mutants are listed in Table 5.1. In general, β-arrestin mutants defective in $PIP_{2/3}$, clathrin and/or AP2 binding are useful in discerning the role of arrestins in trafficking of a particular GPCR. An example of this is depicted in Fig. 5.4 that shows that these mutants can function as effective dominant-negative mutants to inhibit agonist-promoted internalization of the β_2AR (β-arrestin1-ΔLIELD/F391A is the most effective) (Kim & Benovic, 2002). Alternatively, mutants that constitutively localize in CCPs such as the C-terminal minigenes also effectively inhibit GPCR internalization, although these likely function as more general inhibitors and do not necessarily provide insight on whether the process is β-arrestin dependent (Krupnick, Santini et al., 1997; Orsini & Benovic, 1998). To evaluate the role of β-arrestins in GPCR trafficking using these various mutants, one should follow the procedures laid out in Sections 2 and 3 to express the appropriate mutant β-arrestin and analyze the effect on GPCR trafficking.

5. SUMMARY

β-Arrestins bind to GPCRs in a conformationally sensitive manner and are known to regulate: (1) GPCR desensitization by inhibiting GPCR coupling to heterotrimeric G proteins, (2) GPCR endocytosis by promoting

Table 5.1 β-arrestin mutants that regulate GPCR internalization

Subtype	Mutated residues	GPCR	Effect on GPCR trafficking	References
β-Arrestin1	V53D	β_2AR	Constitutive localization in CCPs Reduced agonist-promoted internalization	Ferguson et al. (1996), Krupnick, Goodman et al. (1997), Krupnick, Santini et al. (1997), Zhang et al. (1997)
β-Arrestin2	V54D	β_2AR	Reduced agonist-promoted internalization	Ferguson et al. (1996)
β-Arrestin1	Minigene 319–418	β_2AR	Constitutive localization in CCPs Reduced agonist-promoted internalization	Krupnick, Goodman et al. (1997), Krupnick, Santini et al. (1997)
β-Arrestin2	Minigene 310–410	β_2AR	Reduced agonist-promoted internalization	Laporte et al. (1999)
β-Arrestin2	Minigene 284–409	β_2AR	Reduced agonist-promoted internalization	Orsini and Benovic (1998)
β-Arrestin2	R396A LIEF → AAEA	β_2AR	Agonist-promoted PM localization/no colocalization with AP2	Laporte et al. (2000)
β-Arrestin1	ΔLIELD F391A R395E ΔLIELD/F391A/R395E	β_2AR	Reduced agonist-promoted internalization (maximal effect by combined mutation)	Kim and Benovic (2002)
β-Arrestin2	K233Q/R237Q/K251Q	β_2AR	No agonist-promoted recruitment to CCP Reduced agonist-promoted internalization	Gaidarov et al. (1999)
β-Arrestin2	F389A	TRHR	No agonist-promoted recruitment to PM or CCP	Schmid et al. (2006)
β-Arrestin2	I387A V388A I387A/V388A LIEF → AAEA/R396A	TRHR	Constitutive localization in CCP Reduced agonist-promoted internalization	Burtey et al. (2007)

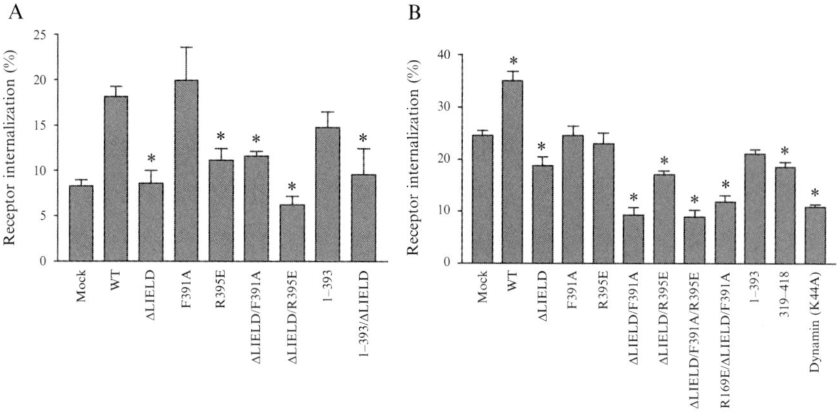

Figure 5.4 Effect of various clathrin and/or adaptin binding deficient β-arrestin mutants on agonist-promoted β_2AR internalization. Internalization of β_2AR in COS-1 cells (A) or HEK293 cells (B) overexpressing wild-type or mutant β-arrestin1 as determined using ELISA to measure cell surface receptor levels. *, $p<0.05$ vs. wild type (WT) (A) or mock (B). Figure is from Kim and Benovic, 2002.

association of GPCR/β-arrestin complexes in CCPs, and (3) arrestin-promoted signaling via the extensive adaptor functions of the β-arrestins (Shukla, Xiao, & Lefkowitz, 2011). Understanding the role of β-arrestins in GPCR trafficking and signaling is critical to better define how these proteins regulate receptor function and ultimately how these proteins function *in vivo* to regulate the biology of an organism.

ACKNOWLEDGMENTS

This work was supported by grants GM44944, GM47417, and CA129626 to J. L. B.

REFERENCES

Audet, M., Lagace, M., Silversides, D. W., & Bouvier, M. (2010). Protein-protein interactions monitored in cells from transgenic mice using bioluminescence resonance energy transfer. *The FASEB Journal*, *24*, 2829–2838.

Barak, L. S., Ferguson, S. S. G., Zhang, J., Martenson, C., & Caron, M. G. (1997). A beta-arrestin/green fluorescent protein biosensor for detecting G protein-coupled receptor activation. *The Journal of Biological Chemistry*, *272*, 27497–27500.

Burtey, A., Schmid, E. M., Ford, M. G. J., Rapport, J. Z., Scott, M. G. H., Marullo, S., et al. (2007). The conserved isoleucine-valine-phenylalanine motif couples activation state and endocytic functions of β-arrestins. *Traffic*, *8*, 914–931.

Charest, P. G., Terrillon, S., & Bouvier, M. (2005). Monitoring agonist-promoted conformational changes of β-arrestin in living cells by intramolecular BRET. *EMBO Reports*, *6*, 1–7.

Ferguson, S. S., Downey, W. E., 3rd, Colapietro, A. M., Barak, L. S., Ménard, L., & Caron, M. G. (1996). Role of β-arrestin in mediating agonist-promoted G protein-coupled receptor internalization. *Science*, *271*, 363–366.

Gaidarov, I., Krupnick, J. G., Falck, J. R., Benovic, J. L., & Keen, J. H. (1999). Arrestin function in G protein-coupled receptor endocytosis requires phosphoinositide binding. *The EMBO Journal, 18*, 871–881.

Goodman, O. B., Jr., Krupnick, J. G., Santini, F., Gurevich, V. V., Penn, R. B., Gagnon, A. W., et al. (1996). β-arrestin acts as a clathrin adaptor in endocytosis of the β_2-adrenergic receptor. *Nature, 383*, 447–450.

Gurevich, V. V., Orsini, M. J., & Benovic, J. L. (2000). Characterization of arrestin expression and function. In J. L. Benovic (Ed.), *Receptor biochemistry and methodology: Regulation of G protein-coupled receptor function and expression*, Vol. 4, (pp. 157–178). New York, NY: Wiley-Liss, Inc.

Hamdan, F. F., Rochdi, M. D., Breton, B., Fessart, D., Michaud, D. E., Charest, P. G., et al. (2007). Unraveling G protein-coupled receptor endocytosis pathways using real-time monitoring of agonist-promoted interaction between β-arrestins and AP-2. *The Journal of Biological Chemistry, 282*, 29089–29100.

Hislop, J. N., & von Zastrow, M. (2011). Analysis of GPCR localization and trafficking. *Methods in Molecular Biology, 746*, 425–440.

Kallal, L., & Benovic, J. L. (2000). Using green fluorescent proteins to study G-protein-coupled receptor localization and trafficking. *Trends in Pharmacological Sciences, 21*, 175–180.

Kallal, L., & Benovic, J. L. (2002). Fluorescence microscopy techniques for the study of G protein-coupled receptor trafficking. *Methods in Enzymology, 343*, 492–506.

Kang, D. S., Kern, R. C., Puthenveedu, M. A., von Zastrow, M., Williams, J. C., & Benovic, J. L. (2009). Structure of an arrestin-2/clathrin complex reveals a novel clathrin binding domain that modulates receptor trafficking. *The Journal of Biological Chemistry, 284*, 8316–8329.

Kang, D. S., & Leeb-Lundberg, L. M. (2002). Negative and positive regulatory epitopes in the C-terminal domains of the human B1 and B2 bradykinin receptor subtypes determine receptor coupling efficacy to $G_{q/11}$-mediated phospholipase Cβ activity. *Molecular Pharmacology, 62*, 281–288.

Kim, Y. M., & Benovic, J. L. (2002). Differential roles of arrestin-2 interaction with clathrin and AP2 in G protein-coupled receptor trafficking. *The Journal of Biological Chemistry, 277*, 30760–30768.

Kohout, T. A., Lin, F. S., Perry, S. J., Conner, D. A., & Lefkowitz, R. J. (2001). β-arrestin 1 and 2 differentially regulate heptahelical receptor signaling and trafficking. *Proceedings of the National Academy of Sciences of the United States of America, 98*, 1601–1606.

Krupnick, J. G., Goodman, O. B., Jr., Keen, J. H., & Benovic, J. L. (1997). Arrestin/clathrin interaction: Localization of the clathrin binding domain of nonvisual arrestins to the C-terminus. *The Journal of Biological Chemistry, 272*, 15011–15016.

Krupnick, J. G., Santini, F., Gagnon, A. W., Keen, J. H., & Benovic, J. L. (1997). Modulation of the arrestin/clathrin interaction in cells: Characterization of beta-arrestin dominant-negative mutants. *The Journal of Biological Chemistry, 272*, 32507–32512.

Lamb, M. E., De Weerd, W. F. C., & Leeb-Lundberg, L. M. (2001). Agonist-promoted trafficking of human bradykinin receptors: Arrestin- and dynamin-independent sequestration of the B2 receptor and bradykinin in HEK293 cells. *The Biochemical Journal, 355*, 741–750.

Laporte, S. A., Oakley, R. H., Zhang, J., Holt, J. A., Ferguson, S. S., Caron, M. G., et al. (1999). The β_2-adrenergic receptor/β-arrestin complex recruits the clathrin adaptor AP-2 during endocytosis. *Proceedings of the National Academy of Sciences of the United States of America, 96*, 3712–3717.

Laporte, S. A., Oakley, R. H., Holt, J. A., Barak, L. S., & Caron, M. G. (2000). The interaction of β-arrestin with the AP-2 adaptor is required for the clustering of β_2-adrenergic receptor into clathrin-coated pits. *The Journal of Biological Chemistry, 275*, 23120–23126.

Luo, J., Busillo, J. M., & Benovic, J. L. (2008). M_3 muscarinic acetylcholine receptor-mediated signaling is regulated by distinct mechanisms. *Molecular Pharmacology, 74*, 338–347.

Masri, B., Salahpour, A., Didriksen, M., Ghisi, V., Beaulieu, J. M., Gainetdinov, R. R., et al. (2008). Antagonism of dopamine D2 receptor/β-arrestin 2 interaction is a common property of clinically effective antipsychotics. *Proceedings of the National Academy of Sciences of the United States of America, 105*, 13656–13661.

Milano, S. K., Kim, Y.-M., Stefano, F., Benovic, J. L., & Brenner, C. (2006). Phosphoinositide-dependent regulation of arrestin oligomerization and function. *The Journal of Biological Chemistry, 281*, 9812–9823.

Moore, C. A., Milano, S. K., & Benovic, J. L. (2007). Regulation of receptor trafficking by GRKs and arrestins. *Annual Review of Physiology, 69*, 451–482.

Mundell, S. J., Orsini, M. J., & Benovic, J. L. (2002). Characterization of arrestin expression and function. *Methods in Enzymology, 343*, 602–613.

Nobles, K. N., Guan, Z., Xiao, K., Oas, T. G., & Lefkowitz, R. J. (2007). The active conformation of β-arrestin1: Direct evidence for the phosphate sensor in the N-domain and conformational differences in the active states of β-arrestins1 and -2. *The Journal of Biological Chemistry, 282*, 21370–21381.

Oakley, R. H., Laporte, S. A., Holt, J. A., Caron, M. G., & Barak, L. S. (2000). Differential affinities of visual arrestin, β-arrestin1, and β-arrestin2 for G protein-coupled receptors delineate two major classes of receptors. *The Journal of Biological Chemistry, 275*, 17201–17210.

Orsini, M. J., & Benovic, J. L. (1998). Characterization of dominant negative arrestins that inhibit β_2-adrenergic receptor internalization by distinct mechanisms. *The Journal of Biological Chemistry, 273*, 34616–34622.

Santini, F., & Keen, J. H. (2000). Characterization of receptor sequestration by immunofluorescence microscopy. In J. L. Benovic (Ed.), *Receptor biochemistry and methodology: Regulation of G protein-coupled receptor function and expression*, Vol. 4, (pp. 231–252). New York, NY: Wiley-Liss, Inc.

Scherrer, G., Tryoen-Toth, P., Filliol, D., Matifas, A., Laustriat, D., Cao, Y. Q., et al. (2006). Knockin mice expressing fluorescent delta-opioid receptors uncover G protein-coupled receptor dynamics in vivo. *Proceedings of the National Academy of Sciences of the United States of America, 103*, 9691–9696.

Schmid, E. M., Ford, M. G. J., Burtey, A., Praefcke, G. J. K., Peak-Chew, S. Y., Mills, I. G., et al. (2006). Role of the AP2 beta-appendage hub in recruiting partners for clathrin-coated vesicle assembly. *PLoS Biology, 4*, 1532–1548.

Shankar, H., Michal, A., Kern, R. C., Kang, D. S., Gurevich, V. V., & Benovic, J. L. (2010). Non-visual arrestins are constitutively associated with the centrosome and regulate centrosome function. *The Journal of Biological Chemistry, 285*, 8316–8329.

Shenoy, S. K., & Lefkowitz, R. J. (2011). β-Arrestin-mediated receptor trafficking and signal transduction. *Trends in Pharmacological Sciences, 32*, 521–533.

Shukla, A. K., Violin, J. D., Whalen, E. J., Gesty-Palmer, D., Shenoy, S. K., & Lefkowitz, R. J. (2008). Distinct conformational changes in β-arrestin report biased agonism at seven-transmembrane receptors. *Proceedings of the National Academy of Sciences of the United States of America, 105*, 9988–9993.

Shukla, A. K., Xiao, K., & Lefkowitz, R. J. (2011). Emerging paradigms of β-arrestin-dependent seven transmembrane receptor signaling. *Trends in Biochemical Sciences, 36*, 457–469.

Temkin, P., Lauffer, B., Jager, S., Cimermancic, P., Krogan, N. J., & von Zastrow, M. (2011). SNX27 mediates retromer tubule entry and endosome-to-plasma membrane trafficking of signalling receptors. *Nature Cell Biology, 13*, 715–721.

Vishnivetskiy, S. A., Gimenez, L. E., Francis, D. J., Hanson, S. M., Hubbell, W. L., Klug, C. S., et al. (2011). Few residues within an extensive binding interface drive receptor interaction and determine the specificity of arrestin proteins. *The Journal of Biological Chemistry*, *286*, 24288–24299.

Vishnivetskiy, S. A., Hosey, M. M., Benovic, J. L., & Gurevich, V. V. (2004). Mapping the arrestin-receptor interface: Structural elements responsible for receptor specificity of arrestin proteins. *The Journal of Biological Chemistry*, *279*, 1262–1268.

Xiao, K., Shenoy, S. K., Nobles, K., & Lefkowitz, R. J. (2004). Activation-dependent conformational changes in β-arrestin 2. *The Journal of Biological Chemistry*, *279*, 55744–55753.

Zhang, J., Barak, L. S., Winkler, K. E., Caron, M. G., & Ferguson, S. S. (1997). A central role for β-arrestins and clathrin-coated vesicle-mediated endocytosis in β_2-adrenergic receptor resensitization. Differential regulation of receptor resensitization in two distinct cell types. *The Journal of Biological Chemistry*, *272*, 27005–27014.

Zhao, Y., & Keen, J. H. (2008). Gyrating clathrin: Highly dynamic clathrin structures involved in rapid receptor recycling. *Traffic*, *9*, 2253–2264.

CHAPTER SIX

Tracking Cell Surface Mobility of GPCRs Using α-Bungarotoxin-Linked Fluorophores

Saad Hannan, Megan E. Wilkins, Philip Thomas, Trevor G. Smart[1]
Department of Neuroscience, Physiology and Pharmacology, University College London, London, United Kingdom
[1]Corresponding author: e-mail address: t.smart@ucl.ac.uk

Contents

1. Introduction 110
   1.1 Rationale for the BBS-tagging technique 111
2. Methodology 113
   2.1 Cloning the BBS site into GPCRs 113
   2.2 HEK-293 and hippocampal cell culture 115
   2.3 Calcium phosphate transfection of cell lines using cDNA 116
3. Validating the BBS Tag 117
   3.1 Expressing $GABA_B$ receptors containing the BBS 117
   3.2 Determining the functional neutrality of the BBS site 118
   3.3 Saturation binding: Apparent affinity of BTX for the BBS 119
4. Experimental Applications for the BBS Tag 120
   4.1 Monitoring receptor insertion at the cell surface 120
   4.2 Intracellular trafficking of internalized receptors 121
   4.3 Live cell imaging and the dynamics of receptor internalization 121
   4.4 Assessing the impact of photobleaching 123
5. Image Analysis 123
   5.1 Measuring rates of internalization and insertion by membrane fluorescence 124
6. Other Applications 126
   6.1 Cysteine modification of the BBS for dual labeling: Determination of GPCR heteromer internalization 126
7. Overview 128
References 128

Abstract

$GABA_B$ receptors are G-protein-coupled receptors (GPCRs) that are activated by GABA, the principal inhibitory neurotransmitter in the central nervous system. Cell surface mobility of $GABA_B$ receptors is a key determinant of the efficacy of slow and prolonged

Methods in Enzymology, Volume 521
ISSN 0076-6879
http://dx.doi.org/10.1016/B978-0-12-391862-8.00006-5

synaptic inhibition initiated by GABA. Therefore, experimentally monitoring receptor mobility and how this can be regulated is of primary importance for understanding the roles of $GABA_B$ receptors in the brain, and how they may be therapeutically exploited. Unusually for a GPCR, heterodimerization between the R1 and R2 subunits is required for the cell surface expression and signaling by prototypical $GABA_B$ receptors. Here, we describe a minimal epitope-tagging method, based on the incorporation of an α-bungarotoxin binding site (BBS) into the $GABA_B$ receptor, to study receptor internalization in live cells using a range of imaging approaches. We demonstrate how this technique can be adapted by modifying the BBS to monitor the simultaneous movement of both R1 and R2 subunits, revealing that $GABA_B$ receptors are internalized as heteromers.

1. INTRODUCTION

$GABA_B$ receptors are classified as class C GPCRs and they couple to three major effector pathways. Two of these involve ion channels—inwardly rectifying K^+ channels (GIRK) in the postsynaptic membranes and voltage-gated Ca^{2+} channels in pre- and postsynaptic terminals, leading to membrane hyperpolarization and the inhibition of neurotransmitter release, respectively (Bettler, Kaupmann, Mosbacher, & Gassmann, 2004). In addition, $GABA_B$ receptors can also couple to adenylyl cyclase increasing or decreasing its activity. Studying the cell surface mobility of $GABA_B$ receptors is considered important because the efficacy of inhibition, caused by the binding of GABA to these receptors, will be influenced by the number of receptors available on the cell surface within signaling microdomains.

To be functionally active, R1 and R2 subunits of the $GABA_B$ receptor need to heterodimerize. This is required for at least four reasons. Firstly, the neurotransmitter (GABA) binding site resides on R1 subunits whereas the G-protein activation domain is located on R2. Secondly, the affinity of agonist binding on R1 increases following dimerization with R2. Thirdly, R1 subunits require R2 subunits to exit the endoplasmic reticulum to gain access to the cell surface membrane. Using this α-bungarotoxin (BTX) binding site (BBS) method, to monitor receptor mobility, we have demonstrated that heterodimerization enables R2 subunits to fulfill a fourth role, in which R2 stabilizes the R1a subunits on the cell surface by slowing their internalization.

The lack of available, highly specific, N-terminal antibodies against $GABA_B$ receptor subunits has meant that most trafficking studies of $GABA_B$ receptors have used cell surface biotinylation or antibodies against the intracellular C-terminus, both of which require the cells to be fixed (Grampp, Sauter, Markovic, & Benke, 2007). The use of the BBS largely obviates the need to fix cells prior to imaging. The strategy is based on a neurotoxin from snake venom, BTX, to bind to and inhibit the function of specific nicotinic acetylcholine receptors (nAChRs) at nanomolar affinity (Corringer, Le Novere, & Changeux, 2000). We have utilized this interaction in the design of a 13 amino acid mimotope of the BBS, WRYYESSLEPYPD, which is sufficient to bind BTX with high affinity (Harel et al., 2001). This mimotope has previously been cloned into ionotropic receptors including, AMPA receptors (Sekine-Aizawa & Huganir, 2004) and $GABA_A$ receptors (Bogdanov et al., 2006), and subsequently metabotropic $GABA_B$ R1a (Wilkins, Li, & Smart, 2008) and R2 receptors (Hannan et al., 2011). The method has also been extended to voltage-gated Ca^{2+} (Tran-Van-Minh & Dolphin, 2010) and K^+ channels (Moise et al., 2010).

In this chapter, we describe the fundamental methods for incorporating BBS tags into GPCRs and include details of how to assess their functional neutrality, and the use of saturation binding on membrane extracts to measure the apparent binding affinity of BTX for the inserted BBS. We then demonstrate the power of the BBS as a minimal reporter method for studying the mobility of GPCRs in live cells and how receptor insertion in the membrane can be studied using fixed cells with confocal microscopy. In addition, we describe a new extension to the method that uses the BBS tag to study the internalization of the two $GABA_B$ receptor subunits, simultaneously.

1.1. Rationale for the BBS-tagging technique

The reason for developing the least invasive-tagging method is based on the premise that to monitor the movement of a protein, we need to interfere with it first by tagging. However, the process of tagging may, of course, affect the movement of the protein. So, how can we faithfully preserve innate protein trafficking behavior from the causality of tagging? This is really another interpretation of Heisenberg's uncertainty principle in quantum mechanics.

Of course, epitope tagging is necessary to monitor protein movement and the BTX-binding strategy has several advantages as a tagging method. Primarily, the BBS is relatively small (13 amino acids) compared to the size of green fluorescent protein (GFP; 238 amino acids), which is the most widely used fluorophore to label proteins. The BBS is comparable in size to myc- (10 amino acids), flag- (8 amino acids), HA- (hemaglutinin, 9 amino acids), AP- (alkaline phosphatase, 15 amino acids) epitope tags, but these tags all require large, specific antibodies to be visualized, whereas the BBS needs only BTX and an organic fluorophore (e.g., rhodamine), which when complexed is still significantly smaller than antibody $F(ab')_2$ fragments and GFP (Table 6.1; Fig. 6.1).

Live cell imaging studies using antibodies have most frequently used primary antibodies allied to a complementary reaction with secondary antibodies or $F(ab')_2$ fragments coupled to organic fluorophores. These complexes are significantly larger in size compared to the BTX–fluorophore combination. Furthermore, BTX labeling is advantageous because, in comparison to fluorescent fusion proteins, it permits the discrimination of cell surface receptors from those located intracellularly. Although pH-sensitive fluorescent fusion proteins have been developed for the detection of only cell surface protein pools (provided the pH-sensitive reporter is only exposed in the extracellular environment: e.g., super-ecliptic pHluorins; Ashby, Ibaraki, & Henley, 2004), often the signal-to-noise ratio at the cell surface is low when compared to BTX–fluorophore conjugates. In addition, labeling of the BBS with BTX is simple and fast and BTX is stable under storage, unlike antibodies which require glycerol and have a short shelf life.

Table 6.1 Physical properties of molecular imaging tags

Molecular tag	Size (no. amino acids)	Dimensions (nm)	Molecular mass (kDa)	Volume (nm^3)	Reference
α-Bungarotoxin	74	4 × 3 × 2	8	24	Love and Stroud (1986)
Antibody	Light chain ~217				
	Heavy chain ~450–576				
	Total ~1334–1586	14.5 × 8.5 × 4	~150 (total)	493	Davies, Padlan, and Segal (1975)
$F(ab')_2$	~450	8 × 5 × 4	50	160	Davies et al. (1975)
GFP	238	4.5 × 3.5 × 3.5	27	55	Ormo et al. (1996)

The dimensions were calculated according to the closest fitting rectangular prism (height × width × depth) that encapsulates the molecular tag. Volumes are calculated according to the dimensions of the rectangular prism.

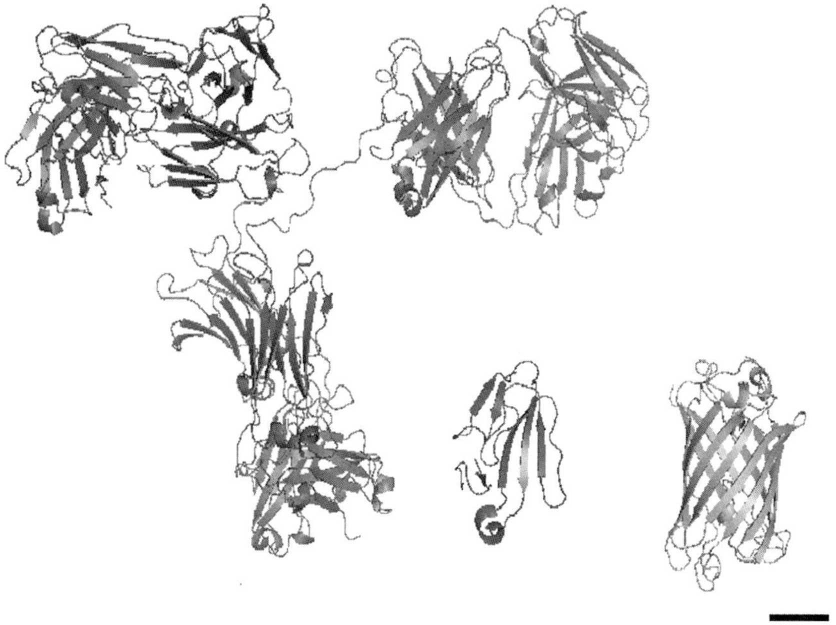

Figure 6.1 Structures of commonly used tags. Model structures are shown for an IgG antibody (magenta; PDB ID—1IGT) and GFP molecule (green; PDB ID—1EMA) in comparison with a molecule of α-bungarotoxin (blue; PDB ID—1HC9). The structures are proportionately scaled to facilitate comparison. Scale bar = 2 nm. (For interpretation of the references to color in this figure legend, the reader is referred to the online version of this chapter.)

2. METHODOLOGY

To use fluorophore-linked BTX as a tag, we first need to insert the BBS into a suitable location in the target protein and ensure that BTX binds to the BBS with high affinity and specificity. In the following section, we provide details of the cloning procedures that are followed for routinely engineering the BBS into receptors. In addition, we describe methods for staining BBS-modified receptors, determining the affinity of the recombinant BBS site for BTX, and testing for functional neutrality of BBS-containing receptors.

2.1. Cloning the BBS site into GPCRs

Previously, we have inserted a BBS tag into the N-termini of R1a (R1a^{BBS}), R1b (R1b^{BBS}), and R2 (R2^{BBS}) subunits of the $GABA_B$ receptor, and in a similar location for the mGluR2 receptor (mGluR2^{BBS}). For this, we adopted an inverse PCR approach allowing the 13 amino acid BBS to be inserted at any position within the protein with just a single polymerase chain reaction (PCR) step. The same technique is applied to GPCRs.

Before cloning, it is necessary to identify a domain in the protein of interest that is not only exposed to the extracellular environment but which will also not disrupt receptor function after the tag is inserted. Terminal signal sequences that are cleaved from the mature protein should obviously be avoided as potential insertion sites. The task of choosing a suitable site is often made easier if previous information is available from the insertion of other epitope tags, such as myc-, flag-, or HA-tags, in the protein. In such cases, the BBS can be inserted adjacent to the existing epitope tag. In addition, the availability of a crystal structure is also very helpful in ascertaining regions of minimal disruption to α-helices and β-sheet structures in the receptor, and for avoiding agonist binding sites and subunit interfaces which may be involved in receptor assembly. The precedent we have followed for placing the BBS in GPCRs comes from experimental observations following insertion of unrelated epitope tags in $GABA_A$ receptor subunits. From this, it is clear that epitope tags are best situated near the N-terminus, for example, between amino acids 4 and 5 (Connolly, Krishek, McDonald, Smart, & Moss, 1996).

Accordingly, we have inserted a BBS in the N-terminus of $GABA_B$ receptor R1a subunits between amino acids 6 and 7 and in the N-terminus of R2 subunits between amino acids 67 and 68 (Fig. 6.2A). Once the BBS insertion site has been chosen, PCR primers containing the BBS site should be designed to anneal to the site of insertion. The minimum Tm sequence encoding for the BBS is 5′-TGGAGATATTATGAAAGTAGTTTA GAACCATATCCAGAT-3′. The last 20 bases of this sequence are added to 21 bases of the coding strand downstream from the site of insertion of the BBS to give the sequence of the forward primer. The first 19 bases of the BBS are added to 21 bases upstream of the coding strand and reverse complemented to give the reverse primer.

We use the Phusion (Finnzymes) DNA polymerase for the PCR, followed by gel extraction (Qiagen Gel extraction Kit), 5′ phosphorylation using polynucleotide kinase (NEB; UK), and ligation at 16 °C using DNA ligase (Roche) to generate the BBS-tagged receptors. After transformation of the ligation product, single colonies are grown up and the DNA is extracted in small quantities using standard protocols. The fidelity of the constructs is checked by DNA sequencing. Once the BTX has been inserted into the receptor subunit, these are used to transfect GIRK cells or hippocampal neurons in culture, as outlined below.

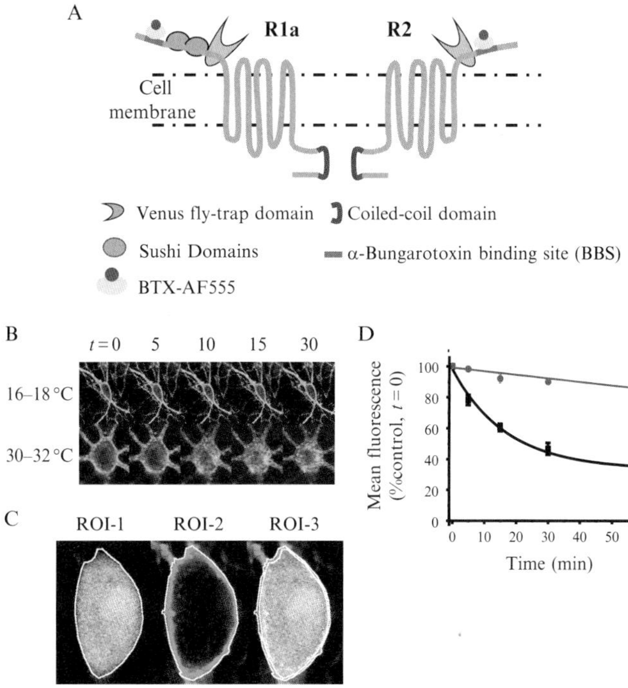

Figure 6.2 Tracking BBS-tagged $GABA_B$ receptors in hippocampal neurons. (A) Schematic diagram depicting the location of the BBS in R1a and R2 subunits. (B) Hippocampal neurons (14–21 DIV) expressing $R1a^{BBS}R2$ and eGFP were incubated in 1 m*M* D-tubocurarine for 5 min followed by 3 μg/ml BTX-AF555 for 10 min at RT and imaged at different time points at 16–18 or 30–32 °C. (C) Defining regions of interest (ROIs, white lines) around a $GABA_BR$ expressing HEK cell tagged with a BBS. (D) Membrane fluorescence decay curve for internalization of BTX-AF555-tagged $R1a^{BBS}R2$ receptors at 30–32 °C (black; $\tau = 17 \pm 3$ min, $n = 3$–6) compared to 16–18 °C (red). (For interpretation of the references to color in this figure legend, the reader is referred to the online version of this chapter.)

2.2. HEK-293 and hippocampal cell culture

Our preferred cell line for cDNA expression is the HEK-293 cells. Use of other cell lines such as CHO, MDCK, and HELA is equally viable. A variant of HEK-293 cells that stably express the inwardly rectifying K^+ channels, Kir3.1 and Kir3.2 (GIRK cells), are maintained at 37 °C in a 95% air/5% CO_2 atmosphere. The culture medium used is Dulbecco's-modified Eagle's medium supplemented with 10% (v/v) fetal calf serum (FCS), penicillin-G/streptomycin (100 U/100 μg/ml), 2 m*M* glutamine, and geneticin (0.5 mg/ml) (all from Invitrogen). Cells are seeded onto poly-L-lysine-coated 22-mm glass coverslips or similar for transfection.

Cultured hippocampal neurons are prepared from E18 Sprague–Dawley rat embryos as described previously (Hannan et al., 2011). This involves enzymatic and mechanical dissociation of the dissected hippocampi into single cells followed by plating onto 18- or 22-mm glass coverslips (Assistence/VWR) coated with poly-D-lysine (Sigma) in a medium containing minimum essential media (Invitrogen), supplemented with 5% (v/v) heat-inactivated FCS, 5% (v/v) heat-inactivated horse serum (Invitrogen), penicillin-G/streptomycin (10 U/10 μg/ml), 2 m*M* glutamine, and 20 m*M* glucose (Sigma). After 1 h, the media is replaced and the cells are maintained until used for experiments in a media containing Neurobasal-A (Invitrogen), supplemented with 1% (v/v) B-27, penicillin-G/streptomycin (50 U/50 μg/ml), 0.5% (v/v) Glutamax (Invitrogen), and 35 m*M* glucose. Neurons are transfected after 7–10 days *in vitro* (DIV).

2.3. Calcium phosphate transfection of cell lines using cDNA

Calcium phosphate methods provide an inexpensive means of delivering cDNA into cells and can routinely achieve high transfection efficiencies. Other methods such as Effectene, Oligofectamine, Lipofectamine, and electroporation could also be used.

GIRK cells are transfected on single 22-mm coverslips as follows:

- cDNAs encoding for $R1a^{BBS}$ (or R1a), $R2^{BBS}$ (or R2), and eGFP are mixed in a ratio of 1:5:1 for a total amount of 4 μg DNA usually in a total volume of 4 μl. We usually adjust our DNA stocks to a concentration of 1 mg/ml.
- Add 20 μl of 340 m*M* $CaCl_2$ to the DNA mixture.
- Add 24 μl HEPES-buffered saline (HBS; 280 m*M* NaCl, 2.8 m*M* Na_2HPO_4, 50 m*M* HEPES (pH 7.2)) to the DNA–$CaCl_2$ mixture and vortex for a few seconds.
- Incubate for 10 min.
- Add the mixture to the cultured cells on coverslips, dropwise.
- Return the cells to the incubator at 37 °C.

If the calcium phosphate treatment compromises cell survival, then the calcium chloride precipitates can be washed off 15–30 min after transfection.

Hippocampal neurons at 7–10 DIV are transfected as follows:

- Into tube “A” place 60 μl HBS (274 m*M* NaCl, 10 m*M* KCl, 1.4 m*M* $Na_2HPO_4{\cdot}7H_2O$, 15 m*M* D-glucose, 42 m*M*HEPES (pH 7.11)).
- Add $R1a^{BBS}$ (or R1a), $R2^{BBS}$ (or R2), and eGFP cDNAs at in a ratio of 1:1:1 to a total amount of 4 μg and a total volume of 54 μl buffer TE in tube “B”.

- Add 2.5 μl of 2.5 *M* $CaCl_2$ to tube B dropwise with continual mixing.
- Hold tube A on a vortexer and add the contents of tube B.
- Protect the mixture from light and incubate at room temperature (RT) for 30 min, vortexing every 5 min.
- During the incubation, make up 2 m*M* kynurenic acid in Neurobasal-A (work from 10 m*M* stocks in Neurobasal-A), filter sterilize, and warm (37 °C). In addition, warm-up sufficient Neurobasal-A to 37 °C for later washing steps.
- Aspirate the media from the neuronal dishes and replace with 1 ml prewarmed 2 m*M* kynurenic acid.
- Add the DNA mixture to the coverslips dropwise.
- Return cells to the incubator at 37 °C for 20 min.
- Wash the cells with 1 ml prewarmed Neurobasal-A (×2) and replace with neuronal maintenance media.
- Cells should be ready for imaging in 96 h when the expression of the constructs should have stabilized.

3. VALIDATING THE BBS TAG

3.1. Expressing $GABA_B$ receptors containing the BBS

The specificity of expression and staining profile of $GABA_B$ receptors tagged with the BBS can be explored in GIRK cells and hippocampal neurons transfected with untagged receptors or receptors containing the BBS. Transfect cells on duplicate coverslips with cDNAs for $R1a^{BBS}$, R2, and eGFP; and on duplicate control coverslips with cDNAs for R1a, R2, and eGFP; and finally, with eGFP only.

- 36–48 h posttransfection for GIRKs and 96 h for neurons, wash cells three times with Krebs containing (m*M*): 140 NaCl, 4.7 KCl, 1.2 $MgCl_2$, 2.5 $CaCl_2$, 11 glucose, and 5 HEPES (pH 7.4).
- Preincubate hippocampal neurons in 1 m*M* D-tubocurarine (D-TC) to prevent BTX-AF555 binding to native nAChRs (Sekine-Aizawa & Huganir, 2004; Wilkins et al., 2008).
- Incubate one coverslip expressing $R1a^{BBS}$, R2, and eGFP; one coverslip expressing R1a, R2, and eGFP; and the coverslip expressing eGFP alone, with 3 μg/ml BTX coupled to Alexa Fluor 555 (BTX-AF555; Molecular Probes) for 10 min.
- Incubate the remaining coverslips expressing $R1a^{BBS}$, R2, and eGFP or R1a, R2, and eGFP with 3 μg/ml unlabeled BTX (Molecular Probes) for 10 min.

- Wash cells three times in Krebs to remove excess BTX.
- Incubate cells in 4% (w/v) PFA in phosphate-buffered saline (PBS) for 10 min followed by washes (×3).
- Incubate cells in 5% (w/v) NH_4Cl followed by washing (×3).
- At this point, the cells can be mounted on glycerol. However, glycerol mounting can give rise to unwanted background fluorescence for imaging and the morphology of the cells can change due to mounting. We therefore image cells without mounting coverslips in glycerol by using a water immersion objective.

We image cells using a Zeiss LSM510 Meta confocal microscope with either a 40 × oil immersion objective or a 40 × water objective. If fluorescence due to the BBS modification is ineffective then the position of insertion for the BBS site can be altered by recloning to allow greater accessibility of BTX to the BBS. In addition, higher concentrations of fluorescent BTX can be tried along with more prolonged incubation times after cell transfection.

3.2. Determining the functional neutrality of the BBS site

The insertion of an epitope can alter the expression efficiency, assembly, and functionality of a receptor; therefore, it is important to ensure that the introduction of the tag is essentially functionally silent before it is used in trafficking experiments. All these facets can be simultaneously examined by assessing the function of the receptor. $GABA_B$ receptors couple to inwardly rectifying K^+ channels via $G_{i/o}$ and this allows us to study the opening of these channels in response to their activation by GABA, comparing wild-type receptors with those in which the BBS tag is inserted. Whole-cell potassium currents in response to GABA application may be obtained from GIRK cells transfected with cDNAs encoding for wild-type or BBS-containing $GABA_B$ receptors, 36–48 h posttransfection. For this, we use electrophysiological techniques.

- Patch pipettes (resistances: 3–5 MΩ) are filled with an internal solution (m*M*): 120 KCl, 2 $MgCl_2$, 11 EGTA, 30 KOH, 10 HEPES, 1 $CaCl_2$, 1 GTP, 2 ATP, 14 creatine phosphate, pH 7.11.
- The GIRK cells are superfused in a Krebs solution and after a seal is established, the KCl concentration in the external solution is increased to 25 m*M*, with a corresponding reduction in the NaCl concentration to 120 m*M*. This shifts the equilibrium potential for K^+ (E_K) from −90 to −47 mV and increases the amplitude and reverses the direction of the $GABA_B$ receptor-activated K^+ currents, a condition convenient for measuring these whole-cell currents.

- Membrane currents are recorded in response to different concentrations of GABA, every 3 min, in the presence or absence of BTX, filtered at 5 kHz (−3 dB, sixth pole Bessel, 36 dB/octave). Any changes greater than 20% in the membrane input conductance or series resistance result in the recording being discarded.
- The GABA concentration–response curves can be generated by measuring the potassium current (I) for each GABA concentration. The current amplitudes are normalized to the maximum GABA response (I_{max}) and the concentration–response relationship fitted with the Hill equation:

$$\frac{I}{I_{max}} = \left[\left(\frac{1}{1+(\mathrm{EC}_{50}/A)^n}\right)\right],$$

where "A" represents GABA concentration, EC_{50} is the GABA concentration activating 50% of the maximum response, and "n" is the Hill slope.

We have found no difference in EC_{50}s between wild-type R1aR2 $GABA_B$ receptors and recombinant $GABA_B$ $R1a^{BBS}R2$ or $R1aR2^{BBS}$ receptors, suggesting that the introduction of the BBS in these receptors at the chosen locations does not alter the function of the receptors (Hannan et al., 2011; Wilkins et al., 2008).

3.3. Saturation binding: Apparent affinity of BTX for the BBS

As well as potentially affecting receptor function, inserting the BBS into new proteins can affect the apparent binding affinity for BTX. Ideally, the affinity should remain unaffected to maintain the fidelity of the mobility measurements for the tagged receptors. Moreover, for live cell imaging, premature dissociation of BTX from the BBS would result in an underestimation of the cell surface fluorescence in live cell internalization studies.

The binding of ^{125}I-BTX (200 Ci/mmol; Perkin Elmer) to $R1a^{BBS}R2$ and $R1aR2^{BBS}$ has been measured using saturation binding assays in the presence or absence of a 1000-fold excess of unlabeled BTX. The apparent binding affinity for BTX with both constructs is in the nanomolar range. Here, we have used the example of $R1aR2^{BBS}$ receptors to describe how the apparent affinity can be measured.

- Transfect two groups of GIRK cells with cDNAs encoding for R1a, $R2^{BBS}$, and eGFP in 6-cm dishes or with cDNAs encoding for the chimeric BTX-sensitive receptor $\alpha7/5HT_{3A}$. This receptor is used because the nACHR α7 subunit retains a high sensitivity to BTX and the chimera with the 5-HT_{3A} receptor enables a higher level of expression in heterologous cell-based systems (Eisele et al., 1993).

- 36–48 h posttransfection, aspirate the media and wash with 10 ml PBS (Sigma). Add 1 ml PBS + 0.5% (w/v) BSA to the dishes and scrape off the transfected cells from the dishes. Gently break-down cell clusters by trituration.
- For each concentration of ^{125}I-BTX, reactions are carried out in triplicates for specific and duplicates for nonspecific binding. Add 50 μl cells per reaction to each tube along with appropriate concentrations of ^{125}I-BTX. Add 50 μl PBS + 0.5% (w/v) BSA for nonspecific reactions and 1000-fold excess unlabeled BTX for specific reactions to make up a total volume of 150 μl.
- Shake for 60 min at RT; during the incubation, soak Whatman GF/A filters in 0.5% (w/v) polyethyleneimine at 4 °C.
- Harvest the cells (Brandel cell harvester) by capturing onto filters, followed by rapid washing with PBS.
- The radiolabel retained on the filters is assayed using a Wallac 1261 gamma counter.
- Subtract nonspecific binding from specific binding.
- Scatchard analysis using a nonlinear regression fitting algorithm is used to obtain values of the B_{max} (maximum number of binding sites) and K_d (binding affinity) from:

$$y = \frac{B_{max} X}{K_d + X},$$

where X is the ^{125}I-BTX concentration.

The same analysis is used for the α7/5HT_{3A} receptor chimera expressed in GIRK cells to compare the specificity of binding.

4. EXPERIMENTAL APPLICATIONS FOR THE BBS TAG

4.1. Monitoring receptor insertion at the cell surface

The cell surface membrane insertion of BBS-tagged $GABA_B$ receptors can be studied in GIRK cells.

- Transfect coverslips with cDNAs encoding for $R1a^{BBS}$, R2, and eGFP.
- 36–48 h posttransfection, wash the coverslips with Krebs (×3).

- Incubate cells in unlabeled BTX (20 μg/ml) for 10 min at RT to label all the cell surface R1a^{BBS}R2 receptors.
- Wash cells in Krebs (×3) to remove excess unlabeled BTX.
- Fix one coverslips in PFA. Incubate the remaining coverslips in a 37 °C incubator in 3 μg/ml BTX-AF555 (in Krebs).
- At the end of 5, 10, 15, 30, and 60 min incubations, remove one coverslip from the incubator and wash cells three times in Krebs to remove the excess BTX-AF555.
- Incubate cells in PFA (4%, w/v) in PBS for 10 min followed by washes (×3).
- Incubate cells in NH_4Cl (5%, w/v) followed by washes (×3).
- Mount coverslips in glycerol if required.

Image and capture the fluorescence of both eGFP and BTX-AF555 using a 488 Argon laser (505–530-nm band-pass filter) for eGFP and a 543-nm Helium-Neon laser (560-nm long-pass filter) for BTX-AF555 for a representative number of cells (e.g., 12).

4.2. Intracellular trafficking of internalized receptors

To follow the fate of the internalized R1a^{BBS}R2 receptors, GIRK cells and hippocampal neurons are cotransfected with appropriate cDNAs for eGFP-tagged markers that identify internal compartments such as: Rab5 (early endosomal compartments), Rab11 (recycling endosomes; Zerial & McBride, 2001), or Rab7 (late endocytic/lysosomal compartments; Meresse, Gorvel, & Chavrier, 1995).

GIRK cells or hippocampal neurons, preincubated in 1 m*M* D-TC, are incubated in 3 μg/ml BTX-AF555 for 10 min at RT to label surface receptors and washed in Krebs to remove the unbound BTX-AF555. Cells are incubated at 37 °C/5% CO_2 in Krebs solution for 30 min for eGFP-Rab5 or eGFP-Rab11 and 60 min for Rab7-GFP to allow the BTX-AF555-tagged receptors to internalize. After the incubation, the cells are fixed and imaged to study the colocalization of BTX-AF555-tagged R1a^{BBS}R2 with eGFP-tagged Rab5-, Rab11-, or Rab7-containing intracellular compartments.

4.3. Live cell imaging and the dynamics of receptor internalization

To determine the rates and extent of receptor internalization, using the BBS approach in live cells, the BBS-tagged receptors are labeled with a fluorophore-tagged BTX and the reduction in cell surface membrane fluorescence is studied over time for single cells. This is advantageous as the controls form part of the same experiment with membrane fluorescence at $t=0$ being used as the initial fluorescence level. Here, as an example, we have used GIRK cells and hippocampal neurons transfected with cDNAs encoding for R1a^{BBS}, R2, and eGFP to demonstrate how internalization of BTX-tagged receptors can be studied in real time. The bath containing the coverslip that is being subjected to imaging should be superfused with Krebs solution at a rate of 1 ml/min to ensure that the media around the cells is constantly replenished and that the cells remain in a viable state, while minimizing mechanical agitation.

- 36–48 h posttransfection for GIRK cells or 7–14 days posttransfection for neurons, wash the cells with Krebs (×3).
- For hippocampal neurons preincubate neurons with 1 m*M* D-TC for 5 min at RT.
- To label BBS-containing receptors with BTX, incubate GIRK cells or hippocampal neurons in 3 μg/ml BTX-AF555 for 10 min at RT followed by washes (×3) in Krebs to remove excess BTX-AF555.
- Load the coverslip on the microscope; quickly find examples of BTX-AF555-labeled cells and select for further analysis. Optimize the settings for imaging as fast as possible and acquire an image for $t=0$ as a mean of 2–4 scans in 8 bits.
- Then, at 5, 10, 15, 30, and 60 min intervals, acquire images of the same cell without changing the field of view or the microscope settings (detector gain, amplifier offset, optical slice thickness, and laser intensity) (Fig. 6.2B). If physical drift becomes noticeable, then the *z*-axis should be adjusted so that images from every time point are from the same optical slice as $t=0$.

The experiments can be extended to study the effect of temperature on receptor internalization by using a Peltier device to cool or heat the perfusion chamber and solutions. We have found that at 16–18 °C there is very little internalization of R1a^{BBS}R2 receptors and internalization proceeds at a fast rate at 22–24 °C, and faster still at 30–32 °C (Fig. 6.2B). During imaging, if the morphology of the cell changes, the osmolarity of the Krebs solution or the flow rate of perfusion should be adjusted. The exposure of cells to the laser should be kept at a minimum while optimizing the confocal settings for imaging at the $t=0$ time point, and also while adjusting subsequent drifts in the *z*-axis. Cells that show classical signs of phototoxicity such as blebbing should be discarded from the analysis.

4.4. Assessing the impact of photobleaching

To ensure that the rate and extent of receptor internalization is being accurately measured using the BTX-linked fluorophores, it is necessary to ascertain that the fluorophore emission spectra are not significantly affected by photobleaching. The extent of photobleaching will vary with the particular fluorophore and experimental conditions used so must be determined under the individual conditions used. To determine the extent of photobleaching for BTX-AF555, GIRK cells can be transfected with cDNAs encoding for the $R1a^{BBS}$, R2, and eGFP. The receptors are then tagged with BTX-AF555 and successive images accrued as quickly as possible.

- 36–48 h posttransfection, wash the cells with Krebs three times.
- Incubate the cells in BTX-AF555 (3 μg/ml) at RT for 10 min to label all cell surface $GABA_B$ receptors.
- Wash the cells three times in Krebs to remove excess BTX-AF555.
- Load coverslip on the microscope with the perfusion running at 16–18 °C, find a suitable labeled cell and quickly optimize the microscope settings for imaging. Take an image at $t = 0$ as a mean of 2–4 scans in 8 bits with a pixel time of 1.6 μs using a 543 nm Helium-Neon laser (560-nm long-pass filter) for BTX-AF555.
- As soon as the image is acquired, acquire another image without changing the confocal settings and continue doing this until the total number of scans reaches 20. Remember, if using an average of 2 scans, this should be a total of 10 scans and for an average of 4, this should be a total of 5 scans.
- Save the images for analysis.

Ten scans of an expressing cell, each one an average of 2 scans, should not take longer than 2–3 min. This process can be made faster by scanning a smaller area. The low temperatures should ensure that minimal receptor internalization take place. If it is necessary to measure photobleaching at higher temperatures, BBS-tagged receptors that do not internalize quickly should be used as the photobleaching levels are measured from cell surface membrane fluorescence and any internalization is likely to affect the measurements of surface fluorescence.

5. IMAGE ANALYSIS

Confocal images can be analyzed using widely available software such as ImageJ, or one of its variants, although other image analysis software such as Metamorph can also be used due to the simplicity of the image analysis process. All versions of Image J that are available offer a region-of-interest (ROI) manager. This is used in the protocol below to accurately define those areas of the membrane that are to be separated for fluorescence quantification.

- Open both the image channels for eGFP and BTX-AF555.
- Open ROI manager (Analyze > Tools > ROI manager) and select the "polygon selection" from the main toolbar.
- Draw the first ROI (ROI-1) using the polygon selection around the cell using the fluorescence of eGFP as a guide (Fig. 6.2C). This fluorescence should be diffuse inside the cell and should indicate the limits of the cell membrane. Click on Add (t) on the ROI manager toolbar to add this ROI (ROI-1) to memory. The program will assign a name for ROI-1 and this will be visible on the ROI manager toolbar.
- Next, choose the BTX-AF555 image and now, draw a second ROI (ROI-2) around the fluorescence on the BTX-AF555 channel (Fig. 6.2C). Add ROI-2 to the ROI manager as in previous step.
- Choose the BTX-AF555 image and click on "show all" on the ROI manager window. This will transfer both ROI-1 and ROI-2 onto the BTX-AF555 image. The nonoverlapping area between ROI-1 and ROI-2 should be the cell surface and represent membrane staining. Use the polygon selection tool to draw a third ROI (ROI-3) between ROI-1 and ROI-2 as a continuous area (Fig. 6.2C). Add this to ROI manager.
- Select an area of the image devoid of any cells that will act as a good representation of the level of background fluorescence and draw a fourth ROI (ROI-4) and add this to the ROI manager.
- While on the BTX-AF555 image, click on "measure" to calculate different fluorescence parameters for the four ROIs. The parameters can be chosen from the main toolbar (Analyze > Set measurements) and should include the area, mean gray value (mean membrane fluorescence per unit area—in this case μm^2), standard deviation, maximum, and minimum fluorescence levels.
- Save the results window and the ROIs. In addition, the results can be copied and pasted to programs like MS Excel where further analysis can be undertaken.

5.1. Measuring rates of internalization and insertion by membrane fluorescence

To calculate the rate of internalization of BBS-tagged $GABA_B$ receptors, we use the rate of change in membrane fluorescence with time (Fig. 6.2D). These data are then fitted with exponential decay or growth functions using standard graphical software such as Microcal Origin. It is important to note that some receptors can recycle at a fast rate after internalization. For such receptors, the membrane fluorescence decay curves for internalization will not give an accurate estimate of the true rate of internalization. When rates of internalization and recycling are fast, the fluorescence observed at the surface membrane at time points after $t=0$ will represent an equilibrium between recycling and internalization.

- Transfer the data from the analyzed ROIs in the previous protocol into MS Excel. Subtract the mean background fluorescence (ROI-4) from the mean cell surface membrane fluorescence (ROI-3) for all the images.
- After background subtraction, the individual mean fluorescence values at each time point are then normalized to the value at $t=0$, which is set to 100%. Thus, the mean fluorescence at subsequent time points should be lower as internalization proceeds.
- Transfer the normalized data for all the cells into Origin.
- Fit the membrane fluorescence values over time according to a mono-exponential decay function where t is time, y is the normalized mean membrane fluorescence at time t, y_0 is the starting fluorescence value at $t=0$ (=100%), and y_{ss} is the steady-state fluorescence after time, $t+\mathrm{d}t$. Thus, for receptor internalization, $y_{ss} < y_0$:

$$y = y_0 + (y_{ss} - y_0)\left(1 - e^{-t/s}\right)$$

The value of τ will provide an estimate of the rate of internalization, noting the caveat mentioned above concerning receptor recycling. The extent of internalization at any time point is essentially a measure of the percentage of receptors internalized up to that time point.

For estimating the rate of receptor insertion into the cell surface membrane, subtract the mean background fluorescence from the mean cell membrane fluorescence in all the captured images.

- Average the resulting mean cell surface membrane fluorescence for all cells at $t=60$ min. This value is then set to 100%.
- Then normalize the mean cell surface membrane fluorescence for every time point, to the value obtained at $t=60$ min.
- Fit the change in the membrane fluorescence values with time according to the above equation for mono-exponential growth, only here, $y_{ss} > y_0$. The value of τ now provides an estimate of the rate of receptor insertion at the cell surface.

6. OTHER APPLICATIONS

6.1. Cysteine modification of the BBS for dual labeling: Determination of GPCR heteromer internalization

Dual labeling using specific fluorophores that bind to different subunits of a GPCR will allow for increased resolution of their relative mobility. We have extended the use of the BBS strategy to study the simultaneous internalization of R1a and R2 subunits in live cells. This is achieved by chemically protecting the BBS on one subunit, for example, R1a, by modifying only this site so that it is reactive to a chemical protectant that can later be removed. To enable one BBS to react with this protectant, we substituted the two serine residues in the center of the BBS (WRYYE**SS**LEPYPD) for cysteines (WRYYE**CC**LEPYPD) using an inverse PCR method. This approach forms R1a$^{BBS\text{-}CC}$ complementing the BBS on R2 which still retains the two serines, R2$^{BBS\text{-}SS}$ for example (Fig. 6.3A). Before applying the chemical protectant, which is a sulfhydryl-specific reagent (MTSES), as the two Cys residues have a tendency to form a disulphide bond that prevents bonding with the protectant. The Cys–Cys bond is therefore first broken by reduction using dithiothreitol (DTT) and after the protection is established using MTSES, the R2$^{BBS\text{-}SS}$ can be labeled with BTX linked to a fluorophore. Then the protectant is removed by reapplying DTT, which does not affect the binding of the first BTX label. Finally, a second application of BTX linked to a new fluorophore binds to the R1a$^{BBS\text{-}CC}$. This provides two BTX-labeled subunits linked to different fluorophores. The internalization of the two subunits can therefore be studied in transfected cells expressing R1a$^{BBS\text{-}CC}$ and R2$^{BBS\text{-}SS}$ (Fig. 6.3B).

- 36–48 h posttransfection for GIRK cells or 7–14 days posttransfection for neurons wash the cells with Krebs.
- For hippocampal neurons, incubate the neurons in 1 m*M* D-TC for 5 min at RT.
- Incubate the cells in 200 μ*M* DTT for 30 min at RT to break the Cys–Cys bond between the two cysteines in the modified BBS-CC on R1a followed by washing in ice-cold Krebs.
- Incubate the cells in 200 μ*M* MTSES (chemical protectant reagent, Toronto research chemicals) for 5 min at 4 °C followed by washing in ice-cold Krebs.
- Incubate the cells with 3 μg/ml BTX coupled to Alexa Fluor 488 (BTX-AF488; Molecular Probes) for 10 min at 4 °C to label R2$^{BBS\text{-}SS}$ subunits followed by washing in ice-cold Krebs.

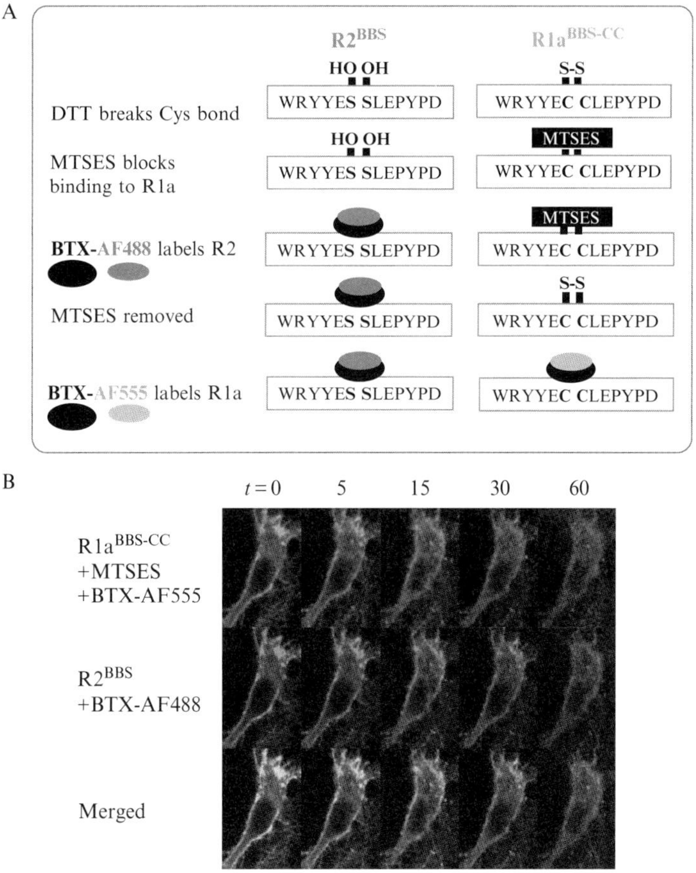

Figure 6.3 $GABA_B$ receptors are constitutively internalized as heterodimers. (A) Schematic diagram detailing the labeling of R2 (wild-type BBS, left) and R1a (mutant BBS, right) subunits with two fluorophores linked to BTX. (B) Hippocampal neurons (14–21 DIV) expressing $R1a^{BBS\text{-}CC}R2^{BBS}$ were incubated in 1 m*M* D-tubocurarine for 5 min followed by 200 μ*M* DTT for 30 min at RT, 20 μ*M* MTSES for 5 min, 3 μg/ml BTX-AF488 for 10 min at 4 °C, 5 m*M* DTT for 5 min at RT, and then 3 μg/ml BTX-AF555 for 10 min at 4 °C before imaging over 60 min at 30–32 °C. *Source: (B) was originally published in Hannan et al. (2011). © The American Society for Biochemistry and Molecular Biology.* (For color version of this figure, the reader is referred to the online version of this chapter.)

- Incubate the cells in 5 m*M* DTT for 5 min at RT to remove the MTSES followed by washing in ice-cold Krebs to remove excess DTT.
- Incubate the cells in 3 μg/ml BTX-AF555 for 10 min at 4 °C to label $R1a^{BBS\text{-}CC}$ subunits. After washing in Krebs, load the coverslip onto the microscope for live cell imaging. Image both BTX-AF488 and

BTX-AF555 channels at different time points to capture the decays in membrane fluorescence. Use these decays to study the relative internalization profiles for the two receptors as previously outlined.

7. OVERVIEW

Overall, inserting a BBS into a membrane protein is a very useful, minimally invasive imaging strategy that can be applied to the study of internalization kinetics of GPCRs and other surface proteins in live cells. The fluorescence data are amenable to quantitatively analyzing the kinetics of receptor internalization from, and insertion into, the cell membrane. Furthermore, this technique can be adapted to study the fate of internalized receptors, and by modifying the BBS, it can also be extended to enable multi-labeling with different fluorophores in the same cells. This allows the internalization of two receptor subunits to be studied simultaneously.

REFERENCES

Ashby, M. C., Ibaraki, K., & Henley, J. M. (2004). It's green outside: Tracking cell surface proteins with pH-sensitive GFP. *Trends in Neurosciences*, *27*, 257–261.

Bettler, B., Kaupmann, K., Mosbacher, J., & Gassmann, M. (2004). Molecular structure and physiological functions of $GABA_B$ receptors. *Physiological Reviews*, *84*, 835–867.

Bogdanov, Y., Michels, G., Armstrong-Gold, C., Haydon, P. G., Lindstrom, J., Pangalos, M., et al. (2006). Synaptic $GABA_A$ receptors are directly recruited from their extrasynaptic counterparts. *The EMBO Journal*, *25*, 4381–4389.

Connolly, C. N., Krishek, B. J., McDonald, B. J., Smart, T. G., & Moss, S. J. (1996). Assembly and cell surface expression of heteromeric and homomeric γ-aminobutyric acid type A receptors. *The Journal of Biological Chemistry*, *271*, 89–96.

Corringer, P. J., Le Novere, N., & Changeux, J. P. (2000). Nicotinic receptors at the amino acid level. *Annual Review of Pharmacology and Toxicology*, *40*, 431–458.

Davies, D. R., Padlan, E. A., & Segal, D. M. (1975). Three-dimensional structure of immunoglobulins. *Annual Review of Biochemistry*, *44*, 639–667.

Eisele, J. L., Bertrand, S., Galzi, J. L., Devillers-Thiery, A., Changeux, J. P., & Bertrand, D. (1993). Chimeric nicotinic-serotonergic receptor combines distinct ligand binding and channel specificities. *Nature*, *366*, 479–483.

Grampp, T., Sauter, K., Markovic, B., & Benke, D. (2007). γ-Aminobutyric acid type B receptors are constitutively internalized via the clathrin-dependent pathway and targeted to lysosomes for degradation. *The Journal of Biological Chemistry*, *282*, 24157–24165.

Hannan, S., Wilkins, M. E., Dehghani-Tafti, E., Thomas, P., Baddeley, S. M., & Smart, T. G. (2011). γ-Aminobutyric acid type B ($GABA_B$) receptor internalization is regulated by the R2 subunit. *The Journal of Biological Chemistry*, *286*, 24324–24335.

Harel, M., Kasher, R., Nicolas, A., Guss, J. M., Balass, M., Fridkin, M., et al. (2001). The binding site of acetylcholine receptor as visualized in the X-ray structure of a complex between α-bungarotoxin and a mimotope peptide. *Neuron*, *32*, 265–275.

Love, R. A., & Stroud, R. M. (1986). The crystal structure of α-bungarotoxin at 2.5 A resolution: Relation to solution structure and binding to acetylcholine receptor. *Protein Engineering*, *1*, 37–46.

Meresse, S., Gorvel, J. P., & Chavrier, P. (1995). The rab7 GTPase resides on a vesicular compartment connected to lysosomes. *Journal of Cell Science, 108*(Pt. 11), 3349–3358.

Moise, L., Liu, J., Pryazhnikov, E., Khiroug, L., Jeromin, A., & Hawrot, E. (2010). K(V)4.2 channels tagged in the S1-S2 loop for α-bungarotoxin binding provide a new tool for studies of channel expression and localization. *Channels (Austin, Tex.), 4*, 115–123.

Ormo, M., Cubitt, A. B., Kallio, K., Gross, L. A., Tsien, R. Y., & Remington, S. J. (1996). Crystal structure of the Aequorea victoria green fluorescent protein. *Science, 273*, 1392–1395.

Sekine-Aizawa, Y., & Huganir, R. L. (2004). Imaging of receptor trafficking by using α-bungarotoxin-binding-site-tagged receptors. *Proceedings of the National Academy of Sciences of the United States of America, 101*, 17114–17119.

Tran-Van-Minh, A., & Dolphin, A. C. (2010). The alpha2delta ligand gabapentin inhibits the Rab11-dependent recycling of the calcium channel subunit α2δ-2. *The Journal of Neuroscience, 30*, 12856–12867.

Wilkins, M. E., Li, X., & Smart, T. G. (2008). Tracking cell surface $GABA_B$ receptors using an α-bungarotoxin tag. *The Journal of Biological Chemistry, 283*, 34745–34752.

Zerial, M., & McBride, H. (2001). Rab proteins as membrane organizers. *Nature Reviews. Molecular Cell Biology, 2*, 107–117.

CHAPTER SEVEN

Regulatory Mechanism of G Protein-Coupled Receptor Trafficking to the Plasma Membrane: A Role for mRNA Localization

Kusumam Joseph, Eleanor K. Spicer, Baby G. Tholanikunnel[1]

Department of Biochemistry and Molecular Biology, Medical University of South Carolina, Charleston, South Carolina, USA

[1]Corresponding author: e-mail address: tholanik@musc.edu

Contents

1. Introduction 132
2. Purification and Identification of β_2-AR mRNA-Binding Proteins 134
   2.1 Purification of β_2-AR mRNA-binding proteins 134
   2.2 Identification of β_2-AR mRNA-binding proteins purified by an RNA-affinity method 135
3. Functional Characterization of β_2-AR mRNA-Binding Proteins in Receptor Expression and Function 137
4. Role of RNA-Binding Protein HuR in Receptor Trafficking to the Plasma Membrane 139
5. β_2-AR mRNA Localization in Cells 141
6. Concluding Remarks and Future Perspectives 146

Acknowledgments 147

References 147

Abstract

Trafficking and localization of G protein-coupled receptors (GPCRs) to the plasma membrane and its retention in the agonist-naive state are critically important for signaling by these receptors. Agonist-induced desensitization of activated GPCRs and their removal from the cell surface have been studied and reviewed extensively. However, less attention has been given to the regulatory mechanisms and different steps that control the trafficking of newly synthesized receptors to the plasma membrane. It is generally believed that the mRNAs encoding GPCRs are targeted to the endoplasmic reticulum by a cotranslational, signal-sequence recognition particle-dependent pathway that results in protein translation and translocation to the plasma membrane. In this chapter, we discuss the importance of *cis*-targeting elements and *trans*-recognition factors in GPCR mRNA translational silencing, trafficking, and localization within the cell and its

Methods in Enzymology, Volume 521
ISSN 0076-6879
http://dx.doi.org/10.1016/B978-0-12-391862-8.00007-7

importance in receptor trafficking to the plasma membrane. Knockdown of the critical *trans*-recognition factors (RNA-binding proteins) resulted in translation of GPCR mRNAs in the perinuclear region and the receptors failed to traffic to the plasma membrane. Thus, a new paradigm is emerging in GPCR trafficking that suggests a fundamental role for mRNA partitioning to specific cytoplasmic regions for efficient plasma membrane localization of the receptors.

1. INTRODUCTION

G protein-coupled receptors (GPCRs) mediate physiological responses to external and internal stimuli that result in intracellular responses (Birnbaumer, Abramowitz, & Brown, 1990; Collins, Lohse, O'Dowd, Caron, & Lefkowitz, 1991; Dohlman, Thorner, Caron, & Lefkowitz, 1991; Malbon & Hadcock, 1993). Thus, receptor abundance and function are critical in normal and pathophysiological states. Many laboratories interested in studying the regulation of receptor expression in cell signaling have used β_2-adrenergic receptors (β_2-ARs) as prototypic GPCRs. Considerable research to date supports the value of β_2-ARs as a model system for the analysis of properties of GPCR superfamily members (Collins et al., 1991; Dohlman et al., 1991). Molecular cloning of hamster and human β_2-AR coding regions (Dixon et al., 1986; Kobilka, Dixon, et al., 1987; Kobilka, Frielle, et al., 1987) and mutagenesis studies have identified consensus sequences that are important in receptor ligand binding (Dixon et al., 1987), GPCR phosphorylation (Campbell et al., 1991; Hausdorff et al., 1991), downregulation, and agonist-mediated sequestration (Barak et al., 1994). Agonist-induced desensitization, sequestration, and downregulation of the GPCR comprise a group of regulatory mechanisms that have been studied extensively (Bunemann, Lee, Pals-Rylaarsdam, Roseberry, & Hosey, 1999; Pitcher, Freedman, & Lefkowitz, 1998).

Compared with the extensive studies performed on the events of the endocytic pathway, molecular mechanisms underlying the synthesis and transport of GPCRs from the endoplasmic reticulum (ER) to the cell surface are relatively less well understood (Achour, Labbe-Jullie, Scott, & Marullo, 2008; Conn, Ulloa-Aguirre, Ito, & Janovick, 2007; Duvernay, Filipeanu, & Wu, 2005). Since the discovery of the signal peptide by Blobel and colleagues (Blobel, 2000; Blobel & Dobberstein, 1975; Lingappa & Blobel, 1980), the targeting of most proteins to various subcellular destinations has been thought to occur after translation. According to this model, mRNAs encoding integral membrane proteins are targeted to the perinuclear ER for signal peptide synthesis. This ER-directed mRNA localization requires both translation of an encoded signal sequence and recognition of the signal sequences by the signal recognition particle (SRP). This event directs mRNA/ribosome/nascent polypeptide complexes to the ER membrane that enables cotranslational protein translocation. Once the protein is inserted into the membrane, the signal peptide is cleaved from the mature protein (Blobel, 2000).

β_2-AR and the vast majority of GPCRs lack a cleavable signal sequence (Guan, Kobilka, & Kobilka, 1992; Singer, 1990) and the molecular mechanisms that lead to the translation and translocation of the amino terminal and first membrane-spanning domain of these proteins are poorly understood. Addition of a cleavable signal peptide to some of the GPCRs (β_2-AR and α1D-adrenoceptor) enhanced the plasma membrane insertion and expression of functional receptors in transfected cells (Guan et al., 1992; Petrovska, Kapa, Klovins, Schioth, & Uhlen, 2005). However, a small group of GPCRs including endothelin B receptors contain cleavable signal peptides (Kochl et al., 2002). Removal of the signal peptide from the endothelin B receptors resulted in retention of the receptor in the ER, suggesting that this peptide is essential for translocation of the N-terminal tail of the receptor across the ER membrane (Kochl et al., 2002). These studies suggest that different GPCRs follow different mechanisms for their synthesis, transport, and integration into the plasma membrane. The above studies also suggest that the amino acid sequence of the encoded protein, rather than the nucleotide sequences of the mRNA, contains information necessary and sufficient for the protein localization to the plasma membrane.

In contrast, recent studies suggest that some mRNAs contain localization elements that govern their partitioning and that protein localization can be controlled by localizing the mRNA transcript prior to translation (Czaplinski & Singer, 2006; St Johnston, 2005). A recent study (Lecuyer et al., 2007) to comprehensively evaluate mRNA localization during early *Drosophila* embryogenesis showed a tight correlation between mRNA distribution and subsequent protein localization. The process of mRNA localization typically utilizes *cis*-targeting elements on the transcript and *trans*-recognition factors to direct specific translationally silenced ribonucleoprotein (RNP) particles to their cellular destinations (Czaplinski & Singer, 2006). Although the list of known localized mRNAs has grown significantly over the past several years (Lecuyer et al., 2007), the prevalence, variety, and importance of mRNA localization events are unknown. Thus, mRNA localization has primarily been considered important in specialized biological processes such as morphogen gradient formation and asymmetric cell division as it occurs in development (St Johnston, 2005).

Relatively little is known, however, about how and where the mRNAs encoding GPCRs are localized in intact cells and the importance of such mRNA localization events, if any, on protein trafficking to the plasma membrane. In the course of our studies on posttranscriptional regulation of β_2-AR, we found that 3′-UTR-mediated control mechanisms are critical in determining the number of receptors that a particular cell type can express (Kandasamy, Joseph, Subramaniam, Raymond, & Tholanikunnel, 2005; Subramaniam, Chen, Joseph, Raymond, & Tholanikunnel, 2004; Subramaniam, Kandasamy, Joseph, Spicer, & Tholanikunnel, 2011). Conservation of the β_2-AR sequences in different species extends beyond the coding region into the 5′- and 3′-UTRs, suggesting that these sequences are important in receptor expression and function (Kobilka, Frielle, et al., 1987; Nakada, Haskell, Ecker, Stadel, & Crooke, 1989; Parola & Kobilka, 1994; Tholanikunnel, Raymond, & Malbon, 1999). Using a highly conserved region of β_2-AR 3′-UTR RNA (Tholanikunnel & Malbon, 1997; Tholanikunnel et al., 1999) as an affinity ligand, we purified and identified multiple RNA-binding proteins that bind specifically to β_2-AR mRNA (Kandasamy et al., 2005). RNAi-mediated knockdown and overexpression of selected β_2-AR RNA-binding proteins identified critical roles for these proteins in receptor trafficking to the plasma membrane by regulating translation and mRNA localization (Tholanikunnel et al., 2010). In this chapter, we focus on the importance of RNA-binding proteins in translation-independent mRNA localization to the cortical ER in GPCR trafficking to the plasma membrane.

2. PURIFICATION AND IDENTIFICATION OF β_2-AR mRNA-BINDING PROTEINS

2.1. Purification of β_2-AR mRNA-binding proteins

We developed a novel RNA-affinity method to purify proteins in DDT_1-MF2 cells that bind to β_2-AR mRNA as detailed by Kandasamy et al. (2005).

1. A highly conserved 21-nt region (5′-CTTTTTTATTTTATTTTTTTA-3′) from the 3′-UTR of β_2-AR cDNA (Tholanikunnel & Malbon, 1997; Tholanikunnel et al., 1999) was cloned into the pSP70 (Promega) vector, downstream from the SP6 promoter. Plasmids containing the T-rich

elements were constructed by using complementary synthetic oligonucleotides flanked by restriction sequences for *Xho*I at the 5′-end and *Cla*I at the 3′-end. The double-stranded oligonucleotides were cloned into pSP70 to generate pSP70-3′-UTR.

2. The vector was linearized by *Cla*I and used for *in vitro* transcription using SP6 RNA polymerase to produce RNA containing the 21-nt U-rich region. The transcribed RNA was extracted with phenol and chloroform and then ethanol precipitated. About 800 μg of U-rich RNA was used as starting material for affinity purification of proteins.
3. The RNA was biotinylated using PHOTOPROBE (Long arm) biotin (Vector Laboratories, Burlingame, CA). Biotin and RNA were mixed and covalently coupled by exposure to light (365 nm, using a hand held UV-lamp) and unincorporated biotin was removed by extracting with 2-butanol. Biotinylated RNA was precipitated with ammonium acetate, washed with 70% ethanol, and dissolved in RNase-free TE (10 m*M* Tris, pH 8.0, 1 m*M* EDTA).
4. Cytosolic extracts were prepared from DDT_1-MF2 cells in hypotonic lysis buffer (10 m*M* Hepes (pH 7.6), 3 m*M* $MgCl_2$, 5 m*M* EDTA, 2 m*M* dithiothreitol, 2.5% glycerol, 0.5% nonidet P-40, 3 mg/ml heparin, 0.5 mg/ml yeast RNA, 40 m*M* KCl, and protease inhibitors) and were precleared to remove any proteins that bind directly to Avidin by incubating with VECTREX Avidin D (Vector laboratories) for 2 h at 4 °C. After this step, avidin beads were pelleted, and cytosolic extracts were recovered. RNase inhibitor was added (final concentration 0.35 units/μl) to the precleared extract, which was then mixed with biotinylated U-rich RNA for 2 h on a rotator at 4 °C. The binding mixture was then incubated with avidin for 2 h at 4 °C on a rotator.
5. The RNA–cytosolic mixture was transferred to a small column and the flow through was collected. The column matrix was washed three times (5 ml each) with lysis buffer containing 40 m*M* KCl. This was followed by two more washes (5 ml each) using the same buffer containing 300 m*M* KCl. Proteins bound to the RNA were eluted with lysis buffer containing 2.0 *M* KCl. All the fractions were dialyzed, concentrated, and then subjected to SDS-PAGE and silver stain analyses (Fig. 7.1A).

2.2. Identification of β_2-AR mRNA-binding proteins purified by an RNA-affinity method

A combination of methods including peptide sequencing, Western blot, and electrophoretic mobility shift assays combined with antibody supershift

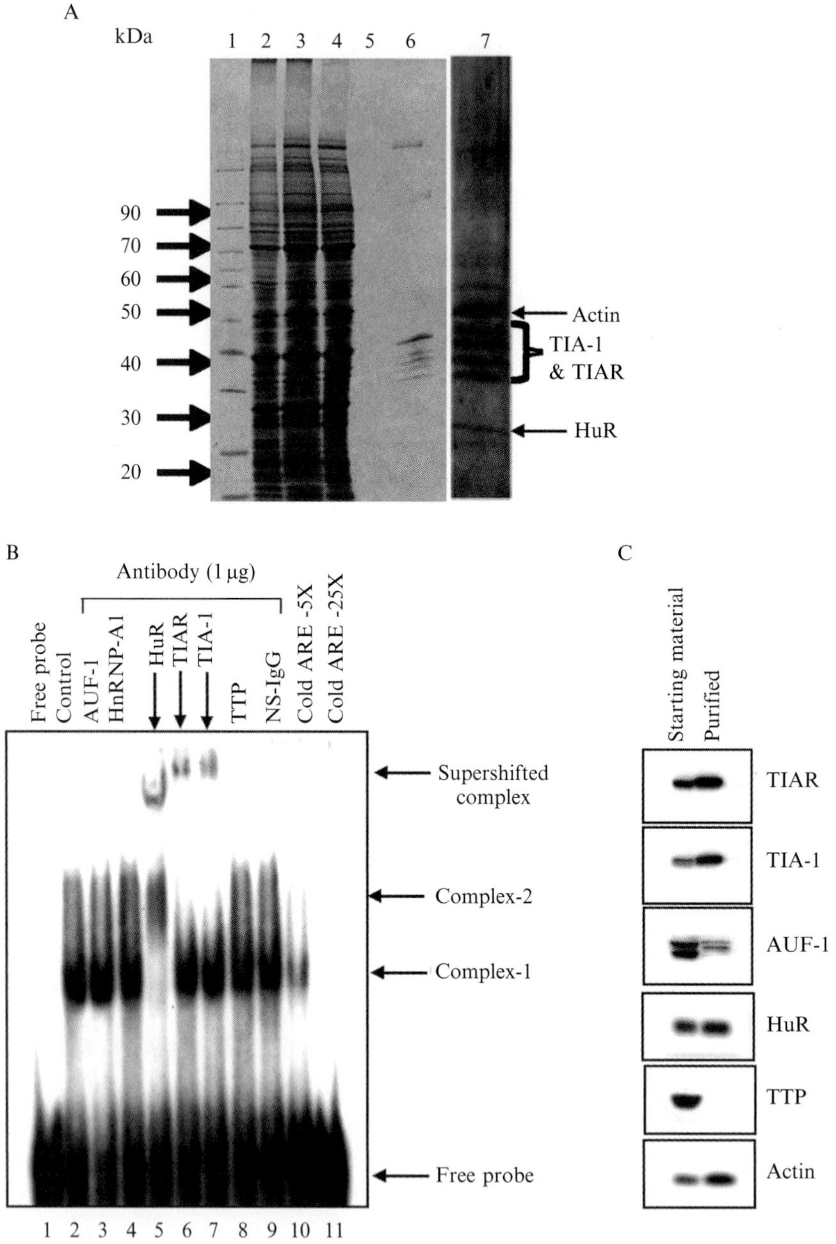

Figure 7.1 Purification and identification of β_2-AR mRNA-binding proteins. (A) SDS-PAGE and silver stain analysis of proteins from each step of the RNA-affinity purification. The protocol used for purification of β_2-AR mRNA-binding proteins from DDT_1-MF2 cells is described in the text. Lane 1, protein molecular weight ladder (Invitrogen). Lane 2,

assays were used to identify the individual proteins that were purified using the RNA-affinity method (Kandasamy et al., 2005). The major proteins that bound to the 3′-UTR of β_2-AR mRNA were identified as T-cell-restricted intracellular antigen-1 (TIA-1), its closely related homologue TIA-1-related protein (TIAR) and HuR. The identification was based on electrophoretic mobility shift and antibody supershift assays (Fig. 7.1B) combined with Western blot analysis (Fig. 7.1C). A distinct protein band in the molecular weight range of ~50 kDa was identified as actin by peptide sequence analysis and confirmed by Western blot (Fig. 7.1A and C).

3. FUNCTIONAL CHARACTERIZATION OF β_2-AR mRNA-BINDING PROTEINS IN RECEPTOR EXPRESSION AND FUNCTION

A combination of methods including RNA interference-mediated knockdown of the β_2-AR mRNA-binding proteins and/or overexpression of the RNA-binding proteins was used to determine the functional significance of the RNA-binding proteins in β_2-AR expression. In addition, we used a reporter gene (Luciferase) containing receptor 3′-UTR sequences to determine the functional significance of the 3′-UTR sequences and the 3′-UTR RNA-binding proteins in control of receptor expression (Kandasamy et al., 2005; Subramaniam et al., 2004, 2011; Tholanikunnel et al., 2010).

Previous studies indicated that deletion of the 3′-UTR results in a two- to threefold increase in receptor expression (Subramaniam et al., 2004), thus, we

cytosolic extract (starting material). Lane 3, precleared cytosolic extract. Lane 4, effluent from the RNA-avidin column. Lane 5, last column wash with binding buffer containing 300 m*M* KCl. Lane 6, proteins eluted with 2.0 *M* KCl, and lane 7, a concentrated aliquot of the affinity purified material, separately subjected to SDS-PAGE and silver stain analysis. (B) Identification of proteins purified by the RNA-affinity purification method by electrophoretic mobility shift and antibody supershift assays. Radiolabeled U-rich RNA (lane 1) were incubated with proteins purified by RNA affinity without (lane 2) or with antibodies raised against AUF-1 (lane 3), hnRNP-A1 (lane 4), HuR (lane 5), TIAR (lane 6), TIA-1 (lane 7), TTP (lane 8), or nonimmune serum (lane 9). Additions of unlabeled ARE RNA (fivefold and 25-fold excess) into samples abolished the gel-shifted complexes, indicating the specificity of the complex formed (lanes 10 and 11). Only antibodies against HuR, TIAR, and TIA-1 produced supershifted complexes. (C) Western blot analysis to confirm the presence of HuR, TIAR, and TIA-1 and actin in the RNA-affinity purified proteins. Cytosolic extract and purified proteins were subjected to SDS-PAGE and Western blot analysis using the antibodies against TIAR, TIA-1, AUF-1, HuR, TTP, and actin. *From Kandasamy et al. (2005).*

anticipated that the 3′-UTR sequences function as negative regulators of receptor expression. Further support for this hypothesis came from the finding that overexpression of the 3′-UTR of β_2-AR mRNA in two different cell lines that endogenously express β_2-AR (DDT$_1$-MF2 and A431 cells), resulted in a twofold increase in receptor expression. The increase in receptor expression was specific to exogenous 3′-UTR RNA sequences because expression of scrambled sequences did not alter receptor expression (Kandasamy et al., 2005). These results suggest that overexpression of β_2-AR 3′-UTR RNA can relieve translational inhibition of endogenous receptor mRNAs by competing for proteins that bind to the receptor 3′-UTR sequences. These decoy experiments also suggest that β_2-AR mRNA is translationally repressed by 3′-UTR sequence-binding proteins in cell lines that endogenously express β_2-AR (Kandasamy et al., 2005).

In addition, cotransfection studies showed that both TIAR and HuR inhibit receptor expression. Cotransfection of β_2-AR cDNAs with TIAR (Kandasamy et al., 2005) or HuR (Tholanikunnel et al., 2010) both resulted in significant reduction in β_2-AR expression that was dependent on the presence of 3′-UTR sequences on β_2-AR cDNA. Polysome profile analysis of β_2-AR mRNA using sucrose density gradient fractionation showed that a substantial amount of endogenously expressed β_2-AR mRNA is associated with low molecular weight fractions, suggesting that these mRNAs are inefficiently translated (Kandasamy et al., 2005; Tholanikunnel et al., 2010). Further, we observed a significant shift in β_2-AR mRNA to the top of sucrose density gradients when TIAR was overexpressed in DDT$_1$-MF2 and A431 cell lines, suggesting that TIAR can influence ribosome association and the rate of β_2-AR mRNA translation (Kandasamy et al., 2005). In addition, knockdown of the RNA-binding protein HuR resulted in a clear shift in β_2-AR mRNA to the heavier polyribosomal fractions suggesting increased translation (Tholanikunnel et al., 2010). These results strongly support the conclusion that the RNA-binding proteins TIAR and HuR negatively regulate translation of β_2-AR mRNA. Further, knockdown of HuR resulted in significant loss of β_2-AR mRNA, suggesting that HuR protein is critically important in maintaining steady-state levels of β_2-AR mRNA. Paradoxically, measurement of receptor expression in HuR knockdown cells showed a two- to threefold increase in receptor protein as compared to controls. Interestingly, inhibition of translation by treatment of DDT$_1$-MF2 cells with cycloheximide resulted in significant accumulation of β_2-AR mRNA in HuR knockdown cells as compared to controls (Tholanikunnel et al., 2010), suggesting that β_2-AR mRNA decay is linked to its translation.

4. ROLE OF RNA-BINDING PROTEIN HuR IN RECEPTOR TRAFFICKING TO THE PLASMA MEMBRANE

Pharmacological and functional properties of β_2-AR expressed in HuR knockdown and control cells were examined by ligand-binding studies using varying concentrations of ^{125}I-CYP. This ligand is cell permeable and measures receptors both on the cell surface and in the cytoplasm. These studies revealed that receptors expressed in HuR knockdown cells have significantly lower affinity for ^{125}I-CYP compared to cells expressing control shRNA (Kusumam Joseph and Baby Tholanikunnel, unpublished results). The lower binding affinity of β_2-AR in HuR knockdown cells as compared to controls was unexpected. To further characterize the properties of receptors expressed in HuR knockdown cells, immunofluorescence staining of β_2-AR and HuR proteins was performed (Tholanikunnel et al., 2010). The β_2-AR staining patterns showed that receptors failed to traffic to the plasma membrane in HuR knockdown cells and appeared instead around the nucleus (Fig. 7.2A, arrows). In cells with significant immunoreactivity for HuR, the receptors preferentially appeared on the plasma membrane (Fig. 7.2A, arrowhead) with no accumulation of receptors around the nucleus. Examination of cells with and without HuR within the same microscopic field confirmed that the accumulation of receptors around the nucleus was specific to HuR knockdown cells (Fig. 7.2A). These results suggest that HuR-mediated translational suppression of β_2-AR mRNA plays a critical role in receptor trafficking to the plasma membrane.

Because knockdown of HuR resulted in overproduction of receptors, it was necessary to rule out the possibility that failed trafficking of β_2-AR in HuR knockdown cells are a result of receptor overexpression perhaps by overwhelming the protein trafficking pathway. This possibility was ruled out by transfecting DDT_1-MF2 cells using full-length and 3′-UTR deleted β_2-AR cDNAs. Such transfection using full-length β_2-AR cDNA resulted in three- to fourfold increase in receptor expression and the receptors were plasma membrane localized (Fig. 7.2B and C). On the contrary, transfection using 3′-UTR deleted β_2-AR cDNA resulted in significant accumulation of receptors around the nucleus (Tholanikunnel et al., 2010). These results suggest a critical role for the 3′-UTR sequences and the RNA-binding protein HuR in β_2-AR trafficking and cell surface expression.

The immunofluorescence staining and confocal studies showing receptor localization to the perinuclear region in HuR knockdown cells were

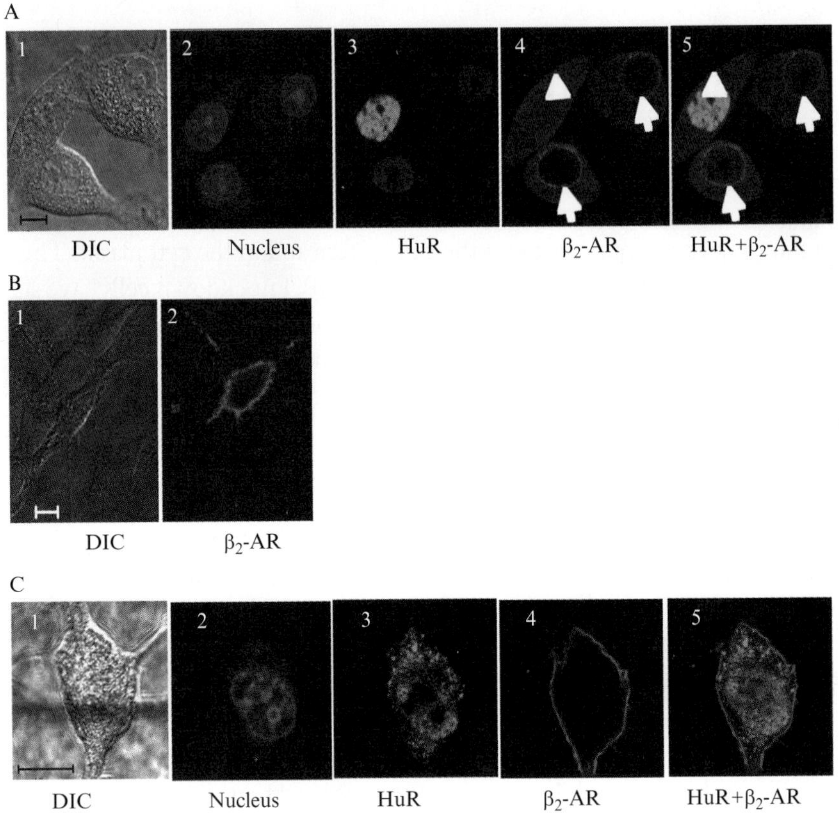

Figure 7.2 Knockdown of the β_2-AR mRNA-binding protein HuR resulted in defective trafficking of receptors to the plasma membrane in DDT_1-MF2 cells. (A) Confocal microscopy images of DDT_1-MF2 cells show immunofluorescence staining of HuR (green) and β_2-AR (red) in HuR knockdown and control cells. The presence of cells with (arrowhead) and without HuR (arrow) was chosen within the same microscopic field to provide an internal control. The accumulation of receptors around the nucleus is seen only in HuR knockdown cells. (B) β_2-AR when overexpressed can traffic to the plasma membrane. Confocal microscopy images show immunofluorescence staining of β_2-AR (red) in DDT_1-MF2 cells transfected with full-length β_2-AR cDNA. (C) Magnified view of a single cell overexpressing β_2-AR cDNA. Confocal microscopy was performed using a Zeiss LSM510META laser-scanning microscope (Carl Zeiss, Inc.). *From Tholanikunnel et al. (2010).* (See Color Insert.)

further confirmed by subcellular fractionation of ^{125}I-CYP binding activity in HuR knockdown and control cells. Isolation of heavy (plasma membrane) and light membrane fractions from HuR knockdown and control cells showed that in control cells more than 80% of the radioligand binding activity was present in the heavy plasma membrane fraction, whereas in HuR knockdown cells less than 40% activity was present in the plasma membrane fraction and more than 60% ligand-binding activity was associated with the light membrane fraction (Tholanikunnel et al., 2010). In addition, HuR knockdown and the appearance of receptors around the nucleus significantly reduced the amount of cAMP induced in response to β_2-AR agonist isoproterenol treatment (Tholanikunnel et al., 2010).

5. β_2-AR mRNA LOCALIZATION IN CELLS

A simple explanation for increased β_2-AR synthesis and its retention around the nucleus in HuR knockdown cells may be the specific roles for this RNA-binding protein in β_2-AR mRNA translational silencing and localization. A well-accepted paradigm of translation-independent mRNA localization is that specific RNA-binding proteins associate with specific mRNA in the nucleus, rendering the mRNA–protein complex translationally inactive when the complex reaches the cytoplasm (Czaplinski & Singer, 2006). Although predominantly nuclear, HuR can shuttle between the nucleus and the cytoplasm (Fan & Steitz, 1998; Keene, 1999), and binding of this protein can variably affect the translational processing of transcripts (Antic & Keene, 1998). Fluorescent *in situ* hybridization (FISH) analysis to localize endogenous mRNA in DDT_1-MF2 cells showed that β_2-AR mRNA is localized to the peripheral cytoplasmic regions of DDT_1-MF2 cells. Also, immunofluorescence staining of HuR protein combined with FISH analysis of β_2-AR mRNA revealed colocalization of cytoplasmic HuR protein with receptor mRNA (Fig. 7.3A). β_2-AR mRNA was restricted to the peripheral cytoplasmic region of DDT_1-MF2 cells. Of particular interest regarding the distribution of β_2-AR mRNA and the nucleocytoplasmic shuttling protein HuR in DDT_1-MF2 cells is the absence of HuR in the bulk of the cytoplasmic regions, but its prominent immunoreactivity in the peripheral cytoplasmic regions of the cells and colocalization with β_2-AR mRNA (Fig. 7.3A, arrow).

β_2-AR mRNA localization by FISH analysis using HuR knockdown cells revealed that the knockdown of HuR decreased β_2-AR mRNA levels significantly and the mRNA failed to traffic to the cell periphery and

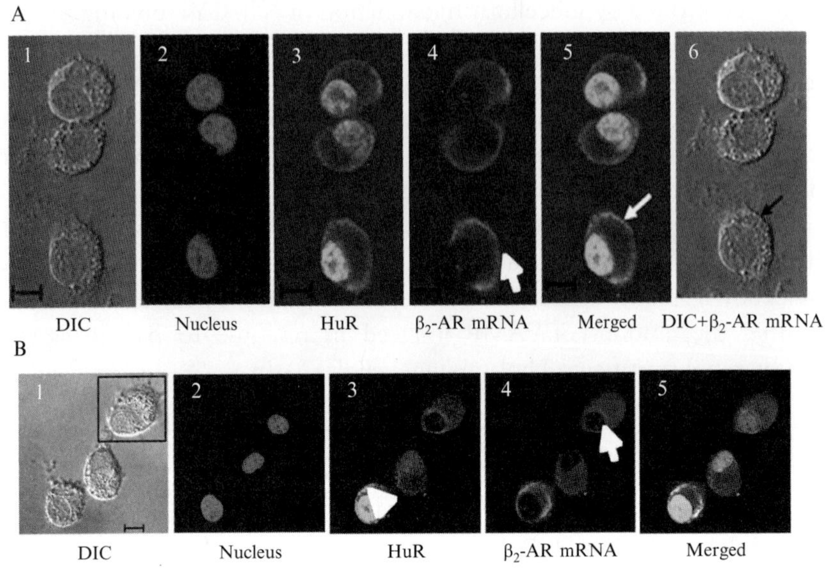

Figure 7.3 β_2-AR mRNA and cytoplasmic HuR are colocalized to the cell periphery and knockdown of HuR results in defective trafficking of receptor mRNA. (A) Confocal images of DDT_1-MF2 cells with fluorescent *in situ* hybridization analysis using digoxigenin-labeled riboprobes directed against β_2-AR mRNA and immunofluorescence staining of HuR protein. Prehybridization and hybridization was performed as described in Tholanikunnel et al. (2010). The RNA-probe signal was amplified using biotinylated antidigoxin followed by streptavidin coupled to Cy3. (B) Knockdown of HuR resulted in decreased levels of β_2-AR mRNA that failed to traffic to plasma membrane and appeared around the nucleus. HuR knockdown (inset) and control cells are shown within the same microscopic field to provide an internal control. *From Tholanikunnel et al. (2010).* (See Color Insert.)

appeared around the nucleus (Fig. 7.3B, arrow). Cells that retained significant quantities of HuR showed higher levels of β_2-AR mRNA that appeared to traffic to the cell periphery (Fig. 7.3B, arrowhead). Thus, HuR protein seems to play a critical role in β_2-AR mRNA stabilization, translational repression, and transport to the cell periphery.

Based on these above studies, we propose that the assembly of the β_2-AR mRNA–HuR protein complex occurs in the nucleus, making the complex translationally inactive when the mRNA reaches the cytoplasm and continued binding of HuR protein prevents translation until β_2-AR mRNA is localized to the peripheral cytoplasmic regions. Because HuR is essential for translational silencing, the knockdown of this protein results in polyribosome association of β_2-AR mRNA upon nuclear exit and translation initiation, making β_2-AR mRNA trafficking and localization problematic. Thus, the appearance of receptors around the nucleus in HuR knockdown cells may be the result of failed mRNA trafficking and premature translational initiation immediately upon nuclear exit. The results of immunofluorescence staining and FISH analysis in control and HuR knockdown cells emphasize the spatiotemporal relationship between β_2-AR mRNA localization and translation and its importance in receptor localization to the plasma membrane.

According to the current model of protein trafficking (Blobel, 2000), the partitioning of the integral membrane protein-encoding mRNA to the ER occurs via a cotranslational, signal sequence/SRP-dependent mechanism. However, most of the GPCRs lack a cleavable signal sequence and have a multiple membrane-spanning topology with their amino terminus on the extracellular side of the membrane (Singer, 1990). In the absence of a cleavable signal sequence, the membrane insertion of these proteins is assisted by internal signal sequences which are inserted in the membrane by the same mechanism that operates for cleavable signal sequences, except that there is no post-insertion cleavage (Blobel, 2000). Thus, the mRNA localization to the ER is dependent on both translation of the mRNA and recognition of these signal sequences by the SRP. Based on this model, it was assumed that the amino acid sequence of the encoded protein, rather than the nucleotide sequence of the RNA, contains the information necessary and sufficient for protein localization to the appropriate sites.

However, a recent study showed that some mRNAs coding for secretory and integral membrane proteins contain localization determinants that can govern their partitioning to the ER membrane in the absence of the signal sequence/SRP pathway (Pyhtila et al., 2008). This suggests that mRNA coding for membrane proteins can be localized to the ER in a translation-dependent or -independent manner (Cohen, 2005; Pyhtila et al., 2008). In the former, it is the amino acid sequence of the signal peptide that determines the localization of mRNA to the ER. On the contrary, translation-independent mRNA localization is RNA based, and thus the information necessary for the localization is contained within the RNA sequence itself (Cohen, 2005). The RNA-based localization pathways require translational silencing mostly facilitated by RNA-binding proteins that recognize mRNAs in a sequence-specific manner. Thus, there are a set of specific proteins for each pathway, although the localization machinery also uses a set of proteins that are common for both pathways, such as the motor proteins (e.g., kinesin) and anchors (actin and tubulin filaments). Relevant to this, it is notable that we isolated cytoskeletal proteins actin, tubulin, and kinesin (Kusumam Joseph and Baby Tholanikunnel, unpublished results) using the 3′-UTR of β_2-AR mRNA as an RNA-affinity ligand.

On the basis of the results obtained from β_2-AR mRNA localization and translational control studies, we propose a new model (Fig. 7.4). Based on this model, GPCR mRNA undergoes direct translation-independent localization to the cell periphery. Of particular interest in this model is the separation of the two processes involving GPCR mRNA localization and protein translocation. These findings describe the first example, to our knowledge, of the importance of translational silencing and mRNA targeting

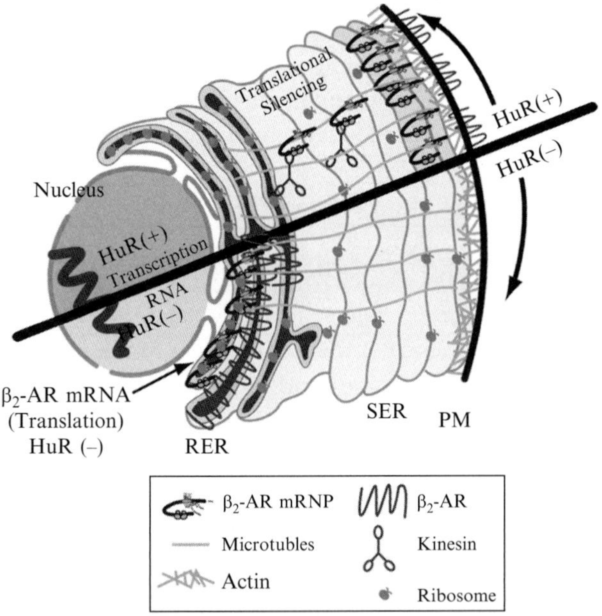

Figure 7.4 Proposed model for β_2-AR mRNP complex formation, its transport and localization to the cell periphery. β_2-AR mRNA is recognized in the nucleus by the nucleocytoplasmic shuttling RNA-binding protein HuR. Upon export, the β_2-AR mRNA–HuR protein complex associates with additional RNA-binding proteins such as Staufen and cytoskeletal elements (actin, tubulin, and kinesin) and is transported to the cell periphery. Continued association of HuR protein silences translational initiation while chaperoning the mRNP complex to the plasma membrane (upper half). The mechanisms involved in β_2-AR mRNA dissociation from TIAR, HuR, and other RNA-binding proteins and translational activation are topics of current interest. When HuR expression is downregulated (lower half) β_2-AR mRNA translation is initiated in perinuclear ER leading to overproduction of receptors but defective trafficking to the plasma membrane. *From Tholanikunnel et al. (2010).* (See Color Insert.)

in GPCR plasma membrane targeting. *In situ* hybridization analysis showed that endogenously expressed β_2-AR mRNA is localized to the peripheral cytoplasmic regions of smooth muscle cells (DDT$_1$-MF2 cells). This provided microscopic evidence that mRNA encoding GPCRs can be localized to the cortical ER network in cells. We also identified HuR as one of the critical RNA-binding proteins that are necessary for translational silencing and mRNA trafficking.

The group of proteins that we isolated by RNA-affinity purification using a conserved region of the 3′-UTR of β_2-AR mRNA also contained the well-characterized double-stranded RNA-binding protein staufen (Kusumam Joseph and Baby Tholanikunnel, unpublished results). Staufen has been identified as a component of the *Oskar* mRNA transport particle and is required for the delivery of this mRNA to the posterior pole during *Drosophila* development (St Johnston, 2005). In addition, staufen has been implicated in the transport of mRNA in mammalian neurons (Macchi et al., 2003). Although these findings suggest that staufen may be a general component of RNA transport particles, a newly identified role of this RNA-binding protein in mRNA decay that is linked to its translation (Kim, Furic, Desgroseillers, & Maquat, 2005; Kim et al., 2007) suggests that staufen protein is capable of participating in two distinct pathways involving mRNA localization and mRNA decay (Meyer & Gavis, 2005). Further work is required to identify the specific roles for staufen protein in β_2-AR mRNA trafficking and its decay.

Long-lasting synaptic plasticity requires the delivery of integral membrane proteins including GPCRs to the postsynaptic membrane (Lledo, Zhang, Sudhof, Malenka, & Nicoll, 1998). The presence of mRNAs encoding integral membrane proteins in dendrites (Bi, Tsai, Lu, Loh, & Wei, 2007) raises the question of how these mRNAs are transported through the ER to the dendrites. Our observation of β_2-AR mRNA localization to the periphery of smooth muscle cells and to the leading edge of growth cones (Tholanikunnel et al., 2010) suggest that GPCR mRNAs can be localized to the distant regions like the leading edge of growth cones through association with specific RNA-binding proteins. Extensive work carried out on pathways of mRNA localization in *Drosophila* (Palacios, 2007; St Johnston, 2005) and *Xenopus* (Cohen, 2005; King, Messitt, & Mowry, 2005) strongly suggest the possibility that mRNA localization and translation necessitates cotrafficking of the translational machinery including the ER (Gerst, 2008). In addition, cell fractionation and detection of RNA in the extrasomatic regions of neurons also support the idea of ER–mRNA cotrafficking (Gerst, 2008). Moreover, it is reported that ER can form flattened sheets and tubules that start from the nuclear periphery and extend not only to the peripheral regions of the cells but can also extend into areas of polarized growth and into the leading edge of motile cells (Shibata, Voeltz, & Rapoport, 2006). Thus, both the movement of mRNA and formation of cortical ER involve dynamic regulation of the cytoskeleton and actin and a microtubule-based transport systems (Gerst, 2008). Using β_2-AR 3′-UTR, actin, tubulin, and kinesin were copurified along with the RNA-binding proteins HuR and staufen, suggesting that cotrafficking of HuR and staufen along with actin and tubulin assisted by kinesin may be involved in β_2-AR mRNA trafficking.

6. CONCLUDING REMARKS AND FUTURE PERSPECTIVES

Localization of GPCR to the cell surface is critically important for signaling by these receptors. However, a number of studies suggest that GPCRs are inefficiently expressed at the plasma membrane (Conn et al., 2007; Guan et al., 1992; Petaja-Repo et al., 2002; Petrovska et al., 2005). Among these, the best-studied example is the gonadotropin-releasing hormone receptor (GnRHR) (Conn, Knollman, Brothers, & Janovick, 2006; Conn et al., 2007; Janovick et al., 2006). The inefficient plasma membrane expression of GnRHR in humans is the result of progressive and convergent evolutionary trends (Janovick et al., 2006) that can bring regulatory advantages to cells (Conn et al., 2006). In other studies, it has been shown that addition (Guan et al., 1992; Petrovska et al., 2005) or deletion (Dunham & Hall, 2009) of amino acid sequences promoted surface expression and functional activity of some GPCRs. Similarly, treatment with agonist or antagonists can facilitate plasma membrane expression of δ opioid receptors (Petaja-Repo et al., 2002). Thus, it appears that inefficient plasma membrane expression is more prevalent in many members of the GPCR family.

Our findings provide evidence for a new regulatory pathway involving RNA-binding proteins in the synthesis and transport of GPCRs to the plasma membrane. Thus, several lines of evidence suggest that different GPCRs use distinct pathways for synthesis, transport, and cell surface expression of the receptors. While inefficient plasma membrane expression and endoplasmic retention of GPCRs can provide a mechanism to produce new receptors without new transcriptional and translational activity (Conn et al., 2007), the pathway involving translational silencing and mRNA localization in the cortical ER can provide cells with a pool of GPCR mRNAs that are ready to be translated. Currently, the mechanisms involved in translational derepression of membrane-localized β_2-AR mRNA are not understood. Further elucidation of the molecular mechanisms involved in translational derepression of β_2-AR mRNA and the associated pathways can provide a foundation for the development of new therapeutic strategies with clinical relevance. It is possible that membrane-associated kinases may control the phosphorylation of RNA-binding proteins resulting in their dissociation from the mRNA and translational activation, as has been suggested for β-actin mRNA (Huttelmaier et al., 2005). If so, then membrane-associated kinases may provide an important regulatory step in GPCR synthesis. Our results also suggest that cortical ER localized microdomains within a particular cell surface can regulate complex signaling events including GPCR mRNA localization, translational derepression, posttranslational modification, and the delivery of mature proteins to the plasma membrane. It is not clear how extensively the pathways involving translational silencing and mRNA localization are used by other members of the GPCR family.

Since GPCRs are valuable drug targets, identification of novel compounds that can act as modulators of GPCR expression may lead to new therapeutic approaches. Compounds currently available are agonists, antagonists, or allosteric modulators of GPCRs (Dunham & Hall, 2009). Identification of RNA-binding proteins with important roles in receptor mRNA stabilization, trafficking, and translation provides an opportunity to design new drugs that are RNA based or that target RNA regulatory proteins.

ACKNOWLEDGMENTS

This work was supported by grants from the National Institute of Health (GM 58740 to B. G. T.) and the American Heart Association grants (0555470U to B. G. T. and 0765356U to K. J.). Imaging facilities for this research were supported, in part, by Cancer Center Support Grant P30 CA138313 to the Hollings Cancer Center, Medical University of South Carolina.

REFERENCES

Achour, L. O., Labbe-Jullie, C., Scott, M. G. H., & Marullo, S. (2008). An escort for GPCRs: Implications for regulation of receptor density at the cell surface. *Trends in Pharmacological Sciences*, *29*, 528–535.

Antic, D., & Keene, J. D. (1998). Messenger ribonucleoprotein complexes containing human ELAV proteins: Interactions with cytoskeleton and translational apparatus. *Journal of Cell Science*, *111*, 183–197.

Barak, L. S., Tiberi, M., Freedman, N. J., Kwatra, M. M., Lefkowitz, R. J., & Caron, M. G. (1994). A highly conserved tyrosine residue in G protein-coupled receptors is required for agonist-mediated beta2-adrenergic receptor sequestration. *The Journal of Biological Chemistry*, *269*, 2790–2795.

Bi, J., Tsai, N. P., Lu, H. Y., Loh, H. H., & Wei, N. (2007). Copb1-facilitated axonal transport and translation of kappa opioid-receptor mRNA. *Proceedings of the National Academy of Sciences of the United States of America*, *104*, 13810–13815.

Birnbaumer, L., Abramowitz, J., & Brown, A. M. (1990). Receptor-effector coupling by G proteins. *Biochimica et Biophysica Acta, 1031*, 163–224.

Blobel, G. (2000). Protein targeting (Nobel lecture). *Chembiochem, 1*, 86–102.

Blobel, G., & Dobberstein, B. (1975). Transfer of proteins across membranes. 1. Presence of proteolytically processed and unprocessed nascent immunoglobulin light chains on membrane-bound ribosomes of murine myeloma. *The Journal of Cell Biology, 67*, 835–851.

Bunemann, M., Lee, K. B., Pals-Rylaarsdam, R., Roseberry, A. G., & Hosey, M. M. (1999). Desensitization of G-protein-coupled receptors in the cardiovascular system. *Annual Review of Physiology, 61*, 169–192.

Campbell, P. T., Hnatowich, M., O'Dowd, B. F., Caron, M. G., Lefkowitz, R. J., & Hausdorff, W. P. (1991). Mutations of the human beta 2-adrenergic receptor that impair coupling to Gs interfere with receptor down-regulation but not sequestration. *Molecular Pharmacology, 39*, 192–198.

Cohen, R. S. (2005). The role of membranes and membrane trafficking in RNA localization. *Biology of the Cell, 97*, 5–18.

Collins, S., Lohse, M. J., O'Dowd, B., Caron, M. G., & Lefkowitz, R. J. (1991). Structure and regulation of G protein-coupled receptors: The beta 2-adrenergic receptor as a model. *Vitamins and Hormones, 46*, 1–39.

Conn, P. M., Knollman, P. E., Brothers, S. P., & Janovick, J. A. (2006). Protein folding as posttranslational regulation: Evolution of a mechanism for controlled plasma membrane expression of a G protein-coupled receptor. *Molecular Endocrinology, 12*, 3035–3041.

Conn, P. M., Ulloa-Aguirre, A., Ito, J., & Janovick, J. A. (2007). G protein-coupled receptor trafficking in health and disease: Lessons learned to prepare for therapeutic mutant rescue in vivo. *Pharmacological Reviews, 59*, 225–250.

Czaplinski, K., & Singer, R. H. (2006). Pathways for mRNA localization in the cytoplasm. *Trends in Biochemical Sciences, 31*, 687–693.

Dixon, R. A., Kobilka, B. K., Strader, D. J., Benovic, J. L., Dohlman, H. G., Frielle, T., et al. (1986). Cloning of the gene and cDNA for mammalian beta-adrenergic receptor and homology with rhodopsin. *Nature, 321*, 75–79.

Dixon, R. A., Sigal, I. S., Rands, E., Register, R. B., Candelore, M. R., Blake, A. D., et al. (1987). Ligand binding to the beta-adrenergic receptor involves its rhodopsin-like core. *Nature, 326*, 73–77.

Dohlman, H. G., Thorner, J., Caron, M. G., & Lefkowitz, R. J. (1991). Model systems for the study of seven-transmembrane-segment receptors. *Annual Review of Biochemistry, 60*, 653–688.

Dunham, J. H., & Hall, R. A. (2009). Enhancement of the surface expression of G protein-coupled receptors. *Trends in Biotechnology, 27*, 541–545.

Duvernay, M. T., Filipeanu, C. M., & Wu, G. (2005). The regulatory mechanisms of export trafficking of G protein-coupled receptors. *Cellular Signalling, 17*, 1457–1465.

Fan, X. C., & Steitz, J. A. (1998). Overexpression of HuR, a nuclear-cytoplasmic shuttling protein, increases the *in vivo* stability of ARE-containing mRNAs. *The EMBO Journal, 17*, 3448–3460.

Gerst, J. E. (2008). Message on the web: mRNA and ER co-trafficking. *Trends in Cell Biology, 18*, 68–76.

Guan, X.-M., Kobilka, T. S., & Kobilka, B. K. (1992). Enhancement of membrane insertion and function in a type IIIb membrane protein following introduction of a cleavable signal peptide. *The Journal of Biological Chemistry, 267*, 21995–21998.

Hausdorff, W. P., Campbell, P. T., Ostrowski, J., Yu, S. S., Caron, M. G., & Lefkowitz, R. J. (1991). A small region of the beta-adrenergic receptor is selectively involved in its rapid regulation. *Proceedings of the National Academy of Sciences of the United States of America, 88*, 2979–2983.

Huttelmaier, S., Zenklusen, D., Lederer, M., Dictenberg, J., Lorenz, M., Meng, X., et al. (2005). Spatial regulation of beta-actin translation by Src-dependent phosphorylation of ZBP1. *Nature, 438*, 512–515.

Janovick, J. A., Knollman, P. E., Brothers, S. P., Ayala-Yanez, R., Aziz, A. S., & Conn, P. M. (2006). Regulation of G protein-coupled receptor trafficking by inefficient plasma membrane expression. *The Journal of Biological Chemistry, 281*, 8417–8425.

Kandasamy, K., Joseph, K., Subramaniam, K., Raymond, J. R., & Tholanikunnel, B. G. (2005). Translational control of beta2-adrenergic receptor mRNA by T-cell-restricted intracellular antigen-related protein. *The Journal of Biological Chemistry, 280*, 1931–1943.

Keene, J. D. (1999). Why is Hu where? Shuttling of early-response-gene messenger RNA subsets. *Proceedings of the National Academy of Sciences of the United States of America, 96*, 5–7.

Kim, Y. K., Furic, L., Desgroseillers, L., & Maquat, L. E. (2005). Mammalian Staufen1 recruits Upf1 to specific mRNA 3'-UTRs so as to elicit mRNA decay. *Cell, 120*, 195–208.

Kim, Y. K., Furic, L., Parisien, M., Major, F., DesGroseillers, L., & Maquat, L. E. (2007). Staufen1 regulates diverse classes of mammalian transcripts. *The EMBO Journal, 26*, 2670–2681.

King, M. L., Messitt, T. J., & Mowry, K. L. (2005). Putting RNAs in the right place at the right time: RNA localization in the frog oocyte. *Biology of the Cell, 97*, 19–33.

Kobilka, B. K., Dixon, R. A., Frielle, T., Dohlman, H. G., Bolanowski, M. A., Sigal, I. S., et al. (1987). cDNA for the human beta 2-adrenergic receptor: A protein with multiple membrane-spanning domains and encoded by a gene whose chromosomal location is shared with that of the receptor for platelet-derived growth factor. *Proceedings of the National Academy of Sciences of the United States of America, 84*, 46–50.

Kobilka, B. K., Frielle, T., Dohlman, H. G., Bolanowski, M. A., Dixon, R. A., Keller, P., et al. (1987). Delineation of the intronless nature of the genes for the human and hamster beta 2-adrenergic receptor and their putative promoter regions. *The Journal of Biological Chemistry, 262*, 7321–7327.

Kochl, R., Alken, M., Rutz, C., Krause, G., Oksche, A., Rosenthal, W., et al. (2002). The signal peptide of the G protein-coupled human endothelin B receptor is necessary for translocation of the N-terminal tail across the endoplasmic reticulum membrane. *The Journal of Biological Chemistry, 277*, 16131–16138.

Lecuyer, E., Yoshida, H., Parthasarathy, N., Alm, C., Babak, T., Cerovina, T., et al. (2007). Global analysis of mRNA localization reveals a prominent role in organizing cellular architecture and function. *Cell, 131*, 174–187.

Lingappa, V. R., & Blobel, G. (1980). Early events in the biosynthesis of secretory and membrane proteins: The signal hypothesis. *Recent Progress in Hormone Research, 36*, 451–475.

Lledo, P. M., Zhang, X., Sudhof, T. C., Malenka, R. C., & Nicoll, R. A. (1998). Postsynaptic membrane fusion and long-term potentiation. *Science, 279*, 399–403.

Macchi, P., Kroening, S., Palacios, I. M., Baldassa, S., Grunewald, B., Ambrosino, C., et al. (2003). Barentsz, a new component of the staufen-containing ribonucleoprotein particles in mammalian cells, interacts with staufen in an RNA-dependent manner. *The Journal of Neuroscience, 23*, 5778–5788.

Malbon, C., & Hadcock, J. (1993). Agonist regulation of gene expression of adrenergic receptors and G proteins. *Journal of Neurochemistry, 60*, 1–9.

Meyer, E. L., & Gavis, E. R. (2005). Staufen does double duty. *Nature Structural and Molecular Biology, 12*, 291–292.

Nakada, M. T., Haskell, K. M., Ecker, D. J., Stadel, J. M., & Crooke, S. T. (1989). Genetic regulation of beta 2-adrenergic receptors in 3T3-L1 fibroblasts. *The Biochemical Journal, 260*, 53–59.

Palacios, I. M. (2007). How does an mRNA find its way? Intracellular localization of transcripts. *Seminars in Cell & Developmental Biology, 18*, 163–170.

Parola, A. L., & Kobilka, B. K. (1994). The peptide product of a 5'leader cistron in the beta2-adrenergic receptor mRNA inhibits receptor synthesis. *The Journal of Biological Chemistry*, *269*, 4497–4505.

Petaja-Repo, U. E., Hogue, M., Bhalla, S., Laperriere, A., Morello, J. P., & Bouvier, M. (2002). Ligands act as pharmacological chaperones and increase the efficiency of delta opioid receptor maturation. *The EMBO Journal*, 7, 1628–1637.

Petrovska, R., Kapa, I., Klovins, J., Schioth, H. B., & Uhlen, S. (2005). Addition of a signal peptide sequence to the alpha1D-adrenoceptor gene increases the density of receptors, as determined by (3H)-prazosin binding in the membranes. *British Journal of Pharmacology*, *144*, 651–659.

Pitcher, J. A., Freedman, N. J., & Lefkowitz, R. J. (1998). G protein-coupled receptor kinases. *Annual Review of Biochemistry*, *67*, 653–692.

Pyhtila, B., Zheng, T., Lager, P. J., Keene, J. K., Reedy, M. C., & Nicchitta, C. V. (2008). Signal sequence- and translation-independent mRNA localization to the endoplasmic reticulum. *RNA*, *14*, 445–453.

Shibata, Y., Voeltz, G. K., & Rapoport, T. A. (2006). Rough sheets and smooth tubules. *Cell*, *126*, 435–439.

Singer, S. J. (1990). The structure and insertion of integral proteins in membranes. *Annual Review of Cell and Developmental Biology*, *6*, 247–296.

St Johnston, D. (2005). Moving messages: The intracellular localization of mRNAs. *Nature Reviews. Molecular Cell Biology*, *6*, 363–375.

Subramaniam, K., Chen, K., Joseph, K., Raymond, J. R., & Tholanikunnel, B. G. (2004). The 3'-untranslated region of the beta2-adrenergic receptor mRNA regulates receptor synthesis. *The Journal of Biological Chemistry*, *279*, 27108–27115.

Subramaniam, K., Kandasamy, K., Joseph, K., Spicer, E. K., & Tholanikunnel, B. G. (2011). The 3'-untranslated region length and AU-rich RNA location modulate RNA-protein interaction and translational control of β_2-adrenergic receptor mRNA. *Molecular and Cellular Biochemistry*, *352*, 125–141.

Tholanikunnel, B. G., Joseph, K., Kandasamy, K., Baldys, A., Raymond, J. R., Luttrell, L. M., et al. (2010). Novel mechanisms in the regulation of G protein-coupled receptor trafficking to the plasma membrane. *The Journal of Biological Chemistry*, *285*, 33816–33825.

Tholanikunnel, B. G., & Malbon, C. C. (1997). A 20-nucleotide (A + U)-rich element of beta2-adrenergic receptor (beta2-AR) mRNA mediates binding to beta2-AR-binding protein and is obligate for agonist-induced destabilization of receptor mRNA. *The Journal of Biological Chemistry*, *272*, 11471–11478.

Tholanikunnel, B. G., Raymond, J. R., & Malbon, C. C. (1999). Analysis of the AU-rich elements in the 3'-untranslated region of beta 2-adrenergic receptor mRNA by mutagenesis and identification of the homologous AU-rich region from different species. *Biochemistry*, *38*, 15564–15572.

CHAPTER EIGHT

Dissecting Trafficking and Signaling of Atypical Chemokine Receptors

Elena Borroni[*,†], **Cinzia Cancellieri**[*,†], **Massimo Locati**[*,†], **Raffaella Bonecchi**[*,†,1]

[*]Humanitas Clinical and Research Center, Rozzano, Italy
[†]Department of Medical Biotechnologies and Translational Medicine, Università degli Studi di Milano, Rozzano, Italy
[1]Corresponding author: e-mail address: raffaella.bonecchi@humanitasresearch.it

Contents

1. Introduction 152
2. ACR Trafficking 153
   2.1 Antibody feeding and cytofluorimetric analysis of D6 internalization 153
   2.2 Confocal microscopy analysis of Rabs-D6 colocalization 155
   2.3 Transient transfection of dominant negative Rab proteins 157
3. Atypical Chemokine Receptor Signaling 158
   3.1 Analysis of Gαi-protein activation by HTRF technology 159
   3.2 Silencing of β-Arrestin1/2 by siRNA technology 161
   3.3 Immunoblotting analysis of Erk1/2 phosphorylation level 162
   3.4 Small G protein activation assay: Affinity precipitation assay using GST-PBD 164
4. Summary 166
References 167

Abstract

Atypical chemokine receptors are a distinct subset of chemokine receptors able to modulate immune responses by acting as chemokine decoy/scavengers or transporters. Intracellular trafficking properties sustained by Gαi-independent signaling have emerged as a major determinant of their biological properties, which support continuous uptake, transport, and/or concentration, of the ligands. Here, we are providing methods to study both trafficking and signaling of this class of chemokine receptors focusing on the atypical chemokine receptor D6 that degrades inflammatory CC chemokines.

Methods in Enzymology, Volume 521
ISSN 0076-6879
http://dx.doi.org/10.1016/B978-0-12-391862-8.00008-9

1. INTRODUCTION

Chemokines are a large family of chemotactic cytokines. Although the major function of chemokines is the coordination of leukocyte recruitment in physiological and pathological conditions, they also mediate other biological activities, including regulation of cell differentiation and proliferation, survival, and senescence (Bonecchi et al., 2009; Charo & Ransohoff, 2006). They are structurally characterized by a conserved protein structure called "chemokine scaffold," strictly dependent on two conserved disulfide bonds connecting conserved cysteine residues whose relative position determines the identification of four subfamilies. CC chemokines, which have the first two cysteine residues adjacent, and CXC chemokines, which have the cysteine residues separated by a single intervening amino acid, account together for the large majority of members (25 and 15, respectively). The C subfamily, with two members have a single cysteine residue in the amino-terminus, and the CX3C chemokine, with three residues separating the cysteine tandem, account for the minority (Murphy et al., 2000).

This structure-based classification of the ligands is reflected in the classification of chemokine receptors, which display, in most cases, significant ligand promiscuity among members of a defined subfamily, but are strictly restricted to members of that given subfamily. Thus, the 10 CC chemokine receptors (CCR1–10) and the 6 CXC chemokine receptors (CXCR1–6) only recognize CC and CXC chemokines, respectively. Similarly, the only receptor for C chemokines (XCR1) and the only receptor for the CX3C chemokine (CX3CR1) are strictly restricted to their respective ligand (Murphy et al., 2000).

Chemokines' biological activities are regulated at several levels. Depending on their expression, chemokines can be distinguished as "homeostatic" chemokines, which control leukocyte homing and lymphocyte recirculation in normal conditions, and "inflammatory" chemokines, produced in response to inflammatory and immune stimuli (Bonecchi et al., 2009). Chemokines are also targets of posttranslational modifications that influence their functional properties, including processing at the amino- and carboxyl-termini by proteases and modifications such as citrullination (Mortier, Van Damme, & Proost, 2008). Finally, a nonredundant role in tuning chemokine biological properties is mediated by their clearance and transport operated by atypical chemokine receptors (ACRs) (Bonecchi, Savino, Borroni, Mantovani, & Locati, 2010). ACRs are a distinct subset of chemokine receptors able to modulate immune responses by acting as chemokine decoy/scavengers or possibly transporters (Mantovani, Bonecchi, & Locati, 2006). These nonconventional chemokine receptors include DARC (Pruenster & Rot, 2006), D6 (Locati et al., 2005), CCRL2 (Zabel et al., 2008), CCX CKR (Comerford, Milasta, Morrow, Milligan, & Nibbs, 2006), and CXCR7 (Boldajipour et al., 2008). Intracellular trafficking properties sustained by Gαi-independent signaling have emerged as a major determinant of ACR biological properties, which support continuous uptake, transport, and/or concentration of the ligand (Borroni & Bonecchi, 2009; Borroni, Bonecchi, Mantovani, & Locati, 2009). Here, we are providing methods to study both trafficking and signaling of one member of this family named D6.

2. ACR TRAFFICKING

Constitutive intracellular trafficking has been proposed as general mechanisms used by receptors, and in particular by scavenger receptors, to rapidly regulate their expression to cope with the immediate needs of the tissue (Prevo, Banerji, Ni, & Jackson, 2004; Schaer et al., 2006). Consistent with this, constitutive ligand-independent internalization and recycling through Rab-dependent pathways have been demonstrated for the ACRs with scavenger function D6 (Bonecchi et al., 2004, 2008; Weber et al., 2004), CCXCKR (Comerford et al., 2006), CXCR7 (Luker, Gupta, Steele, Foerster, & Luker, 2009; Naumann et al., 2010). In this section, we are providing procedures for studying D6 constitutive trafficking by two techniques, fluorescence-activated cell sorting (FACS) and confocal microscopy. All procedures have been performed in the Chinese hamster ovary (CHO)-K1 cell line stable transfected with pcDNA3-expressing human D6, obtained as previously described (Fra et al., 2003).

2.1. Antibody feeding and cytofluorimetric analysis of D6 internalization

Antibody feeding is an immunofluorescence technique that provides a faster analysis of receptor constitutive and ligand-dependent internalization by FACS compared to the traditional biochemical approach based on the covalent modification of surface receptors with biotin (Arancibia-Carcamo, Fairfax, Moss, Kittler, 2006). This technique involves the use of antibodies that recognize receptor-specific extracellular epitopes that can be endogenous signals or engineered extracellular tag on a recombinant receptor. The antibody will label only receptors on the membrane and will follow the receptor until the lysosome if the receptor is targeted to degradation (Fig. 8.1A). If the receptor is targeted to recycling pathways and if the receptor–antibody complex will be stable in the acidic compartments, this technique can be modified to study receptor recycling (Arancibia-Carcamo et al., 2006).

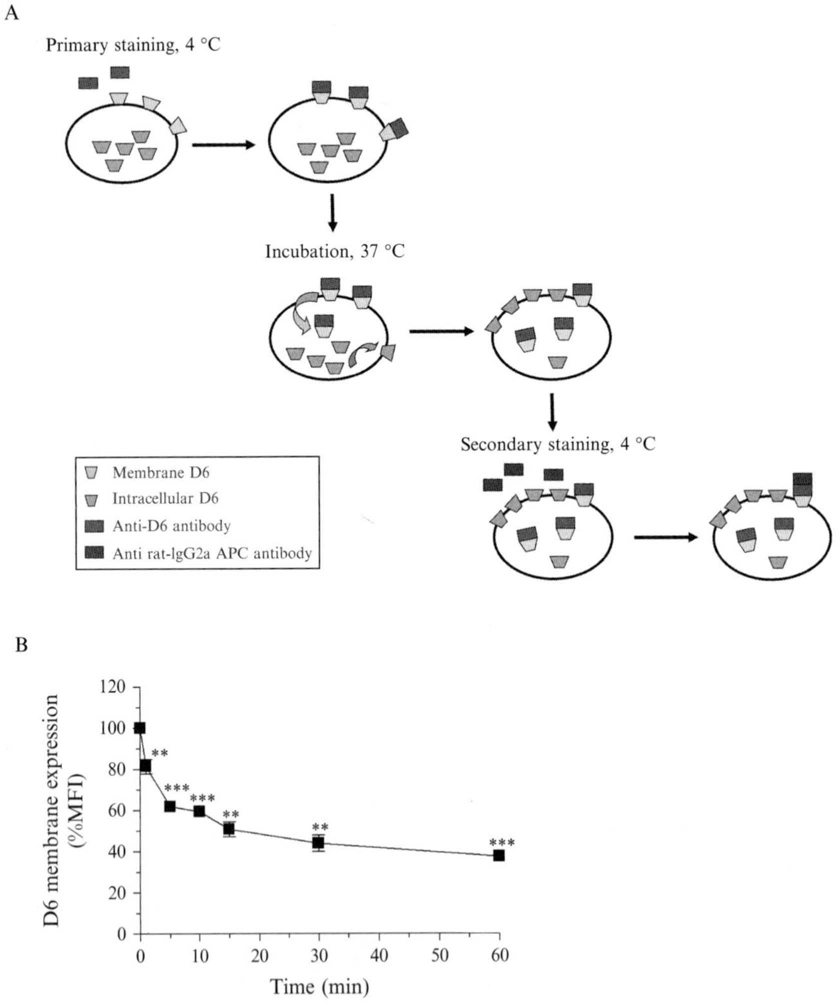

Figure 8.1 Cytofluorimetric analysis of D6 constitutive internalization. (A) Schematic representation of antibody feeding experiment. (B) Quantification of D6 internalization. Data are expressed as percentage of MFI of D6 membrane expression at the indicated time points over $t=0$. Results are mean ± SEM from at least three independent experiments performed. (For color version of this figure, the reader is referred to the online version of this chapter.)

To measure D6 constitutive internalization, the following protocol is used:

1. CHO-K1/D6 cells (5×10^5) are detached by trypsin and washed twice with 500 μl of ice-cold phosphate buffer saline (PBS) (Biosera, East Sussex, UK) + 1% bovin serum albumin (BSA) (GE Healthcare Europe GmbH, Milan, Italy).
2. Cells are resuspended in 100 μl of ice-cold PBS + 1% BSA and stained at 4 °C to inhibit receptor endocytosis with 5 μg/ml of rat anti-hD6 primary antibody or rat IgG2a isotype control (R&D Systems, Minneapolis, MN).
3. Labeled cells are collected after centrifugation at 4 °C and then incubated for increasing time points at 37 °C in 100 μl of prewarmed DMEM/F12 (Cambrex, East Rutherford, New Jersey) + 1% BSA + 25 m*M* HEPES (Gibco, Life Technologies Ltd., Paisley, UK) to allow receptor internalization.
4. Samples are returned to ice to stop endocytosis, washed twice with 500 μl of ice-cold FACS buffer (PBS + 1% BSA + 0.1% NaN_3), and labeled at 4 °C with 4 μg/ml of goat anti-rat 647-conjugated secondary antibody (Molecular Probes, Life Technologies Ltd., Paisley, UK) in 100 μl of ice-cold FACS buffer for 45 min, in order to detect receptors remaining at the cell surface. NaN_3 (sodium azide) is purchased by Merck Chemicals (Darmstadt, Germany).
5. Cells are washed twice with 500 μl of ice-cold FACS buffer and resuspended in 300 μl of FACS fixation buffer: PBS + 1% formaldehyde (Sigma-Aldrich, St. Louis, MO).
6. For each experiment, 10^4 events of viable cells identified by FSC-H and SSC-A parameters are acquired by BD FACSCanto II flow cytometer and analyzed by BD FACSDiva software (BD Biosciences, Franklin Lakes, NJ).
7. The percent of D6 remaining on the cell membrane is calculated from the mean fluorescence intensity (MFI) values as 100 × [MFI (37 °C sample) − MFI (37 °C isotype control)/MFI (4 °C sample) − MFI (4 °C isotype control)]. Figure 8.1B shows quantification of D6 constitutive internalization performed in CHO-K1/D6 cells.

2.2. Confocal microscopy analysis of Rabs-D6 colocalization

Rabs are small GTPases that cycle between GDP-bound (inactive) and GTP-bound (active) states and regulate a number of cellular trafficking events (Neel, Schutyser, Sai, Fan, & Richmond, 2005). Once internalized, receptors are targeted to early endosomes by a process regulated by the small GTPases dynaminII (Dyn) and Rab5 and can be either delivered to the degradative pathway through Rab7 interaction or recycled to the cell surface via a "rapid" pathway involving Rab4 or a "slow" one involving Rab11-positive recycling endosomes (Zerial & McBride, 2001). We set up an immunofluorescence procedure to characterize D6 constitutive internalization and recycling pathways through the analysis of D6-Rabs colocalization by confocal microscopy, described as follows:

1. CHO-K1/D6 cells (1×10^5) are seeded onto Ø 14 mm glass dishes in 24-well plate (BD Biosciences) and grown in 500 μl of DMEM/F12 + 10% fetal cow serum (FCS) (Euroclone, Milan, Italy) + 25 m*M* HEPES + 100 U/ml penicillin/streptomycin (P/S) (Cambrex) at 37 °C for 18 h.
2. Cells are fixed with 4% paraformaldehyde (Merck Chemicals) in PBS for 15 min at room temperature (RT).
3. Fixed cells are washed twice with 500 μl of PBS + 2% BSA and permeabilized with 300 μl PBS + 2% BSA + 0.3% Triton X-100 + 0.1% Glycine (Merck Chemicals) + 5% normal goat serum (Dako, Glostrup, Denmark) for 1 h at RT.
4. Permeabilized cells are extensively washed with 500 μl of washing buffer (PBS + 0.2% BSA + 0.05% Tween 20), incubated in 300 μl of washing buffer with 1 μg/ml of rat anti-hD6 primary antibody for 2 h at RT. Tween 20 is purchased from Merck Chemicals.
5. After washing three times with 500 μl of washing buffer, coverslips are incubated in 300 μl of washing buffer with one of the following antibodies: 2.5 μg/ml of rabbit anti-hRab11 (Invitrogen, Life Technologies Ltd., Paisley, UK), mouse anti-hRab5 or Rab4 (BD Biosciences), 2 μg/ml rabbit anti-hRab7 (Cell Signaling Technology, Beverly, MA) for 1 h at RT.
6. Cells are washed three times with 500 μl of washing buffer and incubated in 300 μl of washing buffer for 1 h at RT with 2 μg/ml of Alexa Fluor secondary antibodies (Molecular Probes), goat anti-rat 594-conjugated to detect D6, goat anti-rabbit 647-conjugated to detect Rab7 and Rab11 or and goat anti-mouse 488-conjugated and Rab4/5. Samples are kept in dark from now on.
7. Stained cells are extensively washed with 500 μl of sterile water (Baxter, Deerfield, IL) and incubated with 300 μl of 300 n*M* DAPI (Invitrogen) for 5 min at RT.
8. After extensive washing in sterile water, specimens are mounted on glass slide (Menzel-Gläser, Thermo Fisher Scientific, Waltham, MA) with 20 μl FluoSave reagent (Calbiochem, Merck4Biosciences) and kept in

dark at RT for 24 h. High-resolution images (1024 × 1024 pixels) are acquired sequentially with a 60 × 1.4 NA. Plan-Apochromat oil immersion objective by using a FV1000 laser scanning confocal microscope (Olympus, Hamburg, Germany). Differential Interference Contrast (DIC, Nomarski technique) is also used. Images are assembled and cropped by Photoshop software (Adobe Systems, San Jose, CA). Quantitative colocalization and statistical analysis are performed by Imaris Coloc software (version 4.2, Bitplane AG, Zurich, Switzerland), FV1000 1.6 colocalization software (Olympus), and ImageJ (free available on http://rsbweb.nih.gov/ij/). D6 colocalization with Rabs is quantified by colocalization volume (CV) and Pearson's Coefficient of Correlation (PCC) parameters and is performed inside a selected region of interest per image, representative of the analyzed cell. For statistic, colocalization analysis is performed on at least 100 cells representative of each experimental condition.

2.3. Transient transfection of dominant negative Rab proteins

Short-term expression of dominant negative Rab proteins is commonly utilized to determine the relative contributions of endocytosis and recycling pathways to their biological activities. Here, we describe a method to generate Rab-dominant negative-transfected CHO-K1/D6 cells through transient transfection of pEGFP-tagged version of Rabs and check D6 expression by FACS described as follows:

1. CHO-K1/D6 cells (4×10^5) are seeded into 6-well plate (BD Biosciences) in 2 ml of DMEM/F12 + 10% FCS + 25 m*M* HEPES without antibiotics (to avoid reduction in transfection efficiency) at 37 °C for 18 h.
2. 50–70% confluent cells are transiently transfected with the pEGFP-tagged dominant negative plasmid Rab4 N121I according to Lipofectamine 2000 procedure as briefly reported in steps 3–6 (Invitrogen).
3. 4 μg of plasmid are diluted in 250 μl of Opti-MEM I Reduced Serum Medium (Invitrogen).
4. Lipofectamine 2000 solution is gently mixed and 10 μl are added to 250 μl of Opti-MEM I. Diluted solution is incubated at RT for 5 min.
5. After incubation, diluted plasmid is combined with diluted Lipofectamine 2000, mixed gently, and incubated for 20–30 min at RT to allow the plasmid–Lipofectamine 2000 complexes to form.

6. Complexes (500 μl total volume) are added carefully drop by drop to each well containing cells and medium. Suspension is mixed gently by rocking the plate back and forth and incubated at 37 °C in a CO_2 incubator for 24 h.
7. Transfected cells are collected through trypsinization (Lonza, Basel, Switzerland) and counted by ADAM-MC automatic cell counter (Digital Bio, Boston, MA). 5×10^5 cells are resuspended in 100 μl of FACS buffer and stained on ice with 5 μg/ml of rat anti-hD6 primary antibody or rat IgG2a isotype control for 1 h.
8. Samples are washed once with 500 μl of FACS buffer and incubated at 4 °C with 4 μg/ml of goat anti-rat 647-conjugated secondary antibody in 100 μl of ice-cold FACS buffer for 45 min.
9. Cells are washed twice with 500 μl of ice-cold FACS buffer and resuspended in 300 μl of fixation buffer.
10. For each experiment, 10^4 events of viable cells identified by FSC-H and SSC-A parameters are acquired by BD FACSCanto II Flow Cytometer and analyzed by BD FACSDiva software (BD Biosciences). Analysis of D6 expression is evaluated in two gates named R1 and R2 which referred to the viable pEGFP^{neg} (untransfected) and pEGFP^{high} (highly transfected) cells, respectively, as shown in Fig. 8.2A. Histograms represent the flow cytometric profiles of D6 expression in gated cell population (light gray = R1 gate and dark gray = R2 gate). Data were expressed as percentage of MFI of R2 over R1 gated population (Fig. 8.2B).

3. ATYPICAL CHEMOKINE RECEPTOR SIGNALING

Upon chemokine receptor engagement, both G protein coupling and β-arrestins recruitment trigger the activation of intracellular signaling cascades that activate signal transduction and prompt directional migration toward the chemokine source (Thelen & Stein, 2008). On the contrary, ACRs show modifications of the conserved DRY motif, which impair the canonical signaling activities typically observed after chemokine receptor triggering and required for cell migration (Bonecchi et al., 2010). Interestingly, it has been reported that some ACRs, including CXCR7 and C5L2, are capable to signaling in Gαi-independent fashion through recruitment and activation of β-arrestin (Rajagopal et al., 2010; Van Lith, Oosterom, Van Elsas, & Zaman, 2009). In this section, we provide the optimal conditions and procedures for studying and dissecting D6 signaling pathways through molecular

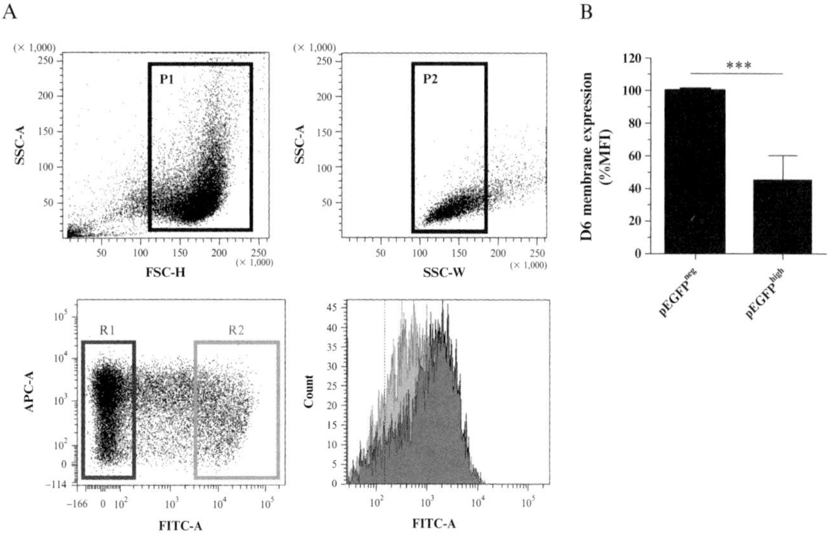

Figure 8.2 Generation of Rabs-dominant negative transient transfected CHO-K1/D6 cells. FACS analysis of CHO-K1/D6 transiently transfected with pEGFP-tagged Rab4 N121I. (A) Based on forward and side scatter parameters, P1 and P2 gates are drawn to exclude dead cells (SSC-A vs. FSC-H) and doublets (SSC-A vs. SSC-W) prior to analysis. R1 and R2 gates referred to the viable $pEGFP^{neg}$ and $pEGFP^{high}$ cells, respectively, in which D6 expression was evaluated. Histograms represent flow cytometric profiles of D6 expression in gated cell population (dark gray = R1 gate and light gray = R2 gate). Dashed lines indicate isotype-matched control mAb. (B) Quantification of D6 membrane expression. Data are expressed as percentage of MFI of R2 over R1 gated population. Results are mean ± SEM from at least three independent experiments performed.

(gene silencing, FRET) and biochemical (immunoprecipitation, immunoblotting) techniques both in CHO-K1 and human embryonic kidney 293 cells (HEK293) stably transfected with pcDNA3-expressing human D6.

3.1. Analysis of Gαi-protein activation by HTRF technology

Recently, a novel high-throughput system to quantify G protein activation based on the homogenous time-resolved fluorescence (HTRF) technique has been developed by CisBio International (Camarillo, CA; 62 AM1PEC). Briefly, HTRF combines standard FRET technology that is based on the transfer of energy between two fluorophores (donor: europium cryptate; acceptor: d2), when in close proximity, with time-resolved measurement of fluorescence, allowing elimination of short-lived background fluorescence. To assess Gαi-protein activation, HTRF assays have been developed for the direct quantitative determination of cyclic AMP based on a competitive immunoassay between native cAMP produced by cells and visualized by a monoclonal antibody anti-cAMP labeled with cryptate (Em: 620 nm), and the cAMP labeled with the dye d2 (Em: 665 nm). We set up a protocol to quantify Gαi-protein activation upon CCL3L1 stimulation in CHO-K1/D6 cells using cAMP cell-based assays (two-step protocol) described as follows:

1. Cells are seeded in 100 μl of DMEM/F12 + 10% FCS + 25 m*M* HEPES + 100 U/ml of P/S at a density of 5×10^4 into 96-well black plates (Greiner Bio-One, Frickenhausen, Germany) previously coated with 100 μl/well of poly-lysine (Sigma-Aldrich) for 30–40 min at 37 °C.
2. After 18 h at 37 °C, cells are washed twice with 100 μl/well of Tris-Krebs (TK) buffer (136 m*M* NaCl + 5 m*M* KCl + 1.2 m*M* $MgCl_2$ + 2.5 m*M* $CaCl_2$ + 10 m*M* glucose + 20 m*M* Trizma base, pH 7.4) and 25 μl of compound buffer (C-buffer: TK buffer + 0.5 m*M* 3-isobutyl-1-methylxanthine (IBMX) to prevent cAMP degradation) are added to each well. Chemicals for buffers preparation are purchased from Merck Chemicals.
3. 25 μl of cAMP standard curve (average range of 0.17–712 n*M* cAMP/well) and cAMP positive control are prepared according to manufacturer's instructions and added to cell-free wells in duplicate. cAMP negative control is done by adding in duplicate 25 μl/well of diluent provided by the kit.
4. 200 n*M* of CCL3L1 (R&D Systems) and 60 μ*M* of forskolin (Sigma-Aldrich) are prepared in C-buffer and 12.5 μl of each compound are added simultaneously in wells corresponding to stimulated cells in triplicate (final concentration of CCL3L1: 100 n*M* and forskolin: 30 μ*M* in a final volume of 25 μl/well). 12.5 μl of C-buffer are added in wells corresponding to unstimulated and negative control cells in triplicate. Twenty-five microliters of C-buffer are added to cAMP standard curve, positive and negative control wells. Plate is gently mixed and incubated in dark for 1 h at RT.
5. cAMP-d2 and anti-cAMP cryptate are prepared according to manufacturer's instructions (for 1 plate: 600 μl of each compound are diluted in 2.4 ml of cAMP and cGMP conjugate and lysis buffer provided by the kit). First, 25 μl of cAMP-d2 are added to all wells with the exception of the ones corresponding to cAMP and cells negative controls. Subsequently, 25 μl of anti-cAMP cryptate are added to all wells and plate is sealed and incubated in dark for 1 h at RT, after gently shaking.
6. Plate is read on the compatible HTRF reader Mithras LB 940 (Berthold Technologies, Bad Wildbad, Germany). Results are calculated from the

665 nm/620 nm ratio × 10^4 (R) and expressed in Delta F% obtained by the following equation: (standard or sample R—negative R)/(negative R) × 100. Standard curve is drawn up by plotting Delta F% versus cAMP concentration, and Delta F% obtained for samples can be reported on the standard curve to deduce respective cAMP concentrations.

3.2. Silencing of β-Arrestin1/2 by siRNA technology

It has become increasingly evident that β-arrestins activity is not only restricted to chemokine receptor desensitization and internalization, but also these adaptor proteins are emerging as signal transducers, able to provide diversity and fine-tuning possibilities to both canonical and ACR signaling activities (Shenoy & Lefkowitz, 2011). The C-terminal sequence of the chemokine receptor D6 constitutively interacts with β-arrestin due to the presence of phosphorilable serine clusters (Galliera et al., 2004). β-Arrestin is required to control D6 stability and recycling properties (McCulloch et al., 2008). Gene silencing by small-interfering RNA (siRNA) technology is particularly suited and widely used to study the role of β-arrestins on chemokine receptors and here we provide a protocol for β-arrestins siRNA optimization in HEK293/D6 described as follows:

1. Cells (5 × 10^5) are seeded into 6-well plate in 2 ml of DMEM + 10% FCS + 25 m*M* HEPES without antibiotics at 37 °C for 18 h.
2. 50–70% confluent cells are transiently transfected with ON-TARGETplus siRNA (Dharmacon, Thermo Scientific Pierce, Rockford, IL) for β-arrestin1, β-arrestin2, and control pool (scramble) according to Lipofectamine RNAiMax procedure as briefly reported in steps 3–6 (Invitrogen).
3. 10× concentrated (corresponding to 500 n*M*) siRNA are diluted in 250 μl of Opti-MEM I. Lipofectamine RNAiMax solution is gently mixed and 7.5 μl are added to 250 μl of Opti-MEM I. Diluted solution is incubated at RT for 5 min.
4. After incubation, diluted siRNA are combined with diluted Lipofectamine RNAiMax, mixed gently and incubated for 20 min at RT to allow the siRNA-Lipofectamine RNAiMax complexes to form.
5. Complexes (500 μl total volume) are added carefully drop by drop to each well containing cells and medium. Suspension is mixed gently by rocking the plate back and forth and incubated at 37 °C in a CO_2 incubator for 72 h.
6. Silenced cells are collected through trypsinization, counted by ADAM-MC automatic cell counter and ready for further experiments. Silencing

ratio of β-arrestin1/2 knocking down is checked by Western blotting. Briefly, cells are lysed for 5 min with 250 μl of ice-cold lysis buffer (50 m*M* Tris–HCl (pH 7.2) + 5 m*M* EDTA + 150 m*M* NaCl + 15 m*M* $MgCl_2$ + 1% Triton + 10% glycerol) freshly prepared and supplemented with protease inhibitors tablet EDTA-free (Roche, Basel, Switzerland) and 100 μg/ml phenylmethylsulfonyl fluoride (PMSF). After 4 °C centrifugation at 13,000 rpm for 20 min, cell supernatants are quantified by DC protein assay (Bio-Rad, Hercules, CA) and 30 μg of total cell lysate is loaded in 10% polyacrilamyde gel. Samples are transferred to polyvinylidene fluoride (PVDF) membrane (Bio-Rad) for 1 h at 100 V. Membrane is blocked with TBS-T (TBS + 0.1% Tween 20. TBS: 25 m*M* Tris–HCl (pH 7.5) + 150 m*M* NaCl) + 5% nonfat dry milk at RT for 1–2 h, rinsed with TBS-T for 5 min and then incubated overnight at 4 °C with 1:1000 dilution in TBS-T + 5% BSA of rabbit anti-hβ-arrestin1/2 primary antibody (Cell Signaling Technology). Membrane is washed five times for 5 min with 10 ml of TBS-T and incubated for 1 h at RT with 1:2000 dilution in TBS-T + 5% BSA of anti-rabbit IgG-horseradish peroxidase-conjugated secondary antibody (Dako). Membrane is washed five times for 5 min with 10 ml of TBS-T and incubated with chemiluminescent HRP substrate (Immobilon Western Millipore, Billerica, MA) for 5 min under gentle mixing. Blot is acquired by ChemiDoc XRS Imaging System (Bio-Rad). Densitometric analysis is performed by Quantity One software (Bio-Rad) and the silencing ratio is calculated as percentage of β-arrestin1/2 band (50 kDa) intensity in silenced cells over scramble-treated cells. Chemicals for TBS and lysis buffer preparation are purchased from Merck Chemicals.

3.3. Immunoblotting analysis of Erk1/2 phosphorylation level

One of the downstream effectors of GPCRs signaling is represented by the family of MAPKs, which regulates cell proliferation and differentiation. These serine/threonine protein kinases are capable of phosphorylating several transcription factors, thereby regulating subsequent transcriptional events, as reported for many chemokine receptors, including CCR5 (Paruch et al., 2007), and for the atypical chemokine receptor CXCR7 (Kumar et al., 2012). There are at least three subtypes of MAPK, the extracellular signal-regulated kinases (ERKs), c-Jun NH2-terminal kinases (JNKs), and p38 MAPK. In this section, we provide detailed description of the protocol set up to study phosphorylation of the Ser/Thr residues of Erk1/2 by Western blotting analysis in CHO-K1/D6 and CCR5 upon CCL3L1 stimulation.

1. Cells (2×10^5) are seeded into 12-well plate (BD Biosciences) in 1 ml of DMEM/F12 + 10% FCS + 25 m*M* HEPES + 100 U/ml of penicillin/streptomycin.
2. Cells are serum-starved by replacing medium with 1 ml of DMEM/F12 + 0.1% BSA + 25 m*M* HEPES + 100 U/ml P/S for 18 h. Cells should reach 70–80% confluence at day of CCL3L1 stimulation.
3. Before starting, medium is replaced with 360 μl of fresh DMEM/F12 + 0.1% BSA + 25 m*M* HEPES + 100 U/ml P/S. By using the same medium, CCL3L1 is prepared at 10 × (corresponding to 1 μ*M*), for a final concentration of 100 n*M*. Lysis buffer (50 m*M* Tris–HCl (pH 8) + 150 m*M* NaCl + 1.5 m*M* $MgCl_2$ + 5 m*M* EDTA + 1% Triton X-100 + 10% glycerol) is freshly prepared and finally supplemented with protease inhibitors tablet EDTA-free, 100 μg/ml PMSF and the phosphatase inhibitors: 0.01 *M* sodium pyrophosphate, 50 m*M* sodium fluoride (NaF), and 0.01 *M* sodium orthovanadate (Na_3VO_4) to neutralize phosphatases action. Chemicals for lysis buffer preparation are purchased from Merck Chemicals.
4. Cells are stimulated at 37° for several time points (1–5–30 min) by adding 40 μl of 10 × CCL3L1. Plates are gently mixed in order to homogenously distribute chemokine. Immediately after stimulation, cells are moved to ice, medium is carefully removed, and cells are gently rinsed once with ice-cold TBS.
5. Stimulated cells are lysed with 100 μl/well of ice-cold lysis buffer for 5 min, scraped, and transferred to 1.5 ml tubes. After 4 °C centrifugation at 13,000 rpm for 20 min, cell supernatant is moved into new microcentrifuge tube and quantified by DC protein assay.
6. 20–30 μg of total cell lysate are loaded into 12% polyacrylamide-precasted mini gel. After running, gel is transferred to PVDF membrane for 1 h at 100 V. Membrane is blocked with TBS + 5% nonfat dry milk at RT for 1 h, rinsed with TBS-T for 5 min, and then incubated overnight at 4 °C with 1:2000 dilution in TBS-T + 5% BSA of rabbit anti-phospho-p44/42 MAPK (hErk1/2) primary antibody (Cell Signaling Technology).
7. Membrane is washed five times for 5 min with 10 ml of TBS-T and incubated for 1 h at RT with 1:2000 dilution in TBS-T + 5% BSA of anti-rabbit IgG-horseradish peroxidase-conjugated secondary antibody.

8. Membrane is washed five times for 5 min with 10 ml of TBS-T and incubated with chemiluminescent HRP substrate (Immobilon Western Millipore, Billerica, MA) for 5 min under gentle mixing. Blot is acquired by ChemiDoc XRS Imaging System. Short exposition time points, starting from 1 s, are recommended in order to visualize pErk1/2 bands (42–44 kDa). α-Tubulin staining is performed on the same blots by using 1:4000 dilution in TBS-T + 5% BSA of mouse anti-hα-tubulin primary antibody (Sigma-Aldrich) for 2 h at RT, followed by incubation at RT for 1 h with 1:2000 dilution in TBS-T + 5% BSA of anti-mouse IgG-horseradish peroxidase-conjugated secondary antibody. α-Tubulin is detected as approximately 50 kDa band. Densitometric analysis is performed by the Quantity One software and Erk1/2 phosphorylation ratio is calculated by pErk1/2 bands normalizing over α-tubulin.

3.4. Small G protein activation assay: Affinity precipitation assay using GST-PBD

Small Rho GTPases have been implicated in many basic cellular processes that influence cell proliferation, motility, chemotaxis, and adhesion (Hall, 1998). The best characterized members of the Rho family GTPases, Rac1, Cdc42, and RhoA, play a central role in the organization of actin filament networks and in membrane ruffling; in particular, they are critical for coupling of extracellular signals to actin cytoskeletal rearrangements, in chemokine-mediated cellular events (Cantrell, 2003). In many canonical chemokine receptors, the activation of Rac1 has been demonstrated to be essential for chemokine-induced lamellipodia formation and fundamental for their chemotactic activity (Di Marzio et al., 2005). In this section, we will describe a protocol to study active GTP-bound form of Rac by affinity precipitation assay with GST-human PAK1 p21-binding domain (PBD) that recognize human, murine, or rat Rac1-GTP.

1. Cells (2.5×10^6) are seeded into poly-lysine-coated 75 cm^2 flasks in 10 ml of DMEM (Cambrex) + 10% FCS + 25 m*M* HEPES + 100 U/ml P/S at 37 °C for 18 h.
2. Cells are serum-starved by replacing medium with 10 ml of DMEM + 0.1 % BSA + 25 m*M* HEPES + 100 U/ml P/S for 18 h. Cells should reach 70–80% confluence at the pull-down assay day.
3. Before starting, medium is replaced with 4.5 ml of fresh DMEM + 0.1% BSA + 25 m*M* HEPES + 100 U/ml P/S. Lysis buffer (25 m*M* Tris–HCl (pH 7.2) + 150 m*M* NaCl + 5 m*M* $MgCl_2$ + 1%

NP-40 + % glycerol) is freshly prepared and finally supplemented with protease inhibitors tablet EDTA-free and 100 μg/ml PMSF.

4. Cells are lysed with 750 μl/flask of ice-cold lysis buffer for 5 min, scraped, and transferred to 1.5 ml tubes (Eppendorf, Hamburg, Germany). After 4 °C centrifugation at 13,000 rpm for 20 min, cell supernatant, representative of total cell lysate (TCL), is moved to new 1.5 ml tube and quantified by DC protein assay. Fifty microliters of total cell lysate are stored at −80 °C and further used to check Western blot procedure and quantify total Rac1.
5. GTPγS (Thermo Scientific Pierce) is added to 500 μg of TCL and used as positive control to ensure that pull-down procedures are working properly. For 500 μl of TCL, 10 μl of 0.5 *M* EDTA (pH 8) (Merck Chemicals) and 5 μl of 10 m*M* GTPγS are added for a final concentration of 10 and 0.1 m*M*, respectively. After vortexing, samples are incubated at 30 °C for 15 min under constant agitation, to ensure the binding between Rac1 and GTP. Reaction is stopped by placing sample on ice and adding 32 μl of 1 *M* $MgCl_2$ for a final concentration of 60 m*M*.
6. New 1.5 ml tubes for each sample are prepared and 100 μl of gluthatione-agarose resin supplied as 50% slurry and 0.05% NaN_3 (Thermo Scientific Pierce) and 400 μl of lysis buffer are added simultaneously. Tubes are gently inverted several times and centrifuged at 4500 rpm for 30 s and supernatant is discarded without touching the resin.
7. 20 μg of GST-human PAK1-PBD (Thermo Scientific Pierce) are added to glutathione-agarose resin. 700 μl of TCL containing at least 500 μg up to 1 mg of proteins are immediately transferred up to the resin. Reaction mixture is vortexed and incubated at 4 °C for 1 h under gentle rocking.
8. Samples are centrifuged at 4500 rpm, 4 °C for 30 s and supernatants are discarded. Rac1-GTP bound to glutathione-agarose resin is washed three times by adding 400 μl of lysis buffer, inverting three or more times and centrifuging at 4500 rpm for 30 s. At the end of the last washing, supernatant is totally removed, without touching the resin. Fifty microliters of 2× reducing sample buffer (125 m*M* Tris–HCl (pH 6.8), 2% glycerol, 4% SDS (w/v) + 0.05% bromophenol blue + 0.71 *M* of β-mercaptoethanol) are added to each sample and incubated at RT for 2 min. After centrifugation at 4500 rpm for 2 min, supernatants containing Rac1-immunoprecipitated lysates (RIL) are

collected carefully without touching the resin and heated for 5 min at 95–100 °C. Samples may be immediately electrophoresed on a gel or stored at −20 °C until use. Chemicals for reducing sample buffer preparation are purchased from Merck Chemicals and β-mercaptoethanol from Bio-Rad.

9. 40 μl/lane of RIL and 25 μg/lane of previously stored TCL are loaded into 12% polyacrylamide-precasted mini gel (Bio-Rad) to detect Rac1-GTP and total Rac1, respectively. After running, gel is transferred to PVDF for 1 h at 100 V. Membrane is blocked with TBS + 3% BSA at RT for 1–2 h, rinsed with TBS-T for 5 min, and then incubated overnight at 4 °C with 1:2000 dilution in TBS-T + 3% BSA + 0.1% NaN_3 of mouse anti-hRac1 primary antibody (Thermo Scientific Pierce).
10. Membrane is washed five times for 5 min with 10 ml of TBS-T and incubated for 1 h at RT with 1:20,000 dilution in TBS-T + 5% nonfat dry milk of anti-mouse IgG-horseradish peroxidase-conjugated secondary antibody (Dako).
11. Membrane is washed five times for 5 min with 10 ml of TBS-T and incubated with 10 ml of chemiluminescent HRP substrate for 5 min under gently mixing. Blot is acquired by ChemiDoc XRS Imaging System. Exposition time of 30–120 s is recommended in order to visualize differences in Rac1 activation between time 0 and 5 min. A GST-PBD band at 35 kDa could be present without interfering with Rac1 band at 22 kDa. Densitometric analysis is performed by Quantity One software and results are obtained by Rac1-GTP normalization over total amount of Rac1 measure in TCL.

4. SUMMARY

Substantial evidence exists confirming that members of the chemokine receptor family have evolved in order to have a regulatory function. They do not mediate chemotactic activity while they function as chemokine presenters, scavengers, or transporters. The exact signal transduction pathway they are coupled to mediate their function still needs to be delineated. Substantial evidence exists indicating that these receptors are uncoupled to the canonical Gαi pathway while they have maintained β-arrestin-dependent signaling to modify intracellular traffic to support continuous uptake, transport, and/or concentration of the ligands.

REFERENCES

Arancibia-Carcamo, I. L., Fairfax, B. P., Moss, S. J., & Kittler, J. T.(2006). Studying the Localization, Surface Stability and Endocytosis of Neurotransmitter Receptors by Antibody Labeling and Biotinylation. In: J. T. Kittler & S. J. Moss (Ed.), The Dynamic Synapse: Molecular Methods in Ionotropic Receptor Biology. Boca Raton (FL): CRC Press; Chapter 6.

Boldajipour, B., Mahabaleshwar, H., Kardash, E., Reichman-Fried, M., Blaser, H., Minina, S., et al. (2008). Control of chemokine-guided cell migration by ligand sequestration. *Cell, 132*, 463–473.

Bonecchi, R., Borroni, E. M., Anselmo, A., Doni, A., Savino, B., Mirolo, M., et al. (2008). Regulation of D6 chemokine scavenging activity by ligand- and Rab11-dependent surface up-regulation. *Blood, 112*, 493–503.

Bonecchi, R., Galliera, E., Borroni, E. M., Corsi, M. M., Locati, M., & Mantovani, A. (2009). Chemokines and chemokine receptors: An overview. *Frontiers in Bioscience, 14*, 540–551.

Bonecchi, R., Locati, M., Galliera, E., Vulcano, M., Sironi, M., Fra, A. M., et al. (2004). Differential recognition and scavenging of native and truncated macrophage-derived chemokine (macrophage-derived chemokine/CC chemokine ligand 22) by the D6 decoy receptor. *Journal of Immunology, 172*, 4972–4976.

Bonecchi, R., Savino, B., Borroni, E. M., Mantovani, A., & Locati, M. (2010). Chemokine decoy receptors: Structure-function and biological properties. *Current Topics in Microbiology and Immunology, 341*, 15–36.

Borroni, E. M., & Bonecchi, R. (2009). Shaping the gradient by nonchemotactic chemokine receptors. *Cell Adhesion & Migration, 3*, 146–147.

Borroni, E. M., Bonecchi, R., Mantovani, A., & Locati, M. (2009). Chemoattractant receptors and leukocyte recruitment: More than cell migration. *Science Signaling, 2*, e10.

Cantrell, D. A. (2003). GTPases and T cell activation. *Immunological Reviews, 192*, 122–130.

Charo, I. F., & Ransohoff, R. M. (2006). The many roles of chemokines and chemokine receptors in inflammation. *The New England Journal of Medicine, 354*, 610–621.

Comerford, I., Milasta, S., Morrow, V., Milligan, G., & Nibbs, R. (2006). The chemokine receptor CCX-CKR mediates effective scavenging of CCL19 in vitro. *European Journal of Immunology, 36*, 1904–1916.

Di Marzio, P., Dai, W. W., Franchin, G., Chan, A. Y., Symons, M., & Sherry, B. (2005). Role of Rho family GTPases in CCR1- and CCR5-induced actin reorganization in macrophages. *Biochemical and Biophysical Research Communications, 331*, 909–916.

Fra, A. M., Locati, M., Otero, K., Sironi, M., Signorelli, P., Massardi, M. L., et al. (2003). Cutting edge: Scavenging of inflammatory CC chemokines by the promiscuous putatively silent chemokine receptor D6. *Journal of Immunology, 170*, 2279–2282.

Galliera, E., Jala, V. R., Trent, J. O., Bonecchi, R., Signorelli, P., Lefkowitz, R. J., et al. (2004). beta-Arrestin-dependent constitutive internalization of the human chemokine decoy receptor D6. *The Journal of Biological Chemistry, 279*, 25590–25597.

Hall, A. (1998). G proteins and small GTPases: Distant relatives keep in touch. *Science, 280*, 2074–2075.

Kumar, R., Tripathi, V., Ahmad, M., Nath, N., Mir, R. A., Chauhan, S. S., et al. (2012). CXCR7 mediated Giα independent activation of ERK and Akt promotes cell survival and chemotaxis in T cells. *Cellular Immunology, 272*, 230–241.

Locati, M., Torre, Y. M., Galliera, E., Bonecchi, R., Bodduluri, H., Vago, G., et al. (2005). Silent chemoattractant receptors: D6 as a decoy and scavenger receptor for inflammatory CC chemokines. *Cytokine & Growth Factor Reviews, 16*, 679–686.

Luker, K. E., Gupta, M., Steele, J. M., Foerster, B. R., & Luker, G. D. (2009). Imaging ligand-dependent activation of CXCR7. *Neoplasia, 11*, 1022–1035.

Mantovani, A., Bonecchi, R., & Locati, M. (2006). Tuning inflammation and immunity by chemokine sequestration: Decoys and more. *Nature Reviews. Immunology, 6*, 907–918.
McCulloch, C. V., Morrow, V., Milasta, S., Comerford, I., Milligan, G., Graham, G. J., et al. (2008). Multiple roles for the C-terminal tail of the chemokine scavenger D6. *The Journal of Biological Chemistry, 283*, 7972–7982.
Mortier, A., Van Damme, J., & Proost, P. (2008). Regulation of chemokine activity by post-translational modification. *Pharmacology and Therapeutics, 120*, 197–217.
Murphy, P. M., Baggiolini, M., Charo, I. F., Hebert, C. A., Horuk, R., Matsushima, K., et al. (2000). International union of pharmacology. XXII. Nomenclature for chemokine receptors. *Pharmacological Reviews, 52*, 145–176.
Naumann, U., Cameroni, E., Pruenster, M., Mahabaleshwar, H., Raz, E., Zerwes, H. G., et al. (2010). CXCR7 functions as a scavenger for CXCL12 and CXCL11. *PLoS One, 5*, e9175.
Neel, N. F., Schutysér, E., Sai, J., Fan, G. H., & Richmond, A. (2005). Chemokine receptor internalization and intracellular trafficking. *Cytokine & Growth Factor Reviews, 16*, 637–658.
Paruch, S., Heinis, M., Lemay, J., Hoeffel, G., Maranon, C., Hosmalin, A., et al. (2007). CCR5 signaling through phospholipase D involves p44/42 MAP-kinases and promotes HIV-1 LTR-directed gene expression. *The FASEB Journal, 21*, 4038–4046.
Prevo, R., Banerji, S., Ni, J., & Jackson, D. G. (2004). Rapid plasma membrane-endosomal trafficking of the lymph node sinus and high endothelial venule scavenger receptor/homing receptor stabilin-1 (FEEL-1/CLEVER-1). *The Journal of Biological Chemistry, 279*, 52580–52592.
Pruenster, M., & Rot, A. (2006). Throwing light on DARC. *Biochemical Society Transactions, 34*, 1005–1008.
Rajagopal, S., Kim, J., Ahn, S., Craig, S., Lam, C. M., Gerard, N. P., et al. (2010). Beta-arrestin- but not G protein-mediated signaling by the "decoy" receptor CXCR7. *Proceedings of the National Academy of Sciences of the United States of America, 107*, 628–632.
Schaer, D. J., Schaer, C. A., Buehler, P. W., Boykins, R. A., Schoedon, G., Alayash, A. I., et al. (2006). CD163 is the macrophage scavenger receptor for native and chemically modified hemoglobins in the absence of haptoglobin. *Blood, 107*, 373–380.
Shenoy, S. K., & Lefkowitz, R. J. (2011). β-Arrestin-mediated receptor trafficking and signal transduction. *Trends in Pharmacological Sciences, 32*, 521–533.
Thelen, M., & Stein, J. V. (2008). How chemokines invite leukocytes to dance. *Nature Immunology, 9*, 953–959.
Van Lith, L. H., Oosterom, J., Van Elsas, A., & Zaman, G. J. (2009). C5a-stimulated recruitment of beta-arrestin2 to the nonsignaling 7-transmembrane decoy receptor C5L2. *Journal of Biomolecular Screening, 14*, 1067–1075.
Weber, M., Blair, E., Simpson, C. V., O'Hara, M., Blackburn, P. E., Rot, A., et al. (2004). The chemokine receptor D6 constitutively traffics to and from the cell surface to internalize and degrade chemokines. *Molecular Biology of the Cell, 15*, 2492–2508.
Zabel, B. A., Nakae, S., Zuniga, L., Kim, J. Y., Ohyama, T., Alt, C., et al. (2008). Mast cell-expressed orphan receptor CCRL2 binds chemerin and is required for optimal induction of IgE-mediated passive cutaneous anaphylaxis. *The Journal of Experimental Medicine, 205*, 2207–2220.
Zerial, M., & McBride, H. (2001). Rab proteins as membrane organizers. *Nature Reviews. Molecular Cell Biology, 2*, 107–117.

SECTION 2

Trafficking Motifs

CHAPTER NINE

Systematic and Quantitative Analysis of G Protein-Coupled Receptor Trafficking Motifs

Carl M. Hurt, Vincent K. Ho, Timothy Angelotti[1]
Department of Anesthesia/CCM, Stanford University Medical School, Stanford, California, USA
[1]Corresponding author: e-mail address: timangel@stanford.edu

Contents

1. Introduction 172
2. Motif Screening by Immunofluorescent Staining 173
   2.1 Epitope tagging of GPCRs 173
   2.2 Identification of trafficking motifs 173
   2.3 Rapid immunofluorescent screening 174
   2.4 Interpretation of total versus surface immunofluorescent staining 175
3. Analysis of Receptor Functionality 175
   3.1 Determination of ligand binding 175
   3.2 Saturation ligand binding assays 177
   3.3 Analysis of receptor signaling 177
   3.4 Interpretation of receptor functionality 179
4. Biochemical Analysis of ER/Golgi Trafficking 179
   4.1 GPCR glycosylation analysis 179
   4.2 Glycolytic analysis of GPCR trafficking 180
   4.3 Interpretation of glycosidic processing 180
5. Quantitative Analysis of GPCR Trafficking 181
   5.1 FACS analysis of receptor expression 181
   5.2 FACS analysis 183
   5.3 Interpretation of FACS analysis 184
6. Summary 184
Acknowledgments 186
References 186

Abstract

Plasma membrane expression of G protein-coupled receptors (GPCRs) is a dynamic process balancing anterograde and retrograde trafficking. Multiple interrelated cellular processes determine the final level of cell surface expression, including endoplasmic reticulum (ER) export/retention, receptor internalization, recycling, and degradation. These processes are highly regulated to achieve specific localization to subcellular

Methods in Enzymology, Volume 521
ISSN 0076-6879
http://dx.doi.org/10.1016/B978-0-12-391862-8.00009-0

domains (e.g., dendrites or basolateral membranes) and to affect receptor signaling. Analysis of potential ER trafficking motifs within GPCRs requires careful consideration of intracellular dynamics, such as protein folding, ER export and retention, and glycosylation. This chapter presents an approach and methods for qualitative and quantitative assessment of these processes to aid in accurate identification of GPCR trafficking motifs, utilizing the analysis of a hydrophobic extracellular trafficking motif in α2C adrenergic receptors as a model system.

1. INTRODUCTION

G protein-coupled receptors (GPCRs) are synthesized and exported along the secretory pathway from the endoplasmic reticulum (ER) to the plasma membrane by a complex process involving multiple mechanisms to insure proper folding, assembly, quality control, selective retention, and transport (Ellgaard & Helenius, 2003). GPCRs are synthesized in the ER, where they are folded, assembled, and then packaged into ER-derived COPII transport vesicles that traverse the Golgi network for glycolytic processing and eventual transport to their membrane localizations (Duvernay, Filipeanu, & Wu, 2005). The resident time within the ER is dependent upon several factors including folding rates and assembly, which may be modified by pharmacological and/or protein chaperones (Brothers, Janovick, & Conn, 2006; Petaja-Repo et al., 2002), as well as specific sequences within the GPCR that dictate ER or Golgi export or retention. Several such trafficking motifs have been identified in various GPCRs, including variations of a carboxy-terminal $F(X)_6LL$ ER export motif (Duvernay et al., 2009; Schulein et al., 1998). Additionally, several arginine-rich ER retention motifs have been identified, including RSRR and RXR motifs in $GABA_B$ R1 and vasopressin type 2 receptors, respectively (Hermosilla & Schulein, 2001; Pagano et al., 2001).

Traditional identification of ER trafficking motifs is often done by mutational analysis and expression; however, interpretation can be complex. For example, mutation or removal of a suspected GPCR ER export motif should lead to ER retention; however, a mutant GPCR that is misfolded could also lead to ER retention. Alternatively, mutation of an ER retention motif should lead to ER export, with an increase in the ratio of plasma membrane/total cellular expression. However, a mutant GPCR with an enhanced intracellular half-life will appear to have increased plasma membrane expression due to increased total cellular receptor expression, but the plasma membrane/total cellular expression ratio may not change. Thus, to ensure proper identification of ER trafficking motifs, a complete analysis must be undertaken to ensure proper interpretation of the data. Three concerns should be addressed prior to identification of a GPCR ER trafficking motif:

1. Removal or mutation of the motif does not affect GPCR function (i.e., it is not misfolded).
2. Alterations in trafficking lead to changes in intracellular processing (e.g., glycosylation).
3. Changes in plasma membrane expression are specific and do not simply reflect changes in total cellular expression only.

α2A and α2C adrenergic receptors (ARs) are highly homologous receptors that target to different neuronal sites (extra- vs. presynaptic, respectively), as well as demonstrate cell-type specific trafficking (Angelotti, Daunt, Shcherbakova, Kobilka, & Hurt, 2010; Brum, Hurt, Shcherbakova, Kobilka, & Angelotti, 2006). Utilizing α2A and α2C ARs as model GPCRs, we describe a systematic approach to identifying GPCR trafficking motifs, culminating in a quantitative membrane expression assay. Using this multiassay approach, we have identified a new extracellular hydrophobic trafficking domain within the amino terminus of α2C ARs (ALAAALAAAAA). These methods are optimized for analysis of ER trafficking; however, they can be utilized to study Golgi or other trafficking motifs.

2. MOTIF SCREENING BY IMMUNOFLUORESCENT STAINING

2.1. Epitope tagging of GPCRs

The majority of methods presented in this chapter require a GPCR-specific antibody for immunofluorescence, immunoblot, and flow-activated cell sorting (FACS) analysis. If a specific antiserum against an extracellular epitope does not exist for the native receptor, it can be tagged at the amino terminus with any one of several epitopes (e.g., FLAG, hemagglutinin, or myc) using standard molecular biological techniques. Epitope-tagged GPCRs need to be tested to ensure no alteration in functionality or biochemical processing; a more detailed methodology can be found elsewhere (Wozniak, Saunders, Schramm, Keefer, & Limbird, 2002).

2.2. Identification of trafficking motifs

Potential trafficking motifs can be identified by any one of several methods, including sequence analysis with known consensus trafficking motifs from other proteins, analysis of convergent and divergent protein sequences between homologous receptors, or analysis based on crystal structure. A comparison of α2A&C ARs reveals that the most divergent sequences are found within the third intracellular loop (iC3) between transmembrane domains V and VI and the amino- and carboxy-terminal regions. Using standard molecular biological methods, these regions were swapped between α2A and α2C ARs (Angelotti et al., 2010) to create a series of chimeric α2A/C ARs for rapid screening of trafficking motifs. Once a region/motif of interest is identified, specific mutations can be constructed within this region prior to further characterization. Thus, this process requires iterative steps of rapid immunofluorescent screening and functional assays to hone in on a specific motif followed by a more complete analysis of membrane trafficking.

2.3. Rapid immunofluorescent screening

Immunofluorescent microscopy is a rapid method to screen for regions or motifs involved in GPCR trafficking and to analyze chimeric and/or mutant receptors with respect to the relative distribution of surface and total receptor expression (Kallal & Benovic, 2002). For example, heterologously expressed α2A ARs demonstrate over 90% plasma membrane expression in nonneuronal cell lines, whereas α2C ARs are predominantly localized to an intracellular site (ER), with only 25% plasma membrane expression (Angelotti et al., 2010; Hurt, Feng, & Kobilka, 2000). Therefore, HA-tagged α2A&C AR chimeras were screened to identify regions involved with this trafficking disparity.

1. Plate HEK293 cells onto 15 × 15-mm poly-D-lysine (70 kDa, Sigma Chemical)-coated glass coverslips placed into 12-well cell culture plates with DMEM + 10% FBS. Plate approximately 100–150,000 cells per well 24 h prior to transfection.
2. Transfect cells with epitope-tagged GPCR chimeras or mutants using Effectene or other transfection reagent (per manufacturer's recommendations) and maintain in culture for 48 h prior to staining.
3. Rinse cells three times with room temperature (RT) PBS and then fix with 4% paraformaldehyde (PFA) in PBS for 5 min at RT. Wash fixed cells three times with PBS at 5-min intervals.
4. For total cellular immunofluorescent staining, permeabilize and block fixed cells at RT for 30 min with blocking solution (40 m*M* HEPES, pH 7.4, 5% dry milk, 2% goat serum (Gemini Bio Products), 0.2% Nonidet P-40 in PBS). For surface immunofluorescent staining, use blocking solution without Nonidet P-40.

5. Label epitope-tagged GPCR chimeras using primary mouse monoclonal antibody (HA epitope = 16B12 [Covance]) at a 1:500 dilution in appropriate blocking solution (with or without NP-40) for 1 h at RT. To counterstain ER or Golgi compartments, add primary antibodies against calreticulin or giantin (respectively). Appropriate dilutions for all antibodies should be determined empirically based upon manufacturer's recommendations.
6. Rinse cells four to five times at RT with PBS over 15 min and reapply the appropriate blocking solution for 30 min at RT.
7. Apply fluorescent-conjugated secondary antibodies in appropriate blocking solution for 1 h at RT (e.g., goat antimouse Alexa Fluor 596, Invitrogen). Dilutions should be based upon manufacturer's recommendations and empiric determination.
8. Rinse cells four to five times with PBS over 15 min at RT. Mount glass coverslips cell-side down onto glass slides using Vectashield (Vector Labs). Allow to air dry at RT in a dark area.
9. Examine and photograph stained cells using a fluorescent microscope with appropriate filter sets for the secondary fluorescent antisera utilized. A magnification over 500 × should be used.

2.4. Interpretation of total versus surface immunofluorescent staining

Rapid visual screening of total versus surface staining of wild-type (WT) and α2A&C AR chimeras will allow for recognition of potential trafficking motifs, based upon observation of altered surface/total expression. Once regions of interest are identified, sequence analysis within these regions can be utilized to refine possible trafficking motifs for further mutational analysis and screening (Fig. 9.1). However, further analysis is necessary to ensure that the increased or decreased immunofluorescent staining observed represents true alterations in trafficking.

3. Analysis of Receptor Functionality

3.1. Determination of ligand binding

As described above, a chimeric or mutant GPCR can lead to ER retention or alterations in plasma membrane expression due to misfolding or changes in protein half-life. To ensure that chimeric or mutated GPCRs are

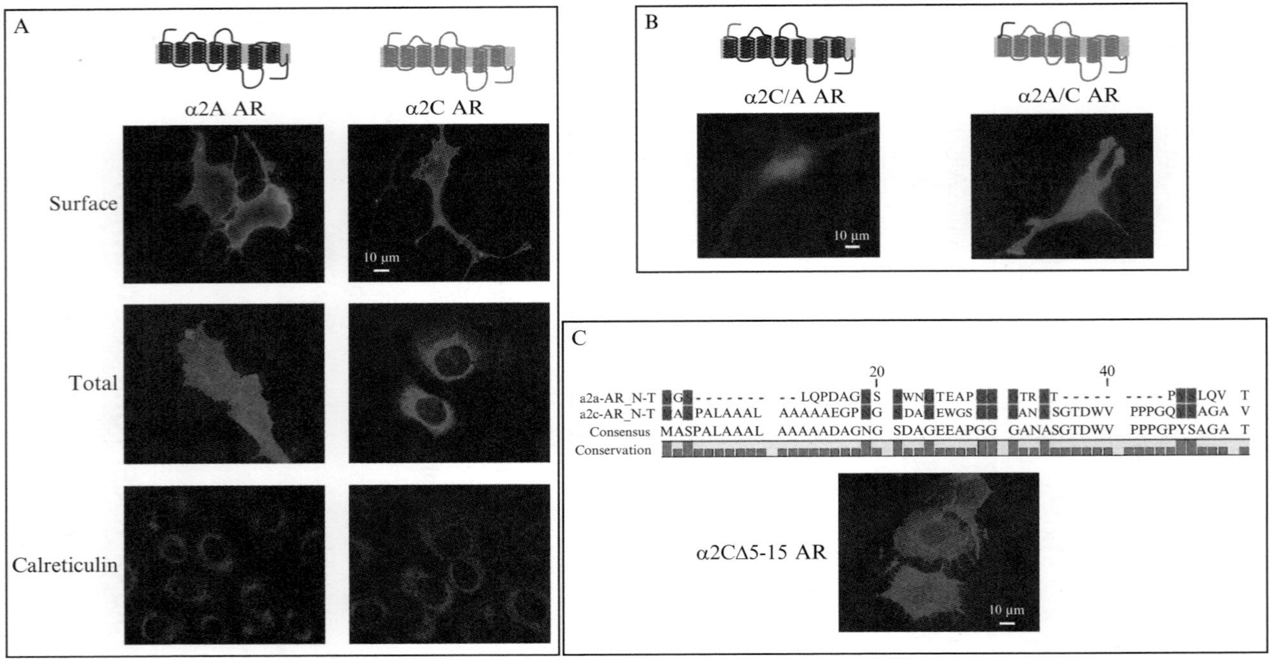

Figure 9.1 Immunofluorescent screening of α2A&C AR trafficking motifs. (A) Surface and total staining of WT HA-α2A&C ARs, co-stained with the ER marker calreticulin, demonstrate predominant intracellular (ER) localization of α2C, compared to α2A, ARs. α2A ARs are primarily found at the plasma membrane. (B) Total cellular staining of α2A/C AR chimeras demonstrates that the extracellular amino terminus of α2A and α2C ARs is a determinant of their relative expression patterns. (C) Removal of an α2C AR amino-terminal hydrophobic region (ALAAALAAAAA = α2CΔ5-15 AR) enhances cell surface expression, as assessed by total cellular staining, suggesting a possible role for this domain in α2C AR membrane expression. *Adapted from Angelotti et al. (2010).* (For color version of this figure, the reader is referred to the online version of this chapter.)

operative prior to further analysis of trafficking, their functionality needs to be examined. Ligand binding can aid in determining if a chimeric or mutant GPCR is folded properly. Membrane binding is preferable to whole cell binding to determine a complete pharmacological profile of binding affinities and expression levels (K_D and B_{max}, respectively).

3.2. Saturation ligand binding assays

1. Split a 15-cm tissue culture dish of confluent HEK293 cells 1:10 in DMEM with 10% FBS. Allow them to replicate at 37 °C until approximately 90% confluent.
2. Transfect selected WT, chimeric, or mutant GPCR constructs into a 15-cm dish of HEK293 cells.
3. After 48 h of growth, lyse cells in 3.0 ml lysis buffer (10 m*M* Tris, 1 m*M* EDTA, pH 7.4 with protease inhibitor cocktail [Roche]) and homogenize 15 times on ice using an appropriately sized dounce homogenizer. Centrifuge at 1000 × *g* for 5 min to remove debris/nuclei and centrifuge the supernatant at 15,000 × *g* for 30 min, both at 16 °C.
4. Resuspend resultant membrane pellet in 1 ml binding buffer (75 m*M* Tris, 12.5 m*M* $MgCl_2$, 1 m*M* EDTA, pH 7.4 with protease inhibitor cocktail) using a 1-ml TB syringe and a 25-g needle on ice. Repeat steps 3 and 4.
5. Determine protein concentration using RC DC assay (Bio-Rad Laboratories) and freeze membranes at −70 °C at approximately 1 mg/ml, dilute with binding buffer as needed.
6. Perform saturation binding using an appropriate ligand, as described previously (Daunt et al., 1997); nonspecific binding is determined by adding excess cold ligand. For α2 AR analysis, $[^3H]$RX-821002 (PerkinElmer) is recommended, with 100 μ*M* yohimbine added to determine nonspecific binding.
7. Calculate equilibrium dissociation constants and B_{max} values from saturation isotherms and competition curves using GraphPAD software (San Diego, CA).

3.3. Analysis of receptor signaling

In addition to ligand binding, the ability of mutant or chimeric GPCRs to activate second messenger pathways should also be determined to ensure functionality. An immunofluorescent MAP kinase assay can be utilized for rapid screening of GPCRs that activate Gi/Go signaling pathways.

For example, following addition of an α2 AR agonist, MAP kinase activation and phosphorylation of extracellular signal-regulated kinase 1 and 2 (ERK1/2) can be detected by specific phospho-ERK1/2 antisera, using an *in cell* immunofluorescent assay (Angelotti et al., 2010).

1. Plate, transfect, and grow Rat1 fibroblasts on poly-D-lysine-coated coverslips, as described above under Section 2.3. Other heterologous cell systems can be utilized.
2. Twenty-four hour after transfection with WT or chimeric GPCRs of interest, serum starve cells in DMEM with 4% bovine serum albumin (BSA) and 20 m*M* HEPES, pH 7.4 for 24 h at 37 °C.
3. Incubate cells in DMEM alone (control) or 1 n*M* dexmedetomidine (agonist) for 10 min at 37 °C in a tissue culture incubator. The appropriate agonist for the GPCR of interest should be chosen.
4. Immediately fix cells with 4% PFA for 5 min at RT.
5. Wash three times with PBS at 5-min intervals, followed by permeabilization and blocking with MAP kinase blocking solution (40 m*M* HEPES, pH 7.4, 3% BSA, 2% goat serum, 0.2% Nonidet P-40 in PBS-CM) for 30 min at RT.
6. Label epitope-tagged GPCR chimeras using primary mouse monoclonal antibody (HA epitope = 16B12 [Covance]) at a 1:500 dilution in MAP kinase blocking solution for 1 h at RT. Detect MAP kinase activity by addition of phospho-ERK1/2 rabbit polyclonal antibody (Millipore) at a 1:1000 dilution. Appropriate dilutions for all antibodies should be determined empirically based upon manufacturer's recommendations.
7. Rinse cells four to five times at RT with PBS over 15 min and reapply appropriate blocking solution for 30 min at RT.
8. Apply fluorescent-conjugated secondary antibodies (e.g., goat antimouse Alexa Fluor 596 and donkey anti-rabbit Alexa Fluor 488 antibodies [Invitrogen]) in appropriate blocking solution for 1 h at RT. Dilutions should be based upon manufacturer's recommendations and empiric determination.
9. Rinse cells four to five times with PBS over 15 min at RT. Mount glass coverslips cell-side down onto glass slides using Vectashield (Vector Labs). Allow to air dry at RT in a dark area.
10. Examine and photograph stained cells using a fluorescent microscope with appropriate filter sets for the secondary fluorescent antisera utilized. A magnification over 500× should be used.

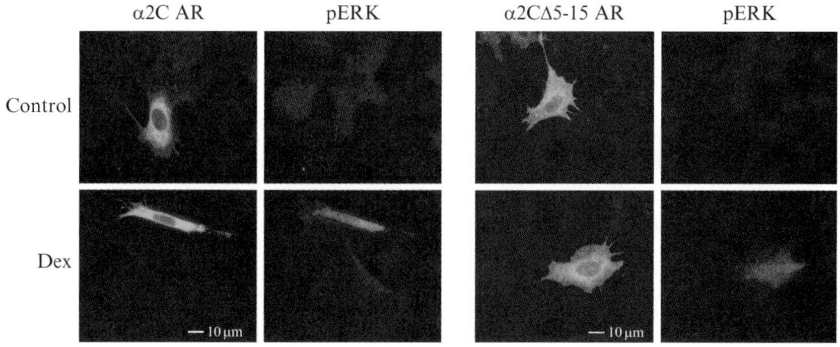

Figure 9.2 Functional screening of α2C AR mutations by MAP kinase. An immunofluorescent assay for MAP kinase (pERK) in response to agonist (Dex = dexmedetomidine) allows for screening of GPCR chimeras and mutations to ensure pharmacological function prior to further trafficking analysis. Permeabilized cells were stained with monoclonal antibody 16B12 to examine total cellular HA-α2 AR expression. Note minimal pERK staining in the absence of agonist. Both α2CΔ5-15 and α2CA7D ARs can activate MAP kinase and bind RX-821002 (data not shown) suggesting that they are not misfolded and are functional. *Adapted from Angelotti et al. (2010).* (See Color Insert.)

3.4. Interpretation of receptor functionality

Ligand binding and second messenger screening assays can be utilized to ensure that mutant or chimeric GPCRs are functional. The relative expression levels determined by binding may vary; however, ligand affinity should not be altered drastically. If ligand binding is drastically altered, consider using a different ligand (agonist vs. antagonist). In a similar manner, MAP kinase activation may not occur or be relevant to a specific GPCR, thus another second messenger assay should be considered (e.g., cyclase, phosphoinositol turnover). Prior to going forward with trafficking analysis, all mutant or chimeric GPCRs should retain ligand binding and second messenger activation to demonstrate functionality (Fig. 9.2).

4. BIOCHEMICAL ANALYSIS OF ER/GOLGI TRAFFICKING

4.1. GPCR glycosylation analysis

As GPCRs move through the secretory pathway from ER to Golgi, glycosidic processing/maturation occurs (Hurt et al., 2000). The glycosylation state can be determined by digestion of GPCR-containing membrane preparations with endoglycosidase H (Endo H) and peptide:N-glycosidase

F (PNGase F). Endo H removes asparagine-linked high-mannose-content glycans from glycoproteins that have not trafficked to the *cis*-medial Golgi apparatus, and thus are residing within the ER as immature proteins. PNGase F removes all asparagine-linked glycans regardless of their level of processing as immature or mature proteins. Therefore, analysis of the relative immature and mature glycoprotein content of a mutant or chimeric GPCR can determine its effect on secretory pathway trafficking.

4.2. Glycolytic analysis of GPCR trafficking

1. Endo H and PNGase F can be purchased from New England Biolabs. All necessary buffers and accessory reagents are supplied.
2. Thaw GPCR membrane preparations on ice (see Section 3.2).
3. Add 50–100 μg of membrane to a 1.5-ml microcentrifuge tube. Three reactions are necessary for each GPCR studied, control (no enzyme), Endo H, and PNGase F.
4. Manufacturer's instructions should be followed for the enzymatic digestion; however, due to aggregation of HA-α2A&C ARs at temperatures above 55 °C, detergent and Laemmli loading buffer denaturation reactions were carried out at 55 °C.
5. Incubate the final glycosidic reaction for 4 h at 37 °C. Stop reaction by adding one-fifth volume of 5 × SDS sample buffer (Hurt et al., 2000) and load each sample onto a 10% SDS PAGE gel. Large format gels will enhance detection of reaction products.
6. Following standard immunoblotting protocols, transfer the electrophoresed proteins onto nitrocellulose (or other membrane) and probe the blot with an appropriate epitope-specific monoclonal antisera and appropriate secondary antibody at dilutions appropriate for your detection methods (e.g., LiCOR, ECL).

4.3. Interpretation of glycosidic processing

Immunoblot analysis of GPCRs demonstrates multiple cell-specific patterns representing mature and immature glycosylation (Hurt et al., 2000). For example, WT-α2A ARs are predominantly found with mature glycosylation (Endo H insensitive), whereas α2C ARs are predominantly found with immature glycosylation (Endo H sensitive) (Fig. 9.3). Removal of the putative extracellular hydrophobic trafficking motif (α2CΔ5-15) enhances mature glycosylation, consistent with the enhanced plasma membrane staining seen by immunofluorescence. In addition, disruption of this domain by a charged

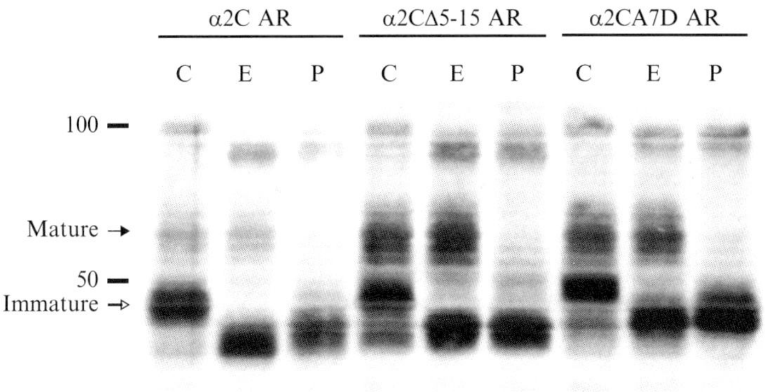

Figure 9.3 Glycosylation analysis of α2C AR trafficking. Cellular processing and maturation of α2C ARs and various trafficking mutations can be assessed by biochemical analysis of N-linked glycosylation (C = control, E = Endo H, and P = PNGase F). Mature and immature represent Endo H-resistant and -sensitive GPCR forms, respectively. Note that α2CΔ5-15 and α2CA7D ARs both have enhanced mature forms, suggestive of enhanced ER to Golgi trafficking and glycosidic processing. These results are consistent with a role for an extracellular hydrophobic domain in trafficking and plasma membrane expression. *Adapted from Angelotti et al. (2010).*

point mutation (α2CA7D) has the same effect. The ratio of mature to immature glycosylation (ER retained) correlates with the ratio of surface to intracellular expression seen by immunocytochemistry and FACS analysis. WT, chimeric, and mutant GPCR glycosylation analysis can assist in determining if a putative trafficking motif truly affects movement through the secretory pathway.

5. QUANTITATIVE ANALYSIS OF GPCR TRAFFICKING

If a GPCR trafficking motif is functional, then mutating it should alter the ratio of cell surface to total receptor expression. For example, removal of an ER retention motif should lead to ER export and thus an increase in cell surface/total GPCR expression (Angelotti et al., 2010). Qualitative immunofluorescent microscopy cannot readily discern such changes, especially if transient transfection methods are utilized and/or there is a large intracellular pool of receptor. Therefore, quantification of GPCR expression by FACS can assess the effects of mutating a potential trafficking motif.

5.1. FACS analysis of receptor expression

FACS can be utilized to measure the relative surface and total cell receptor levels for a given WT, chimeric, or mutant GPCR. From these data, the % surface and % intracellular expression can then be calculated to determine if a trafficking motif has affected this ratio. A proper FACS analysis requires measurements of nontransfected cells and determination of cell death to ensure accurate measurements of relative surface and total cell receptor levels. A more complete description of FACS analysis of transfected cells can be found elsewhere (Adams, Lopez, Sellers, & Kaelin, 1997).

1. Split a 10-cm tissue culture dish of confluent HEK293 cells 1:10 in DMEM with 10% FBS. Allow them to replicate at 37 °C until approximately 50% confluent.
2. Transfect cells with epitope-tagged GPCR chimeras or mutants using Effectene or other transfection reagent (per manufacturer's recommendations) and maintain in culture for 72 h prior to analysis. In addition, three separate dishes of nontransfected cells will be required for controls (unstained, stained, and dead cell assessment).
3. Treat transfected cells with 10 μ*M* cycloheximide (protein synthesis inhibitor) for 90 min at 37 °C to reduce background intracellular retention of GPCRs.
4. Wash plates two times with Hanks buffered saline solution (HBSS). Lift cells off the dish with 10 ml HBSS + 5 m*M* EDTA.
5. Centrifuge cells at 200 × *g* for 5 min at 4 °C. Aspirate supernatant and resuspend cell pellet in 10 ml PBS at 4 °C. Repeat centrifugation and resuspend pellet in 1 ml PBS.
6. Stain one control set of cells with propidium iodide (PI, Sigma) to determine cell death in the preparation. Incubate the sample preparation with PI in PBS (1 mg/ml) at 4 °C for 15 min on ice. Centrifuge and resuspend cell pellet in 250 μl PBS without PI and keep sample on ice in the dark. PI cannot be used on cells that will subsequently undergo fixation, as this will lead to dissociation from DNA of dead cells and binding to DNA of fixed cells. Fluorescent Live/Dead fixable dyes (Becton Dickinson) can be used with fixed cells.
7. Fix all other cells (transfected and nontransfected controls) by adding 1 ml of 4% PFA (2% final concentration) for 10 min at RT.
8. Place two 1-ml fractions (nonpermeabilized and permeabilized) into Eppendorf tubes. Centrifuge cells at 2500 × *g* for 5 min at RT to remove PFA and resuspend cell pellet in 1 ml PBS. Repeat three times at 10-min intervals.
9. For surface expression analysis (nonpermeabilized), centrifuge cells as above and resuspend in 1 ml nonpermeabilized blocking solution (PBS + 2% FBS) for 45 min prior to fluorescent labeling.

10. For total expression analysis (permeabilized), centrifuge cells as above and resuspend in 1 ml permeabilized blocking solution (PBS + 2% FBS + 0.1% Triton X-100) for 15 min and wash once with 1 ml nonpermeabilized blocking solution for 30 min.
11. Set aside the permeabilized and nonpermeabilized, nontransfected cells for unstained controls.
12. Centrifuge both cell preparations at 2500 × *g* for 5 min at RT. Resuspend cell pellets in 250 μl blocking solution (in appropriate permeabilized or nonpermeabilized blocking solution) containing fluorescein-conjugated 16B12 antibody (Covance) at 0.75 μg/ml. Incubate for 30 min at RT in the dark.
13. Similarly, permeabilized and nonpermeabilized, nontransfected cells should be prepared as stained controls in the same manner.
14. Centrifuge cell preparations at 2500 × *g* for 5 min at RT and wash cell pellets three times at 5-min intervals with 500 μl PBS + 2% FBS. Repeat once more and resuspend in 200–400 μl PBS. Cover tubes with foil.

5.2. FACS analysis

1. FACS analysis can be performed using a LSR-1 (Becton Dickinson Bioscience) or other similar device.
2. Prior to running cell samples, dissociate cells by passing through a 35-μm pore size, 5-ml round-bottom tube with a cell strainer cap (Becton Dickinson).
3. Following FACS device recommendations, gate the instrument to remove cell debris and clumps of aggregated cells using the nonpermeabilized, unstained control cells.
4. Dead cells can autofluoresce and have increased nonspecific immunofluorescent staining. Using the PI-stained control cells, the fraction of dead cells in the population being studied should be determined. PI is excited using a blue 488-nm laser and detected using Cy5-PE/PI setting at 665/40 nm for the photomultiplier tube. The fraction of dead cells in these preparations should be low (less than a few percent of the population) so as to not adversely affect data quality.
5. Depending upon the transfection efficiency, the total number of cells needed to be analyzed can vary between 10,000 to 100,000 cells. Permeabilized and nonpermeabilized, nontransfected, stained control cells are used to set the gates for determination of transfected (stained) versus untransfected (minimally stained) GPCR-expressing cells. Fluorescent

intensities should be obtained from 50,000 to 100,000 cells for each WT, mutant, or chimeric construct under permeabilized and nonpermeabilized conditions.

6. For a good representation of the transfected cell population, a minimum of 2000 transfected cells should be used to determine median fluorescent intensity.
7. A minimum of three independent experiments should be carried out to calculate average median fluorescent intensity for each construct under each condition (permeabilized and nonpermeabilized).
8. Calculate the percentage of GPCR surface expression relative to total cellular expression by dividing the average median surface (nonpermeabilized) fluorescent intensity by the average median total (permeabilized) fluorescent intensity and report as average ± standard error of the mean (SEM):

$$\%\text{surface expression} = \frac{\text{average median surface fluorescent intensity}}{\text{average median total fluorescent intensity}} \quad [9.1]$$

$$\%\text{intracellular expression} = 100 - \%\text{surface expression} \quad [9.2]$$

5.3. Interpretation of FACS analysis

FACS analysis allows quantitative measurement of relative surface and total expression for a given WT or mutant GPCR (Fig. 9.4). With this information, the fraction of GPCR at the plasma membrane can be calculated (% surface) and the relative effect of a given trafficking motif on surface expression can be delineated with more certainty over qualitative immunofluorescent microscopic methods. It is hard to discern if an increase in surface expression seen by immunofluorescent microscopy is due to an overall increase in total protein expression or truly represents enhanced trafficking, especially in transiently transfected cells. For example, deletion or mutation of the extracellular hydrophobic domain (α2CΔ5-15 and α2CA7D AR) leads to increased surface expression compared to WT-α2C ARs, demonstrating a change in plasma membrane trafficking.

6. SUMMARY

Appropriate trafficking and localization of GPCRs to specific subdomains within a cell or neuron can be an important aspect of their physiological function (Edwards, Tan, & Limbird, 2000). A more thorough

A

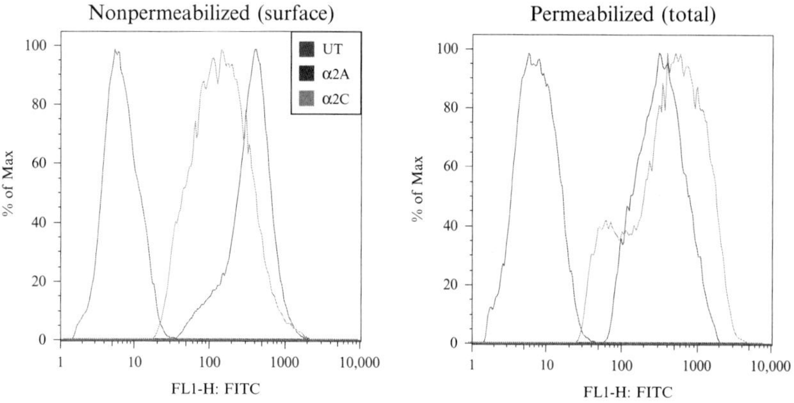

B

	Average surface median fluorescent intensity	Average total median fluorescent intensity	Average % surface expression	Average % intracellular expression
α2A AR	347.3 ± 50.2	376.3 ± 52.2	92 ± 3.9	8 ± 3.9
α2C AR	126 ± 16.6	495.9 ± 59.1	25 ± 1	75 ± 1
α2CΔ5-15 AR	305.8 ± 50	582.8 ± 115.8	52 ± 7.3	48 ± 7.3
α2CA7D AR	271 ± 60.4	558 ± 131.8	49 ± 5.8	51 ± 5.8

Fluorescent intensity represented as arbitrary units. Values given are the average ± SEM.

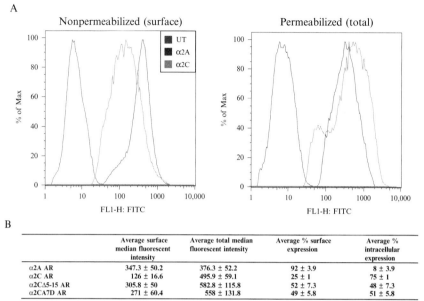

Figure 9.4 Quantitative FACS analysis of plasma membrane trafficking. (A) Relative plasma membrane expression can be easily quantified by measuring GPCR levels in permeabilized (total) and nonpermeabilized (surface) cells using a one-step fluorescent labeling procedure (UT = untransfected). Following proper FACS gating to remove dead cells, measurement of single cell fluorescence allows for the determination of median fluorescent intensity for each GPCR under both conditions. Notice the shift in α2C AR median fluorescence intensity following permeabilization, reflecting the larger intracellular pool of receptor B. Calculation of relative surface and intracellular expression levels can be performed using median fluorescent measurements, demonstrating that both α2CΔ5-15 and α2CA7D ARs enhance plasma membrane expression. *Adapted from Angelotti et al. (2010).* (See Color Insert.)

understanding of the biochemical machinery underlying transport of cargo proteins is ongoing (De Matteis & Luini, 2011), and an important aspect of trafficking is understanding protein motifs involved with GPCR transport. However, given the complexities of protein synthesis, folding, assembly, and transport through the ER/Golgi/plasma membrane, determination of trafficking motifs is not as straightforward as it may seem.

We have applied the above series of experimental methods to understand and identify a hydrophobic extracellular trafficking motif within α2C ARs, which is cell-type specific and transferable to α2A ARs (Angelotti et al., 2010). By initially constructing α2A&C AR chimeras and screening them with immunofluorescent microscopy, we were able to determine that a potential ER retention motif resided within the amino terminus of α2C ARs. Sequence analysis of multiple divergent species of α2C ARs suggested that a highly conserved motif may be responsible. Through a reiterative process of immunofluorescent staining, an extracellular hydrophobic domain (ALAAALAAAAA) was identified. Furthermore, alternative explanations such as alterations in receptor folding or half-life were removed from consideration by use of functional assays, including ligand binding and MAP kinase activation. Glycosidic processing clearly demonstrated that the motif regulated trafficking from ER to Golgi, resulting in altered glycosylation. Lastly, quantitative FACS analysis demonstrated that this motif truly regulated cell surface expression, not just total cellular expression. By applying a series of rigorous cell biological and biochemical methods, a novel trafficking motif was identified in α2C ARs. These methods should be applicable to other GPCRs.

ACKNOWLEDGMENTS

This work was support in part by research MSTP grant GM07365 from the National Institutes of General Medical Sciences (C. M. H.) and NINDS K08NS050654-01A1 (T. A.). The authors would like to thank Dr. Brian Kobilka for his support during the early stages of this research.

REFERENCES

Adams, P. D., Lopez, P., Sellers, W. R., & Kaelin, W. G., Jr. (1997). Fluorescence-activated cell sorting of transfected cells. *Methods in Enzymology*, *283*, 59–72.

Angelotti, T., Daunt, D., Shcherbakova, O. G., Kobilka, B., & Hurt, C. M. (2010). Regulation of G-protein coupled receptor traffic by an evolutionary conserved hydrophobic signal. *Traffic*, *11*, 560–578.

Brothers, S. P., Janovick, J. A., & Conn, P. M. (2006). Calnexin regulated gonadotropin-releasing hormone receptor plasma membrane expression. *Journal of Molecular Endocrinology*, *37*, 479–488.

Brum, P. C., Hurt, C. M., Shcherbakova, O. G., Kobilka, B., & Angelotti, T. (2006). Differential targeting and function of alpha(2A) and alpha(2C) adrenergic receptor subtypes in cultured sympathetic neurons. *Neuropharmacology*, *51*, 397–413.

Daunt, D. A., Hurt, C., Hein, L., Kallio, J., Feng, F., & Kobilka, B. K. (1997). Subtype-specific intracellular trafficking of alpha2-adrenergic receptors. *Molecular Pharmacology*, *51*, 711–720.

De Matteis, M. A., & Luini, A. (2011). Mendelian disorders of membrane trafficking. *The New England Journal of Medicine*, *365*, 927–938.

Duvernay, M. T., Dong, C., Zhang, X., Zhou, F., Nichols, C. D., & Wu, G. (2009). Anterograde trafficking of G protein-coupled receptors: Function of the C-terminal F(X)6LL motif in export from the endoplasmic reticulum. *Molecular Pharmacology*, *75*, 751–761.

Duvernay, M. T., Filipeanu, C. M., & Wu, G. (2005). The regulatory mechanisms of export trafficking of G protein-coupled receptors. *Cellular Signalling*, *17*, 1457–1465.

Edwards, S. W., Tan, C. M., & Limbird, L. E. (2000). Localization of G-protein-coupled receptors in health and disease. *Trends in Pharmacological Sciences*, *21*, 304–308.

Ellgaard, L., & Helenius, A. (2003). Quality control in the endoplasmic reticulum. *Nature Reviews. Molecular Cell Biology*, *4*, 181–191.

Hermosilla, R., & Schulein, R. (2001). Sorting functions of the individual cytoplasmic domains of the G protein-coupled vasopressin V(2) receptor in Madin Darby canine kidney epithelial cells. *Molecular Pharmacology*, *60*, 1031–1039.

Hurt, C. M., Feng, F. Y., & Kobilka, B. (2000). Cell-type specific targeting of the alpha 2c-adrenoceptor. Evidence for the organization of receptor microdomains during neuronal differentiation of PC12 cells. *The Journal of Biological Chemistry*, *275*, 35424–35431.

Kallal, L., & Benovic, J. L. (2002). Fluorescence microscopy techniques for the study of G protein-coupled receptor trafficking. *Methods in Enzymology*, *343*, 492–506.

Pagano, A., Rovelli, G., Mosbacher, J., Lohmann, T., Duthey, B., Stauffer, D., et al. (2001). C-terminal interaction is essential for surface trafficking but not for heteromeric assembly of GABA(b) receptors. *The Journal of Neuroscience*, *21*, 1189–1202.

Petaja-Repo, U. E., Hogue, M., Bhalla, S., Laperriere, A., Morello, J. P., & Bouvier, M. (2002). Ligands act as pharmacological chaperones and increase the efficiency of delta opioid receptor maturation. *The EMBO Journal*, *21*, 1628–1637.

Schulein, R., Hermosilla, R., Oksche, A., Dehe, M., Wiesner, B., Krause, G., et al. (1998). A dileucine sequence and an upstream glutamate residue in the intracellular carboxyl terminus of the vasopressin V2 receptor are essential for cell surface transport in COS. M6 cells. *Molecular Pharmacology*, *54*, 525–535.

Wozniak, M., Saunders, C., Schramm, N., Keefer, J. R., & Limbird, L. E. (2002). Morphological and biochemical strategies for monitoring trafficking of epitope-tagged G protein-coupled receptors in agonist-naive and agonist-occupied states. *Methods in Enzymology*, *343*, 530–544.

CHAPTER TEN

Identification of Endoplasmic Reticulum Export Motifs for G Protein-Coupled Receptors

Guangyu Wu[1]

Department of Pharmacology and Toxicology, Georgia Health Sciences University, Augusta, Georgia, USA

[1]Corresponding author: e-mail address: guwu@georgiahealth.edu

Contents

Abstract

Coat protein complex II (COPII) vesicle-mediated protein export from the endoplasmic reticulum (ER) can be controlled by linear, independent motifs embedded within the cargo. ER export motifs directly interact with selective components of COPII vesicles and enhance cargo recruitment onto COPII vesicles. Moreover, ER export motifs are able to confer their transport abilities to other proteins. We have recently identified a novel ER export motif for α_{2B}-adrenergic receptor (α_{2B}-AR). This motif selectively interacts with Sec24C/D isoforms of COPII vesicles and facilitates α_{2B}-AR export from the ER as well as transport to the cell surface. This motif can also mediate CD8 glycoprotein transport. These studies indicate that ER export of nascent G protein-coupled receptors (GPCRs) may be directed by specific codes that mediate receptor interaction with the ER-derived COPII vesicles. In this chapter, I discuss experimental approaches to identify ER export motifs for GPCRs by using α_{2B}-AR as a model.

Methods in Enzymology, Volume 521
ISSN 0076-6879
http://dx.doi.org/10.1016/B978-0-12-391862-8.00010-7

1. INTRODUCTION

Similar to many other plasma membrane proteins, the life of GPCRs begins at the endoplasmic reticulum (ER), where they are synthesized, folded, and assembled. Correctly folded and properly assembled receptors are able to pass through the ER quality control mechanism and to exit the ER. It is well known that exit of nascent cargo from the ER is mediated through coat protein complex II (COPII) transport vesicles that can be formed from three components: the small GTPase Sar1, the Sec23/24 heterodimer, and the Sec13/31 heterodimer (Barlowe et al., 1994; Bickford, Mossessova, & Goldberg, 2004; Gurkan, Stagg, Lapointe, & Balch, 2006; Kuge et al., 1994). A number of recent studies have also demonstrated that the recruitment of cargo onto COPII vesicles is a selective process and can be mediated by ER export motifs.

Several ER export motifs have been identified with diacidic and dihydrophobic motifs being well characterized. The diacidic motifs have been found in the cytoplasmic C-termini of a number of membrane proteins, including vesicular stomatitis viral glycoprotein, cystic fibrosis transmembrane conductance regulator, ion channels, the yeast membrane proteins Sys1p and Gap1p, and G protein-coupled angiotensin II receptors (Ma et al., 2001; Mikosch, Kaberich, & Homann, 2009; Nishimura & Balch, 1997; Nishimura et al., 1999; Votsmeier & Gallwitz, 2001; Wang et al., 2004; Zhang, Dong, Wu, Balch, & Wu, 2011; Zuzarte et al., 2007). The DxE motif (Nishimura & Balch, 1997; Nishimura et al., 1999; Zuzarte et al., 2007), together with neighboring residues (Sevier, Weisz, Davis, & Machamer, 2000), directs the concentration of the cargo molecule during export from the ER, thereby enhancing the rate of its exit from the ER. The dihydrophobic motifs are required for efficient transport of the ERGIC-53, p24 family of proteins, and the Erv41–Erv46 complex from the ER to the Golgi (Dominguez et al., 1998; Fiedler, Veit, Stamnes, & Rothman, 1996; Nufer, Kappeler, Guldbrandsen, & Hauri, 2003; Otte & Barlowe, 2002).

The function of ER export motifs to enhance cargo transport is mediated through direct interaction with components of COPII vesicles, particularly the Sec24 subunit (Farhan et al., 2007; Mancias & Goldberg, 2008; Miller et al., 2003; Mossessova, Bickford, & Goldberg, 2003; Votsmeier & Gallwitz, 2001; Wang et al., 2004). There are four Sec24 isoforms (Sec24A, Sec24B, Sec24C, and Sec24D) identified in human cells, and based on the sequence homology, they can be further divided into Sec24A/B and Sec24C/D subclasses. Three cargo recognition sites have been mapped in Sec24 (Miller et al., 2003; Mossessova et al., 2003), demonstrating that different ER export motifs interact with the same COPII component to facilitate ER export. Meanwhile, different ER export motifs may have selectivity toward Sec24 isoforms (Mancias & Goldberg, 2008). Based on these studies, ER export motifs should have the following characteristics: (1) modulation of cargo transport at the level of the ER, (2) linear sequence, (3) direct interaction with components of the COPII vesicles, (4) transferable transport properties.

Recent studies from our and other laboratories have identified several motifs essential for GPCR export from the ER and transport from the ER to the cell surface (Bermak, Li, Bullock, & Zhou, 2001; Dong, Filipeanu, Duvernay, & Wu, 2007; Dong & Wu, 2006; Donnellan, Kimbembe, Reid, & Kinsella, 2011; Duvernay, Dong, Zhang, Robitaille, et al., 2009, Duvernay, Dong, Zhang, Zhou, et al., 2009; Duvernay, Zhou, & Wu, 2004; Robert, Clauser, Petit, & Ventura, 2005). However, the molecular mechanisms underlying the function of these motifs in controlling GPCR export remain elusive (Duvernay, Dong, Zhang, Zhou, et al., 2009; Margeta-Mitrovic, Jan, & Jan, 2000; Robert et al., 2005). By using progressive deletion, site-directed mutagenesis, and protein–protein interaction assays, we have recently revealed that a three Arg (3R) motif in the third intracellular loop (ICL3) mediates α_{2B}-adrenergic receptor (α_{2B}-AR) interaction with Sec24C/D isoforms. Mutation of the 3R motif attenuates α_{2B}-AR exit from the ER and transport to the cell surface. Interestingly, the 3R motif can facilitate α_{2B}-AR export trafficking when transferred to the C-terminus. Furthermore, the 3R motif can also enhance CD8 glycoprotein transport to the cell surface (Dong et al., 2012). These data demonstrate that the 3R motif represents a novel ER export motif. In this chapter, I discuss experimental approaches to identify specific sequences/motifs involved in the regulation of ER export of newly synthesized GPCRs.

2. EXPERIMENTAL APPROACHES TO IDENTIFY ER EXPORT MOTIFS FOR GPCRs

2.1. Manipulation of GPCRs

In order to identify the motifs essential for export from the ER, the first step is to select a domain to be manipulated in a given GPCR. The C-terminus and three intracellular loops are positioned toward the cytoplasm which may provide docking sites for components of the ER-derived COPII transport vesicles. In contrast, the N-terminus and three extracellular loops are located in the lumen of the ER and are spatially impossible to directly interact with the transport machinery in the cytoplasm. Therefore, the C-terminus and the intracellular loops should be considered in searching for ER export motifs that are able to physically interact with cytosolic transport machinery.

Once a specific domain is selected to be studied, the next step is to generate the receptor mutant in which the domain of interest is removed and the transport properties of the truncated receptor mutant are compared with those of its wild-type counterpart.

If the deleted mutant indeed is defective in the transport, the deleted domain may contain signals important for receptor transport. If the domain is not too long, such as the C-terminus of α_{2B}-AR which contains only 20 amino acid residues, site-directed mutagenesis can be directly applied to define specific residues that modulate receptor transport. Each of the residues in the domain of interest is mutated to alanine and then the effect of mutations on receptor transport is determined. If site-directed mutagenesis is cumbersome for a domain which is relatively large, the progressive deletion strategy can be utilized to search for a subdomain containing the export signal.

2.2. Measurement of receptor expression at the cell surface

If the cell surface expression of a receptor mutant, in which the domain of interest is deleted or individual amino acid residue mutated, is significantly lower than that of its wild-type counterpart, the deleted domain or the mutated residue is important for receptor export to the cell surface. The cell surface expression of GPCRs can be quantitated by a number of well-established methods, such as ligand binding, flow cytometry, and biotinylation. Our laboratory has established a ligand binding assay in intact living cells for quantifying the cell surface numbers of adrenergic and angiotensin II receptors (Dong & Wu, 2007; Dong, Zhou, Fugetta, Filipeanu, & Wu, 2008; Duvernay, Dong, Zhang, Robitaille, et al., 2009; Duvernay, Dong, Zhang, Zhou, et al., 2009; Filipeanu, Zhou, Fugetta, & Wu, 2006). This method has been used to study the role of Ras-like small GTPases and specific motifs in modulating the intracellular trafficking of GPCRs (Dong & Wu, 2006; Duvernay, Dong, Zhang, Robitaille, et al., 2009; Duvernay, Dong, Zhang, Zhou, et al., 2009; Wang & Wu, 2012). The measurement of the cell surface expression of α_1-AR, β-AR and angiotensin II receptor by intact cell ligand binding is described elsewhere

(Dong & Wu, 2007; Filipeanu, Zhou, Claycomb, & Wu, 2004; Filipeanu et al., 2006; Zhang et al., 2011). The following describes the intact cell ligand binding assay to measure the cell surface expression of α_{2B}-AR using the radioligand [^{3}H]-RX821002.

HEK293 cells are cultured on 6-well dishes in DMEM and transfected with 1.0 μg of wild-type and mutated α_{2B}-AR plasmids. After 6 h, the cells are split into 12-well dishes precoated with poly-L-lysine. After 24–36 h transfection, the cells are incubated with DMEM plus [^{3}H]-RX821002 (PerkinElmer) at a concentration of 20 n*M* in a total of 400 μl for 90 min at room temperature. For the nonspecific binding, the cells are incubated with [^{3}H]-RX821002 plus rauwolscine (Sigma-Aldrich) at a concentration of 10 μ*M*. The cells are washed twice with 1 ml ice-cold DMEM to remove the excess radioligand. The remained ligands are extracted by digesting the cells with 1 *M* NaOH for 2 h at room temperature. The liquid phase is collected and 5 ml of Ecoscint A scintillation fluid (National Diagnostics Inc., Atlanta, GA) added. The amount of radioactivity retained is measured by liquid scintillation spectrometry.

Ligand binding of membrane preparations has been long used to quantify GPCR expression. The major advantage of intact cell ligand binding is that it has the ability to accurately measure the numbers of receptors at the plasma membrane, compared with ligand binding of total membrane preparations which contain the receptors expressed in the intracellular compartments, such as the ER and the Golgi. This is particularly important when the mutation or deletion disrupts receptor export, resulting in intracellular retention of the receptor, but does not alter receptor–ligand binding properties.

As mutation of the receptors may alter receptor–ligand binding affinity, other alternative methods such as flow cytometry should be used to confirm the intact cell ligand binding results. For this purpose, the receptors are tagged with an epitope such as FLAG or HA at their N-termini and their expression at the cell surface is measured by flow cytometry following staining with epitope antibodies in nonpermeabilized cells.

In addition to the cell surface expression, it is very important to determine if the mutation or deletion influences the overall synthesis of the receptor, which can be accomplished by a number of methods. If the receptor is tagged with fluorescence proteins such as GFP, the simplest way to measure the overall receptor expression is to directly measure the total fluorescent signal of GFP by flow cytometry or the total GFP expression by immunoblotting using GFP antibodies (Duvernay, Dong, Zhang, Robitaille, et al., 2009).

It is also necessary to exclude the possibility that the reduction in the cell surface expression of a mutated receptor is caused by constitutive receptor internalization. For this goal, receptor internalization blockers can be used to inhibit endocytic transport. If the cell surface expression of a mutated receptor is restored upon coexpression with dominant-negative internalization mutants or treatments with pharmacological blockers, defective export of the mutated receptor is likely caused by facilitated endocytic transport (Dong & Wu, 2006).

In addition to measuring the cell surface expression of the receptor, it is also necessary to define if the mutation of the traffic motif could attenuate signaling elicited by that particular receptor. Although it is expected that the effects of the mutation on the cell surface expression and signaling will be parallel, dysfunction of the mutated receptor will not only confirm the defect in the anterograde transport but also indicate the importance of export process in regulating receptor signal propagation in cells. The selection of functional readouts depends on the type of receptor studied. In our studies, the agonist-mediated activation of the mitogen-activated protein kinases ERK1/2 has been used as a functional readout for several family A GPCRs (Dong & Wu, 2006; Duvernay, Dong, Zhang, Robitaille, et al., 2009; Duvernay, Dong, Zhang, Zhou, et al., 2009).

2.3. Analysis of receptor export from the ER

Once a domain or a specific motif is identified to be important for receptor export, the next step is to directly visualize the subcellular distribution of the mutated receptor by microscopy. If a mutated receptor is extensively localized inside the cell but not on the cell surface, this, together with the reduced cell surface expression measured by intact cell ligand binding, will further indicate that the mutated residue is essential for receptor transport to the cell surface. Microscopic analysis of receptor localization is also able to precisely define the intracellular compartment where the mutated residue modulates receptor export. This could be accomplished by colocalization of the mutated receptors with various intracellular organelle markers (Dong & Wu, 2006; Duvernay, Dong, Zhang, Robitaille, et al., 2009; Duvernay, Dong, Zhang, Zhou, et al., 2009; Duvernay et al., 2004). Analysis of the subcellular colocalization of GFP-tagged α_{2B}-AR with ER markers in fixed cells is described below.

HEK293 cells cultured on 6-well plates are transfected with 100 ng of wild type or mutated α_{2B}-AR tagged with GFP at its C-terminus. Six hours later, the cells are split onto cover slips precoated with poly-L-lysine in 6-well plates. Twenty-four hours following transfection, the cells are washed with ice-cold phosphate buffer saline (PBS) twice with gentle shaking. The cells are fixed with fresh 2 ml of 4% paraformaldehyde–4% sucrose mixture in PBS for 15 min. After washing the cells with 2 ml of ice-cold PBS for 3×5 min, the cells are permeabilized with 2 ml of PBS containing 0.2% Triton X-100 for 5 min. The cells are incubated with 0.24% normal donkey serum (Jackson Immuno) in PBS for 1 h and then with primary antibodies again the ER marker calregulin or calnexin (Santa Cruz) at 1:50 dilution on parafilm with the cover slips upside down for 2 h. Antibodies against other intracellular markers should be used as controls, such as ERGIC53 (Alexis Biochemicals)—an ERGIC marker; GM130 (BD Biosciences)—a *cis*-Golgi marker; p230 (BD Biosciences)—a TGN marker. The cells are washed with cold PBS and incubated with 1 ml of Alexa Fluor 594-labeled secondary antibody (Invitrogen) at a dilution of 1:2000 for 1 h. Following washing as above, the cells are incubated with 4,6-diamidino-2-phenylindole for 5 min to stain the nuclei. The cover slips are mounted with the mounting media and the fluorescent signal is detected by using an appropriate microscope.

Analysis of the cell surface expression and subcellular distribution of receptors as described above will provide important information about receptor anterograde transport from the ER to the cell surface at the steady status. As receptor transport passes through a number of intracellular compartments such as the ERGIC, Golgi, and TGN en route from the ER to the cell surface, the influences of a specific motif on the cell surface expression and export from the ER may not be identical. It is possible that mutation of an ER export motif significantly slows down the rate of receptor export from the ER, without dramatically reducing receptor cell surface expression and disrupting subcellular localization at the steady status. Therefore, after identifying a motif that modulates receptor export from the ER as discussed above, one should consider measuring the effect of mutating the motif on the kinetics of receptor export from the ER.

The kinetics of receptor export from the ER can be analyzed by fluorescence recovery after photobleaching (FRAP) experiments or by metabolic pulse-chase labeling experiments measuring the formation of complex glycosylated receptors (Bermak et al., 2001; Petaja-Repo, Hogue, Laperriere, Walker, & Bouvier, 2000). The following describes a FRAP protocol that was used in our laboratory to determine the role of the 3R motif in the ER export of α_{2B}-AR in COS7 cells (Dong et al., 2012). As α_{2B}-AR is not accumulated in the Golgi under the normal condition, we took advantage of a previously characterized Golgi-localized YS mutant of α_{2B}-AR (Dong & Wu, 2006) and determined if mutation of the 3R motif could influence the transport of the YS mutant to the Golgi.

COS7 cells are grown on 29-mm glass-bottom culture dishes (In Vitro Scientific) and transfected with either YS mutant in which the YS motif is mutated to Ala or the YS/3R mutant in which both the YS and the 3R motifs are mutated to Ala. Both mutants are tagged with GFP at their C-termini. Prior to imaging, standard media are replaced with CO_2-independent media. FRAP experiments are performed on a Leica SP2-TCS confocal microscope using a 63 × water immersion objective and the FRAP Wizard application. During imaging, culture dishes are maintained at 37 °C using a Harvard Apparatus (Holliston, MA) temperature controller and PDMI-2 Micro Incubator. Scan settings are 256 × 256 pixel resolution at 1400 Hz. For FRAP, 20 prebleach frames are acquired (0.287 s acquisition time/frame), followed by 20 bleach frames (0.287 s/frame) with the 488 laser, and then by 200 postbleach frames acquired at 5.0-s intervals. The region of interest for bleach/recovery measurement is manually drawn around the Golgi of each cell, and the mean pixel value determined for the region of interest at each recovery time point.

By using this protocol, we found that after photobleaching the intensities of both mutants were markedly reduced and unable to recover to the prebleach intensities, suggesting a large number of the immobilized α_{2B}-AR in the Golgi. The recovery of the YS mutant after photobleaching was much faster than the YS/3R mutant, and the half-time of the recovery of the YS mutant was much shorter than the YS/3R mutant (Dong et al., 2012). These data demonstrate that the mutation of the 3R motif in α_{2B}-AR induces defective transport from the ER to the Golgi.

2.4. Assays for receptor interaction with components of COPII-coated vesicles

Once a motif is identified to modulate receptor export from the ER, the next step is to delineate the possible molecular mechanisms. Because a number of factors, such as proper folding, interactions with accessory proteins, and dimerization, potentially modulate the ER export of GPCRs, disruption of each of these processes could lead to defective export from the ER. For example, misfolding is the most common cause for the loss of ability of GPCRs to transport from the ER to the cell surface. Indeed, a number of loss-of-function GPCRs, which are associated with many hereditary diseases, are induced by their misfolding, resulting in intracellular accumulation of the receptors. Circular dichroism, treatments with chemical and pharmacological chaperones, and low temperature culture have been used to determine if a GPCR mutant is correctly folded. However, these methods may not work for all kinds of misfolding. Therefore, the folding problem has been very difficult to be excluded in studying GPCR export from the ER.

One important function of ER export motifs is to mediate cargo interaction with components of COPII vesicles. Therefore, it is crucial to determine if the motif is able to interact with the COPII vesicles. We have used glutathione *S*-transferase (GST) fusion protein pull-down, coimmunoprecipitation, and peptide-conjugated agarose affinity matrix to measure the interaction of the 3R motif with four different Sec24 isoforms. GST fusion protein pull-down and coimmunoprecipitation assays have been essentially described in chapter 6 in volume 522 of this series. The following section briefly describes the methods to measure the 3R motif interaction with Sec24 isoforms using the ICL3 peptide-conjugated agarose as affinity matrix (Dong et al., 2012).

The α_{2B}-AR ICL3 peptide NH2-GKNVGVASGQWWRRRTQLSRE-COOH and its mutant NH2-GKNVGVASGQWWAAATQLSRE-COOH were synthesized, purified by HPLC to >75% and directly conjugated to agarose beads by Biosynthesis Inc. (Lewisville, TX). The peptide-conjugated agarose beads (10 μl, approximately 13 μmol peptides) were incubated with 300 μg cell lysates prepared from HEK293 cells transfected with GFP-tagged Sec24 in 300 μl of binding buffer containing 20 m*M* Tris–HCl, pH 7.4, 2% NP-40, 120 m*M* NaCl at 4 °C overnight. The resin was washed for four times each with 1 ml of binding buffer. The bound GFP–Sec24 were solubilized in 1 × SDS gel loading buffer, separated by SDS-PAGE, and detected by immunoblotting using GFP antibodies.

We have demonstrated in both GST fusion protein pull-down and peptide-conjugated affinity matrix that the ICL3 interacted with Sec24C and Sec24D more strongly than with Sec24A and Sec24B (Fig. 10.1A and B) and mutation of the 3R motif markedly reduced their interaction with Sec24 (Fig. 10.1B) (Dong et al., 2012).

2.5. Analysis of the ability of ER export motifs to transport other proteins

Another important property of ER export motifs is to confer their transport abilities to other proteins. For this propose, CD8 glycoprotein has been used in many studies. The following section describes a protocol used in our laboratory to define if the 3R motif could facilitate the transport of CD8

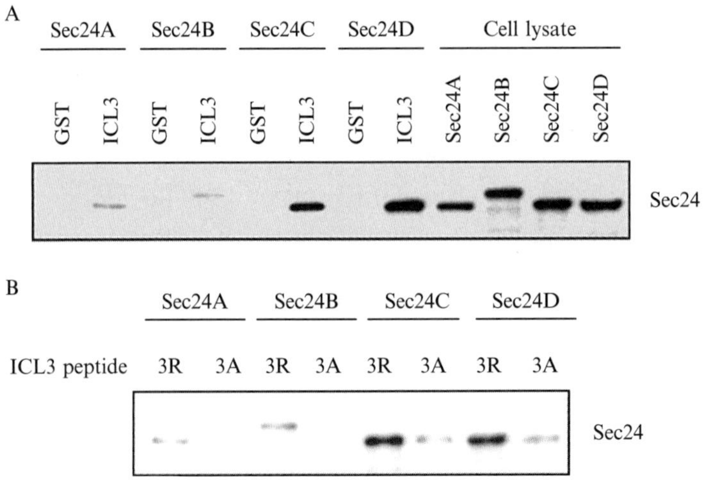

Figure 10.1 Interaction of the third intracellular loop of α_{2B}-AR with different Sec24 isoforms. (A) The interaction of the third intracellular loop (ICL3) of α_{2B}-AR with Sec24A, Sec24B, Sec24C, and Sec24D isoforms as determined in GST fusion protein pull-down assay. The ICL3 of α_{2B}-AR was generated as GST fusion proteins. Sec24 isoforms tagged with GFP were transiently expressed in HEK293 cells and total cell homogenates were incubated with GST or GST–ICL3 fusion proteins. (B) The interaction of the ICL3 with Sec24 isoforms as measured by ICL3 peptide-conjugated agarose matrix. The ICL3 fragment G349-E369 containing the 3R motif (3R) or its mutant in which the 3R motif was mutated to three alanines (3A) was synthesized and conjugated to agarose. The ICL3-conjugated agarose beads were incubated with total cell homogenates containing GFP-tagged Sec24 isoforms. In both (A) and (B), bound Sec24 isoforms were revealed by immunoblotting using GFP antibodies (Dong et al., 2012).

glycoprotein (Dong et al., 2012). To generate the α_{2B}-AR and CD8 chimeras (CD8-WT), the extracellular domains of human CD8α glycoprotein were generated by PCR using primers (forward primer, 5′-GTCACTCGAGACCATGGCCTTACCAGTGACCGCC-3′ and reverse primer 5′-GTTAGAATTCCCCCGCCGCTGGCCGGCAC-3′) and the α_{2B}-AR fragment containing the seventh TM, the C-terminus (from residues 404 to 453) and the ICL3 peptide WWWRRRTQ generated by using primers (forward primer, 5′-GTCAGAATTCTGCAAGGTACCGCATGGC-3′ and reverse primer, 5′-GTAC AAGCTT TCA CTG TGT CCG TCT GCG CC-3′). The two PCR products were digested with *Xho*I/*Eco*RI and *Eco*RI/*Hin*dIII, respectively, and then ligated into the pcDNA3.1(−) vector (Fig. 10.2A). To measure cell surface expression of α_{2B}-AR–CD8, HEK293 cells were transfected with α_{2B}-AR–CD8 chimeric proteins. The cells were incubated with FITC-conjugated human CD8

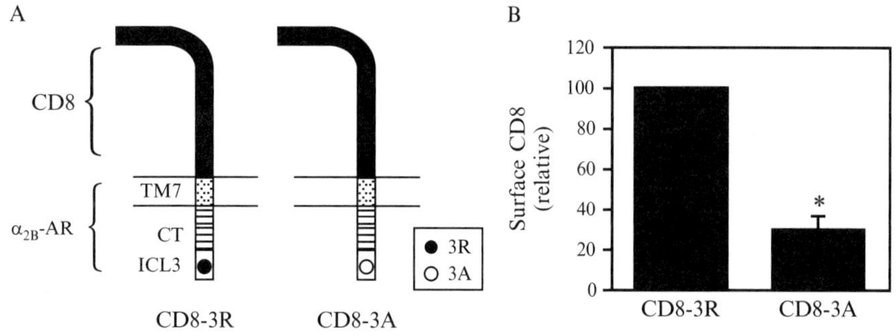

Figure 10.2 Effect of the 3R motif of α_{2B}-AR on the cell surface export of CD8 glycoprotein. (A) A diagram showing generation of chimeric α_{2B}-AR-CD8 proteins. The truncated α_{2B}-AR containing the seventh transmembrane (TM7) and the C-terminus (CT) of α_{2B}-AR was conjugated with the N-terminal portion of CD8 glycoprotein and an ICL3 fragment containing the 3R motif (CD8-3R) or the 3A motif (CD8-3A) at the C-terminus. (B) Cell surface expression of CD8-3R and CD8-3A. CD8-3R and CD8-3A were transiently expressed in HEK293 cells and their cell surface expression was measured by flow cytometry following staining using anti-CD8 antibodies in nonpermeabilized cells. $^*P < 0.05$ versus CD8-3R (Dong et al., 2012).

antibodies (eBioscience) at 4 °C for 30 min. To measure the total expression level of α_{2B}-AR–CD8, the cells were permeabilized with 0.2% Triton X-100 in PBS for 5 min on ice before staining with anti-CD8 antibodies.

We found that the cell surface expression of the chimera CD8-3A was markedly lower than that of the chimera CD8-3R (Fig. 10.2B). Consistently, immunostaining studies revealed cell surface expression of the chimeric protein CD8-3R, whereas CD8-3A was almost undetectable at the cell surface (Dong et al., 2012). These data demonstrate that the 3R motif in α_{2B}-AR is sufficient to direct the export of CD8 glycoprotein to the cell surface.

3. CONCLUSIONS

It is becoming increasingly appreciated that membrane proteins may use specific codes to exit from intracellular compartments, particularly the ER. A number of motifs have also been identified to be required for GPCR export from the ER. However, the molecular mechanisms remain poorly understood. Therefore, continuing search for linear, independent ER export motifs that not only mediate GPCR interaction with components of the COPII vesicles but also confer the transport ability to facilitate the ER exit of other proteins will provide important insights into how newly synthesized GPCRs exit from the ER compartment.

ACKNOWLEDGMENT

This work was supported by National Institutes of Health Grant R01GM076167.

REFERENCES

Barlowe, C., Orci, L., Yeung, T., Hosobuchi, M., Hamamoto, S., Salama, N., et al. (1994). COPII: A membrane coat formed by Sec proteins that drive vesicle budding from the endoplasmic reticulum. *Cell*, *77*, 895–907.

Bermak, J. C., Li, M., Bullock, C., & Zhou, Q. Y. (2001). Regulation of transport of the dopamine D1 receptor by a new membrane-associated ER protein. *Nature Cell Biology*, *3*, 492–498.

Bickford, L. C., Mossessova, E., & Goldberg, J. (2004). A structural view of the COPII vesicle coat. *Current Opinion in Structural Biology*, *14*, 147–153.

Dominguez, M., Dejgaard, K., Fullekrug, J., Dahan, S., Fazel, A., Paccaud, J. P., et al. (1998). gp25L/emp24/p24 protein family members of the cis-Golgi network bind both COP I and II coatomer. *The Journal of Cell Biology*, *140*, 751–765.

Dong, C., Filipeanu, C. M., Duvernay, M. T., & Wu, G. (2007). Regulation of G protein-coupled receptor export trafficking. *Biochimica et Biophysica Acta*, *1768*, 853–870.

Dong, C., Nichols, C. D., Guo, J., Huang, W., Lambert, N. A., & Wu, G. (2012). A triple Arg motif mediates alpha2B-adrenergic receptor interaction with Sec24C/D and export. *Traffic*, *13*, 857–868.

Dong, C., & Wu, G. (2006). Regulation of anterograde transport of alpha2-adrenergic receptors by the N termini at multiple intracellular compartments. *The Journal of Biological Chemistry*, *281*, 38543–38554.

Dong, C., & Wu, G. (2007). Regulation of anterograde transport of adrenergic and angiotensin II receptors by Rab2 and Rab6 GTPases. *Cellular Signalling*, *19*, 2388–2399.

Dong, C., Zhou, F., Fugetta, E. K., Filipeanu, C. M., & Wu, G. (2008). Endoplasmic reticulum export of adrenergic and angiotensin II receptors is differentially regulated by Sar1 GTPase. *Cellular Signalling*, *20*, 1035–1043.

Donnellan, P. D., Kimbembe, C. C., Reid, H. M., & Kinsella, B. T. (2011). Identification of a novel endoplasmic reticulum export motif within the eighth alpha-helical domain (alpha-H8) of the human prostacyclin receptor. *Biochimica et Biophysica Acta*, *1808*, 1202–1218.

Duvernay, M. T., Dong, C., Zhang, X., Robitaille, M., Hebert, T. E., & Wu, G. (2009). A single conserved leucine residue on the first intracellular loop regulates ER export of G protein-coupled receptors. *Traffic*, *10*, 552–566.

Duvernay, M. T., Dong, C., Zhang, X., Zhou, F., Nichols, C. D., & Wu, G. (2009). Anterograde trafficking of G protein-coupled receptors: Function of the C-terminal F(X)6LL motif in export from the endoplasmic reticulum. *Molecular Pharmacology*, *75*, 751–761.

Duvernay, M. T., Zhou, F., & Wu, G. (2004). A conserved motif for the transport of G protein-coupled receptors from the endoplasmic reticulum to the cell surface. *The Journal of Biological Chemistry*, *279*, 30741–30750.

Farhan, H., Reiterer, V., Korkhov, V. M., Schmid, J. A., Freissmuth, M., & Sitte, H. H. (2007). Concentrative export from the endoplasmic reticulum of the gamma-aminobutyric acid transporter 1 requires binding to SEC24D. *The Journal of Biological Chemistry*, *282*, 7679–7689.

Fiedler, K., Veit, M., Stamnes, M. A., & Rothman, J. E. (1996). Bimodal interaction of coatomer with the p24 family of putative cargo receptors. *Science*, *273*, 1396–1399.

Filipeanu, C. M., Zhou, F., Claycomb, W. C., & Wu, G. (2004). Regulation of the cell surface expression and function of angiotensin II type 1 receptor by Rab1-mediated endoplasmic reticulum-to-Golgi transport in cardiac myocytes. *The Journal of Biological Chemistry, 279*, 41077–41084.

Filipeanu, C. M., Zhou, F., Fugetta, E. K., & Wu, G. (2006). Differential regulation of the cell-surface targeting and function of beta- and alpha1-adrenergic receptors by Rab1 GTPase in cardiac myocytes. *Molecular Pharmacology, 69*, 1571–1578.

Gurkan, C., Stagg, S. M., Lapointe, P., & Balch, W. E. (2006). The COPII cage: Unifying principles of vesicle coat assembly. *Nature Reviews. Molecular Cell Biology, 7*, 727–738.

Kuge, O., Dascher, C., Orci, L., Rowe, T., Amherdt, M., Plutner, H., et al. (1994). Sar1 promotes vesicle budding from the endoplasmic reticulum but not Golgi compartments. *The Journal of Cell Biology, 125*, 51–65.

Ma, D., Zerangue, N., Lin, Y. F., Collins, A., Yu, M., Jan, Y. N., et al. (2001). Role of ER export signals in controlling surface potassium channel numbers. *Science, 291*, 316–319.

Mancias, J. D., & Goldberg, J. (2008). Structural basis of cargo membrane protein discrimination by the human COPII coat machinery. *The EMBO Journal, 27*, 2918–2928.

Margeta-Mitrovic, M., Jan, Y. N., & Jan, L. Y. (2000). A trafficking checkpoint controls GABA(B) receptor heterodimerization. *Neuron, 27*, 97–106.

Mikosch, M., Kaberich, K., & Homann, U. (2009). ER export of KAT1 is correlated to the number of acidic residues within a triacidic motif. *Traffic, 10*, 1481–1487.

Miller, E. A., Beilharz, T. H., Malkus, P. N., Lee, M. C., Hamamoto, S., Orci, L., et al. (2003). Multiple cargo binding sites on the COPII subunit Sec24p ensure capture of diverse membrane proteins into transport vesicles. *Cell, 114*, 497–509.

Mossessova, E., Bickford, L. C., & Goldberg, J. (2003). SNARE selectivity of the COPII coat. *Cell, 114*, 483–495.

Nishimura, N., & Balch, W. E. (1997). A di-acidic signal required for selective export from the endoplasmic reticulum. *Science, 277*, 556–558.

Nishimura, N., Bannykh, S., Slabough, S., Matteson, J., Altschuler, Y., Hahn, K., et al. (1999). A di-acidic (DXE) code directs concentration of cargo during export from the endoplasmic reticulum. *The Journal of Biological Chemistry, 274*, 15937–15946.

Nufer, O., Kappeler, F., Guldbrandsen, S., & Hauri, H. P. (2003). ER export of ERGIC-53 is controlled by cooperation of targeting determinants in all three of its domains. *Journal of Cell Science, 116*, 4429–4440.

Otte, S., & Barlowe, C. (2002). The Erv41p-Erv46p complex: Multiple export signals are required in trans for COPII-dependent transport from the ER. *The EMBO Journal, 21*, 6095–6104.

Petaja-Repo, U. E., Hogue, M., Laperriere, A., Walker, P., & Bouvier, M. (2000). Export from the endoplasmic reticulum represents the limiting step in the maturation and cell surface expression of the human delta opioid receptor. *The Journal of Biological Chemistry, 275*, 13727–13736.

Robert, J., Clauser, E., Petit, P. X., & Ventura, M. A. (2005). A novel C-terminal motif is necessary for the export of the vasopressin V1b/V3 receptor to the plasma membrane. *The Journal of Biological Chemistry, 280*, 2300–2308.

Sevier, C. S., Weisz, O. A., Davis, M., & Machamer, C. E. (2000). Efficient export of the vesicular stomatitis virus G protein from the endoplasmic reticulum requires a signal in the cytoplasmic tail that includes both tyrosine-based and di-acidic motifs. *Molecular Biology of the Cell, 11*, 13–22.

Votsmeier, C., & Gallwitz, D. (2001). An acidic sequence of a putative yeast Golgi membrane protein binds COPII and facilitates ER export. *The EMBO Journal, 20*, 6742–6750.

Wang, X., Matteson, J., An, Y., Moyer, B., Yoo, J. S., Bannykh, S., et al. (2004). COPII-dependent export of cystic fibrosis transmembrane conductance regulator from the ER uses a di-acidic exit code. *The Journal of Cell Biology, 167*, 65–74.

Wang, G., & Wu, G. (2012). Small GTPase regulation of GPCR anterograde trafficking. *Trends in Pharmacological Sciences*, *33*, 28–34.

Zhang, X., Dong, C., Wu, Q. J., Balch, W. E., & Wu, G. (2011). Di-acidic motifs in the membrane-distal C termini modulate the transport of angiotensin II receptors from the endoplasmic reticulum to the cell surface. *The Journal of Biological Chemistry*, *286*, 20525–20535.

Zuzarte, M., Rinne, S., Schlichthorl, G., Schubert, A., Daut, J., & Preisig-Muller, R. (2007). A di-acidic sequence motif enhances the surface expression of the potassium channel TASK-3. *Traffic*, *8*, 1093–1100.

CHAPTER ELEVEN

Amino Acid Residues of G-Protein-Coupled Receptors Critical for Endoplasmic Reticulum Export and Trafficking

Motonao Nakamura[*,1], **Daisuke Yasuda**[*], **Nobuaki Hirota**[*], **Teruyasu Yamamoto**[†], **Satoshi Yamaguchi**[†], **Takao Shimizu**[*], **Teruyuki Nagamune**[†]

[*]Department of Biochemistry and Molecular Biology, Graduate School of Medicine, The University of Tokyo, Tokyo, Japan

[†]Department of Chemistry and Biotechnology, Graduate School of Engineering, and Center for NanoBio Integration, The University of Tokyo, Tokyo, Japan

[1]Corresponding author: e-mail address: moto-nakamura@umin.net

Contents

1. Introduction 204
2. Generation of Mutant GPCRs 205
3. Examination of Mutant GPCR Trafficking 205
   3.1 ER retention of mutant GPCRs 206
   3.2 Requirement of conserved residues and helix 8 for ER export of GPCRs 208
4. ER Export of Mutant GPCRs by Specific Ligands 209
   4.1 Receptor-specific ligands facilitate the ER export of mutant GPCRs 209
   4.2 Export of mutant GPCRs from ER and surface trafficking following treatment with ligands 210
5. Functional Analysis of Surface-Trafficked Mutant GPCRs in Living Cells 210
   5.1 Analysis of surface-trafficked mutant GPCRs 210
   5.2 Properties of surface-trafficked mutant GPCRs 213
6. Conclusion 214
Acknowledgments 215
References 215

Abstract

Analysis of the structural features of rhodopsin-type G-protein-coupled receptors (GPCRs) revealed the existence of an additional α-helix, termed helix 8, in the C-terminal tail. Furthermore, these GPCRs were determined to possess several conserved residues in their transmembrane domains. The functional deficiencies of receptors in which these domains or residues have been mutated have not been examined in living cells due to their accumulation in the endoplasmic reticulum (ER), although the ligand

Methods in Enzymology, Volume 521
ISSN 0076-6879
http://dx.doi.org/10.1016/B978-0-12-391862-8.00011-9

affinities of these receptors have been tested in membrane preparations. Recent studies have demonstrated that ER-accumulated receptors are effectively exported from ER using membrane permeable ligands as pharmacological chaperones. Here, we identified several residues of the platelet-activating factor receptor and leukotriene B_4 type-II receptor that are crucial for export from ER. Moreover, we used their specific ligands as pharmacological chaperones to traffic ER-accumulated GPCRs to the cell surface in order to examine the functional deficiencies of each mutant receptor. Here, we introduce the novel technique of site-specific N-terminal labeling of cell surface proteins in living cells with Sortase-A, a transpeptidase isolated from *Staphylococcus aureus*, to evaluate the trafficking of receptors after agonist stimulation.

1. INTRODUCTION

G-protein-coupled receptors (GPCRs) are synthesized and modified in the endoplasmic reticulum (ER). During this process, mutations in the proteins are recognized by the ER quality control system, which prevents the receptors from trafficking to the surface. Small molecules known as "pharmacological chaperones or pharmacoperones" were recently identified (Bernier, Lagace, Bichet, & Bouvier, 2004; Morello, Petaja-Repo, Bichet, & Bouvier, 2000). These compounds bind to newly synthesized proteins and facilitate their folding, resulting in enhanced export of these proteins from ER. Various specific synthetic ligands for GPCRs have already been developed, making these ligands ideal for use as chaperones (Bernier, Bichet, & Bouvier, 2004; Conn & Ulloa-Aguirre, 2010; Conn, Ulloa-Aguirre, Ito, & Janovick, 2007; Dunham & Hall, 2009).

Leukotrienes, prostaglandins, platelet-activating factor (PAF), lysophosphatidic acid (LPA), endocannabinoids, etc. are known as lipid mediators. These molecules exert their biological effects via the activation of cognate GPCRs (Shimizu, 2009). Previously, we identified several GPCRs bound by these lipids, including PAF receptor (Honda et al., 1991; Nakamura et al., 1991), leukotriene B_4 receptors BLT1 (Yokomizo, Izumi, Chang, Takuwa, & Shimizu, 1997) and BLT2 (Yokomizo, Kato, Terawaki, Izumi, & Shimizu, 2000), and LPA receptors LPA4 (Noguchi, Ishii, & Shimizu, 2003) and P2Y5/LPA6 (Yanagida et al., 2009). Because these lipid mediators play crucial roles in the immune response and maintain cellular homeostasis, numerous synthetic ligands have been generated in an effort to control these lipid–GPCR pathways. During the structural and functional analyses of these GPCRs, several receptors with mutations in crucial domains were found to be accumulated in ER. Because some of the synthetic ligands for these receptors are membrane permeable, we were able to use these ligands as chaperones to successfully transport these mutants to the surface and examine their signaling and intracellular trafficking.

2. GENERATION OF MUTANT GPCRs

Rhodopsin-type GPCRs possess conserved amino acid residues in their transmembrane domains (TMs), such as Asp and Leu in TM2; Phe, Cys, and Pro in TM6; and Asp/Asn, Pro, and Tyr in TM7 (Bockaert & Pin, 1999; Lagerstrom & Schioth, 2008), as well as an amphipathic helix, helix 8, in the C-terminal tail (Cherezov et al., 2007; Jaakola et al., 2008; Murakami & Kouyama, 2008; Nakamura, Yasuda, Hirota, & Shimizu, 2010; Palczewski et al., 2000; Warne et al., 2008). The structural features of GPCRs revealed that mutation of these residues causes structural and functional defects in these proteins, resulting in their accumulation in ER. To confirm the importance of these residues for the correct folding and function of GPCRs in living cells, we generated several mutant GPCRs: the human PAF receptor (hPAFR), in which conserved residues were mutated (Ala substitution of Leu^{59}, Asp^{63}, Pro^{247}, Asp^{289}, or Pro^{290}), and human BLT2 (hBLT2) lacking helix 8 (truncation of Pro^{301}-Arg^{302}-Phe^{303}-Leu^{304}-Thr^{305}-Arg^{306}-Leu^{307}-Phe^{308}, hBLT2/ΔH8). HA-tagged hPAFR (HA-hPAFR) and hBLT2 (HA-hBLT2) were used as templates to construct these mutant receptors using the QuikChange Site-Directed Mutagenesis Kit (Stratagene, La Jolla, CA) following the manufacturer's instructions. The primer sets utilized have been listed in previous reports (Hirota et al., 2010; Yasuda et al., 2009). All mutant receptors were subcloned into a pcDNA3 vector.

3. EXAMINATION OF MUTANT GPCR TRAFFICKING

To evaluate the ER export of these mutant GPCRs, we performed flow cytometric analysis, glycosidase treatment followed by Western blot, and confocal microscopy. All of our results demonstrated that mutation of the conserved residues or helix 8 resulted in the ER retention of these receptors, confirming the importance of these residues and domain for correct folding during biosyntheses. The experimental procedures are detailed below.

3.1. ER retention of mutant GPCRs

3.1.1 Analysis of surface expression: Flow cytometry

Cell surface expression of the mutant receptors was semi-quantitatively measured by flow cytometry.

1. HeLa cells (1×10^6 cells/60-mm dish) were cultured in 5 ml of Dulbecco's modified Eagle's medium (DMEM; Sigma-Aldrich, St. Louis, MO) supplemented with 10% (v/v) fetal bovine serum at 37 °C in 5% CO_2.
2. The cells were transfected with a plasmid harboring wild-type (WT) or mutated receptors using Lipofectamine 2000 (Invitrogen, Carlsbad, CA) according to the manufacturer's protocol and cultured for 24 h.
3. For staining, cells were incubated with an anti-HA antibody (1:500, clone 3F10; Roche Applied Science, Penzberg, Germany) in 100 μl of phosphate-buffered saline (PBS) (calcium chloride and magnesium chloride free) supplemented with 2% goat serum (PBS/GS) at room temperature for 1 h, followed by staining with phycoerythrin (PE)-conjugated anti-rat IgG (1:1000; Beckman Coulter Electronics Ltd., Fullerton, CA) at room temperature for 30 min.
4. The cells were then washed twice with PBS/GS and analyzed with an EPICS XL flow cytometer (Beckman Coulter Electronics Ltd., Fullerton, CA).

3.1.2 Analysis of glyco-modification: Glycosidase treatment followed by Western blot

The accumulation of mutant GPCRs in ER results in the immature glyco-modification of these proteins.

1. Two days after transfection of HeLa cells (1×10^6 cells/60-mm dish), they were harvested in PBS containing 2 m*M* EDTA and disrupted by sonication (30 s × 20 times, 30-s intervals, on ice) in ice-cold sonication buffer [25 m*M* HEPES–NaOH (pH 7.4), 250 m*M* sucrose, 10 m*M* $MgCl_2$, protease inhibitor mixture (Roche Applied Science, Penzberg, Germany)].
2. Cell debris was removed by centrifugation at 1000 × *g* for 10 min at 4 °C, and the resultant supernatants were used as protein samples.
3. Protein concentration was determined by the Bradford method using a Protein Assay Kit (Bio-Rad, Hercules, CA) with bovine serum albumin (BSA; Sigma-Aldrich, St. Louis, MO) as a standard.
4. Protein samples were treated with endoglycosidase-H (Endo-H; 0.005 units; Roche Applied Science, Penzberg, Germany) in 50 μl

of buffer (11.7 mM Na_2HPO_4, 168.3 mM NaH_2PO_4, 0.4% sodium dodecyl sulfate (SDS), 20 mM EDTA, 2% 2-mercaptoethanol) or with *N*-glycosidase F (PNGase-F; 1 unit; Roche Applied Science, Penzberg, Germany) in 50 μl of buffer (139.2 mM Na_2HPO_4, 40.8 mM NaH_2PO_4, 0.4% SDS, 20 mM EDTA, 2% 2-mercaptoethanol) for 16 h at 4 °C.

5. The resultant samples were suspended in sample buffer [25 mM Tris–HCl (pH 6.5), 5% glycerol, 1% SDS, and 0.05% bromophenol blue], then separated on 10% SDS-polyacrylamide gels, and transferred to a nitrocellulose membrane.
6. For Western blot analyses, nitrocellulose membranes were incubated with 5% skim milk in TBS-T [20 mM Tris-buffered saline (pH 7.4), 0.1% Tween 20] and probed with primary antibody (anti-HA antibody; 1:500, clone 3F10) for 1 h at room temperature.
7. The membranes were washed twice with TBS-T and incubated with an HRP-conjugated anti-rat IgG antibody (1:1000; Santa Cruz Biotechnology, Santa Cruz, CA) for 1 h.
8. Chemiluminescent signals were visualized using an ECL Western Blotting Detection System (GE Healthcare, Little Chalfont, UK).

3.1.3 ER localization: Confocal microscopy

1. HeLa cells (5×10^5 cells) were seeded on collagen-coated glass-bottomed 35-mm dishes (MatTek Corporation, Ashland, MA).
2. After transfection and incubation for 6, 12, or 24 h at 37 °C in 5% CO_2, the cells were fixed with 2% paraformaldehyde for 10 min at room temperature and rinsed twice with ice-cold PBS.
3. The cells were then incubated with 1/4× permeabilization reagent (Beckman Coulter, Marseille, France) for 10 min at room temperature.
4. Primary antibodies were then added [anti-HA (1:500, 3F10) and anti-calreticulin (1:1000; Stressgen Bioreagents, Ann Arbor, MI) or anti-golgin-97 (1:1000; Invitrogen, Carlsbad, CA)] and the mixture was incubated for 1 h at room temperature.
5. After the cells were washed twice with ice-cold PBS, cells were incubated with secondary antibodies [Alexa Fluor 488-conjugated anti-rat IgG (1:1000; Invitrogen, Carlsbad, CA) and Alexa Fluor 546-conjugated anti-rabbit IgG (1:1000; Invitrogen, Carlsbad, CA) or Alexa Fluor 546-conjugated anti-mouse IgG (1:1000; Invitrogen, Carlsbad, CA)] for 1 h at room temperature.

6. Images were obtained using an LSM510 Laser Scanning Confocal Microscope (Carl Zeiss, Jena, Germany) equipped with argon and helium/neon lasers and a 100 × oil immersion objective lens.

3.2. Requirement of conserved residues and helix 8 for ER export of GPCRs

All mutants generated in Section 2 (HA-hPAFR/D63A, HA-hPAFR/P247A, and HA-hBLT2/ΔH8) showed significantly decreased surface expression compared with the WT receptor in flow cytometric analyses (data not shown; see Hirota et al., 2010; Yasuda et al., 2009). In Western blot analyses, HA-hPAFR/WT showed several bands: a single broad band (approx. 40–50 kDa) and two narrow bands (30 and 27 kDa) (data not shown; see Hirota et al., 2010; Yasuda et al., 2009). The broad species disappeared following the treatment of the sample with PNGase-F, but not after treatment with Endo-H, indicating that these proteins possess mature glyco-chains. On the other hand, treatment with either PNGase-F or Endo-H reduced the intensity of the 30-kDa band, leading to an increase in the intensity of the 27-kDa protein. These data indicate that the 30- and 27 kDa proteins correspond to a core glyco-chain-conjugated and non-glycosylated form of the receptor, respectively, both of which could be retained in ER. By contrast, the mature glycosylated form (broad band) was not detected in HA-hPAFR/D63A- and HA-hPAFR/P247A-expressing cells, although the core glyco-chain-conjugated and non-glycosylated forms of the receptors were observed (data not shown; see Hirota et al., 2010). Similar results were obtained in HA-hBLT2/ΔH8-expressing cells (data not shown; see Yasuda et al., 2009), suggesting that the mutant receptors are not exported from ER. Confocal microscopy analysis revealed that at 6 and 12 h after transfection, HA-hPAFR/WT localized in the ER and the Golgi apparatus, and at 24 h, it was trafficked to the plasma membrane. By contrast, there was no overlap between HA-hPAFR/D63A or HA-hPAFR/P247A and golgin-97, a Golgi apparatus marker, even though these mutant receptors colocalized with calreticulin, an ER marker, indicating that these receptors were accumulated in ER (data not shown; see Hirota et al., 2010). Similar results were obtained from analysis of HA-hBLT2 with regard to the surface localization of HA-hBLT2/WT and the accumulation of HA-hBLT2/ΔH8 in ER (data not shown; see Yasuda et al., 2009). Taken together, these data indicated that mutation of the conserved residues or helix 8 leads to an aberrant protein structure, which is recognized by the ER quality control system and prevented from trafficking to the surface.

4. ER EXPORT OF MUTANT GPCRs BY SPECIFIC LIGANDS

Several reports demonstrate that mutant receptors are rescued from ER accumulation by pharmacological chaperones that enhance correct folding and stability (Fan, Perry, Gao, Schwarz, & Maki, 2005; Janovick, Maya-Nunez, & Conn, 2002; Morello, Salahpour, et al., 2000; Noorwez et al., 2003; Petaja-Repo et al., 2002; Robert, Auzan, Ventura, & Clauser, 2005). In this section, we show that the specific ligands for hPAFR and hBLT2 act as chaperones for their mutant receptors.

4.1. Receptor-specific ligands facilitate the ER export of mutant GPCRs

Y-24180, a PAFR antagonist, was generously donated by Yoshitomi Pharmaceutical Industries, Ltd. (Osaka, Japan). Methylcarbamyl (mc)-PAF C-16, a PAFR agonist, was purchased from Cayman Chemical (Ann Arbor, MI). A BLT2 antagonist, ZK 158252, was generously donated by Bayer Schering Pharma AG (Berlin, Germany). To evaluate the trafficking of mutant GPCRs to the cell surface, flow cytometric analysis and immunofluorescence confocal microscopy were performed.

Flow cytometric analysis

1. Four hours after the transfection of HeLa cells (1×10^6 cells/60-mm dish), 1 μ*M* Y24180 or 1 μ*M* mc-PAF for hPAFR and 10 μ*M* ZK 158252 for hBLT2 were added to the culture medium, and the cells were incubated for 20 h.
2. The cells were then incubated with anti-HA antibody in 100 μl of PBS/GS at room temperature for 1 h, followed by staining with PE-conjugated anti-rat IgG at room temperature for 30 min.
3. The cells were then washed twice with PBS/GS and analyzed with an EPICS XL flow cytometer.

Immunofluorescence confocal microscopy

1. Cells (5×10^5 cells) were seeded on collagen-coated glass-bottomed 35-mm dishes and transfected.
2. Four hours after transfection, 1 μ*M* Y24180 or 1 μ*M* mc-PAF for hPAFR and 10 μ*M* ZK 158252 for hBLT2 were added to the culture medium, and the cells were incubated for 20 h.

3. The cells were then fixed with 2% paraformaldehyde for 10 min at room temperature and rinsed twice with ice-cold PBS.
4. The cells were incubated with 1/4 × permeabilization reagent for 10 min at room temperature. Primary antibodies [anti-HA (3F10) and anti-calreticulin] were added, and the mixture was incubated for 1 h at room temperature.
5. After the cells were washed twice with ice-cold PBS, secondary antibodies (Alexa Fluor 488-conjugated anti-rat IgG and Alexa Fluor 546-conjugated anti-rabbit IgG) were added, and the cells were incubated for 1 h at room temperature.
6. Images were obtained using an LSM510 Laser Scanning Confocal Microscope equipped with argon and helium/neon lasers and a 100 × oil immersion objective lens.

4.2. Export of mutant GPCRs from ER and surface trafficking following treatment with ligands

Although the mutation of the conserved residues or helix 8 of GPCRs resulted in their accumulation in ER, all of these mutants, including HA-hPAFR/D63A, HA-hPAFR/P247A, and HA-hBLT2/ΔH8, were exported from ER following treatment with their specific ligands. The surface localization of mutant receptors after treatment with specific ligands was confirmed by flow cytometric analysis and immunofluorescence confocal microscopy (data not shown; see Hirota et al., 2010; Yasuda et al., 2009).

5. FUNCTIONAL ANALYSIS OF SURFACE-TRAFFICKED MUTANT GPCRs IN LIVING CELLS

In Section 4, we successfully transported mutant GPCRs to the cell surface using their specific ligands as pharmacological chaperones. Next, we examined signaling and internalization of the surface-trafficked mutant GPCRs. In this section, we describe several methods used to evaluate the function of the surface-trafficked mutant GPCRs, including agonist-elicited intracellular Ca^{2+} responses and internalization analysis using Sortase-A-mediated labeling (Tsukiji & Nagamune, 2009; Yamamoto & Nagamune, 2009).

5.1. Analysis of surface-trafficked mutant GPCRs

5.1.1 Agonist-elicited intracellular Ca^{2+} response

1. HeLa cells were plated on a 100-mm dish (3×10^6 cells/dish) and incubated for 16 h before transfection.

2. Four hours after transfection of HeLa cells with plasmids bearing WT or mutant GPCRs using Lipofectamine 2000, 1 μ*M* Y-24180 for HA-hPAFR or 10 μ*M* ZK 158252 for HA-hBLT2 was added.
3. The cells were incubated for 20 h, followed by extensive washing with PBS.
4. Intracellular Ca^{2+} increases were elicited by mc-PAF or 12*S*-hydroxy-5-*cis*-8,10-*trans*-heptadecatrienoic acid (12-HHT), a strong agonist for BLT2, and evaluated as follows: the transfected cells were loaded with 3 μ*M* Fura-2 AM (Dojin, Kumamoto, Japan) in a modified HEPES-Tyrode's BSA buffer [25 m*M* HEPES–NaOH (pH 7.4), 140 m*M* NaCl, 2.7 m*M* KCl, 1 m*M* $CaCl_2$, 12 m*M* $NaHCO_3$, 5.6 m*M* D-glucose, 0.37 m*M* NaH_2PO_4, 0.49 m*M* $MgCl_2$, 0.1% (w/v) BSA] containing 0.01% pluronic acid (Molecular Probes, Eugene, OR) at 37 °C for 1 h.
5. The cells were washed and resuspended in HEPES-Tyrode's BSA buffer at a density of 1×10^6 cells/ml.
6. The cell suspension (0.5 ml) was applied to a CAF-110 system (JASCO Corp., Tokyo, Japan), and ligand solution (5 μl) was added to activate the receptor.
7. The intracellular Ca^{2+} concentration was determined by measuring the ratio of emission fluorescence at 510 nm by excitation at 340 and 380 nm. The free Ca^{2+} concentration was calculated from the equation:

$$\left[Ca^{2+}\right]_i = K_d(R_{free}/R_{bound})[(F - F_{min})/(F_{max} - F)]$$

K_d: Ca^{2+}-binding dissociation constant (224 n*M* for Fura-2)

F: 510 nm fluorescence intensity ratio

F_{max}: maximal fluorescence intensity ratio determined after the addition of 0.1% Triton X-100 to permeabilize the cells in the presence of 1 m*M* Ca^{2+}

F_{min}: minimal fluorescence intensity ratio determined after permeabilization and the addition of 5 m*M* EGTA

R_{bound}: fluorescence at an excitation of 380 nm determined after the addition of 0.1% Triton X-100 to permeabilize the cells in the presence of 1 m*M* Ca^{2+}

R_{free}: fluorescence at an excitation of 380 nm determined after permeabilization and the addition of 5 m*M* EGTA

5.1.2 Sortase-A-mediated labeling of surface GPCRs for analysis of internalization

1. An N-terminally 6 × His-tagged Sortase-A (His_6-Sortase-A)-expression plasmid was introduced into the *Escherichia coli* BL21(DE3) strain.

2. The cells were grown in Luria-Bertani (LB) medium to an OD (600 nm) value of 0.8, and the production of His_6–Sortase-A was then induced by the addition of 0.3 m*M* isopropyl-β-D-thiogalactopyranoside.
3. The cells were grown for 16 h at 27 °C and harvested by centrifugation. The cell pellets were resuspended in 50 m*M* phosphate, 300 m*M* NaCl (pH 7.0) and lysed by sonication on ice.
4. His_6–Sortase-A was purified from the soluble fraction of the lysate by His-Trap chelating column (GE Healthcare) according to the manufacturer's protocol and dialyzed against a buffer containing 50 m*M* Tris–HCl (pH 8.0) and 150 m*M* NaCl. Alexa Fluor 488-Leu-Pro-Glu-Thr-Gly-Gly (AF488-LPETGG) was generated by reaction of the synthesized peptide with Alexa Fluor 488-carboxylic acid, succinimidyl ester (Invitrogen, Carlsbad, CA) (2 equiv) in dry DMSO with diisopropylethylamine for 2 h, followed by purification with reversed-phase HPLC.
5. The Sortase-A recognition sequence, Met-Leu-Pro-Glu-Thr-Gly-Gly-Gly-Gly-Gly ($MLPETG_5$), was created at the N-terminus of WT and mutant HA-hPAFRs by PCR, and the resulting constructs were subcloned into pcDNA3. The primer sets utilized are presented in our previous paper (Hirota et al., 2010).
6. Labeling of the cell surface receptors by His_6-Sortase-A was achieved as shown in Fig. 11.1. The transpeptidation reaction was performed by incubation of the transfected HeLa cells in DMEM (0.1% BSA) containing

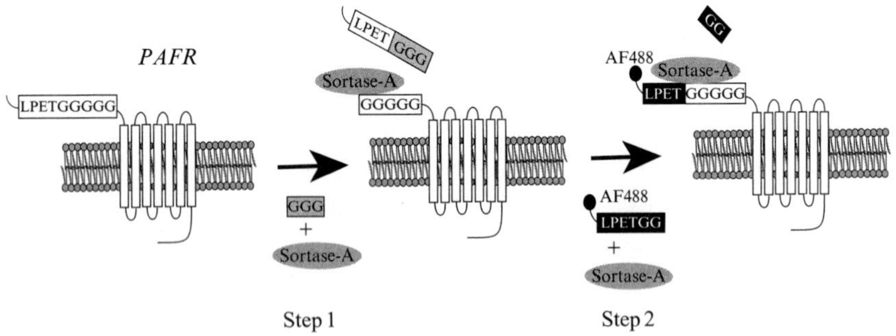

Figure 11.1 N-terminal labeling of $LPETG_5$-tagged cell surface PAFRs with LPETGG-tethered Alexa Fluor-488 (AF488) by Sortase-A. In Step 1, the cells expressing $LPETG_5$-tagged PAFRs were incubated with His_6-Sortase-A and triglycine, resulting in the cleavage between the Thr and Gly of the $LPETG_5$-tag and generation of the pentaglycine-tagged PAFRs. In the following step (Step 2), the labeling of the target receptors was achieved by incubation of the cells with His_6-Sortase-A and LPETGG-tethered Alexa Fluor-488.

30 μ*M* His_6-Sortase-A and 1 m*M* triglycine at 37 °C in 5% CO_2 for 30 min (Fig. 11.1, Step 1).

7. After the cells were washed twice with PBS, they were incubated in fresh DMEM (0.1% BSA) containing 30 μ*M* His_6-Sortase-A and 10 μ*M* AF488-LPETGG peptide at 37 °C in 5% CO_2 for 15 min (Fig. 11.1, Step 2).
8. The cells were then washed with PBS and observed using a confocal laser microscope.

5.2. Properties of surface-trafficked mutant GPCRs

5.2.1 Surface-trafficked mutant GPCRs elicit intracellular Ca^{2+} signaling

HA-hPAFR/P247A was translocated to the cell surface by the treatment with Y-24180. Cells were washed with PBS to remove the compound, and agonist (mc-PAF)-induced intracellular Ca^{2+} mobilization was examined. In these cells, mc-PAF induced an intracellular Ca^{2+} increase with a dose dependency similar to that of HA-hPAFR/WT-expressing cells. These results suggested that HA-hPAFR/P247A is functional, even though it is recognized as a misfolded protein by the ER quality control system (data not shown; see Hirota et al., 2010). By contrast, no mc-PAF-elicited Ca^{2+} mobilization was detected in HA-hPAFR/D63A-expressing cells, even though this mutant was successfully trafficked to the cell surface by Y-24180 (data not shown; see Hirota et al., 2010). Thus, D63 is important for the function of PAFR. Interestingly, the chaperones enable mutant receptors to circumvent quality control in ER, regardless of whether or not they are functional. In hBLT2/ΔH8-expressing cells, which were not pretreated with ZK 158252, an increase in intracellular Ca^{2+} was not observed in response to 12-HHT, a strong agonist for BLT2. However, after treatment with ZK 158252, 12-HHT-elicited Ca^{2+} mobilization was detected in these cells (data not shown; see Yasuda et al., 2009).

5.2.2 Accumulation of hPAFR/P247A in early endosome after agonist-mediated ER export

When mc-PAF, a stable PAFR agonist, was used as a chaperone for the rescue of the HA-hPAFR/P247A mutant receptor, the surface expression of this mutant was not increased. Intriguingly, HA-hPAFR/P247A localized predominantly to the early endosomes (data not shown; see Hirota et al., 2010). Thus, it is possible that HA-hPAFR/P247A was trafficked to the cell surface by the help of mc-PAF, but subsequently disappeared from the surface due to aberrant trafficking, for example, enhanced internalization, deficiency in recycling, and/or accelerated degradation. Indeed, this was confirmed using Sortase-A-mediated labeling of cell surface proteins.

5.2.3 Recycling deficiency of hPAFR/P247A after agonist stimulation

After HA-hPAFR/WT- and HA-hPAFR/P247A-expressing cells were treated with Y-24180 (antagonist) to facilitate surface expression, the cells were stimulated with mc-PAF and internalization of the receptors was evaluated. Using a Sortase-A-mediated labeling technique, surface HA-hPAFR/WT and HA-hPAFR/P247A receptors were specifically labeled with Alexa Fluor-488, and translocation of these receptors induced by mc-PAF was examined by confocal microscopy. Although a small amount of Alexa Fluor-488-labeled HA-hPAFR/WT was detected in the early endosomes, a considerable portion of this wild-type HA-hPAFR was located on the cell surface due to effective recycling (data not shown; see Hirota et al., 2010). On the other hand, Alexa Fluor-488-labeled HA-hPAFR/P247A receptors were predominantly detected in the early endosomes after stimulation with mc-PAF. These results suggested that this mutant exhibits abnormal trafficking, for example, through enhanced internalization, deficiency in recycling, and/or accelerated degradation.

6. CONCLUSION

This study demonstrates that some mutant GPCRs are able to bypass the ER quality control system by the action of pharmacological chaperones. However, it is still unclear how this system discriminates between correctly folded and misfolded receptors. We describe here a novel technique, site-specific N-terminal labeling of cell surface proteins on living cells by Sortase-A, that can be used to evaluate the trafficking of mutant receptors after agonist stimulation. The speed of this method is comparable to that of other labeling methods, and it is useful for pulse–chase labeling with high time resolution. Furthermore, enzymatic labeling by Sortase-A can be performed in serum-containing medium, thereby decreasing cell damage. The use of this technique allows not only the study of the trafficking of cell surface proteins, including GPCRs, but also the opportunity to study the dynamics of GPCR movement in real time.

ACKNOWLEDGMENTS

This work was supported in part by a Grants-in Aid from the Ministry of Education, Culture, Sports, Science and Technology of Japan (to M. N. and T. S.), a grant from the Japan Society for the Promotion of Science (Global COE program) (to T. S.), and the Center for NanoBio Integration at the University of Tokyo (T. S. and T. N.).

REFERENCES

Bernier, V., Bichet, D. G., & Bouvier, M. (2004). Pharmacological chaperone action on G-protein-coupled receptors. *Current Opinion in Pharmacology, 4*, 528–533.

Bernier, V., Lagace, M., Bichet, D. G., & Bouvier, M. (2004). Pharmacological chaperones: Potential treatment for conformational diseases. *Trends in Endocrinology and Metabolism, 15*, 222–228.

Bockaert, J., & Pin, J. P. (1999). Molecular tinkering of G protein-coupled receptors: An evolutionary success. *The EMBO Journal, 18*, 1723–1729.

Cherezov, V., Rosenbaum, D. M., Hanson, M. A., Rasmussen, S. G., Thian, F. S., Kobilka, T. S., et al. (2007). High-resolution crystal structure of an engineered human beta2-adrenergic G protein-coupled receptor. *Science, 318*, 1258–1265.

Conn, P. M., & Ulloa-Aguirre, A. (2010). Trafficking of G-protein-coupled receptors to the plasma membrane: Insights for pharmacoperone drugs. *Trends in Endocrinology and Metabolism, 21*, 190–197.

Conn, P. M., Ulloa-Aguirre, A., Ito, J., & Janovick, J. A. (2007). G protein-coupled receptor trafficking in health and disease: Lessons learned to prepare for therapeutic mutant rescue in vivo. *Pharmacological Reviews, 59*, 225–250.

Dunham, J. H., & Hall, R. A. (2009). Enhancement of the surface expression of G protein-coupled receptors. *Trends in Biotechnology, 27*, 541–545.

Fan, J., Perry, S. J., Gao, Y., Schwarz, D. A., & Maki, R. A. (2005). A point mutation in the human melanin concentrating hormone receptor 1 reveals an important domain for cellular trafficking. *Molecular Endocrinology, 19*, 2579–2590.

Hirota, N., Yasuda, D., Hashidate, T., Yamamoto, T., Yamaguchi, S., Nagamune, T., et al. (2010). Amino acid residues critical for endoplasmic reticulum export and trafficking of platelet-activating factor receptor. *The Journal of Biological Chemistry, 285*, 5931–5940.

Honda, Z., Nakamura, M., Miki, I., Minami, M., Watanabe, T., Seyama, Y., et al. (1991). Cloning by functional expression of platelet-activating factor receptor from guinea-pig lung. *Nature, 349*, 342–346.

Jaakola, V. P., Griffith, M. T., Hanson, M. A., Cherezov, V., Chien, E. Y., Lane, J. R., et al. (2008). The 2.6 angstrom crystal structure of a human A2A adenosine receptor bound to an antagonist. *Science, 322*, 1211–1217.

Janovick, J. A., Maya-Nunez, G., & Conn, P. M. (2002). Rescue of hypogonadotropic hypogonadism-causing and manufactured GnRH receptor mutants by a specific protein-folding template: Misrouted proteins as a novel disease etiology and therapeutic target. *The Journal of Clinical Endocrinology and Metabolism, 87*, 3255–3262.

Lagerstrom, M. C., & Schioth, H. B. (2008). Structural diversity of G protein-coupled receptors and significance for drug discovery. *Nature Reviews. Drug Discovery, 7*, 339–357.

Morello, J. P., Petaja-Repo, U. E., Bichet, D. G., & Bouvier, M. (2000). Pharmacological chaperones: A new twist on receptor folding. *Trends in Pharmacological Sciences, 21*, 466–469.

Morello, J. P., Salahpour, A., Laperriere, A., Bernier, V., Arthus, M. F., Lonergan, M., et al. (2000). Pharmacological chaperones rescue cell-surface expression and function of misfolded V2 vasopressin receptor mutants. *The Journal of Clinical Investigation, 105*, 887–895.

Murakami, M., & Kouyama, T. (2008). Crystal structure of squid rhodopsin. *Nature, 453*, 363–367.
Nakamura, M., Honda, Z., Izumi, T., Sakanaka, C., Mutoh, H., Minami, M., et al. (1991). Molecular cloning and expression of platelet-activating factor receptor from human leukocytes. *The Journal of Biological Chemistry, 266*, 20400–20405.
Nakamura, M., Yasuda, D., Hirota, N., & Shimizu, T. (2010). Specific ligands as pharmacological chaperones: The transport of misfolded G-protein coupled receptors to the cell surface. *IUBMB Life, 62*, 453–459.
Noguchi, K., Ishii, S., & Shimizu, T. (2003). Identification of p2y9/GPR23 as a novel G protein-coupled receptor for lysophosphatidic acid, structurally distant from the Edg family. *The Journal of Biological Chemistry, 278*, 25600–25606.
Noorwez, S. M., Kuksa, V., Imanishi, Y., Zhu, L., Filipek, S., Palczewski, K., et al. (2003). Pharmacological chaperone-mediated in vivo folding and stabilization of the P23H-opsin mutant associated with autosomal dominant retinitis pigmentosa. *The Journal of Biological Chemistry, 278*, 14442–14450.
Palczewski, K., Kumasaka, T., Hori, T., Behnke, C. A., Motoshima, H., Fox, B. A., et al. (2000). Crystal structure of rhodopsin: A G protein-coupled receptor. *Science, 289*, 739–745.
Petaja-Repo, U. E., Hogue, M., Bhalla, S., Laperriere, A., Morello, J. P., & Bouvier, M. (2002). Ligands act as pharmacological chaperones and increase the efficiency of delta opioid receptor maturation. *The EMBO Journal, 21*, 1628–1637.
Robert, J., Auzan, C., Ventura, M. A., & Clauser, E. (2005). Mechanisms of cell-surface rerouting of an endoplasmic reticulum-retained mutant of the vasopressin V1b/V3 receptor by a pharmacological chaperone. *The Journal of Biological Chemistry, 280*, 42198–42206.
Shimizu, T. (2009). Lipid mediators in health and disease: Enzymes and receptors as therapeutic targets for the regulation of immunity and inflammation. *Annual Review of Pharmacology and Toxicology, 49*, 123–150.
Tsukiji, S., & Nagamune, T. (2009). Sortase-mediated ligation: A gift from Gram-positive bacteria to protein engineering. *Chembiochem, 10*, 787–798.
Warne, T., Serrano-Vega, M. J., Baker, J. G., Moukhametzianov, R., Edwards, P. C., Henderson, R., et al. (2008). Structure of a beta1-adrenergic G-protein-coupled receptor. *Nature, 454*, 486–491.
Yamamoto, T., & Nagamune, T. (2009). Expansion of the sortase-mediated labeling method for site-specific N-terminal labeling of cell surface proteins on living cells. *Chemical Communications (Cambridge, England)*, 1022–1024.
Yanagida, K., Masago, K., Nakanishi, H., Kihara, Y., Hamano, F., Tajima, Y., et al. (2009). Identification and characterization of a novel lysophosphatidic acid receptor, p2y5/LPA6. *The Journal of Biological Chemistry, 284*, 17731–17741.
Yasuda, D., Okuno, T., Yokomizo, T., Hori, T., Hirota, N., Hashidate, T., et al. (2009). Helix 8 of leukotriene B4 type-2 receptor is required for the folding to pass the quality control in the endoplasmic reticulum. *The FASEB Journal, 23*, 1470–1481.
Yokomizo, T., Izumi, T., Chang, K., Takuwa, Y., & Shimizu, T. (1997). A G-protein-coupled receptor for leukotriene B4 that mediates chemotaxis. *Nature, 387*, 620–624.
Yokomizo, T., Kato, K., Terawaki, K., Izumi, T., & Shimizu, T. (2000). A second leukotriene B(4) receptor, BLT2. A new therapeutic target in inflammation and immunological disorders. *The Journal of Experimental Medicine, 192*, 421–432.

SECTION 3

GPCR Oligomerization

CHAPTER TWELVE

G-Protein-Coupled Heteromers: Regulation in Disease

Ivone Gomes[*], **Achla Gupta**[*], **Lakshmi A. Devi**[*,†,1]

[*]Department of Pharmacology and Systems Therapeutics, Mount Sinai School of Medicine, New York, New York, USA

[†]The Friedman Brain Institute, Mount Sinai School of Medicine, New York, New York, USA

[1]Corresponding author: e-mail address: lakshmi.devi@mssm.edu

Contents

1. Introduction 220
2. Generation of Heteromer-Selective mAbs 221
   2.1 Preparation of antigen for immunization 222
   2.2 Subtractive immunization 223
   2.3 Fusion protocol to generate hybridoma-secreting clones 224
   2.4 Purification of mAb from hybridoma supernatant 226
   2.5 Serum preparation 227
   2.6 Trouble-shooting and precautions 228
3. ELISA for Detection of Receptor Heteromers 228
4. Immunofluorescence for Visualization of Receptor Heteromers 230
5. Immunoprecipitation and Western Blotting 232
   5.1 Lysis of cells/tissues 233
   5.2 Problems and troubleshooting 235
6. Summary and Perspectives 235
Acknowledgment 236
References 236

Abstract

Over the past decade, an increasing number of studies have shown that G-protein-coupled receptors including opioid and cannabinoid receptors associate to form heteromers. Moreover, G-protein-coupled receptor heteromerization leads to the modulation of the binding, signaling, and trafficking properties of individual receptors. Although very little information is available about the physiological role of receptor heteromers, some studies have shown that the levels of some heteromers are upregulated in disease states such as preeclamptic pregnancy, schizophrenia, Parkinson's, ethanol-induced liver fibrosis, and development of tolerance to morphine. The recent generation of antibodies that selectively recognize distinct heteromers and, of peptides that selectively disrupt them, have started to elucidate the contribution of heteromers to the disease state. Here, we describe the methods for the generation of heteromer-selective antibodies and elucidation of their levels and localization under normal and pathological conditions.

Methods in Enzymology, Volume 521
ISSN 0076-6879
http://dx.doi.org/10.1016/B978-0-12-391862-8.00012-0

1. INTRODUCTION

Drugs of abuse such as opioids and cannabinoids act through G-protein-coupled receptors (GPCRs), the opioid and cannabinoid receptors. Three opioid (μ, δ, and κ) and cannabinoid (CB_1R, CB_2R, and GPR55) receptor subtypes have been identified (Balenga, Henstridge, Kargl, & Waldhoer, 2011; Dietis, Rowbotham, & Lambert, 2011; Di Marzo, Piscitelli, & Mechoulam, 2011). Both receptors signal via $G_{\alpha i/o}$ proteins to activate similar signal transduction cascades leading to decreases in intracellular cyclic AMP levels, inhibition of neurotransmitter release, and to increases in mitogen-activated protein kinase phosphorylation (Bushlin, Rozenfeld, & Devi, 2010; Cichewicz, 2004; Howlett et al., 2002; Vigano, Rubino, & Parolaro, 2005). Moreover, activation of either receptor induces similar physiological responses such as antinociception, sedation, reward, and emotional responses (Maldonado, Valverde, & Berrendero, 2006; Manzanares et al., 1999). This similarity in mechanisms of action and physiological responses suggests the possibility of interactions between the opioid and cannabinoid systems.

Opioid receptor subtypes can associate to form higher-order structures, a process known as heteromerization. For example, μ (μOR) and δ (δOR) opioid receptors heteromerize and these modulate binding, signaling, and morphine-mediated analgesia (Gomes et al., 2004, 2000; Gomes, Ijzerman, Ye, Maillet, & Devi, 2011; Kabli et al., 2010; Levac, O'Dowd, & George, 2002; Rozenfeld & Devi, 2007). Heteromerization between δOR and κ opioid receptors (κOR) leads to novel pharmacology and alteration of individual receptor-trafficking properties (Berg et al., 2012; Bhushan, Sharma, Xie, Daniels, & Portoghese, 2004; Jordan & Devi, 1999). Furthermore, opioid receptors can heteromerize with other family A GPCRs such as α2A adrenergic (Jordan, Gomes, Rios, Filipovska, & Devi, 2003; Rios, Gomes, & Devi, 2004), β2 adrenergic (Jordan, Trapaidze, Gomes, Nivarthi, & Devi, 2001), chemokine (Chen et al., 2004; Hereld & Jin, 2008; Pello et al., 2008), substance P (Pfeiffer et al., 2003), or somatostatin receptors (Pfeiffer et al., 2002). Interestingly, heteromerization between CB_1R and μOR, δOR, or angiotensin AT1 receptors (AT1Rs) leads to alterations in signaling and localization of CB_1R (Rios, Gomes, & Devi, 2006; Rozenfeld et al., 2012, 2011). However, little information is available about the physiological role of GPCR heteromers due to a lack of appropriate tools to study them in endogenous tissues and to distinguish from receptor homomers. Studies using mainly coimmunoprecipitation techniques suggest the involvement of some GPCR heteromers in disease. Heteromers between dopamine D1–D2 receptors have been implicated in major depression (Pei et al., 2010), between AT1R and adrenergic α1D or AT1R and bradykinin B2 receptors with preeclamptic pregnancy (AbdAlla, Abdel-Baset, Lother, el Massiery, & Quitterer, 2005; Gonzalez-Hernandez Mde, Godinez-Hernandez, Bobadilla-Lugo, & Lopez-Sanchez, 2010) and between dopamine receptor subtypes as well as dopamine D2 and adenosine 2A receptors in schizophrenia (Dziedzicka-Wasylewska, Faron-Gorecka, Gorecki, & Kusemider, 2008; Faron-Gorecka, Gorecki, Kusmider, Wasylewski, & Dziedzicka-Wasylewska, 2008; Fuxe et al., 2005; Maggio & Millan, 2010; Perreault, O'Dowd, & George, 2011). However, direct demonstration of heteromers *in vivo* has not been possible due to a lack of appropriate reagents.

We recently generated monoclonal antibodies (mAbs) that selectively recognize heteromers over individual receptor homomers using a subtractive immunization strategy. This enabled studies to directly explore the physiological role of GPCR heteromers. For example, these antibodies can be used to detect the presence of a heteromer in a specific tissue/region. A case in point is the detection of δOR–κOR heteromers in peripheral sensory neurons using δOR–κOR selective antibodies (Berg et al., 2012). Alternatively, the antibodies could implicate the heteromer in a disease state. μOR–δOR heteromer-selective antibodies detect increased heteromer levels in brain regions involved in pain processing following chronic morphine administration under conditions leading to the development of tolerance (Gupta et al., 2010), suggesting that they may play a role in tolerance. This is supported by studies showing that μOR–δOR heteromer disruption leads to enhanced morphine analgesia with a concomitant decrease in tolerance (He et al., 2011). CB_1R–AT1R heteromer-selective antibodies detect a significant heteromer upregulation in hepatic stellate cells of rats chronically treated with ethanol (Rozenfeld et al., 2011), suggesting its involvement in ethanol-induced liver fibrosis. Here, we describe the generation of heteromer-selective antibodies and their use with enzyme-linked immunosorbent assays (ELISAs), immunofluorescence, immunoprecipitation, and Western blotting to detect levels and localization of heteromers in native tissues under normal and pathological conditions.

2. GENERATION OF HETEROMER-SELECTIVE mAbs

The advantages of mAbs to probe for heteromer levels in normal and disease states are that they recognize a single epitope, are highly specific, and can be produced in large quantities. The challenge in the generation of heteromer-selective mAbs is that the "heteromer-specific" epitope may be present in a cell or tissue at relatively low levels thus limiting the chances of being detected by antibody-producing cells. This limitation can be overcome through the use of a subtractive immunization strategy (Salata et al., 1992; Sleister & Rao, 2001, 2002). The first step in subtractive immunization involves tolerizing mice with membrane preparations that either do not express the receptor of interest (usually HEK-293 membranes) or express one of the two receptors (e.g., Neuro2A neuroblastoma cell membranes used to generate CB_1R–AT1R heteromer-selective mAb (Rozenfeld et al., 2011) endogenously express CB_1R). Tolerization is achieved by killing activated antibody-producing cells through the use of a chemical agent such as cyclophosphamide. Once a consistently low titer, as ascertained by ELISA, is obtained with the membranes used for the tolerization step, the mice are immunized with membranes from cells coexpressing the receptors forming the targeted heteromer. The immunizations are repeated until a high titer is obtained by ELISA at which time the mice are killed and spleens are removed for the generation of mAbs. Once the mAbs are generated, their heteromer selectivity is characterized by ELISA, immunofluorescence, or Western blot analysis using cells that express the individual or a combination of receptors or tissues from wild type and animals lacking one of the receptors. These steps are described below using the μOR–δOR heteromer as a model system.

2.1. Preparation of antigen for immunization

For the generation of μOR–δOR heteromer-selective mAbs (μOR–δOR mAb), we used membranes from human embryonic kidney 293 (HEK-293) cells (American Type Culture Collection, Manassas, VA) for the tolerization step and from cells coexpressing Flag-μOR and myc-δOR (HEK-μδ) for the second immunogenic step.

1. Grow HEK-293 cells in 10-cm^2 dishes (Becton-Dickinson, Falcon) in Dulbecco's modified Eagle's medium (DMEM; Gibco-BRL, Gaithersburg, MD) containing 10% fetal bovine serum (FBS; Gibco-BRL) and 1% penicillin–streptomycin (Gibco-BRL).
2. When ~80–85% confluent, cotransfect cells with Flag-μOR and myc-δOR (6 μg each) using Lipofectamine 2000 according to the manufacturer's protocol (Invitrogen).
3. Thirty-six to forty-eight hours after transfection, remove growth media and rinse cells twice with 5 ml phosphate-buffered saline (PBS).

4. Add 1.5 ml of prechilled buffer A (50 m*M* Tris–Cl, pH 7.4, containing protease inhibitor cocktail (Sigma-Aldrich)) to the dish, collect cells in an Eppendorf tube with the help of a rubber policemen, and centrifuge at 3000 × *g* for 3 min at 4 °C.
5. Resuspend the cell pellet in 0.2 ml of freshly prepared ice-cold buffer B (20 m*M* Tris–Cl buffer, pH 7.4, containing 2 m*M* EGTA, 1 m*M* $MgCl_2$, 250 m*M* sucrose, and protease inhibitor cocktail).
6. Homogenize cells on ice using a Teflon pestle for Eppendorf tubes (10–20 strokes), adjust the volume to 0.5 ml with ice-cold buffer B and pass through an insulin syringe three times.
7. Adjust the volume to 10 ml with ice-cold buffer B and centrifuge at 27,000 × *g* for 15 min at 4 °C.
8. Resuspend the pellet in ice-cold buffer B and repeat steps 6 and 7.
9. Resuspend the pellet in minimum volume of PBS containing 10% glycerol and pass through insulin syringe.
10. Determine membrane protein concentration using Pierce BCA protein assay reagent (Thermo Scientific) according to manufacturer's protocol.
11. Mix membranes from either HEK or HEK-μδ cells (50–100 μg protein in 0.2 ml PBS) with 0.2 ml Freund's complete adjuvant (Gibco).
12. Pass the mixture through an insulin syringe 10 times.
13. Place the suspension in a rotating shaker at moderate speed for 20 min at 4 °C and repeat step 11.
14. Confirm the formation of an emulsion by placing a small drop in 10 ml of water in a petridish. If the drop disperses in water, the emulsification is incomplete in which case repeat steps 11 and 12. If the emulsification is complete, the drop does not disperse in water. For booster injections, prepare the emulsion in Freund's incomplete adjuvant.

2.2. Subtractive immunization

Subtractive immunization is carried out in Balb/c mice (6-week-old female). Use at least three mice for immunization. Prior to immunization, collect blood from the tail vein to prepare preimmune serum (described in Section 2.6).

1. Inject mice intraperitoneally with 50–100 μg of HEK-293 membranes emulsified in complete Freund's adjuvant.
2. Administer cyclophosphamide (100 mg/kg body weight; Sigma-Aldrich) daily for the next 3 days.
3. Every 15 days, administer an intraperitoneal booster injection with membranes emulsified in incomplete Freund's adjuvant and repeat step 2.

4. Collect blood from the tail vein on the 4th day after each booster injection, prepare serum (described in Section 2.6), and monitor antibody production by ELISA (described in Section 3).
5. Continue booster injections until a consistent low titer is observed with HEK-293 membranes.
6. Inject the mice intraperitoneally with HEK-μδ membranes emulsified in Freund's complete adjuvant.
7. Repeat steps 3 and 4 with HEK-μδ membranes emulsified in incomplete Freund's adjuvant.
8. Continue booster injections until a consistent high titer is observed with HEK-μδ membranes and not with HEK membranes alone or expressing individual receptors. Generate mAbs as described below.

2.3. Fusion protocol to generate hybridoma-secreting clones

To generate hybridoma-secreting clones, splenocytes from immunized mice are fused with mouse myeloma cells (Gupta et al., 2010; Kohler & Milstein, 1975). We use SP2/0-Ag-14 myeloma cells (American Type Culture Collection, Manassas, VA) that have the machinery necessary for antibody secretion although on their own they do not secrete antibodies (Shulman et al., 1978). SP2/0-Ag-14 cells have a mutation in one of the enzymes of the salvage pathway for purine nucleotide biosynthesis rendering the latter nonfunctional (Shulman et al., 1978). The fusion process is relatively inefficient, given that only 1 in 10^5 of starting cells form viable hybrids (Shulman et al., 1978). The unfused splenocytes do not grow in culture and eventually die. To kill unfused myeloma cells, compounds that block *de novo* purine biosynthesis such as aminopterin, methotrexate, or azaserine are used (Shulman et al., 1978). Fused myeloma cells do not die under these conditions since they have a functional salvage pathway derived from the splenocyte fusion partner (Shulman et al., 1978).

1. Thaw a vial of SP2/0-Ag-14 cells and grow them in DMEM containing 10% FBS and 1% penicillin–streptomycin. At the time of fusion, these cells should be under the exponential growth phase.
2. Three days before fusion, inject mice intravenously with HEK-μδ membranes (50–100 μg).
3. On the day of the fusion, kill an immunized mouse using CO_2 from compressed gas.
4. In a sterile hood, place the mouse on its right side and swab the body with 70% ethanol.
5. Cut open the mouse to expose the spleen taking care to not puncture the gut or surrounding splenic arteries and veins.

6. Remove the spleen and place it in a sterile Petridish containing 5 ml of sterile serum-free Iscove's modified Dulbecco's media (IMDM; Gibco-BRL).
7. Wash the spleen three to five times in serum-free IMDM.
8. Remove any adherent fat and transfer the spleen into a Petridish containing 5 ml of serum-free IMDM.
9. Tease apart the spleen with the help of a 21-G needle to release the splenocytes. The medium becomes cloudy as the cells are released from the spleen.
10. Transfer the cell suspension into a 50-ml centrifuge tube; let it stand for 2 min to allow the tissue clumps to settle down.
11. Transfer the supernatant containing the cell suspension minus the tissue clumps into a fresh 50-ml tube.
12. Make up the volume to 10 ml with serum-free IMDM and centrifuge at $400 \times g$ for 10 min.
13. Resuspend cell pellet in 10 ml serum-free IMDM and count the number of live cells using 0.2% Trypan blue (Gibco-BRL) and an improved Neubauer counting chamber. A spleen from an immunized mouse yields $0.5–2 \times 10^8$ splenocytes.
14. Wash the splenocytes twice in 10 ml of serum-free IMDM by centrifuging at $400 \times g$ for 10 min.
15. Harvest the SP2/0-Ag-14 cells and wash them twice as described in step 14 for splenocytes.
16. Resuspend the SP2/0-Ag-14 cells in 10 ml serum-free IMDM and count the number of cells as described in step 13.
17. Mix the spleen and myeloma cells in a 50-ml centrifuge tube at a ratio of 10:1 (splenocytes:myeloma) and centrifuge at $800 \times g$ for 5 min. Discard the supernatant.
18. Gently dislodge the cell pellet by tapping and place the tube in a beaker of water at 37 °C.
19. Slowly add 1 ml of 50% PEG 4000 (Sigma-Aldrich) prewarmed to 37 °C. This addition should be complete within a minute, continually stir the mixture with the pipette tip. The cells must not be in contact with the concentrated PEG solution for more than 2 min.
20. Slowly add (over a period of 1 min) 1 ml of serum-free IMDM prewarmed to 37 °C. Continuously stir the cell suspension with the pipette tip while media is being added.
21. Add 4 ml of serum-free IMDM (prewarmed to 37 °C) over a period of 3–4 min, without stirring, sliding it slowly down the side of the tube.

22. Add 20 ml of serum-free IMDM (prewarmed to 37 °C) followed by 20 ml of IMDM containing 15% FBS, again by sliding slowly down the side of the tube.
23. Cap the tube, invert it once, and leave it at 37 °C for 2 h.
24. Centrifuge the cells (400 × *g* for 10 min), discard the supernatant, and resuspend the pellet in IMDM containing 15% FBS and 1 × solution of hypoxanthine, aminopterin, and thymidine (HAT; Gibco-BRL).
25. Plate the cells in 24-well plates at 2 ml/well.
26. Put the plates in a 10% CO_2 incubator at 37 °C.
27. After a week, remove 1 ml of medium from each well and add 1 ml of complete media containing HAT supplement. Repeat this step once more. After 15 days in culture, screen the wells under a microscope for the presence of hybridoma-secreting clones.
28. Subclone wells containing clones into 96-well plates so as to obtain 1 cell/well.
29. Allow the clones to grow in 200 μl of complete media containing HAT supplement.
30. Change the media every 3 days by removing 100 μl of media from each well and adding 100 μl of complete media containing HAT supplement.
31. Check the wells after 1 week for the presence of hybridoma clones.
32. After 15 days in culture, test for the presence of hybridoma-secreting clones by ELISA (described under Section 3) using 50 μl of supernatant from each well and membranes from HEK-293, HEK-μδ, HEK-μ, or HEK-δ cells.
33. Clones that give a high titer with only HEK-μδ but not with HEK-μ or HEK-δ membranes are transferred into 25-cm^2 flasks along with 5 ml of complete growth media containing HAT supplement.
34. Determine the isotype of the antibody secreted by each hybridoma-secreting clone using the IsoStrip Mouse Monoclonal Antibody Isotyping Kit (Roche) and 150 μl of hybridoma supernatant (1:10 diluted in PBS containing 1% BSA).
35. For large-scale preparation of mAbs, either grow hybridomas in large tissue culture flasks (Integra Celline 1000; Integra Biosciences AG, Switzerland) or generate ascites fluid as described by Hoogenraad and Wraight (1986). Either procedure will yield ~10 mg/ml of antibodies.

2.4. Purification of mAb from hybridoma supernatant

We routinely purify mAbs using a protocol described by Ey, Prowse, and Jenkin (1978) that takes advantage of protein A-Sepharose CL-4B beads.

1. Pass the hybridoma supernatant through a 0.2-μm filter to remove aggregates; adjust the pH to 8.0 by adding 1/10 volume of 1.0 *M* Tris.
2. Swell 1 g of protein A-Sepharose CL-4B beads (GE Healthcare) in 50 ml distilled water.
3. Transfer the slurry into a column (4–5 ml of gel); equilibrate with 1.0 *M* Tris, pH 8.0. Check that the pH of the eluate is 8.0.
4. Pass the antibody solution through the protein A column. About 10–20 mg of antibody/ml wet beads can be bound.
5. Wash the column with 10 volumes of 100 m*M* Tris (pH 8.0) followed by 10 volumes of 10 m*M* Tris (pH 8.0).
6. Elute bound antibodies by adding 500 μl aliquots of 100 m*M* glycine (pH 3.0) and collecting the eluate in 1.5-ml Eppendorf tubes containing 50 μl of 1 *M* Tris (pH 8.0). Mix each tube gently to bring the pH back to neutral. Do not shake vigorously to avoid frothing and bubbling. Collect ~25 fractions.
7. Identify immunoglobulin-containing fractions using a spectrophotometer at 280 nm; determine protein levels in positive fractions using Pierce BCA protein assay reagent (Thermo Scientific) according to manufacturer's protocol.
8. Store in aliquots at −20 °C till use.

2.5. Serum preparation

Serum is prepared from blood collected from the tail vein of mice before (preimmune serum) and after (immune serum) immunization.

1. Place the mice in a small restraining cage.
2. Warm the tail with an infrared lamp or hot water (50–60 °C) to increase the flow of blood to the tail.
3. Disinfect the tail with ethanol, make a small incision at the tip with a sterile blade, and collect blood (~75–100 μl) into a sterile Eppendorf tube. Apply pressure or use a styptic pencil to stop the bleeding.
4. Allow blood to clot for 30–60 min at 37 °C for 2 h.
5. Separate the clot from the sides of the tube using a gel-loading tip.
6. Keep the tube at 4 °C overnight to allow the clot to contract.
7. Collect the serum by centrifugation at 10,000 × *g* for 10 min at 4 °C.
8. Store the serum in aliquots at −20 °C.
9. For working dilutions of preimmune and immune serum, add 50 μl of 0.5 *M* EDTA (pH7.4) and 2 μl of 50% thimerosal to prevent microbial growth and store at 4 °C.

2.6. Trouble-shooting and precautions

1. Instead of membranes, whole cells expressing the receptors of interest can be used to generate mAbs. In this case, cells should be washed free of phenol red with PBS since it is toxic and can lead to liver damage and jaundice in mice. Approximately, 10^6–10^8 cells should suffice for primary immunization and booster injections.
2. The most common problem during mAb production is that of bacterial, fungal, yeast, or mycoplasma contamination. Bacteria, seen under the light microscope as rods or cocci, can be treated with antibiotics. If the growth media contains penicillin and streptomycin, add 50 μg/ml gentamycin. Alternatively, subclone cells into media containing antibiotics in an effort to dilute out the bacteria. However, it is best to discard the contaminated cells and sterilize the contaminated wells with 70% ethanol. Fungi or yeast can be detected under the microscope and this type of contamination prevented by adding fungizone (2.5 μg/ml) to the growth media. However, once fungal or yeast contamination appears, it is best to discard the cells and sterilize the wells. Wells are sterilized by removing the media by suction, carefully avoiding touching adjacent wells, adding Chloros (sodium hypochlorite) for 2–3 min, rinsing with sterile water, and leaving the wells empty. In addition, the lid of the plate should be replaced to minimize the possibility of further contamination. Mycoplasma contamination is difficult to diagnose as the organism cannot be seen under the light microscope but can be the cause of slow cell growth. It is possible to rescue cells contaminated with mycoplasma by treating them with a mycoplasma removal agent (ICN Biomedicals, Inc.) according to manufacturer's protocol.

3. ELISA FOR DETECTION OF RECEPTOR HETEROMERS

ELISA, a technique commonly used to screen for the presence on an antigenic epitope on membrane preparations, in whole cells or tissue sections, can also detect GPCRs by using antibodies that recognize endogenous receptors or epitope tags (Flag, *myc*, or HA) present in the N-terminal region of the receptors. We use ELISA to screen for and determine the selectivity of heteromer-selective antibodies (Gupta et al., 2010; Rozenfeld et al., 2011). In the case of μ-δ mAb, we used membranes from (i) HEK cells, (ii) HEK cells expressing individual receptors, and (iii) HEK-μδ cells to detect hybridoma clones that gave a high titer with only HEK-μδ (Gupta et al., 2010). We also examined the selectivity of μ-δ mAb clones using membranes from the brain of wild type and mice lacking μOR, δOR, or both. We found that the μ-δ mAb recognized an epitope present only in membranes from wild type but not from knockout animals (Gupta et al., 2010). We examined whether the μ-δ mAb exhibited cross-reactivity with other heteromers involving μOR or δO using membrane preparations from cells coexpressing μOR–α2A adrenergic receptor (α2AR), μOR–CB_1R, δOR–α2AR, or δOR–CB_1R (Gupta et al., 2010). We found that the μ-δ mAb did not exhibit a significant degree of cross-reactivity with other heteromers involving μOR or δOR (Gupta et al., 2010). In addition, we used ELISA to determine the effect of a pathological condition such as development of tolerance to morphine (μOR–δOR heteromers) or chronic ethanol treatment (CB_1R–AT1R heteromers) on heteromer levels (Gupta et al., 2010; Rozenfeld et al., 2011). Thus, ELISA is a versatile technique that can be used to determine heteromer levels under pathological techniques. Below, we describe a routinely used ELISA protocol.

1. Add 10–100 μg of HEK, HEK-μδ, HEK-μ, or HEK-δ membranes (in PBS) to each well of an ELISA plate (96-well, flat-bottom immuno plate; Nalgene Nunc International, USA) and air-dry the plates.
2. Incubate with 200 μl of 3% BSA (Sigma) in PBS for 1 h at 37 °C.
3. Discard the BSA solution.
4. Add 100 μl of μOR–δOR-selective hybridoma supernatant (or 1:500 to 1:1000 dilution of purified mAb) to the wells. Add 100 μl of PBS to wells used to determine nonspecific binding.
5. Incubate for 16–18 h at 4 °C.
6. Wash wells four times (5 min each wash at RT) with 200 μl of 1% BSA (Sigma) in PBS.
7. Incubate with 100 μl of peroxidase-conjugated anti-mouse antibody (1:1000 in PBS containing 1% BSA) for 1 h at 37 °C.
8. Repeat step 6.
9. Incubate with 100 μl of substrate (0.5 mg/10 ml of *ortho*-phenylenediamine in 0.15 *M* citrate-phosphate buffer, pH 5 containing 10 μl of H_2O_2) for 10 min at RT; terminate the reaction by adding 50 μl of 5N H_2SO_4.
10. Transfer the supernatant to another 96-well plate and read at 490 nm in an ELISA reader (Bio-Rad model 550).

This ELISA protocol can be adapted to monitor surface heteromer levels following chronic treatment with drugs. For example, cells ($\sim 2.5 \times 10^5$ cells/well) coexpressing μOR and δOR are seeded in growth media in a 24-well plate. After cells have attached, they are treated with morphine or any lipophilic opiate, preferably an antagonist (1 μ*M* final concentration in growth media) for different time periods (0–48 h). Controls are treated with same amount of growth media. At the end of the incubation period, media are removed, plates are placed on ice, and wells are rinsed with 200 μl of cold PBS. Cells are fixed with ice-cold methanol for 5 min followed by two washes with 500 μl PBS. Cells are then subjected to ELISA using primary and secondary antibodies as described above.

ELISA can also be carried out in tissue sections, to compare regional differences in antibody recognition in control and pathological conditions and to examine the effect of drugs/ligands on receptor levels. For this, 10-μm tissue sections are placed on Fisher Brand Superfrost Plus slides (Fisher Scientific, PA) and circled immediately with ImmEdge PAP pen (Vector Laboratories, Inc., CA) to form a water-proof barrier; the resulting wells hold approximately 200 μl solution. ELISA is then carried out as described above.

The most common problems encountered during the ELISAs are elevated background, poor, or absence of a signal. Elevated background signal can be due to inadequate washing and draining of the wells, contamination of the substrate solution with metal ions or oxidizing reagents. The latter can be avoided by using only distilled/deionized water in the preparation of the different solutions and by using clean plastic-ware. Prior substrate exposure to light leads to poor signal. The substrate should be prepared about 10 min before use and should be kept in the dark till use. Sometimes, the signal develops very rapidly (i.e., in less than a minute), giving very high titers that are outside the range of the plate reader. For this reason, pilot experiments using different dilutions of primary and secondary antibodies should be carried out to establish conditions where optimum signal is obtained after 10 min of incubation with substrate.

4. IMMUNOFLUORESCENCE FOR VISUALIZATION OF RECEPTOR HETEROMERS

Immunofluorescence uses fluorescently labeled secondary antibodies to visualize proteins in cells and tissues and can provide information about the tissue distribution of a given protein as well as its subcellular distribution. We used immunofluorescence to show that coexpression of μOR with δOR leads to intracellular retention of μOR–δOR heteromers in the Golgi apparatus, and this is rescued by the expression of RTP4 (Decaillot, Rozenfeld, Gupta, & Devi, 2008).

We used this technique to visualize μOR–δOR heteromers in the brains of wild-type mice (but not μOR or δDOR knockout mice) treated chronically with morphine under a paradigm that leads to the development of tolerance. The increases in μOR–δOR heteromer levels were robust in the medial nucleus of the trapezoid body and in the rostral ventral medulla, brain regions involved in the processing of painful stimuli (Gupta et al., 2010). In the case of CB_1R–AT1R heteromers, immunofluorescence studies showed that heteromerization leads to surface expression of CB_1R and colocalization of CB_1R and AT1R at the cell surface of hepatic stellate cells that were chronically treated with ethanol (Rozenfeld et al., 2011). The protocol that we use for immunofluorescence is as follows:

1. If using cells, place poly-D-lysine-coated coverslips (Fisher brand, Fisher Scientific, USA) in 12-well plates (one/well). Seed the plates using complete growth media with cells (1×10^4) expressing the receptors of interest. Next day, remove media by suction and wash wells twice with sterile PBS prior to fixation and immunofluorescence.
2. If using tissue sections, collect brains, dissect them into different regions, snap freeze in liquid nitrogen, and store at −80 °C until use. Cut cryostat sections (5–8 μm) and mount them in Superfrost Plus slides. Slides can be stored at −80 °C until use. Before staining, warm the slides at RT for 30 min.
3. Fix cells/tissue sections with either acetone (5 min) or 4% paraformaldehyde, pH 7.4 (15 min) at RT.
4. Wash wells/slides three to four times (5 min each) with Tris-buffered saline, pH 7.4 (TBS).
5. Block nonspecific binding sites with 1% goat serum in TBS for 30 min.
6. Incubate cells/tissue sections overnight at 4 °C with the appropriate dilution of the primary antibody.
7. Wash wells/slides four times (5 min each) with gentle shaking in TBS.
8. Incubate wells/slides with the appropriate dilution of fluorescently labeled secondary antibody for 60–90 min in the dark with gentle shaking.
9. Repeat step 7.
10. Counterstain wells/slides with DAPI for nuclear visualization if desired.
11. Remove coverslips containing cells and mount them on slides using Vector shield (Vector Laboratories, Inc., Burlingame, CA); similarly coverslip tissue sections using Vector shield.

12. Examine cells/tissue sections under a fluorescence microscope (Leica DM 6000B).

Common problems encountered with immunofluorescence are high background and sensitivity particularly of heteromer-selective antibodies to fixatives. High background can be reduced by adding 0.1% Tween-20 to washes or increasing the concentration of normal serum (5–10%) used for blocking. If antibodies are sensitive to fixatives, then changing the fixation conditions (glutaraldehyde, methanol, or 10% formalin) followed by extensive washing to remove all traces of the fixative is recommended.

5. IMMUNOPRECIPITATION AND WESTERN BLOTTING

Immunoprecipitation is a useful technique to demonstrate heteromerization between differentially epitope-tagged GPCRs. After cell lysis, one of the receptors is immunoprecipitated using antibodies recognizing the epitope tag present on the receptor (*myc*-tagged receptors are immunoprecipitated using anti-*myc* antisera). The immunoprecipitates are then subjected to SDS-PAGE under nonreducing conditions, and Western blots are probed with antibodies to the epitope tag present to the other receptor. In order to avoid cross-reactivity, antibodies from two different species are used. A signal is detected in the blots only if there is an association between the two epitope-tagged receptors. We used this strategy to demonstrate heteromerization between μOR and δOR (Gomes et al., 2004, 2000), δOR and κOR (Jordan & Devi, 1999), μOR and α2AR (Jordan et al., 2003), δOR and α2AR (Rios et al., 2004), δOR and CB_1R (Rozenfeld et al., 2012), as well as CB_1R and AT1R (Rozenfeld et al., 2011). When examining heteromerization in endogenous tissue, antibodies that selectively recognize each receptor or the heteromer are needed (Gomes et al., 2004; Gupta et al., 2010; Rozenfeld et al., 2012, 2011). In the case of the heteromer, the selective antibody could be used to immunoprecipitate the heteromer, and antibodies to individual receptors to probe the Western blots (Gupta et al., 2010). Coimmunoprecipitation and Western blot analysis can also be used to examine interactions between GPCR heteromers and signaling molecules such as β-arrestins, or chaperones such as RTP-4, AP-3, or AP-2. Such studies showed that μOR–δOR heteromers are constitutively associated with β-arrestin (Rozenfeld & Devi, 2007), and that RTP-4 interacts with μOR–δOR heteromers (Decaillot et al., 2008). Thus, immunoprecipitation and Western blotting could be used to (i) probe heteromer levels under healthy and pathological conditions, (ii) identify differences in the levels of heteromer-associated proteins, or (iii) help identify novel proteins associated with the heteromer under pathological conditions. The following sections describe the different steps involved in coimmunoprecipitation and Western blotting.

5.1. Lysis of cells/tissues

Different buffers can be used to lyse transfected cells or tissues. Two important considerations in the choice of lysis buffer are the efficient solubilization of the desired receptor without affecting receptor associations and whether the lysis buffer interferes with receptor recognition by the antibody used for immunoprecipitation. Variables that can drastically affect the solubilization of proteins are salt concentration, pH, and type of detergent used. Lysis buffers commonly used to examine receptor interactions include

Buffer G: 50 m*M* Tris–Cl, pH 7.4 containing 300 m*M* NaCl, 1% Triton X-100, 10% glycerol, 1.5 m*M* $MgCl_2$, and 1 m*M* $CaCl_2$.

RIPA buffer: 50 m*M* Tris–Cl, pH 8 containing 150 m*M* NaCl, 1% NP-40, 0.5% deoxycholate, 0.1% SDS, and 1 m*M* $CaCl_2$.

NP-40 buffer: 10 m*M* Tris–Cl, pH 8 containing 1% NP-40, 150 m*M* NaCl, 1 m*M* EDTA, 10% glycerol, and 1 m*M* $CaCl_2$.

CHAPS buffer: 1% in 50 m*M* Tris–Cl, pH 7.4.

Dodecyl maltoside buffer: 0.5% dodecyl maltoside in 50 m*M* Tris–Cl, pH 7.4.

1. Take a 10-cm dish containing cells expressing the receptors of interest. For tissues, use 20–50 mg of fresh or frozen tissue.
2. Collect cells from plate by addition of 3 ml enzyme-free Hank's-based cell dissociation buffer (Invitrogen) followed by incubation for 3–5 min in a CO_2 incubator at 37 °C and up-and-down pipetting.
3. Spin down the cells at 3000 × *g* for 3 min.
4. Resuspend the cell pellet in 1 ml PBS. Transfer to a 1.5-ml Eppendorf tube.
5. Repeat steps 3–4. Discard the supernatant.
6. Add 200 μl of prechilled lysis buffer containing protease inhibitor cocktail and 10 m*M* iodoacetamide to the Eppendorf tube containing cells/tissue.
7. Homogenize cells/tissue on ice with a Teflon pestle.
8. Make the volume to 1.5 ml with prechilled lysis buffer.
9. Pass the homogenate through an insulin syringe two to three times and transfer to a fresh Eppendorf tube.
10. Incubate for 1 h at 4 °C using a rotating shaker.

11. Centrifuge at 14,000 rpm for 20 min at 4 °C.
12. Transfer supernatant to a fresh Eppendorf tube.
13. Take an aliquot for protein estimation using BCA assay reagent.
14. To an Eppendorf tube on ice, add 150 μg of protein lysate, 5 μg of antibody (polyclonal antibody to one receptor or to the epitope tag present on the receptor or monoclonal heteromer-selective antibody), and 12 μl of protease inhibitor cocktail. Make the volume to 1.2 ml with prechilled lysis buffer.
15. Incubate overnight on a rotating shaker at 4 °C.
16. Equilibrate protein A beads in prechilled lysis buffer and add 150 μl to each tube.
17. Incubate for 2 h at 4 °C on a rotating shaker.
18. Centrifuge at 14,000 rpm for 1 min at 4 °C.
19. Wash the beads three times with 500 μl of lysis buffer containing protease inhibitors.
20. Remove the lysis buffer completely with the help of an insulin syringe.
21. Add 70 μl of 2 × sample buffer (120 m*M* Tris–Cl, pH 6.8 containing 4% SDS, 20% glycerol, and 0.002% bromophenol blue) to the pellet and incubate for 15 min at 60 °C.
22. Spin down the samples and run 10–15 μl on 8% SDS-PAGE gels.
23. Transfer the separated proteins to nitrocellulose membranes overnight at 30 V.
24. Rinse membranes briefly in TBS-T (50 m*M* Tris–Cl, pH 7.4 containing 150 m*M* NaCl, 1 m*M* $CaCl_2$, and 0.1% Tween 20; prepared fresh).
25. Block the membranes for 1 h at RT with 10 ml of 5% nonfat dried milk in TBS-T.
26. Wash the membranes in TBS-T four times (15 min/wash) in a shaker.
27. Incubate with 1:1000 to 1:5000 primary antibody diluted in 30% Odyssey buffer (Li-cor) in TBS-T containing 0.01% sodium azide for 16 h at 4 °C.
28. Wash membranes four times (15 min each wash) with 15 ml TBS-T.
29. Incubate with 1:10,000 dilution of IRDye 680 or IRDye 800 secondary antibody (Li-cor) in 30% Odyssey buffer in TBS-T containing 0.01% sodium azide for 1–2 h at RT on a shaker.
30. Wash membranes four times (15 min each wash) with 15 ml TBS-T.
31. Visualize the signal and densitize it using the Odyssey Imaging system (Li-cor).

5.2. Problems and troubleshooting

Artifactual receptor aggregation during solubilization/immunoprecipitation due to the inherent hydrophobic nature of GPCRs is a major concern when examining receptor heteromerization. To rule this out, a variety of solubilization conditions including different combinations of detergents have to be used. In addition, controls where cells expressing individual receptors are mixed prior to solubilization and immunoprecipitation have to be used. If the heteromers are observed only in cells coexpressing both receptors and not in the mixed cells, this would suggest that they are not the result of artifactual aggregation.

Harsh solubilization procedures can disrupt receptor associations. Cross-linking reagents have been used to address this concern (Cvejic & Devi, 1997). The presence of receptor heteromers in the presence of cross-linking reagents, irrespective of the latter's functional properties, would then suggest that they stabilize the interactions and do not induce receptor heteromerization.

To prevent the masking of the antigenic epitope due to the presence of nonspecific proteins, it is advisable to preclear the cell lysate with normal serum (e.g., rabbit serum if using rabbit polyclonal antibody for immunoprecipitation) followed by binding to protein A beads. This removes proteins that bind nonspecifically to the antibody or to the beads.

The presence of a high background or nonspecific bands in Western blots can be minimized through the use of a different blocking buffer, reducing the time of incubation with primary/secondary antibody, using harsher conditions for washing membranes after antibody incubation (e.g., 50 m*M* Tris–Cl, pH 7.5 containing 150 m*M* NaCl, 1% NP-40, 0.5% deoxycholate, and 0.1% sodium dodecyl sulfate), or adding detergent to the primary/secondary antibody preparation up to a concentration of 1%.

6. SUMMARY AND PERSPECTIVES

Recent evidence indicates that some GPCR heteromers are upregulated in disease states. However, the physiological role of these heteromers in pathology is not clearly understood due to the lack of tools to distinguish between heteromer- and homomer-mediated effects. The development of heteromer-selective antibodies and of specific heteromer disrupting TAT peptides could help elucidate their role in pathological conditions. This is clearly shown in the case of μOR–δOR heteromers where selective antibodies detect increased heteromer levels in brain regions involved in pain perception following chronic treatment with morphine. Studies with TAT peptides that selectively disrupt μOR–δOR heteromers suggest that the latter may keep morphine-mediated signaling via μOR in the desensitized state. Thus, GPCR heteromers could be novel targets for the development of therapeutics to treat diseases. This, however, would require an understanding of the functional role of GPCR heteromers during pathology. The different protocols described in this review used in combination with the heteromer-selective antibodies or with peptides that selectively disrupt the heteromers could help in elucidating these roles.

ACKNOWLEDGMENT

L. A. D. is supported by NIH grants DA008863 and DA019521.

REFERENCES

AbdAlla, S., Abdel-Baset, A., Lother, H., el Massiery, A., & Quitterer, U. (2005). Mesangial AT1/B2 receptor heterodimers contribute to angiotensin II hyperresponsiveness in experimental hypertension. *Journal of Molecular Neuroscience*, *26*, 185–192.

Balenga, N. A., Henstridge, C. M., Kargl, J., & Waldhoer, M. (2011). Pharmacology, signaling and physiological relevance of the G protein-coupled receptor 55. *Advances in Pharmacology*, *62*, 251–277.

Berg, K. A., Rowan, M. P., Gupta, A., Sanchez, T. A., Silva, M., Gomes, I., et al. (2012). Allosteric interactions between delta and kappa opioid receptors in peripheral sensory neurons. *Molecular Pharmacology*, *81*, 264–272.

Bhushan, R. G., Sharma, S. K., Xie, Z., Daniels, D. J., & Portoghese, P. S. (2004). A bivalent ligand (KDN-21) reveals spinal delta and kappa opioid receptors are organized as heterodimers that give rise to delta(1) and kappa(2) phenotypes. Selective targeting of delta-kappa heterodimers. *Journal of Medicinal Chemistry*, *47*, 2969–2972.

Bushlin, I., Rozenfeld, R., & Devi, L. A. (2010). Cannabinoid-opioid interactions during neuropathic pain and analgesia. *Current Opinion in Pharmacology*, *10*, 80–86.

Chen, C., Li, J., Bot, G., Szabo, I., Rogers, T. J., & Liu-Chen, L. Y. (2004). Heterodimerization and cross-desensitization between the mu-opioid receptor and the chemokine CCR5 receptor. *European Journal of Pharmacology*, *483*, 175–186.

Cichewicz, D. L. (2004). Synergistic interactions between cannabinoid and opioid analgesics. *Life Sciences*, *74*, 1317–1324.

Cvejic, S., & Devi, L. A. (1997). Dimerization of the delta opioid receptor: Implication for a role in receptor internalization. *Journal of Biological Chemistry*, *272*, 26959–26964.

Decaillot, F. M., Rozenfeld, R., Gupta, A., & Devi, L. A. (2008). Cell surface targeting of mu-delta opioid receptor heterodimers by RTP4. *Proceedings of the National Academy of Sciences of the United States of America*, *105*, 16045–16050.

Dietis, N., Rowbotham, D. J., & Lambert, D. G. (2011). Opioid receptor subtypes: Fact or artifact? *British Journal of Anaesthesia*, *107*, 8–18.

Di Marzo, V., Piscitelli, F., & Mechoulam, R. (2011). Cannabinoids and endocannabinoids in metabolic disorders with focus on diabetes. *Handbook of Experimental Pharmacology*, *203*, 75–104.

Dziedzicka-Wasylewska, M., Faron-Gorecka, A., Gorecki, A., & Kusemider, M. (2008). Mechanism of action of clozapine in the context of dopamine D1–D2 receptor hetero-dimerization—A working hypothesis. *Pharmacological Reports*, *60*, 581–587.

Ey, P. L., Prowse, S. J., & Jenkin, C. R. (1978). Isolation of pure Igg1, Igg2a and Igg2b immunoglobulins from mouse serum using protein A-sepharose. *Immunochemistry, 15*, 429–436.

Faron-Gorecka, A., Gorecki, A., Kusmider, M., Wasylewski, Z., & Dziedzicka-Wasylewska, M. (2008). The role of D1-D2 receptor hetero-dimerization in the mechanism of action of clozapine. *European Neuropsychopharmacology, 18*, 682–691.

Fuxe, K., Ferre, S., Canals, M., Torvinen, M., Terasmaa, A., Marcellino, D., et al. (2005). Adenosine A2A and dopamine D2 heteromeric receptor complexes and their function. *Journal of Molecular Neuroscience, 26*, 209–220.

Gomes, I., Gupta, A., Filipovska, J., Szeto, H. H., Pintar, J. E., & Devi, L. A. (2004). A role for heterodimerization of mu and delta opiate receptors in enhancing morphine analgesia. *Proceedings of the National Academy of Sciences of the United States of America, 101*, 5135–5139.

Gomes, I., Ijzerman, A. P., Ye, K., Maillet, E. L., & Devi, L. A. (2011). G protein-coupled receptor heteromerization: A role in allosteric modulation of ligand binding. *Molecular Pharmacology, 79*, 1044–1052.

Gomes, I., Jordan, B. A., Gupta, A., Trapaidze, N., Nagy, V., & Devi, L. A. (2000). Heterodimerization of mu and delta opioid receptors: A role in opiate synergy. *Journal of Neuroscience, 20*, RC110.

Gonzalez-Hernandez Mde, L., Godinez-Hernandez, D., Bobadilla-Lugo, R. A., & Lopez-Sanchez, P. (2010). Angiotensin-II type 1 receptor (AT1R) and alpha-1D adrenoceptor form a heterodimer during pregnancy-induced hypertension. *Autonomic & Autacoid Pharmacology, 30*, 167–172.

Gupta, A., Mulder, J., Gomes, I., Rozenfeld, R., Bushlin, I., Ong, E., et al. (2010). Increased abundance of opioid receptor heteromers after chronic morphine administration. *Science Signal, 3*, ra54.

He, S. Q., Zhang, Z. N., Guan, J. S., Liu, H. R., Zhao, B., Wang, H. B., et al. (2011). Facilitation of mu-opioid receptor activity by preventing delta-opioid receptor-mediated codegradation. *Neuron, 69*, 120–131.

Hereld, D., & Jin, T. (2008). Slamming the DOR on chemokine receptor signaling: Heterodimerization silences ligand-occupied CXCR4 and delta-opioid receptors. *European Journal of Immunology, 38*, 334–337.

Hoogenraad, N. J., & Wraight, C. J. (1986). The effect of pristane on ascites tumor formation and monoclonal antibody production. *Methods in Enzymology, 121*, 375–381.

Howlett, A. C., Barth, F., Bonner, T. I., Cabral, G., Casellas, P., Devane, W. A., et al. (2002). International Union of Pharmacology. XXVII. Classification of cannabinoid receptors. *Pharmacological Reviews, 54*, 161–202.

Jordan, B. A., & Devi, L. A. (1999). G-protein-coupled receptor heterodimerization modulates receptor function. *Nature, 399*, 697–700.

Jordan, B. A., Gomes, I., Rios, C., Filipovska, J., & Devi, L. A. (2003). Functional interactions between mu opioid and alpha 2A-adrenergic receptors. *Molecular Pharmacology, 64*, 1317–1324.

Jordan, B. A., Trapaidze, N., Gomes, I., Nivarthi, R., & Devi, L. A. (2001). Oligomerization of opioid receptors with beta 2-adrenergic receptors: A role in trafficking and mitogen-activated protein kinase activation. *Proceedings of the National Academy of Sciences of the United States of America, 98*, 343–348.

Kabli, N., Martin, N., Fan, T., Nguyen, T., Hasbi, A., Balboni, G., et al. (2010). Agonists at the delta-opioid receptor modify the binding of micro-receptor agonists to the micro-delta receptor hetero-oligomer. *British Journal of Pharmacology, 161*, 1122–1136.

Kohler, G., & Milstein, C. (1975). Continuous cultures of fused cells secreting antibody of predefined specificity. *Nature, 256*, 495–497.

Levac, B. A., O'Dowd, B. F., & George, S. R. (2002). Oligomerization of opioid receptors: Generation of novel signaling units. *Current Opinion in Pharmacology, 2*, 76–81.

Maggio, R., & Millan, M. J. (2010). Dopamine D2-D3 receptor heteromers: Pharmacological properties and therapeutic significance. *Current Opinion in Pharmacology, 10*, 100–107.

Maldonado, R., Valverde, O., & Berrendero, F. (2006). Involvement of the endocannabinoid system in drug addiction. *Trends in Neurosciences, 29*, 225–232.

Manzanares, J., Corchero, J., Romero, J., Fernandez-Ruiz, J. J., Ramos, J. A., & Fuentes, J. A. (1999). Pharmacological and biochemical interactions between opioids and cannabinoids. *Trends in Pharmacological Sciences, 20*, 287–294.

Pei, L., Li, S., Wang, M., Diwan, M., Anisman, H., Fletcher, P. J., et al. (2010). Uncoupling the dopamine D1-D2 receptor complex exerts antidepressant-like effects. *Nature Medicine, 16*, 1393–1395.

Pello, O. M., Martinez-Munoz, L., Parrillas, V., Serrano, A., Rodriguez-Frade, J. M., Toro, M. J., et al. (2008). Ligand stabilization of CXCR4/delta-opioid receptor heterodimers reveals a mechanism for immune response regulation. *European Journal of Immunology, 38*, 537–549.

Perreault, M. L., O'Dowd, B. F., & George, S. R. (2011). Dopamine receptor homooligomers and heterooligomers in schizophrenia. *CNS Neuroscience and Therapeutics, 17*, 52–57.

Pfeiffer, M., Kirscht, S., Stumm, R., Koch, T., Wu, D., Laugsch, M., et al. (2003). Heterodimerization of substance P and mu-opioid receptors regulates receptor trafficking and resensitization. *Journal of Biological Chemistry, 278*, 51630–51637.

Pfeiffer, M., Koch, T., Schroder, H., Laugsch, M., Hollt, V., & Schulz, S. (2002). Heterodimerization of somatostatin and opioid receptors cross-modulates phosphorylation, internalization, and desensitization. *Journal of Biological Chemistry, 277*, 19762–19772.

Rios, C., Gomes, I., & Devi, L. A. (2004). Interactions between delta opioid receptors and alpha-adrenoceptors. *Clinical and Experimental Pharmacology and Physiology, 31*, 833–836.

Rios, C., Gomes, I., & Devi, L. A. (2006). mu opioid and CB1 cannabinoid receptor interactions: Reciprocal inhibition of receptor signaling and neuritogenesis. *British Journal of Pharmacology, 148*, 387–395.

Rozenfeld, R., Bushlin, I., Gomes, I., Tzavaras, N., Gupta, A., Neves, S., et al. (2012). Receptor heteromerization expands the repertoire of cannabinoid signaling in rodent neurons. *PLoS One*, 7, e29239.

Rozenfeld, R., & Devi, L. A. (2007). Receptor heterodimerization leads to a switch in signaling: Beta-arrestin2-mediated ERK activation by mu-delta opioid receptor heterodimers. *The FASEB Journal, 21*, 2455–2465.

Rozenfeld, R., Gupta, A., Gagnidze, K., Lim, M. P., Gomes, I., Lee-Ramos, D., et al. (2011). AT1R-CBR heteromerization reveals a new mechanism for the pathogenic properties of angiotensin II. *EMBO Journal, 30*, 2350–2363.

Salata, R. A., Malhotra, I. J., Hampson, R. K., Ayers, D. F., Tomich, C. S., & Rottman, F. M. (1992). Application of an immune-tolerizing procedure to generate monoclonal antibodies specific to an alternate protein isoform of bovine growth hormone. *Analytical Biochemistry, 207*, 142–149.

Shulman, M., Wilde, C. D., & Kohler, G. (1978). A better cell line for making hybridomas secreting specific antibodies. *Nature, 276*, 269–270.

Sleister, H. M., & Rao, A. G. (2001). Strategies to generate antibodies capable of distinguishing between proteins with >90% amino acid identity. *Journal of Immunological Methods, 252*, 121–129.

Sleister, H. M., & Rao, A. G. (2002). Subtractive immunization: A tool for the generation of discriminatory antibodies to proteins of similar sequence. *Journal of Immunological Methods, 261*, 213–220.

Vigano, D., Rubino, T., & Parolaro, D. (2005). Molecular and cellular basis of cannabinoid and opioid interactions. *Pharmacology, Biochemistry, and Behavior, 81*, 360–368.

CHAPTER THIRTEEN

Hetero-oligomerization and Specificity Changes of G Protein-Coupled Purinergic Receptors: Novel Insight into Diversification of Signal Transduction

Tokiko Suzuki[*], **Kazunori Namba**[†], **Natsumi Mizuno**[*,‡], **Hiroyasu Nakata**[§,1]

[*]Department of Cellular Signaling, Graduate School of Pharmaceutical Sciences, Tohoku University, Sendai, Japan
[†]Otolaryngology/Laboratory of Auditory Disorders, National Institute of Sensory Organs, National Tokyo Medical Center, Tokyo, Japan
[‡]Department of Pharmacotherapy of Lifestyle Related Diseases, Graduate School of Pharmaceutical Sciences, Tohoku University, Sendai, Japan
[§]Department of Brain Development and Neural Regeneration, Tokyo Metropolitan Institute of Medical Science, Tokyo, Japan
[1]Corresponding author: e-mail address: nakata-hy@igakuken.or.jp

Contents

1. Introduction 240
2. Measurement of GPCR Dimerization 242
   2.1 Coimmunoprecipitation 242
   2.2 Bioluminescence resonance energy transfer[2] 244
   2.3 Immunoelectron microscopy 247
3. Receptor Pharmacology 250
   3.1 Ligand binding assay 250
   3.2 cAMP assay 251
   3.3 Ca^{2+} assay 253
4. Conclusion 255
References 256

Abstract

The formation of homo- and hetero-oligomers between various G protein-coupled receptors (GPCRs) has been demonstrated over the past decade. In most cases, GPCR heterodimerization increases the diversity of intracellular signaling. GPCR-type purinergic receptors (adenosine and P2Y receptors) are actively reported to form hetero-oligomers with each other, with GPCRs belonging to the same group (type 1,

Methods in Enzymology, Volume 521
ISSN 0076-6879
http://dx.doi.org/10.1016/B978-0-12-391862-8.00013-2

rhodopsin-like), and even with GPCRs from another group. This chapter describes common strategies to identify dimerization of purinergic receptors (coimmunoprecipitation, bioluminescence resonance energy transfer (BRET), and immunoelectron microscopy) and to assess the alteration of their pharmacology (ligand binding, intracellular cAMP, and intracellular Ca^{2+} assays). We have reported dimerization of purinergic receptors using these strategies in transfected human embryonic kidney 293T cells and native brain tissue. Our data suggest that homo- and hetero-oligomerization between purinergic receptors exert unique pharmacology in this receptor group. According to these discoveries, heterodimerization is likely to be employed for the "fine-tuning" of purinergic receptor signaling.

1. INTRODUCTION

Posttranslational processing of G protein-coupled receptors (GPCRs), including multimerization, generates diversity of cellular signaling transduction that cannot be obtained from genomic sequences. In the past decade, many GPCRs have been reported to form homo- and hetero-oligomers, which generate various pharmacological effects (Birdsall, 2010). Information concerning interacting partners and predicted dimerization interfaces are available in GRIPDB (G Protein-Coupled Receptor Interaction Partner Database; http://grip.cbrc.jp/; Nemoto, Fukui, & Toh, 2011). This database presents information about experimentally identified GPCR oligomers and their annotations, experimentally suggested interfaces for the oligomerization, and the oligomerization interfaces predicted using computational methods.

Though increasing evidence has accumulated to indicate that various GPCRs form homo- and hetero-oligomers, there are limited data that reveal the physiological significance of oligomerization *in vivo*. Most studies are focused on the exploration of pharmacological changes following GPCR dimerization *in vitro*. As a result, the significance of heterodimerization can be classified into three categories: (1) diversification of signal transduction, (2) requirement for membrane trafficking, and (3) altering ligand specificity.

1. Diversification of signal transduction: Most GPCR heterodimers alter signal transduction by changing ligand binding affinity, activation of G proteins, or internalization and desensitization. For example, the adenosine A_1 (A_1R) and $P2Y_1$ ($P2Y_1R$) receptors form a heterodimer with each other in cotransfected HEK (human embryonic kidney)

293T cells and in native rat brain (Yoshioka, Hosoda, Kuroda, & Nakata, 2002; Yoshioka, Saitoh, & Nakata, 2001, 2002). Upon heterodimerization, the $P2Y_1R$ agonist was capable of inducing signaling via $G_{i/o}$ protein coupling to A_1R (Yoshioka et al., 2001). Another example is the adenosine $A_{2A}R$ and dopamine D_2R heterodimer, where both receptors are suggested to be involved in Parkinson's disease. Long-term stimulation with D_2R agonist led to the internalization of $A_{2A}R/D_2R$ oligomers, and opposing effects were observed in cells treated with D_2R antagonist or with the $A_{2A}R$ agonist (Vidi, Chemel, Hu, & Watts, 2008).

2. Requirement for membrane trafficking: Some GPCRs have been reported to need heterodimerization for proper membrane trafficking. The $GABA_B$ receptor is composed of two subunits ($GABA_{B1}$ and $GABA_{B2}$) and is nonfunctional when the monomers are expressed individually but functional when they are coexpressed (Jones et al., 1998; Kaupmann et al., 1998). This phenomenon also occurs with M71 olfactory receptors that require the formation of heterodimers with $A_{2A}R$, $P2Y_1R$, and $P2Y_2R$ for functional expression at the cell membrane (Bush et al., 2007).
3. Altering ligand specificity: An intriguing example of the significance of GPCR heterodimerization is changes in ligand specificity. There are three subtypes of taste receptor T1R, that is, T1R1, T1R2, and T1R3. The T1R2/T1R3 heterodimer recognizes diverse natural and synthetic sweeteners. By contrast, the human T1R1/T1R3 heterodimer responds to the *umami* taste stimulus L-glutamate, and this response is enhanced by 5′-ribonucleotides, a hallmark of *umami* taste (Palmer, 2007).

Adenosine and nucleotides exert various physiological effects via purinergic receptors. ATP is hydrolyzed extracellularly into ADP, AMP, and adenosine. ATP and ADP bind to P2 receptors, whereas adenosine binds to the adenosine receptor (Ralevic & Burnstock, 1998). This intriguing relationship between these ligands suggests some interaction between adenosine and P2 receptors. Adenosine receptors are further subclassified into A_1, A_{2A}, A_{2B}, and A_3, all of which are GPCRs. The P2 receptors are further subclassified into ligand-gated ion channel-type $P2X_{1-7}$ receptors and G protein-coupled $P2Y_{1, 2, 4, 6, 11, 12, 13, 14}$ receptors. P2Y receptors respond to a wider range of agonists including purines, pyrimidines, and UDP glucose.

Evidence showing the existence of homo- and hetero-oligomerization between purinergic receptors and other GPCRs has been accumulating.

According to our data and GRIPDB, homodimerization occurs in the A_1R, $A_{2A}R$, and $P2Y_2Rs$ (Kamiya, Saitoh, Yoshioka, & Nakata, 2003; Suzuki et al., 2009; Yoshioka, Saitoh, et al., 2002). Heterodimerization is observed in various combinations including $A_1R/P2Y_1R$ (Yoshioka et al., 2001); $A_1R/P2Y_2R$ (Suzuki, Namba, Tsuga, & Nakata, 2006); $A_{2A}R/P2Y_1R/P2Y_{12}R$ (Nakata, Suzuki, Namba, & Oyanagi, 2010); A_1R/thromboxane A_2 receptor TPα (Mizuno, Suzuki, Hirasawa, & Nakahata, 2011); A_1R/D_1R (Ginés et al., 2000); $A_{2A}R/D_2R$ (Hillion et al., 2002); $A_1R/A_{2A}R$ (Ciruela et al., 2006); $P2Y_4R/P2Y_6R$ (D'Ambrosi, Iafrate, Saba, Rosa, & Volonté, 2007); M71/$A_{2A}R$, $P2Y_1R$, and $P2Y_2R$ (Bush et al., 2007); A_1R/metabotropic glutamate receptor 1 (Ciruela et al., 2001); and $A_{2A}R$/metabotropic glutamate receptor 5 (Ferre et al., 2002).

There are many biochemical techniques to measure protein–protein interactions for soluble proteins, whereas different specific strategies are required for membrane-bound GPCRs. In this chapter, we describe some of the useful procedures to investigate GPCR oligomerization, especially hetero-oligomerization between purinergic receptors.

2. MEASUREMENT OF GPCR DIMERIZATION

2.1. Coimmunoprecipitation

Coimmunoprecipitation of detergent-solubilized receptors is the most common and classical technique for research into possible GPCR dimerization. The key points of this strategy are the selection of antibodies for two different GPCR targets and membrane solubilization conditions. Therefore, exploration of the optimum antibodies and optimum condition for receptor solubilization may be required.

We explored the coimmunoprecipitation between HA-tagged A_1R and Myc-tagged $P2Y_2R$ using cotransfected HEK293T cells (Suzuki et al., 2006). In addition, we performed immunoprecipitation of nontagged A_1R and Myc-$P2Y_2R$ using solubilized membranes from nontagged A_1R/Myc-$P2Y_2R$-cotransfected HEK293T cells. We detected nontagged A_1R in the complex precipitated with anti-Myc (Fig. 13.1). This result indicates that a complex between nontagged A_1R and Myc-$P2Y_2R$ was formed in the membranes of the cotransfected cells, and coimmunoprecipitates of HA-A_1R/Myc-$P2Y_2R$ (Suzuki et al., 2006) were not due to nonspecific interactions between both tags.

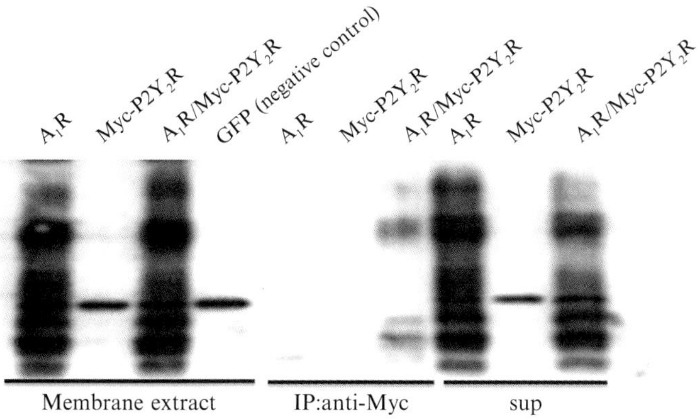

Figure 13.1 Coimmunoprecipitation of cells coexpressing A_1R/Myc-P2Y_2R. Anti-Myc antibody was used to coimmunoprecipitate A_1R with Myc-P2Y_2R WT (seventh lane from left). The first to fourth lanes represent expression of receptor proteins and reactivity of the antibodies used. Membrane extract, solubilized membrane extract; IP, immunoprecipitation; sup, supernatant following immunoprecipitation; IB, immunoblotting.

2.1.1 Protocol

On day 1, detach HEK293T cells grown in Dulbecco's modified Eagle's medium (DMEM) supplemented with 10% fetal bovine serum using trypsin–EDTA solution and seed into 100-mm dishes at a density of 3.5×10^6 cells/dish.

The following day, transfect cells with a mixture of 5 μg of cDNA plasmids, nontagged A_1R, and Myc-tagged P2Y_2R at the N-terminus and 15 μl of FuGENE HD (Roche Applied Bioscience, Manheim, Germany).

On day 4, collect cells ($\sim 1 \times 10^7$ cells), wash cells twice with DPBS (Dulbecco's phosphate-buffered saline) (137 m*M* NaCl, 2.7 m*M* KCl, 1.5 m*M* KH_2PO_4, and 8.1 m*M* Na_2HPO_4), and sonicate in hypotonic lysis buffer (50 m*M* Tris-acetate, pH 7.4) containing protease inhibitor cocktail (Roche Applied Bioscience). After centrifugation at $17{,}400 \times g$ for 30 min, solubilize cell membranes by incubating with 300 μl of Tx buffer (50 m*M* Tris-acetate, pH 7.4, 300 m*M* NaCl, and 1% Triton X-100) containing protease inhibitor cocktail and incubate for 1 h at 4 °C. After centrifugation, preclear the supernatant containing the solubilized membranes by incubating, for 30 min at 4 °C, with 30 μl of 50% Protein G-Sepharose™ 4 Fast Flow (GE Healthcare, Piscataway, NJ) in Tx buffer, followed by centrifugation at $17{,}400 \times g$ for 10 s. Then collect the supernatant and mix with 1 μg of anti-Myc 9E10 antibody (Roche Applied Bioscience) and incubate for 1 h at 4 °C, followed by incubation for 2 h with 30 μl of 50% Protein G-Sepharose™ 4 Fast Flow (GE Healthcare). Wash the immune complex twice with 300 μl of Tx buffer, and elute the bound proteins with 30 μl of SDS-PAGE (sodium dodecyl sulfate–polyacrylamide gel electrophoresis) sample buffer. Then separate the solubilized proteins using SDS-PAGE and transfer electrophoretically onto a nitrocellulose membrane. After blocking with 5% skimmed milk in 20 m*M* Tris–HCl, pH 7.6, 137 m*M* NaCl, and 0.1% Tween 20, probe for the receptors using rabbit polyclonal anti-A_1R antibody (Sigma-Aldrich, St. Louis, MO), followed by horseradish peroxidase-conjugated anti-rabbit IgG antibody (MP Biomedicals, Irvine, CA).

2.2. Bioluminescence resonance energy transfer[2]

The coimmunoprecipitation strategies described above are simple and do not require special equipment. However, the above techniques may seem limited because the solubilization of hydrophobic GPCRs can lead to aggregation that could be mistaken for oligomerization. Furthermore, the solubilization process with detergents can inhibit the association between GPCRs. Biophysical assays based on light resonance energy transfer (RET) are therefore quite valuable for overcoming these issues.

Bioluminescence resonance energy transfer (BRET) is a biophysical technique that represents a powerful tool with which to measure protein–protein interaction in living cells (Pfleger & Eidne, 2006). BRET is a natural process involving nonradiative energy transfer between donor and suitable acceptor fluorophores after oxidation of substrate. An appropriate distance between fluorophores causing BRET is 10–100 Å, which is consistent with the distance where two molecules can interact with each other. $BRET^2$ is an advanced BRET technology, which uses Renilla luciferase (Rluc) as the donor, codon-humanized form of wild-type GFP (green fluorescent protein) (GFP^2) as the acceptor, and DeepBlueC (Coelenterazine 400A) as the substrate. The use of this substrate results in a 395-nm Rluc emission peak and an emission maximum at 510 nm for GFP^2, which offers greater separation of the emission spectra of the donor and acceptor moieties. We employed this technique to measure homodimerization of A_1R (Suzuki et al., 2009 and Fig. 13.2A and B) and $P2Y_2R$ (Fig. 13.2A and C) and heterodimerization of $A_1R/P2Y_1R$ (Yoshioka, Saitoh, et al., 2002), $A_{2A}R/D_2R$ (Kamiya et al., 2003), and $A_1R/P2Y_2R$ (Fig. 13.2A).

Although a powerful strategy, BRET studies can be misleading if not conducted using the appropriate controls. Nonspecific BRET can result from the overexpression of noninteracting proteins when acceptor

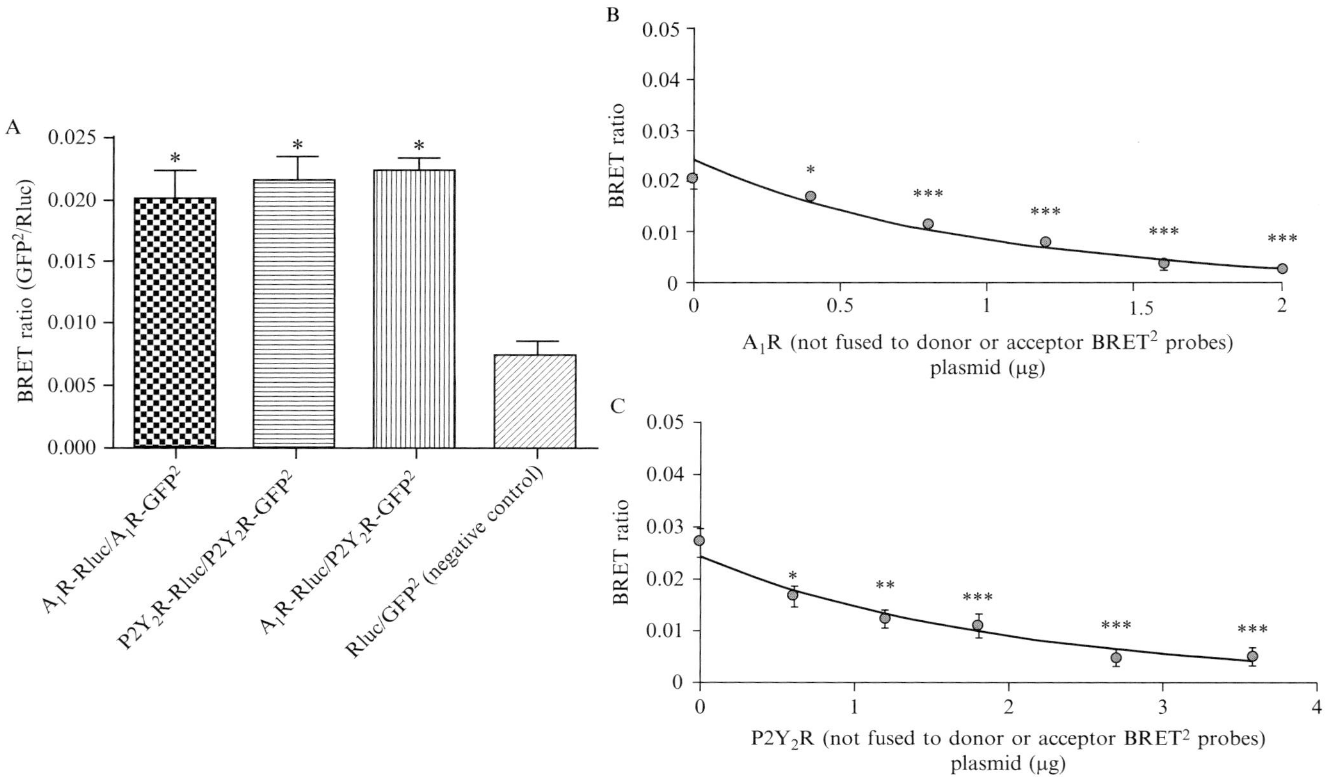

Figure 13.2 Single-point BRET^2 (A) and competitive BRET^2 (B and C) analysis. (A) HEK293T cells were cotransfected with an equal concentration of indicated plasmids. For nonspecific control, constant amounts of Rluc-N3 and GFP^2-N3 vectors were cotransfected (Rluc/GFP^2). BRET signals were measured using cells cultured for 48 h after transfection. The BRET^2 ratio was plotted as a function of the expression ratio of receptor-GFP^2 over receptor-Rluc. Experiments were performed three times using six replicates. *$P < 0.05$ versus negative control by Dunnett's test. HEK293T cells were cotransfected with a fixed concentration of (B) HA-A$_1$R-Rluc and HA-A$_1$R-GFP^2 plasmids and increasing amounts of HA-A$_1$R not fused to donor or acceptor BRET^2 probes. HEK293T cells were also cotransfected with a fixed concentration of (C) Myc-P2Y$_2$R-Rluc and Myc-P2Y$_2$R-GFP^2 plasmids and Myc-P2Y$_2$R not fused to donor or acceptor BRET^2 probes. Data are expressed as mean ± S.E.M. of three experiments. *$P < 0.05$, **$P < 0.01$, ***$P < 0.001$ versus 0 (Dunnett's test). (For color version of this figure, the reader is referred to the online version of this chapter.)

molecules are in close proximity to donor molecules. Thus, it is important to select proper negative controls that can be compared at similar expression levels. According to the guidelines for RET experiments (Marullo & Bouvier, 2007), three BRET methods have been recommended including competitive BRET experiments, as mentioned below.

Results of single-point $BRET^2$ and competitive $BRET^2$ analysis are shown in Fig. 13.2A–C. For single-point $BRET^2$, HEK293T cells were cotransfected with equal amounts of two kinds of plasmids, A_1R-Rluc/A_1R-GFP^2, $P2Y_2R$-Rluc/$P2Y_2R$-GFP^2, A_1R-Rluc/$P2Y_2R$-GFP^2, and soluble-free Rluc/GFP^2 for negative control. All three combinations resulted in significantly high $BRET^2$ signal in the negative control (Fig. 13.2A), indicating formation of A_1R homodimer, $P2Y_2R$ homodimer, and A_1R/$P2Y_2R$ heterodimer in living cells. For competitive $BRET^2$ analysis, fixed amounts of A_1R-Rluc/A_1R-GFP^2 or $P2Y_2R$-Rluc/$P2Y_2R$-GFP^2 and increasing amounts of nonlabeled A_1R or nonlabeled $P2Y_2R$ were coexpressed in HEK293T cells. As shown in Fig. 13.2B and C, BRET signal was decreased as the concentration of nonlabeled A_1R or $P2Y_2R$ increased. In addition, $BRET^2$ of $P2Y_2R$-Rluc/$P2Y_2R$-GFP^2 was inhibited as cotransfection of nonlabeled A_1R was increased (data not shown). These data confirmed the specific formation of A_1R homodimer, $P2Y_2R$ homodimer, and A_1R/$P2Y_2R$ heterodimer.

2.2.1 Protocol

2.2.1.1 Plasmid construction and transient transfection

For $BRET^2$ experiments, construct plasmids that have Rluc and GFP^2 at the C-terminus of the cDNA of target GPCRs, without stop codon. We constructed HA-A_1R-GFP^2, HA-A_1R-Rluc, Myc-$P2Y_2R$-Rluc, and Myc-$P2Y_2R$-GFP^2 by amplification of the HA-tagged rat A_1R or Myc-tagged rat $P2Y_2R$ coding sequence, without stop codon, using sense and antisense primers containing distinct restriction enzyme sites at the 5′ and 3′ ends, respectively (Mizuno et al., 2011; Suzuki et al., 2006). The fragments were then subcloned in-frame into the appropriate sites of the pGFP^2-N3 and pRluc-N3 expression vectors (Perkin Elmer Life Sciences, Boston, MA).

For single-point BRET experiments, cotransfect HEK293T cells (5×10^6 cells per 35-mm dish) with a fixed amount (e.g., 1 μg) of Rluc-tagged and GFP^2-tagged plasmids. For competitive BRET experiments, cotransfect HEK293T cells (5×10^5 per 35-mm dish) with a fixed amount (e.g., 1 μg) of Rluc-tagged and GFP^2-tagged plasmids and increasing concentrations of unlabeled receptor plasmids (e.g., 0, 0.4, 0.8, 1.2, 1.6, 2.0 μg). Transfection was done using 3 μl of FuGENE HD (Roche Applied Bioscience). Nontransfected cells or cells transfected with only Rluc-tagged plasmids can be used for control.

2.2.1.2 $BRET^2$ measurement

Forty-eight hours after transfection, harvest cells and suspend in assay buffer (DPBS containing 0.1 mg/ml $CaCl_2$, 0.1 mg/ml $MgCl_2$, and 1 mg/ml D-glucose). Next, distribute the cells into white-walled 96-well plates (Opti-Plate; Perkin Elmer Life Sciences) at a density of 1×10^6 cells per well, and incubate for 20 min at 37 °C. Then add DeepBlueC (Perkin Elmer Life Sciences) or Coelenterazine 400A (Biotium Inc., CA) to cells at a final concentration of 5 μM, and immediately determine BRET signal using a Fusion α universal microplate analyzer (Perkin Elmer Life Sciences) for the detection of Rluc at 410 nm and GFP^2 at 515 nm. The BRET ratio was calculated as emission at 515 nm over emission at 410 nm. Correct the values by subtracting the background signal detected using non-transfected cells.

2.3. Immunoelectron microscopy

Immunoelectron microscopy is another tool for the more direct assessment of GPCR heterodimerization. In this chapter, we introduce this method to examine heterodimerization of transfected and endogenous purinergic receptors using HEK293T and intact rat brain, respectively.

To assess the hetero-oligomerization of endogenous GPCRs in brain tissue, three experimental steps are typically required. The first step is to obtain immunoelectron microscopic data acquired from preembedding methods using transfected culture cells. The second step is to acquire data using transfected cells with postembedding methods. The last step is to acquire data from tissues using postembedding methods.

As shown in Fig. 13.3A and B, immunogold particles were localized individually or in clusters, indicating that both transfected HA-A_1R and Myc-P2Y_2R form monomers and homo-oligomers in HEK293T cells. When Myc-P2Y_2R-transfected HEK293T cells were incubated with both anti-HA and anti-Myc antibodies, we detected single particles scattered all over the cells (Fig. 13.3A). A similar observation was also observed in HA-A_1R-transfected HEK293T cells incubated with both anti-HA and anti-Myc antibodies (Fig. 13.3B). By contrast, HEK293T cells cotransfected with both HA-A_1R and Myc-P2Y_2R displayed clusters of different-sized particles, mainly at the cell surface (Fig. 13.3C), suggesting the formation of hetero-oligomers. Similarly, we observed clusters of different-sized gold particles at cytoplasmic membranes in cell bodies in the rat cerebellum, indicating the presence of heteromeric complexes of endogenous A_1R and P2Y_2R in the rat cerebellum (Fig. 13.3D).

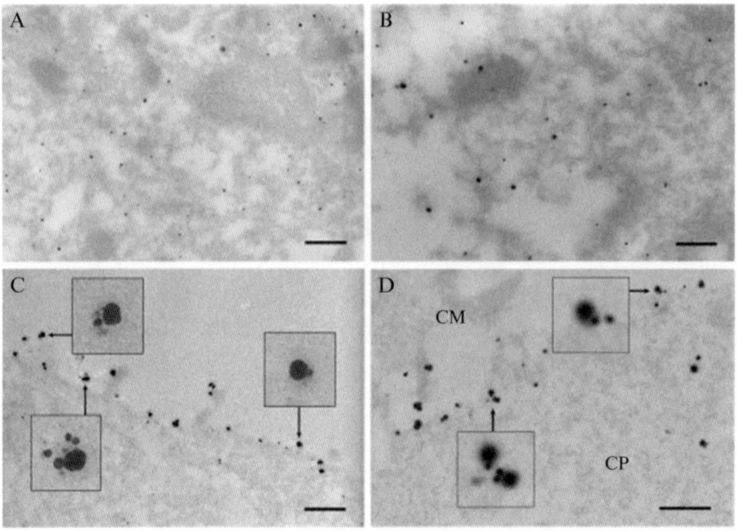

Figure 13.3 Immunogold electron microscopy (postembedding) method to detect A_1R and $P2Y_2R$ in transfected HEK293T cells using nanogold particles. (A) Localization of Myc-$P2Y_2R$ (small particles) detected with anti-Myc and nonspecific immunoreaction with anti-HA (large particles) in Myc-$P2Y_2R$-transfected HEK293T cells. (B) Localization of HA-A_1R detected with anti-HA and nonspecific immunoreaction with anti-Myc in HA-A_1R-transfected HEK293T cells. (C) Anti-HA and anti-Myc immunolocalization in HEK293T cells cotransfected with HA-A_1R and Myc-$P2Y_2R$. (D) Visualization of A_1R and $P2Y_2R$ in rat brain. Localization of A_1R (small particles) and $P2Y_2R$ (large particles) on the cell surface of Purkinje cells detected with both anti-A_1R and anti-$P2Y_2R$. Arrows indicate two adjacent receptors on the cell membrane. CM, cell membrane; CP, cytoplasm. Bars represent 100 nm.

In our earlier experiments, under the same experimental conditions, oligomerization of A_1R and $P2Y_2R$ in rat brain tissue occurred in hippocampal pyramidal cells, cerebellum, and pyramidal cells in the frontal cortex (Namba et al., 2010). Hetero-oligomers of $A_1R/P2Y_2R$ were detected in significant numbers at the cell surface. From these studies, it can be speculated that the $A_1R/P2Y_2R$ hetero-oligomer responsible for unique signal transduction (Suzuki et al., 2006) also occurs in the rat brain.

2.3.1 *Protocol*

2.3.1.1 Preembedding method of transfected HEK293T cells

Forty-eight hours after transfection of HA-A_1R and/or Myc-$P2Y_2R$ plasmids, briefly wash the cells with 0.1 *M* phosphate-buffered saline (PBS) and fix with 4% paraformaldehyde and permeabilize with 0.25% Triton X-100 (PBST) for 15 min. Incubate the cells with anti-HA rat antibody (1 μg/ml) and anti-Myc mouse antibody (1 μg/ml) for 3 h at 4 °C. After washing with PBS, incubate the cells with 10-nm gold particle-conjugated goat anti-rat IgG antibody (rat IgG-10, 1:1000; BBI International, Lakewood, CO) and 5-nm gold particle-conjugated goat anti-mouse IgG antibody (mouse IgG-5, 1:1000; BBI International) for 4 h at 4 °C. After washing, fix the cells with 2.5% glutaraldehyde in 0.15 *M* sodium cacodylate (pH 7.4) for 2 h. Postfix the cells with 1% osmium tetroxide for 4 h at room temperature. After washing twice with 50% ethanol, stain the cells with 1% uranyl acetate in 50% ethanol for 30 min. Then dehydrate the cells on culture dishes with 50%, 70%, and 95% ethanol for 10 min each and in 100% ethanol for 2 × 15 min. Further dehydrate the cells in 1:1 mixture of ethanol and propylene oxide for 1 min. Following this step, rinse the cells on culture dishes with 100% propylene oxide, immediately add 1:1 mixture of propylene oxide and Epon 812 resin (NISSIN EM, Tokyo, Japan) and leave for 8 min. Embed the cells on the culture dishes with straight Epon 812 resin for 2 × 5 h. Flat-embed the culture dishes with another straight Epon 812 resin, and bake at 60 °C for 48 h. Freeze the dishes in liquid nitrogen for 1 min, and subsequently submerge the dishes in room temperature water. Then, crack the frozen dishes to remove the dish from the resin completely. Cut the polymerized resin into 1.5 × 1.5-mm blocks using a razor blade on a hot plate, and fix to the tip of the foundation capsule, using instant glue, with the cellular surface lying face up. Cut the block into thin sections (40 nm) using a diamond knife in an ultramicrotome. Observe the specimens with an H7500 electron microscope (Hitachi, Japan).

2.3.1.2 Postembedding method of transfected HEK293T cells and brain tissue

Before transfection, prepare a polyethyleneimine-coated plain microscope slide (25 × 75 × 1 mm). Perform fixation and dehydration as described above. Remove the ethanol with the 100% propylene oxide for 10 min. Incubate the slide in 1:1 mixture of propylene oxide and LR-white resin (NISSIN EM) for 30 min. Embed the cells on the slide with straight LR-white for 2 × 5 h. Put the polymerized LR-white capsule on cells embedded with straight LR-white, and bake at 60 °C for 48 h. Detach the polymerized LR-white from the slide. Cut the block into thin sections using a diamond knife in an ultramicrotome. Cut the dissected brain tissue into 1.0-mm^3 blocks that are then incubated with lead(II) acetate (Sigma-Aldrich) buffer for 1 h at room temperature. Following this, dehydrate the brain tissue with 50%, 70%, and 95% ethanol for 30 min each and in 100% ethanol for 2 × 2 h. Further dehydrate in 1:1 mixture of ethanol and propylene oxide for 2 h.

Exchange with 100% propylene oxide for 2 × 2 h, and add 1:1 mixture of propylene oxide and LR-white to the brain tissue for 2 h. Embed the brain tissues with straight LR-white for 2 × 5 h. Place specimen at the bottom of a gelatin capsule and fill with LR-white. Polymerize in a 60 °C oven for 48 h. Cut out the specimen in polymerized resin and place the tip in a polymerized LR-white capsule. Mount the ultra-thin sections (40 nm) on 200-mesh nickel grids (NISSIN EM) and incubate in PBST containing 1% bovine serum albumin for 10 min. After immunoreactions, rinse sections for 2 × 2 min by placing grids on large droplets of PBST. Stain sections with uranyl acetate for 10 min and rinse in PBS for 2 × 2 min.

3. RECEPTOR PHARMACOLOGY

Hetero-oligomerization of GPCRs can affect various aspects of receptors, including altering ligand binding specificity, signal transduction, and cellular trafficking. We therefore explored purinergic receptor activity in cotransfected HEK293T cells by ligand binding assay, adenylate cyclase assay, and intracellular Ca^{2+} assay.

3.1. Ligand binding assay

We revealed that A_1R ligand binding was attenuated by $P2Y_2R$ ligands in the $A_1R/P2Y_2R$ coexpressing HEK293T cells (Suzuki et al., 2006). To analyze what caused the inhibition of A_1R agonist binding by $P2Y_2R$ activation in the $A_1R/P2Y_2R$-coexpressing cells, we examined [^{3}H]CCPA (2-chloro-N^6-cyclopentyladenosine), a selective A_1R agonist, saturation binding in the presence or absence of 100 μ*M* UTP, a $P2Y_2R$ agonist (Fig. 13.4). In the absence of UTP, the K_d and B_{max} values were 0.23 ± 0.03 n*M* and 0.84 ± 0.04 pmol/mg protein, respectively. Conversely, in the presence of UTP, these values were estimated to be 1.61 ± 0.11 n*M* and 0.34 ± 0.02 pmol/mg protein, respectively. These K_d and B_{max} values are significantly different from the original values. Therefore, inhibition of A_1R agonist binding by the $P2Y_2R$ agonist UTP is due to a decrease in both the affinity and the number of binding sites of the A_1R in the membranes from $A_1R/P2Y_2R$-coexpressing cells.

3.1.1 Protocol

Perform transfection into HEK293T as above using FuGENE HD (Roche Applied Bioscience) and harvest these cells 48 h later for use in the ligand binding assay. Collect transfected cells (approx. 10^7 cells) and wash twice with DPBS. Disrupt the cells by sonication in hypotonic lysis buffer

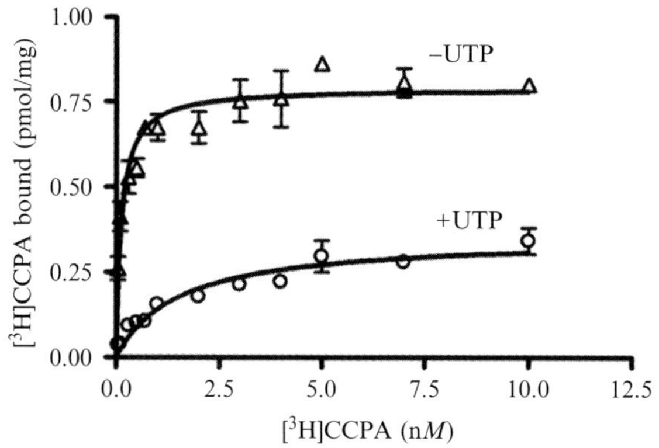

Figure 13.4 [^{3}H]CCPA saturation binding in the presence or absence of 100 μ*M* UTP in membranes from cells coexpressing HA-A_1R and Myc-P2Y_2R. Saturation binding of [^{3}H] CCPA to HEK293T cell membranes coexpressing HA-A_1R and Myc-P2Y_2R was determined in the presence (circle) and absence (triangle) of 100 μ*M* UTP. Results are shown as the mean of three independent experiments performed in duplicate.

containing protease inhibitor cocktail. Centrifuge at 17,400 × *g* for 30 min and suspend the pellet using the same buffer. Repeat this three times. Incubate the membranes (15–30 μg) for 1.5 h at 25 °C with various concentrations of radiolabeled ligand [^{3}H]CCPA (42.6 Ci/mmol; PerkinElmer Life Sciences, Boston, MA), 50 m*M* Tris-acetate (pH 7.4), 1 m*M* EDTA, 5 m*M* $MgCl_2$, and 4 units/ml of adenosine deaminase (ADA; Sigma-Aldrich). Nonspecific binding can be determined by examining the binding in the presence of xanthine amine congener (10 μ*M*), an adenosine receptor antagonist. After filtration through GF/B glass fiber filters, quantify radioactive ligand bound to the receptors by counting the radioactivity of the filters. Determine K_D and B_{max} values by nonlinear regression analysis using one-site binding model (GraphPad Prism 4; GraphPad, San Diego, CA).

3.2. cAMP assay

We examined whether heteromerization affects cellular signaling transduction. For this purpose, we studied A_1R agonist-induced inhibition of adenylyl cyclase, a key index of A_1R function (Fig. 13.5). As expected, CPA (N^6-cyclopentyladenosine), a selective A_1R agonist, decreased forskolin (FSK)-evoked cAMP generation in cells expressing HA-A_1R alone and in cells coexpressing HA-A_1R and Myc-P2Y_2R (Fig. 13.5A and B). Interestingly, we found that UTP inhibited the CPA-mediated signaling in the HA-A_1R/Myc-P2Y_2R-coexpressing cells (Fig. 13.5B, black bar).

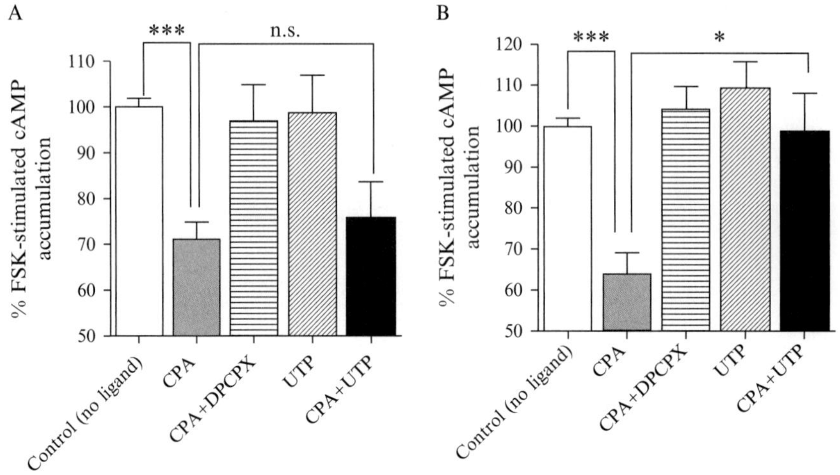

Figure 13.5 Effects of A_1R and $P2Y_2R$ agonists and antagonists on FSK-stimulated cAMP accumulation in cells expressing HA-A_1R alone (A) and coexpressing HA-A_1R and Myc-$P2Y_2R$ (B). The results were normalized by the cAMP concentration in control cells (without ligands). Results shown are the means of more than two independent experiments performed in quadruplicate. Values were compared by Dunnett's test. n.s., no significant difference ($P > 0.05$) where indicated. $^*P < 0.05$, $^{***}P < 0.001$ versus where indicated, between groups. *Part of this figure was reprinted from Suzuki et al. (2006), with permission from Elsevier.*

This result is consistent with the inhibitory effect of UTP on [^{3}H]CCPA binding to the membranes from cells coexpressing A_1R and $P2Y_2R$ (Fig. 13.4, open circles). The antagonist-like effect of UTP on A_1R-mediated signaling in cells expressing A_1R alone was negligible (Fig. 13.5A, black bars). UTP alone did not significantly affect FSK-induced cAMP production in cells expressing HA-A_1R alone and in cells coexpressing HA-A_1R and Myc-$P2Y_2R$ (Fig. 13.5A and B).

3.2.1 Protocol

Transfect HEK293T cells plated on collagen-coated 24-well plates at 1×10^5 cells/well as described above and culture for another 24 h. Next, treat the cells for 1 h at 37 °C with 1 unit/ml ADA and incubate for 15 min at 37 °C with an equal volume of serum-free DMEM containing 100 μ*M* of Ro-20-1724 and 1 unit/ml ADA with or without receptor antagonist. Subsequently, incubate the cells with receptor agonist and FSK (10 μ*M* final concentration) for $G_{i/o}$-coupled receptor for a further 10 min. Terminate the reaction by adding ice-cold 0.1N HCl for 10 min, followed by neutralization with 1N NaOH. Extract the intracellular cAMP with 65% ethanol for 10 min on ice. Determine the intracellular cAMP concentration using a cAMP Biotrak Enzyme Immunoassay System (GE Healthcare).

3.3. Ca^{2+} assay

Measurement of intracellular Ca^{2+} concentration is needed for analysis of signal transduction of $G_{q/11}$-coupled GPCRs. Changes in intracellular free Ca^{2+} concentration ($[Ca^{2+}]_i$) were monitored by the intensity of Fura-2 fluorescence at 37 °C using a spectrofluorometer. Fura-2-acetoxymethyl ester (Fura-2-AM) is a membrane-permeable derivative of the ratiometric calcium indicator Fura-2, used to measure cellular calcium concentrations by fluorescence. When added to cells, Fura-2-AM crosses cell membranes, and once inside the cell, the acetoxymethyl groups are removed by cellular esterases. Removal of the acetoxymethyl esters regenerates "Fura-2," the pentacarboxylate calcium indicator. Measurement of Ca^{2+}-induced fluorescence at both 340 and 380 nm allows for calculation of calcium concentration-based 340/380 ratios.

We focused on three purinergic receptor subtypes, $A_{2A}R$, $P2Y_1R$, and $P2Y_{12}R$, which are expressed in platelets and are thought to play individual roles in platelet aggregation (Jin & Kunapuli, 1998). Formation of a heteromeric complex composed of these three receptors has been demonstrated using coimmunoprecipitation techniques in HEK293T cells coexpressing these three receptors (Nakata et al., 2010). For further exploration, we examined the possibility of functional interaction between these purinergic receptors (Suzuki, Obara, Moriya, Nakata, & Nakahata, 2011). We analyzed the effect of ligands for $A_{2A}R$ and $P2Y_{12}R$ on the mobilization of intracellular Ca^{2+} as a key index of $P2Y_1R$-mediated signaling. These observations were performed using HEK293T cells transfected with $A_{2A}R$ and $P2Y_{12}R$, which endogenously express $P2Y_1R$ (Schachter, Sromek, Nicholas, & Harden, 1997). As expected, stimulation of the $P2Y_1R$ with 2MeSADP (2-methylthio ADP), an agonist of $P2Y_1R$ and $P2Y_{12}R$, caused an increase in Ca^{2+} release (Fig. 13.6A), which was absolutely blocked by pretreatment with MRS2179, a $P2Y_1R$-specific antagonist (Fig. 13.6B). These findings indicated that normal $P2Y_1R$ signaling via $G_{q/11}$ in HEK293T cells existed. Pretreatment with CGS21680, an $A_{2A}R$ specific agonist, did not affect Ca^{2+} mobilization induced by 2MeSADP (Fig. 13.6C). By contrast, treatment with ZM241385 and ARC69931MX, which are specific antagonists for

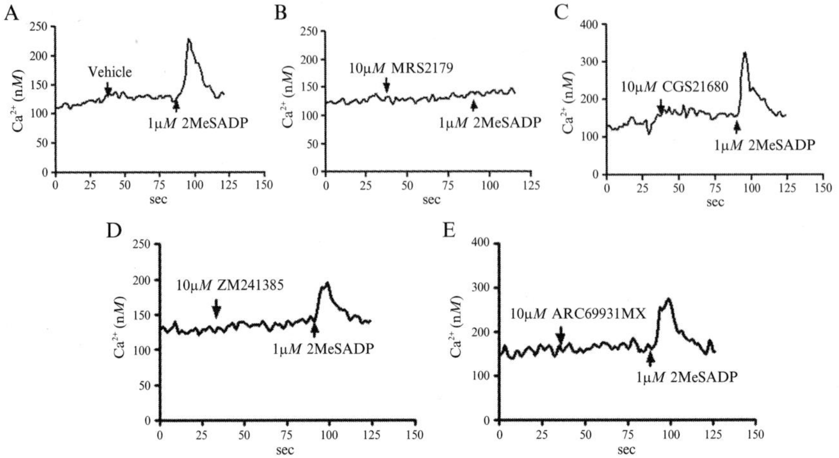

Figure 13.6 Mobilization of Ca^{2+} evoked by 2MeSADP prestimulated with various ligands. Fura 2-loaded HEK293T cells transfected with $A_{2A}R$ and $P2Y_{12}R$ were preincubated for 1 min with either vehicle as control (A), 10 μ*M* MRS2179 (B), 10 μ*M* CGS21680 (C), 10 μ*M* ZM241385 (D), or 10 μ*M* ARC69931MX (E) as indicated. Cells were then stimulated with 1 μ*M* 2MeSADP (arrow), and fluorescence intensity was measured. Traces shown are representative of at least three separate experiments.

$A_{2A}R$ and $P2Y_{12}R$, respectively, caused decreased Ca^{2+} release (Fig. 13.6D and E). These results indicate negative regulation of $P2Y_1R$ function by $A_{2A}R$- and $P2Y_{12}R$-specific antagonists.

3.3.1 Protocol

Wash the transfected cells twice with Tyrode's solution (137 m*M* NaCl, 2.7 m*M* KCl, 1.0 m*M* $MgCl_2$, 1.8 m*M* $CaCl_2$, 20 m*M* HEPES, and 5.6 m*M* glucose, pH 7.4). After centrifugation at $250 \times g$ for 2 min, treat the cells with Fura-2-AM (1 μ*M*; Dojindo Laboratories, Kumamoto, Japan) at 37 °C for 15 min. Then centrifuge the cells at $250 \times g$ for 2 min and wash twice with Tyrode's buffer. Suspend the cells in modified Tyrode's buffer (137 m*M* NaCl, 2.7 m*M* KCl, 1.0 m*M* $MgCl_2$, 0.18 m*M* $CaCl_2$, 20 m*M* HEPES, and 5.6 m*M* glucose, pH 7.4) at a density of 2×10^5 cells/ml. Use 1.5 ml of the cell suspension for Fura-2 assays at 37 °C, with gentle stirring, in a spectrofluorometer with excitation at 340 and 380 nm and emission at 510 nm. Calibration of the signal should be performed for each sample by adding 0.15% Triton X-100 to obtain maximal fluorescence and then 2.5 m*M* EGTA to obtain minimal fluorescence. The ratio of fluorescence at 340 nm to that at 380 nm is a measure of $[Ca^{2+}]_i$, assuming a K_D of 244 n*M* Ca^{2+} for Fura-2-AM.

4. CONCLUSION

As mentioned above, purinergic receptors frequently form heterodimers, which exert specific alterations in signal transduction. It is likely that the physiological effects of adenine nucleotides and nucleosides rely on a membrane network of appropriate receptors and enzymes rather than the presence of isolated proteins. Depending on the partners included in the network, a previously unforeseen number of different responses can be expected to arise in response to ATP, ADP, and adenosine. These results suggest that a large number of combinations for direct interaction may occur to form a membrane network of receptors and enzymes for adenine nucleosides and nucleotides. Purinergic receptors are known to be widely expressed in various cells. Frequently, the output by the addition of nucleotides or nucleosides to cells is not always what is expected from independent molecularly defined cloned purinergic receptor types (Burnstock & Knight, 2004). Although other mechanisms may exist, it is quite feasible that such functional modification can be produced by the direct interaction between purinergic receptors of other colocalized subtypes. As reviewed in this chapter, purinergic receptors tend to form heterodimers with other GPCRs, and therefore, if purinergic receptors are located in close proximity on the cell membrane, it is natural to speculate that the resulting heterodimers between purinergic receptors produce various changes or modification in the purinergic signaling.

GPCRs are known to be proteins that are difficult to crystallize. Efforts to elucidate the crystal structures of GPCRs bound to diffusible ligands have recently yielded structures for five rhodopsin-like GPCRs: the human β2 adrenergic receptor (Rosenbaum et al., 2011), turkey β1 adrenergic receptor (Warne et al., 2011), human A_{2A} adenosine receptor (Xu et al., 2011), human CXCR4 chemokine receptor (Wu et al., 2010), and the D3 dopamine receptor (Chien et al., 2010). GPCR hetero-oligomerization may increase the diversity of GPCR-responsive signals by altering receptor pharmacology, and such oligomers thus represent a potential set of novel therapeutic targets. Therefore, agents that promote or disrupt oligomerization, or that selectively induce heterodimerization, may be of clinical and research value. Future clarification of the crystal structure of heterodimerized GPCRs will dramatically facilitate new drug discovery.

REFERENCES

Birdsall, N. J. (2010). Class A GPCR heterodimers: Evidence from binding studies. *Trends in Pharmacological Sciences*, *31*, 499–508.

Burnstock, G., & Knight, G. E. (2004). Cellular distribution and functions of P2 receptor subtypes in different systems. *International Review of Cytology*, *240*, 31–304.

Bush, C. F., Jones, S. V., Lyle, A. N., Minneman, K. P., Ressler, K. J., & Hall, R. A. (2007). Specificity of olfactory receptor interactions with other G protein-coupled receptors. *The Journal of Biological Chemistry*, *282*, 19042–19051.

Chien, E. Y., Liu, W., Zhao, Q., Katritch, V., Han, G. W., Hanson, M. A., et al. (2010). Structure of the human dopamine D_3 receptor in complex with a D_2/D_3 selective antagonist. *Science*, *330*, 1091–1095.

Ciruela, F., Casadó, V., Rodrigues, R. J., Luján, R., Burgueño, J., Canals, M., et al. (2006). Presynaptic control of striatal glutamatergic neurotransmission by adenosine A_1-A_{2A} receptor heteromers. *The Journal of Neuroscience*, *26*, 2080–2087.

Ciruela, F., Escriche, M., Burgueno, J., Angulo, E., Casado, V., Soloviev, M. M., et al. (2001). Metabotropic glutamate 1α and adenosine A_1 receptors assemble into functionally interacting complexes. *The Journal of Biological Chemistry*, *276*, 18345–18351.

D'Ambrosi, N., Iafrate, M., Saba, E., Rosa, P., & Volonté, C. (2007). Comparative analysis of $P2Y_4$ and $P2Y_6$ receptor architecture in native and transfected neuronal systems. *Biochimica et Biophysica Acta*, *1768*, 1592–1599.

Ferre, S., Karcz-Kubicha, M., Hope, B. T., Popoli, P., Burgueno, J., Gutierrez, M. A., et al. (2002). Synergistic interaction between adenosine A_{2A} and glutamate mGlu5 receptors: Implications for striatal neuronal function. *Proceedings of the National Academy of Sciences of the United States of America*, *99*, 11940–11945.

Ginés, S., Hillion, J., Torvinen, M., Le Crom, S., Casadó, V., Canela, E. I., et al. (2000). Dopamine D_1 and adenosine A_1 receptors form functionally interacting heteromeric complexes. *Proceedings of the National Academy of Sciences of the United States of America*, *97*, 8606–8611.

Hillion, J., Canals, M., Torvinen, M., Casado, V., Scott, R., Terasmaa, A., et al. (2002). Coaggregation, cointernalization, and codesensitization of adenosine A_{2A} receptors and dopamine D_2 receptors. *The Journal of Biological Chemistry*, *277*, 18091–18097.

Jin, J., & Kunapuli, S. P. (1998). Coactivation of two different G protein-coupled receptors is essential for ADP-induced platelet aggregation. *Proceedings of the National Academy of Sciences of the United States of America*, *95*, 8070–8074.

Jones, K. A., Borowsky, B., Tamm, J. A., Craig, D. A., Durkin, M. M., Dai, M., et al. (1998). $GABA_B$ receptors function as a heteromeric assembly of the subunits $GABA_BR1$ and $GABA_BR2$. *Nature*, *396*, 674–679.

Kamiya, T., Saitoh, O., Yoshioka, K., & Nakata, H. (2003). Oligomerization of adenosine A_{2A} and dopamine D_2 receptors in living cells. *Biochemical and Biophysical Research Communications*, *306*, 544–549.

Kaupmann, K., Malitschek, B., Schuler, V., Heid, J., Froestl, W., Beck, P., et al. (1998). $GABA_B$-receptor subtypes assemble into functional heteromeric complexes. *Nature*, *396*, 683–687.

Marullo, S., & Bouvier, M. (2007). Resonance energy transfer approaches in molecular pharmacology and beyond. *Trends in Pharmacological Sciences*, *28*, 362–365.

Mizuno, N., Suzuki, T., Hirasawa, N., & Nakahata, N. (2011). Hetero-oligomerization between adenosine A_1 and thromboxane A_2 receptors and cellular signal transduction on stimulation with high and low concentrations of agonists for both receptors. *European Journal of Pharmacology*, *677*, 5–14.

Nakata, H., Suzuki, T., Namba, K., & Oyanagi, K. (2010). Dimerization of G protein-coupled purinergic receptors: Increasing the diversity of purinergic receptor signal responses and receptor functions. *Journal of Receptor and Signal Transduction Research, 30*, 337–346.

Namba, K., Suzuki, T., & Nakata, H. (2010). Immunogold electron microscopic evidence of in situ formation of homo- and heteromeric purinergic adenosine A1 and P2Y2 receptors in rat brain. *BMC Res. Notes*. 3, 323.

Nemoto, W., Fukui, K., & Toh, H. (2011). GRIPDB—G protein coupled Receptor Interaction Partners DataBase. *Journal of Receptor and Signal Transduction Research, 31*, 199–205.

Palmer, R. K. (2007). The pharmacology and signaling of bitter, sweet, and umami taste sensing. *Molecular Interventions*, 7, 87–98.

Pfleger, K. D., & Eidne, K. A. (2006). Illuminating insights into protein-protein interactions using bioluminescence resonance energy transfer (BRET). *Nature Methods, 3*, 165–174.

Ralevic, V., & Burnstock, G. (1998). Receptors for purines and pyrimidines. *Pharmacological Reviews, 50*, 413–492.

Rosenbaum, D. M., Zhang, C., Lyons, J. A., Holl, R., Aragao, D., Arlow, D. H., et al. (2011). Structure and function of an irreversible agonist-β_2 adrenoceptor complex. *Nature, 469*, 236–240.

Schachter, J. B., Sromek, S. M., Nicholas, R. A., & Harden, T. K. (1997). HEK293 human embryonic kidney cells endogenously express the $P2Y_1$ and $P2Y_2$ receptors. *Neuropharmacology, 36*, 1181–1187.

Suzuki, T., Namba, K., Tsuga, H., & Nakata, H. (2006). Regulation of pharmacology by hetero-oligomerization between A_1 adenosine receptor and $P2Y_2$ receptor. *Biochemical and Biophysical Research Communications, 351*, 559–565.

Suzuki, T., Namba, K., Yamagishi, R., Kaneko, H., Haga, T., & Nakata, H. (2009). A highly conserved tryptophan residue in the fourth transmembrane domain of the A_1 adenosine receptor is essential for ligand binding but not receptor homodimerization. *Journal of Neurochemistry, 110*, 1352–1362.

Suzuki, T., Obara, Y., Moriya, T., Nakata, H., & Nakahata, N. (2011). Functional interaction between purinergic receptors: Effect of ligands for A_{2A} and $P2Y_{12}$ receptors on $P2Y_1$ receptor function. *FEBS Letters, 585*, 3978–3984.

Vidi, P. A., Chemel, B. R., Hu, C. D., & Watts, V. J. (2008). Ligand-dependent oligomerization of dopamine D_2 and adenosine A_{2A} receptors in living neuronal cells. *Molecular Pharmacology, 74*, 544–551.

Warne, T., Moukhametzianov, R., Baker, J. G., Nehmé, R., Edwards, P. C., Leslie, A. G., et al. (2011). The structural basis for agonist and partial agonist action on a β_1-adrenergic receptor. *Nature, 469*, 241–244.

Wu, B., Chien, E. Y., Mol, C. D., Fenalti, G., Liu, W., Katritch, V., et al. (2010). Structures of the CXCR4 chemokine GPCR with small-molecule and cyclic peptide antagonists. *Science, 330*, 1066–1071.

Xu, F., Wu, H., Katritch, V., Han, G. W., Jacobson, K. A., Gao, Z. G., et al. (2011). Structure of an agonist-bound human A_{2A} adenosine receptor. *Science, 332*, 322–327.

Yoshioka, K., Hosoda, R., Kuroda, Y., & Nakata, H. (2002). Hetero-oligomerization of adenosine A_1 receptors with $P2Y_1$ receptors in rat brains. *FEBS Letters, 531*, 299–303.

Yoshioka, K., Saitoh, O., & Nakata, H. (2001). Heteromeric association creates a P2Y-like adenosine receptor. *Proceedings of the National Academy of Sciences of the United States of America, 98*, 7617–7622.

Yoshioka, K., Saitoh, O., & Nakata, H. (2002). Agonist-promoted heteromeric oligomerization between adenosine A_1 and $P2Y_1$ receptors in living cells. *FEBS Letters, 523*, 147–151.

CHAPTER FOURTEEN

Bimolecular Fluorescence Complementation Analysis of G Protein-Coupled Receptor Dimerization in Living Cells

Karin F.K. Ejendal, Jason M. Conley, Chang-Deng Hu, Val J. Watts[1]
Department of Medicinal Chemistry and Molecular Pharmacology, College of Pharmacy, Purdue University, West Lafayette, Indiana, USA
[1]Corresponding author: e-mail address: wattsv@purdue.edu

Contents

1. Introduction 260
2. Generation of GPCR–BiFC Fusion Proteins 262
   2.1 Required materials 263
   2.2 Planning the generation of GPCR–BiFC fusion constructs 263
   2.3 Construction of GPCR–BiFC fusion vectors 265
   2.4 Verification of expression and function of GPCR–BiFC fusions 265
3. Detection of GPCR Interactions using BiFC and Fluorescence Microscopy 266
   3.1 Required materials and equipment 266
   3.2 Detection of BiFC fluorescent signal using microscopy 267
   3.3 Localization of complemented receptor signal 267
   3.4 Quantitative image analysis 268
   3.5 Evaluation of results 269
4. Microscopic Detection of GPCR Interactions by mBiFC 270
   4.1 Required materials and equipment 271
   4.2 Construction of mBiFC fusion proteins 272
   4.3 Transfection and drug treatment 272
   4.4 Image acquisition 272
   4.5 Quantitative analysis and results assessment 272
5. Fluorometric Detection of GPCR Dimerization using BiFC and mBiFC 273
   5.1 Required materials and equipment 274
   5.2 Detection of BiFC fluorescent signal using fluorometry 274
   5.3 Analysis of receptor interactions and interpretation of results 275
6. Summary 275
Acknowledgments 276
References 276

Methods in Enzymology, Volume 521
ISSN 0076-6879
http://dx.doi.org/10.1016/B978-0-12-391862-8.00014-4

Abstract

Emerging evidence indicates that G protein-coupled receptor (GPCR) signaling is mediated by receptor–receptor interactions at multiple levels. Thus, understanding the biochemistry and pharmacology of those receptor complexes is an important part of delineating the fundamental processes associated with GPCR-mediated signaling in human disease. A variety of experimental approaches have been used to explore these complexes, including bimolecular fluorescence complementation (BiFC) and multicolor BiFC (mBiFC). BiFC approaches have recently been used to explore the composition, cellular localization, and drug modulation of GPCR complexes. The basic methods for applying BiFC and mBiFC to study GPCRs in living cells are the subject of the present chapter.

ABBREVIATIONS

BiFC bimolecular fluorescence complementation
BiLC bimolecular luminescence complementation
BRET bioluminescence resonance energy transfer
FRET fluorescence resonance energy transfer
GFP green fluorescent protein
GPCR G protein-coupled receptor
VC155 or CC155 C-terminal fragments of Venus or Cerulean split between amino acid residues 154–155
VN173 or CrN173 N-terminal fragment of Venus or Cerulean split between amino acid residues 172–173

1. INTRODUCTION

In response to an extracellular stimulus, G protein-coupled receptors (GPCRs) convey the signal across the cell membrane to the intracellular signaling machinery. To achieve the appropriate response to the stimulus, the activity of the receptor is tightly regulated on several levels. Specific patterns of receptor expression, localization, and oligomerization offer modes of regulation. A large number of recent reviews have been devoted to the study of receptor oligomerization and the physiological implications of such complexes (e.g., see Albizu, Moreno, Gonzalez-Maeso, & Sealfon, 2010; Milligan, 2009; Pin et al., 2007; Rozenfeld & Devi, 2011). Receptor complexes may be homomeric or heteromeric and multiple types of GPCRs are expressed in a given cell. However, coexpression does not necessitate receptor interaction and oligomerization because the receptor proteins may be localized in separate organelles or partitions of the cell membrane. It is also possible that receptors interact only under specific conditions and that interaction can be modulated by neurotransmitters or receptor ligands (e.g., see Kearn, Blake-Palmer, Daniel, Mackie, & Glass, 2005; Stockton & Devi, 2011; Vidi, Chemel, Hu, & Watts, 2008; Vidi, Ejendal, Przybyla, & Watts, 2011). Understanding the regulation and physiological implications of GPCR oligomerization may benefit the development of pharmaceutical therapies targeting specific receptor oligomers.

Fluorescent approaches have been extensively used to study and visualize cellular functions. Molecular biology in combination with the discovery of green fluorescent protein (GFP; reviewed by Tsien, 1998) has made it possible to construct and express fusion proteins with a fluorescent protein moiety or "tag." In the past decade, bimolecular fluorescence complementation (BiFC) has emerged as a simple and elegant approach to study and visualize protein–protein interactions (Kerppola, 2006; Shyu & Hu, 2008; Wilson, Magliery, & Regan, 2004). In BiFC, the fluorescent protein, such as GFP, is "split" into two nonfluorescent fragments. When the fragments are brought into close proximity, the two fragments reconstitute the fluorophore (Fig. 14.1). Thus, when two proteins carrying BiFC fusion tags interact, the BiFC fragments form a fluorescent protein that can be detected with microscopy or fluorometry.

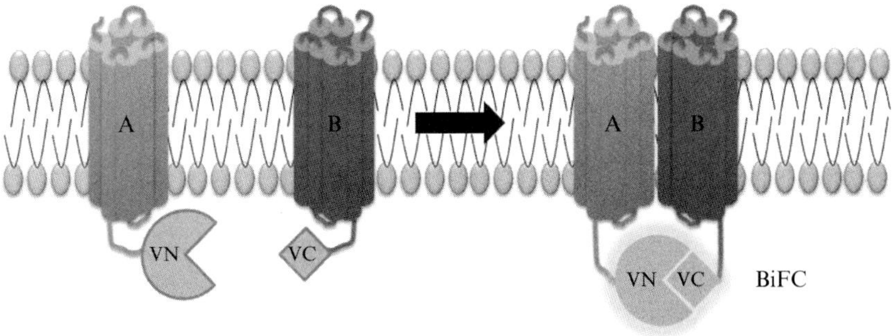

Figure 14.1 Schematic representation of receptors fused to split BiFC fragments of Venus. If receptor A-VN and receptor B-VC do not interact, no complementation of the fluorophore occurs, thus, no fluorescence is detected (left). In contrast, when receptors A and B interact, the -VN and -VC fragments are brought in close proximity to allow for fluorescence complementation of yellow Venus (right). The fluorescence can be visualized by fluorescence microscopy or fluorometry. Alternatively, the receptors may also be fused with -CrN and -CC fragments of Cerulean, a cyan-shifted variant of GFP. (For interpretation of the references to color in this figure legend, the reader is referred to the online version of this chapter.)

BiFC is one of several techniques available to study protein–protein interactions. When selecting the most suitable technology to use, a number of factors must be taken into consideration. First and foremost, what type of information about the interaction is desired? The majority of techniques utilize fusion proteins, where the interactions between two fusion tags imply interaction between the two proteins of interest. Therefore, the nature of the tags may be important for the method of choice. Another factor is the robustness of the signal, that is, the background signal and signal-to-noise ratio, the limitations in the distance between the target proteins, and stability as well as interactions over time. Moreover, some techniques permit analysis of subcellular localization of the protein complex, whereas others do not (Kerppola, 2006; Shyu & Hu, 2008; Wilson et al., 2004). Compared to other available techniques such as fluorescence and bioluminescence energy transfer (FRET and BRET), bimolecular luminescence complementation (BiLC), and various split enzyme technologies, the BiFC technology has a few key strengths worth highlighting (Kerppola, 2006; Shyu & Hu, 2008; Vidi & Watts, 2009; Wilson et al., 2004). A major advantage is the relative simplicity of the assay; the protein complex has endogenous fluorescence (i.e., does not depend on addition of a cofactor) and can be visualized in living cells using an epifluorescence microscope. BiFC can be used to gain spatial and temporal information in living cells, and the technology is suitable for detection of both weak and transient interactions. Further, using multicolor BiFC (mBiFC), two distinct dimeric complexes can be simultaneously visualized (see Section 4). However, it is also important to point out that there are limitations with the BiFC technology, including the potential for false positives, which make the design of appropriate controls critical. Detailed descriptions of both advantages and limitations can be found in several BiFC reviews and protocols (Kerppola, 2008; Shyu & Hu, 2008; Shyu, Suarez, & Hu, 2008). In addition, a number of recent articles have discussed the specific use of BiFC in the study of GPCR signaling (Rose, Briddon, & Holliday, 2010; Vidi et al., 2011; Vidi, Przybyla, Hu, & Watts, 2010; Vidi & Watts, 2009). The methodology described below will provide the basics for applying BiFC and mBiFC to study GPCR interactions in living cells.

2. GENERATION OF GPCR–BiFC FUSION PROTEINS

Prior to performing a BiFC experiment, several experimental design parameters must be considered, including selection of fluorescent proteins (e.g., Venus or Cerulean), the location of fluorescent protein fragment for fusion (e.g., C-terminus of receptor), and the composition of the linker region between the receptor and the fluorescent protein fragment. Once an experimental strategy is defined, molecular cloning techniques are used to generate the receptor–BiFC fusion constructs, followed by initial verification of expression and native function. Section 2 serves as a guide to the preparation of BiFC-tagged receptor constructs and preliminary experimental optimization steps.

2.1. Required materials

- Plasmid vectors containing BiFC fragment sequences (see Fig. 14.2)
- cDNAs encoding the receptor coding sequences
- Oligonucleotide primers

2.2. Planning the generation of GPCR–BiFC fusion constructs

BiFC experiments are performed by measuring the resulting fluorescence intensity upon complementation of two nonfluorescent fragments of fluorescent proteins including spectral variants of GFP, YFP, CFP (Hu & Kerppola, 2003; Shyu, Liu, Deng, & Hu, 2006), and also red fluorescent proteins (Fan et al., 2008; Jach, Pesch, Richter, Frings, & Uhrig, 2006). The choice of fluorescent proteins can be guided by properties such as peak excitation and emission wavelengths, fluorescence intensity, temperature sensitivity, and maturation time. Here, we describe the use of fragments of the enhanced YFP variant Venus and the enhanced CFP variant Cerulean, but other fluorescent proteins may also be suitable. There are several possible sites at which to split the fluorescent protein for BiFC fragments, such as between amino acid residues 154–155 and 172–173 (Shyu et al., 2006) or 210–211 (Ohashi, Kiuchi, Shoji, Sampei, & Mizuno, 2012). In previous studies from our laboratory (Przybyla & Watts, 2010; Vidi, Chemel, et al., 2008; Vidi, Chen, Irudayaraj, & Watts, 2008), we successfully used overlapping -VN173/-VC155 fragments to reconstitute the Venus signal. However, nonoverlapping fragments, such as -VN155/-VC155 fragments can also be used (Hu & Kerppola, 2003). Recent studies in BiFC fragment optimization have identified a number of mutations resulting in an enhanced signal-to-noise ratio compared to the original fragments (Kodama & Hu, 2010; Lin et al., 2011; Nakagawa, Inahata, Nishimura, & Sugimoto, 2011; Zhou, Lin, Zhou, Deng, & Xia, 2011). The desired properties noted above should be used as a guide for the selection of appropriate pBiFC vectors (Fig. 14.2).

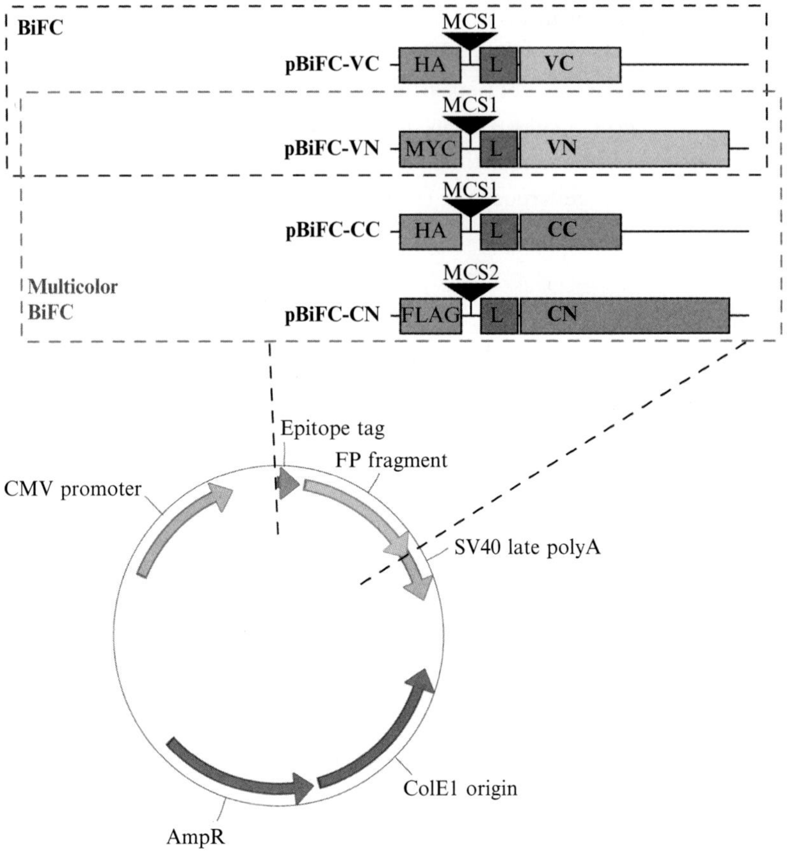

Figure 14.2 Schematic of plasmids containing the BiFC fusion tags. Complete plasmid sequences and maps can be found at the Hu lab Web site (http://people.pharmacy.purdue.edu/~hu1/). *This material is reproduced with permission of* Current Protocols in Neuroscience *and John Wiley & Sons, Inc.* (For color version of this figure, the reader is referred to the online version of this chapter.)

Previous studies from our laboratory have utilized GPCRs that have been fused with the BiFC fragments at the C-terminus (Przybyla & Watts, 2010; Vidi, Chemel, et al., 2008; Vidi, Chen, et al., 2008); however, N-terminal fusions with BiFC tags are also possible. The efficiency of fluorescence complementation may be affected by the linker region between the C-terminus of the receptor and the N-terminus of the fluorescent protein fragment. It is recommended to have a linker of 15–20 residues in length that is composed of small, uncharged amino acids, but the length and composition of the linker region may need to be optimized for the most favorable fluorescence complementation and/or GPCR function (Kerppola, 2006; Shyu et al., 2008).

When heteromers between two different GPCRs are studied (e.g., A + B), it may be useful to construct the reciprocal combinations of GPCR–BiFC fusions (e.g., A-VN + B-VC as well as A-VC + B-VN). Due to possible self-complementation of the nonfluorescent fragments, it is essential to include negative controls for dimerization. If the interaction interface is known, a GPCR construct with a mutated or deleted interaction surface is suitable (e.g., A′-VN). If the interaction surface is not known, a noninteracting and structurally homologous receptor that is localized to the same cellular compartment may serve as a negative control for interaction. However, caution must be taken as the formation of BiFC complexes is also dependent on the topology of the two fusion proteins.

2.3. Construction of GPCR–BiFC fusion vectors

Once the pBiFC vectors have been chosen, molecular cloning techniques should be used to insert the receptor coding sequence into the pBiFC vector in frame with both the N-terminal epitope tag and the C-terminal fluorescent protein fragment. Use the cDNA of the receptor as a PCR template and design primers that (i) disrupt the stop codon of the GPCR coding sequence and (ii) contain appropriate restriction endonuclease sites to subclone the PCR product into pBiFC vectors. Verify the GPCR–BiFC fusion constructs by DNA sequencing.

2.4. Verification of expression and function of GPCR–BiFC fusions

It is important to establish that the GPCR–BiFC fusion proteins retain the expression, ligand binding, and functional properties of the wild-type (untagged) GPCR prior to studying receptor interactions by BiFC methods. Receptor expression and the ligand-binding properties of the tagged receptor can be verified by radioligand binding experiments. The pBiFC plasmids (Fig. 14.2) contain N-terminal epitope tags, allowing receptor expression to be examined by immunoblot or dot blot analysis. Experiments that measure the functional activation of the specific GPCRs (e.g., cAMP accumulation, ERK1/2 phosphorylation, GIRK channel current, etc.) should be carried out to confirm the retention of receptor function after fusion to fluorescent protein or fluorescent protein fragment.

A variety of cell lines (e.g., HEK293, CAD, COS-1, HeLa, etc.) may be suitable for receptor expression, microscopy, and functional assays. Studies in our lab have examined the expression of dopamine, adenosine, and cannabinoid receptors (Przybyla & Watts, 2010; Vidi, Chemel, et al., 2008; Vidi, Chen, et al., 2008) in neuronally derived Cath. A differentiated (CAD) cells (Qi, Wang, McMillian, & Chikaraishi, 1997). Thus, although the protocols in this unit can be used as initial guides for receptor expression, the transfection protocols in Sections 3–5 need to be optimized for each receptor, cell model, and assay format.

3. DETECTION OF GPCR INTERACTIONS USING BiFC AND FLUORESCENCE MICROSCOPY

The interactions between two BiFC-tagged receptors of interest can be visualized by fluorescence microscopy in live cells. Here, we describe a basic protocol for detection of GPCR dimers in CAD cells. For detection of interactions between multiple receptors, please refer to Section 4, mBiFC.

3.1. Required materials and equipment

- Plasmids containing the receptor sequence fused to the sequences of the BiFC fragments of Venus (e.g., pBiFC-A-VN173, pBiFC-B-VC155, pBiFC-A′-VN173) or Cerulean variant (Shyu et al., 2006).
- Plasmids containing a normalization protein with distinct spectral properties such as mCherry-Mem (Yost, Mervine, Sabo, Hynes, & Berlot, 2007). Plasmids containing fluorescent markers for specific cellular compartments (i.e., Golgi apparatus, endosomes, and endoplasmic reticulum).
- Cell expression system (e.g., CAD cells) and basic tissue culture supplies
- Transfection reagents (e.g., Lipofectamine2000 from Invitrogen).
- Supplies for microscopy (e.g., coverglass slides from LabTek, Nunc, or glass coverslips added to tissue culture cluster plate).
- Phosphate-buffered saline (PBS) or cell culture medium without pH indicator.
- Inverted epifluorescence microscope (e.g., Nikon TE2000-U) equipped with objectives (20 × and 60 ×), excitation source (e.g., mercury lamp), filter sets for Venus/eYFP (excitation at 500/20 nm; emission at 535/30 nm), filter sets for Cerulean/eCFP (excitation at 430/25 nm; emission at 470/30 nm) and filter sets for mCherry/DsRed (excitation at 572/23 nm; emission at 640/50 nm), and a sensitive CCD camera (e.g., Photometrics CoolSNAP-ES Digital Monochrome CCD).
- Image analysis software (e.g., ImageJ; available at http://rsbweb.nih.gov/ij/).

3.2. Detection of BiFC fluorescent signal using microscopy

3.2.1 Cells and transfection

Seed CAD cells in 4-well LabTek slides and incubate in a humidified incubator at 37 °C with 5% CO_2 to approximately 60–80% confluency. Transfect the cells following the manufacturer's recommended steps. A good starting point may be 1 μl/well Lipofectamine2000 in combination with 100–500 ng DNA/well of receptors A-VN, B-VC, and negative controls as appropriate. A transfection control (normalizer), preferably expressed in the same cellular compartment but with spectral properties distinct from the complemented BiFC fluorophore, may be cotransfected with the BiFC fusion proteins (e.g., the membrane marker mCherry-Mem, 10 ng DNA/well). Incubate the cells for 24 h in a humidified incubator at 37 °C and 5% CO_2 to allow for protein expression. Prior to imaging, gently rinse the cells with prewarmed (37 °C) PBS or colorless culture media to remove cell debris that may interfere with the visualization of the fluorescence signal. The cell density, quantities, ratios of DNA and Lipofectamine2000, and the expression time may need optimization, depending on receptor of interest and the cell system.

3.2.2 Imaging of GPCR oligomerization

The appropriate filter sets are used to visualize the complemented Venus (500/20, 535/30) or Cerulean (430/25, 470/30) signals. The BiFC signal may be visible at lower magnification (20×, air); however, a weaker complemented signal may require imaging with a 60 × oil immersion objective. Further, the higher magnification is used to determine subcellular localization as well as image acquisition.

3.3. Localization of complemented receptor signal

The localization of the BiFC signal may be examined by cotransfection of the complementing receptors with fluorescent markers targeting cellular organelles as well as the cell membrane (Przybyla & Watts, 2010; Vidi, Chemel, et al., 2008; Vidi, Chen, et al., 2008). Cells are seeded, transfected, and prepared as described above with the modifications or additions described below. Generally, we have used the Cerulean (i.e., A-CrN and B-CC) complementing receptor pair with YFP-labeled intracellular markers or mCherry-Mem for membrane localization. The BiFC receptor fusion constructs are cotransfected with the transmedial Golgi marker (YFP-Golgi; YFP fusion to residues 1–81 of the β1,4-galactosyltransferase), the ER marker (YFP-ER; YFP fused to the ER targeting sequence of calreticulin and the ER retrieval sequence KEDL), the endosomal marker RhoB fused to YFP (YFP-Endo), or the membrane marker mCherry-Mem (N-terminal fragment of Gap43 fused to mCherry).

To acquire images, select and image the field based on the normalizer (e.g., mCherry-Mem or YFP-Endo), followed by the same field in the channel for the complemented BiFC signal. Fluorescent localization studies are typically only qualitative in nature and are replicated at least three times. Some degree of overlap is anticipated in all cellular compartments; however, we have observed the following patterns with subtypes from the dopamine, cannabinoid, and adenosine receptor families. Robust membrane localization is typically observed consistent with plasma membrane targeting of the receptor dimer. Overlap with both the ER and endosome markers has also been observed, supporting a role for the ER in dimer assembly (Herrick-Davis, Weaver, Grinde, & Mazurkiewicz, 2006) as well as suggesting that the dimer undergoes proper endosomal trafficking (Seachrist & Ferguson, 2003). More limited expression has been observed in the Golgi apparatus with exception of the A_{2A} adenosine receptor homodimers (Przybyla & Watts, 2010; Vidi, Chemel, et al., 2008; Vidi, Chen, et al., 2008). The expression patterns appear to be cell-type dependent, and results should be compared to endogenous expression where data are available.

3.4. Quantitative image analysis

To compare the interactions between receptors A+B and the control A′+B, the complemented fluorescence signals should be quantified. It is crucial to image and quantify a large number of cells (>100) from at least three independent experiments to offset any perceived differences that may be due to experimental variations, such as transfection efficiency. Utilize the fluorescent normalizer to select the field to be imaged in order to avoid user bias.

For acquisition of images, select and image the field based on the normalizer (e.g., mCherry-Mem), followed by the same field in the channel for the complemented BiFC signal. For proper analysis and comparison, it is important to keep the exposure time for all the images in one channel consistent (i.e., exposure time for A+B should be the same as for A′+B). Using the image analysis software (e.g., ImageJ), create an image stack of the two images. For quantification, the fluorescent intensities will be determined in each microscopic field.

i. To establish the background fluorescence intensity in each channel, select a region of interest (ROI) devoid of any cells and use the values from each channel.
ii. Select a ROI in each cell (either the whole cell or a specific region of the cell) and determine the intensities in both channels.
iii. Subtract the background fluorescence values in both channels.
iv. Calculate the BiFC/normalizer ratio for each cell (or subsection of cell, ROI).
v. Determine the median of the BiFC/normalizer ratio for the experiment.
vi. Average the median values from three or more independent experiments.

A quantitative analysis of the subcellular distribution of BiFC signals (e.g., signal at plasma membrane vs. intracellular signal) can also be performed. The methodology for doing this is described below for mBiFC analysis (see Section 4.5 and Fig. 14.4).

3.5. Evaluation of results

Fluorescence complementation of the nonfluorescent VN and VC (or CrN and CC) fragments suggests an interaction between A and B. Compare the intensity and localization of the A+B signal to coexpression with B+the negative control A′ (i.e., the fusion protein with the noninteracting or mutant receptor). The experimental findings can either support or contradict the occurrence of a specific interaction between receptors A and B. Any of the following outcomes can be expected:

i. The data indicate that there is a specific interaction between A and B if a complemented Venus signal is detected with A-VN+B-VC, and if the signal is absent or significantly reduced with the negative control A′-VN and B-VC. This conclusion relies on the assumption that the expression levels of A-VN and the control A′-VN are comparable and that their cellular localization is the same.
ii. The data show that complemented Venus signals are detected in both experimental and control experiments (e.g., A-VN+B-VC and A′-VN+B-VC). If this is the scenario, then BiFC (or the current BiFC assay design) may not be suitable to study interactions between receptors A and B.
iii. The experiments fail to show any complemented Venus signal upon coexpression of A-VN and B-VC. Although these observations indicate that the two BiFC fragments are not brought together by the tagged receptors, it does not necessarily demonstrate that the (native)

receptors do not interact. A negative result may be the consequence of steric hindrance, for example, the interaction surface between the receptors may not allow the fragments to interact. It is also possible that trafficking and expression of the fusion construct is different from the untagged receptor. Strategies to overcome these issues may include modifying the linker between the receptor and BiFC fragment or changing the position or identity of the BiFC tag.

The BiFC experiments only provide evidence that the GPCRs may form dimers under the conditions tested. Given that BiFC complexes are likely irreversible (Kerppola, 2006; Shyu & Hu, 2008), additional biochemical and functional data should be provided to support the observed interaction.

4. MICROSCOPIC DETECTION OF GPCR INTERACTIONS BY mBiFC

Many GPCRs are able to form both homo- and heteromeric complexes with a variety of GPCR interaction partners such as A_{2A}–A_{2A} and A_{2A}–D_2 (Vidi, Chemel, et al., 2008). As BiFC experiments can be performed with several different fluorescent proteins (Section 2, Table 14.1), selection of fluorescent proteins with distinct spectral properties allows for

Table 14.1 Properties of fluorescent proteins and select BiFC fragments

Protein or protein fragment	Application(s)	Excitation peak (nm)	Emission peak (nm)
Venus[a] (eYFP variant)		515	528
Cerulean[b] (eCFP variant)	Normalizer, colocalization	439	479
VN173+VC155[c]	BiFC	515	528
CrN173+CC155[c]	BiFC/mBiFC	439	479
VN173+CC155[c]	mBiFC	504	513
mCherry[d]	Normalizer, colocalization	587	610

A two letter code is used to identify the BiFC fragments, where the first letter indicates the parent fluorescent protein and the second indicates to which portion of the fluorophore it corresponds. Thus, the BiFC fragments of the Venus variant of YFP are designated VN or VC, for Venus N-terminal and Venus C-terminal fragments, respectively.

[a]Nagai et al. (2002).

[b]Rizzo, Springer, Granada, and Piston (2004).

[c]Shyu et al. (2006).

[d]Shaner et al. (2004).

multiple GPCR–GPCR interactions to be measured simultaneously in living cells by mBiFC technology (Hu & Kerppola, 2003). The basis for mBiFC is that residues of the N-terminal fragment primarily determine the color of GFP-derived fluorescent proteins such as yellow-shifted Venus or blue-shifted Cerulean (Hu & Kerppola, 2003; Shyu et al., 2006). This observation allows for experiments designed to exploit the promiscuous nature of the -CC BiFC fragment. Complementation of -CC with -VN will result in fluorescence in the Venus channel, whereas complementation with—CrN will result in fluorescence in the Cerulean channel. These different fluorescent signals correspond to the coexistence of two distinct dimeric GPCR complexes, as shown in Fig. 14.3. mBiFC techniques have been applied to study changes in the relative abundance (and subcellular localization) of two dimeric GPCR complexes in response to pharmacological modulation (Przybyla & Watts, 2010; Vidi, Chemel, et al., 2008; Vidi, Chen, et al., 2008).

4.1. Required materials and equipment

- All materials from Section 3.1.
- Plasmids containing the receptor sequence fused to the sequences of BiFC fragments -VN, -CrN, and -CC.
- Receptor ligands for pharmacological modulation.

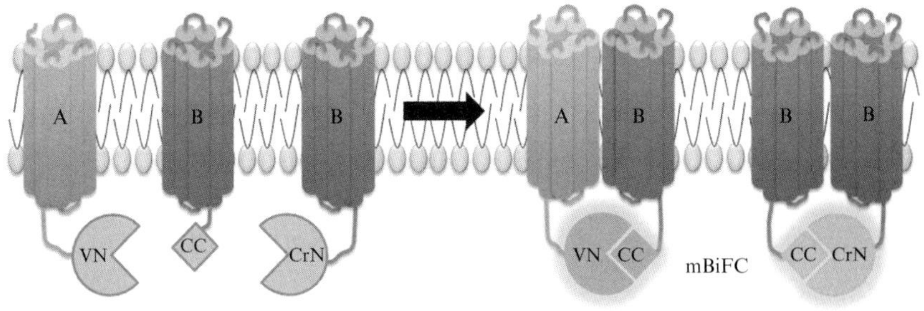

Figure 14.3 Depiction of the principle of multicolor BiFC (mBiFC). Simultaneous expression of three receptor BiFC fusions allows for the visualization and detection of two distinct dimers. The interaction between receptors A and B results in complementation of a Venus signal, whereas the receptor B homodimer results in a Cerulean signal. As a consequence of unique fluorescent properties of VN + CC and CrN + CC (see Table 14.1), the two receptor dimers can be readily distinguished. If receptor B neither forms homodimers nor interacts with A, no fluorescence complementation occurs (left). (For color version of this figure, the reader is referred to the online version of this chapter.)

4.2. Construction of mBiFC fusion proteins

Construct GPCR–BiFC fusions as described in Section 2. It is important to generate appropriate combinations of GPCR–BiFC fusions for the specific GPCR interactions to be studied. For example, to simultaneously detect heterodimers of receptors A and B and homodimers of B, the combination of A-VN, B-CC, and B-CrN would be suitable (Fig. 14.3).

4.3. Transfection and drug treatment

Seed cells at appropriate cell densities and cotransfect with the desired BiFC constructs as described above for microscopic detection of BiFC (Section 3). The pharmacological modulation of specific GPCR interactions can be measured by mBiFC. The temporal aspects of transfection and drug treatment should be optimized for each cell system and receptor of interest. Studies from our laboratory show that the receptor interactions can be modulated successfully by treatment with receptor ligands for 10–30 h beginning at 4 h posttransfection (Przybyla & Watts, 2010; Vidi, Chemel, et al., 2008). For pharmacological modulation, 100 × stock concentrations of drugs are added to the culture media (vehicle controls are treated with only the diluent) and incubated at 37 °C and 5% CO_2 for the appropriate incubation time.

4.4. Image acquisition

Follow the general BiFC assay protocol from Section 3. Collect fluorescent signals using filter sets that are specific for the Venus signal (-VN/-CC complementation), Cerulean signal (-CrN/-CC complementation), and normalizer fluorescent signal (e.g., mCherry-Mem marker). For quantitative analysis, image at least 100 cells for each condition and convert the images from the three channels (i.e., settings to detect signals for Venus, Cerulean, and mCherry) into an image stack.

4.5. Quantitative analysis and results assessment

Use image analysis software, such as ImageJ or Metamorph, to perform the quantitative analysis of images from the mBiFC experiment. Subtract the background fluorescence (see Section 3.3) and determine the Venus/Cerulean fluorescence ratio for each cell. Calculate the median Venus/Cerulean fluorescence ratio for each condition. An average of the median fluorescence ratios should be calculated from at least three-independent experiments.

Quantitative analysis of the subcellular distribution of BiFC signals (e.g., signal at plasma membrane vs. intracellular signal) can also be

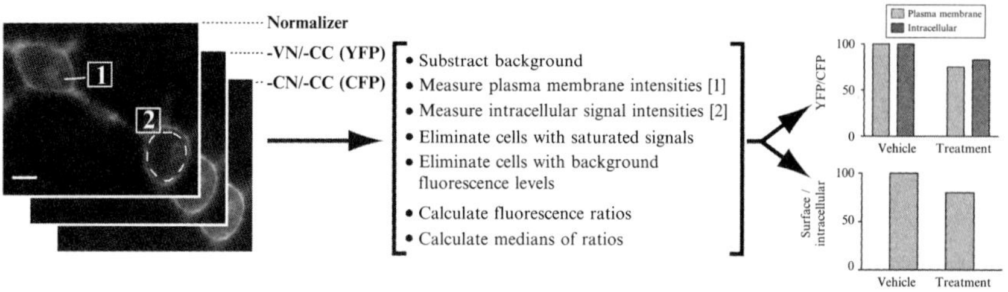

Figure 14.4 Illustration of steps involved in ratiometric analysis of multicolor BiFC measurements discussed in Section 4.5. *This material is reproduced with permission of* Current Protocols in Neuroscience *and John Wiley & Sons, Inc.* (For color version of this figure, the reader is referred to the online version of this chapter.)

performed. The plasma membrane signal intensity can be estimated by obtaining the maximum pixel intensity along lines drawn perpendicular to the cell surface (Fig. 14.4). Intracellular signals can be estimated by drawing an ROI that includes the area inside of the mCherry-Mem cell surface marker (Fig. 14.4). A ratio of the surface to intracellular signal for each cell is calculated from the average pixel intensities of the ROIs. The median value for each experimental condition is averaged from at least three-independent experiments.

Compare BiFC ratios (either Venus/Cerulean or Surface/Intracellular signal) across the different treatments. When interpreting changes in fluorescence ratios, it is important to take into account changes in individual receptor levels upon drug/ligand treatment. Therefore, measure receptor expression levels in response to ligand exposure (analogous to those used for the mBiFC assay) by techniques such as radioligand receptor binding or immunoblotting.

5. FLUOROMETRIC DETECTION OF GPCR DIMERIZATION USING BiFC AND mBiFC

In addition to microscopic detection of BiFC signals, fluorometric detection of BiFC signals as a measurement of GPCR dimerization can also be used. Fluorometric detection of BiFC signals offers several advantages over microscopic detection. For example, fluorometric detection of BiFC is relatively higher throughput than microscopic detection (which requires time-consuming image acquisition and analyses), thus allowing for more conditions to be examined per experiment. Also, fluorometry-based BiFC measures large populations of cells, reducing the risk for experimental bias introduced by performing microscopic analysis of small subsets of cells chosen by the user. Fluorometric detection of BiFC, however, cannot distinguish spatial aspects of GPCR interactions (e.g., membrane vs. intracellular localization). Therefore, fluorometric detection of BiFC is a complementary approach to microscopic BiFC detection.

5.1. Required materials and equipment

- All materials from Section 3.1
- 12-well tissue culture-treated cluster plates (e.g., BD Falcon)
- Microcentrifuge tubes
- Protein concentration kit (BCA assay, Pierce)
- Black 96-well plates (e.g., Nunc 96-well round bottom)
- A fluorescence plate reader with appropriate filter sets to distinguish Venus and Cerulean fluorescent signals

5.2. Detection of BiFC fluorescent signal using fluorometry

5.2.1 *Transfection*

Seed cells into a 12-well plate at a density that will provide 60–80% confluency after 24 h of growth in a humidified incubator at 37 °C and 5% CO_2. Transfect cells with desired GPCR–BiFC fusion constructs (see Sections 3 and 4 for BiFC and mBiFC, respectively). To normalize the BiFC Venus signal to expression levels in the specific experiment, cotransfect cells with a fluorescent protein that has distinct spectral properties, such as Cerulean or mCherry-Mem. Perform experiments in triplicate (transfection condition or treatment), including mock-transfected cells for background signal measurement and cells transfected with Venus and Cerulean alone for calculation of bleed-through corrected fluorescence signals (Vidi, Chemel, et al., 2008).

5.2.2 *Preparation of cell suspensions*

At 24 h posttransfection, remove transfection mix and gently wash cells with warm PBS (500 μl/well). Detach and resuspend cells in 330 μl PBS and transfer to a microcentrifuge tube. Use 30 μl of each cell suspension (10 μl in triplicate) to estimate protein concentrations by Pierce BCA assay, while maintaining the remaining cell suspension at 37 °C. Using the results of the protein assay, equalize protein concentrations (e.g., 200 ng protein/μl) by adding an appropriate volume of PBS. Transfer 200–400 μl of each cell suspension into a well of a black 96-well plate. Alternatively, the normalization of cell numbers can be carried out using DAPI staining of nuclear DNA (C.W. Dessauer, personal communication).

5.2.3 Fluorescence intensity measurements

Measure fluorescence signals using a multiwell fluorescence plate reader. Use excitation and emission filters specific for the wavelengths that correspond to the fluorescent proteins selected for the experiment (see Table 14.1).

5.3. Analysis of receptor interactions and interpretation of results

To analyze the fluorescence intensity data, subtract the background fluorescence by averaging the signal from the mock-transfected cells and subtracting the resulting value from each fluorescent signal. There may be bleed-through fluorescence from Venus into the Cerulean signal (and vice versa) in cells coexpressing Venus and Cerulean. Fluorescence bleed-through corrected values (i.e., V_{cor} and C_{cor}) can be calculated for cells coexpressing Venus and Cerulean or BiFC constructs by applying fluorescence intensity values and bleed-through coefficients to the following equations:

$$V_{\text{cor}} = (V - y\text{C})/1 - xy$$
$$C_{\text{cor}} = (C - x\text{V})/1 - xy$$

V and C represent fluorescence intensity values from Venus and Cerulean signals collected in cells coexpressing Venus and Cerulean. The "x coefficient" is the C/V fluorescence ratio in cells expressing only Venus. The "y coefficient" is the V/C fluorescence ratio in cells expressing only Cerulean.

Calculate the average of replicate values from each condition and determine the Venus/normalizer ratios. The fluorescence values for BiFC correspond to receptor interaction relative to a normalizer fluorescent protein. For mBiFC experiments, calculate the average of replicate values from each condition for the Venus signal (-VN/- CC) and Cerulean signal (-CrN/-CC) and determine the Venus/Cerulean ratio. The fluorescence ratios correspond to the relative efficiency of complementation and receptor–receptor interaction (see Section 4).

6. SUMMARY

In this chapter, we described the use of BiFC to study dimerization as well as global localization of GPCRs using basic instrumentation such as an epifluorescence microscope and fluorescence plate reader. Using confocal microscopy, the applications of this assay could be expanded to include more detailed studies of GPCR trafficking and localization. The BiFC technology has shown to be suitable to detect specific interactions between G protein subunits (Hynes, Mervine, Yost, Sabo, & Berlot, 2004; Hynes, Yost, Yost, & Berlot, 2011) as well as receptors and adenylyl cyclase (Ejendal, Przybyla, & Watts, 2011). In addition, BiFC has been used to detect drug-induced interactions between receptors and β-arrestin (Kilpatrick, Briddon, Hill, & Holliday, 2010). This application is similar to the framework used in the PathHunter technology (DiscoveRx, Corp.) that uses enzyme complementation to measure either direct or heterologous receptor activation via β-arrestin recruitment. Thus, potentially endless possibilities exist to use the BiFC assay for studying interactions between different signaling components and the composition and design of receptor (or effector-)-specific signaling complexes (Vidi et al., 2011). mBiFC techniques have also been applied to study changes in the relative abundance (and subcellular localization) of two dimeric GPCR complexes in response to pharmacological modulation (Przybyla & Watts, 2010; Vidi, Chemel, et al., 2008; Vidi, Chen, et al., 2008). In addition, mBiFC is useful when studying G protein βγ complex formation and localization (Hynes et al., 2011). BiFC can also be combined with FRET (BiFC-FRET) or BRET (BiFC-BRET) to measure higher-ordered receptor complexes (Guo et al., 2008; Vidi, Chen, et al., 2008; Vidi & Watts, 2009). The BiFC technology is straightforward, robust, and applicable across many proteins and cellular systems for the study of GPCR dimerization, localization, and signaling. In summary, when used appropriately in conjunction with additional techniques to study receptor oligomerization, the BiFC technology serves as a valuable tool for research scientists.

ACKNOWLEDGMENTS

This work was supported by Purdue University, the National Institute of Mental Health (MH60397), and the Brain & Behavior Research Foundation. We would like to thank Dr. Pierre-Alexandre Vidi and Julie A. Przybyla for previous research contributions using BiFC and mBiFC. Further, we would like to thank Mary MacDowell for proofreading and improving the present chapter.

REFERENCES

Albizu, L., Moreno, J. L., Gonzalez-Maeso, J., & Sealfon, S. C. (2010). Heteromerization of G protein-coupled receptors: Relevance to neurological disorders and neurotherapeutics. *CNS & Neurological Disorders Drug Targets*, *9*, 636–650.

Ejendal, K. F. K., Przybyla, J. A., & Watts, V. J. (2011). Adenylyl cyclase isoform-specific signaling of GPCRs. In S. Siehler & G. Milligan (Eds.), *G protein-coupled receptors: Structure, signaling, and physiology* (pp. 189–216). New York, USA: Cambridge University Press.

Fan, J. Y., Cui, Z. Q., Wei, H. P., Zhang, Z. P., Zhou, Y. F., Wang, Y. P., et al. (2008). Split mCherry as a new red bimolecular fluorescence complementation system for visualizing protein–protein interactions in living cells. *Biochemical and Biophysical Research Communications*, *367*, 47–53.

Guo, W., Urizar, E., Kralikova, M., Mobarec, J. C., Shi, L., Filizola, M., et al. (2008). Dopamine D2 receptors form higher order oligomers at physiological expression levels. *The EMBO Journal*, *27*, 2293–2304.

Herrick-Davis, K., Weaver, B. A., Grinde, E., & Mazurkiewicz, J. E. (2006). Serotonin 5-HT2C receptor homodimer biogenesis in the endoplasmic reticulum: Real-time visualization with confocal fluorescence resonance energy transfer. *The Journal of Biological Chemistry*, *281*, 27109–27116.

Hu, C. D., & Kerppola, T. K. (2003). Simultaneous visualization of multiple protein interactions in living cells using multicolor fluorescence complementation analysis. *Nature Biotechnology*, *21*, 539–545.

Hynes, T. R., Mervine, S. M., Yost, E. A., Sabo, J. L., & Berlot, C. H. (2004). Live cell imaging of Gs and the beta2-adrenergic receptor demonstrates that both alphas and beta1gamma7 internalize upon stimulation and exhibit similar trafficking patterns that differ from that of the beta2-adrenergic receptor. *The Journal of Biological Chemistry*, *279*, 44101–44112.

Hynes, T. R., Yost, E. A., Yost, S. M., & Berlot, C. H. (2011). Multicolor BiFC analysis of G protein betagamma complex formation and localization. *Methods in Molecular Biology*, *756*, 229–243.

Jach, G., Pesch, M., Richter, K., Frings, S., & Uhrig, J. F. (2006). An improved mRFP1 adds red to bimolecular fluorescence complementation. *Nature Methods*, *3*, 597–600.

Kearn, C. S., Blake-Palmer, K., Daniel, E., Mackie, K., & Glass, M. (2005). Concurrent stimulation of cannabinoid CB1 and dopamine D2 receptors enhances heterodimer formation: A mechanism for receptor cross-talk? *Molecular Pharmacology*, *67*, 1697–1704.

Kerppola, T. K. (2006). Visualization of molecular interactions by fluorescence complementation. *Nature Reviews. Molecular Cell Biology*, *7*, 449–456.

Kerppola, T. K. (2008). Bimolecular fluorescence complementation (BiFC) analysis as a probe of protein interactions in living cells. *Annual Review of Biophysics*, *37*, 465–487.

Kilpatrick, L. E., Briddon, S. J., Hill, S. J., & Holliday, N. D. (2010). Quantitative analysis of neuropeptide Y receptor association with beta-arrestin2 measured by bimolecular fluorescence complementation. *British Journal of Pharmacology*, *160*, 892–906.

Kodama, Y., & Hu, C. D. (2010). An improved bimolecular fluorescence complementation assay with a high signal-to-noise ratio. *Biotechniques*, *49*, 793–805.

Lin, J., Wang, N., Li, Y., Liu, Z., Tian, S., Zhao, L., et al. (2011). LEC-BiFC: A new method for rapid assay of protein interaction. *Biotechnic & Histochemistry*, *86*, 272–279.

Milligan, G. (2009). G protein-coupled receptor hetero-dimerization: Contribution to pharmacology and function. *British Journal of Pharmacology*, *158*, 5–14.

Nagai, T., Ibata, K., Park, E. S., Kubota, M., Mikoshiba, K., & Miyawaki, A. (2002). A variant of yellow fluorescent protein with fast and efficient maturation for cell-biological applications. *Nature Biotechnology*, *20*, 87–90.

Nakagawa, C., Inahata, K., Nishimura, S., & Sugimoto, K. (2011). Improvement of a Venus-based bimolecular fluorescence complementation assay to visualize bFos-bJun interaction in living cells. *Bioscience, Biotechnology, and Biochemistry*, *75*, 1399–1401.

Ohashi, K., Kiuchi, T., Shoji, K., Sampei, K., & Mizuno, K. (2012). Visualization of cofilin-actin and Ras-Raf interactions by bimolecular fluorescence complementation assays using a new pair of split Venus fragments. *Biotechniques*, *52*, 45–50.

Pin, J. P., Neubig, R., Bouvier, M., Devi, L., Filizola, M., Javitch, J. A., et al. (2007). International Union of Basic and Clinical Pharmacology. LXVII. Recommendations for

the recognition and nomenclature of G protein-coupled receptor heteromultimers. *Pharmacological Reviews*, *59*, 5–13.
Przybyla, J. A., & Watts, V. J. (2010). Ligand-induced regulation and localization of cannabinoid CB1 and dopamine D2L receptor heterodimers. *The Journal of Pharmacology and Experimental Therapeutics*, *332*, 710–719.
Qi, Y., Wang, J. K., McMillian, M., & Chikaraishi, D. M. (1997). Characterization of a CNS cell line, CAD, in which morphological differentiation is initiated by serum deprivation. *The Journal of Neuroscience*, *17*, 1217–1225.
Rizzo, M. A., Springer, G. H., Granada, B., & Piston, D. W. (2004). An improved cyan fluorescent protein variant useful for FRET. *Nature Biotechnology*, *22*, 445–449.
Rose, R. H., Briddon, S. J., & Holliday, N. D. (2010). Bimolecular fluorescence complementation: Lighting up seven transmembrane domain receptor signalling networks. *British Journal of Pharmacology*, *159*, 738–750.
Rozenfeld, R., & Devi, L. A. (2011). Exploring a role for heteromerization in GPCR signalling specificity. *The Biochemical Journal*, *433*, 11–18.
Seachrist, J. L., & Ferguson, S. S. (2003). Regulation of G protein-coupled receptor endocytosis and trafficking by Rab GTPases. *Life Sciences*, *74*, 225–235.
Shaner, N. C., Campbell, R. E., Steinbach, P. A., Giepmans, B. N., Palmer, A. E., & Tsien, R. Y. (2004). Improved monomeric red, orange and yellow fluorescent proteins derived from Discosoma sp. red fluorescent protein. *Nature Biotechnology*, *22*, 1567–1572.
Shyu, Y. J., Hiatt, S. M., Duren, H. M., Ellis, R. E., Kerppola, T. K., & Hu, C. D. (2008). Visualization of protein interactions in living Caenorhabditis elegans using bimolecular fluorescence complementation analysis. *Nature Protocols*, *3*, 588–596.
Shyu, Y. J., & Hu, C. D. (2008). Fluorescence complementation: An emerging tool for biological research. *Trends in Biotechnology*, *26*, 622–630.
Shyu, Y. J., Liu, H., Deng, X., & Hu, C. D. (2006). Identification of new fluorescent protein fragments for bimolecular fluorescence complementation analysis under physiological conditions. *Biotechniques*, *40*, 61–66.
Shyu, Y. J., Suarez, C. D., & Hu, C. D. (2008). Visualization of ternary complexes in living cells by using a BiFC-based FRET assay. *Nature Protocols*, *3*, 1693–1702.
Stockton, S. D., Jr., & Devi, L. A. (2011). Functional relevance of mu-delta opioid receptor heteromerization: A role in novel signaling and implications for the treatment of addiction disorders. *Drug and Alcohol Dependence*, *121*, 167–172.
Tsien, R. Y. (1998). The green fluorescent protein. *Annual Review of Biochemistry*, *67*, 509–544.
Vidi, P. A., Chemel, B. R., Hu, C. D., & Watts, V. J. (2008). Ligand-dependant oligomerization of dopamine D2 and adenosine A2A receptors in living neuronal cells. *Molecular Pharmacology*, *74*, 544–551.
Vidi, P. A., Chen, J., Irudayaraj, J. M., & Watts, V. J. (2008). Adenosine A(2A) receptors assemble into higher-order oligomers at the plasma membrane. *FEBS Letters*, *582*, 3985–3990.
Vidi, P. A., Ejendal, K. F. K., Przybyla, J. A., & Watts, V. J. (2011). Fluorescent protein complementation assays: New tools to study G protein-coupled receptor oligomerization and GPCR-mediated signaling. *Molecular and Cellular Endocrinology*, *331*, 185–193.
Vidi, P. A., Przybyla, J. A., Hu, C. D., & Watts, V. J. (2010). Visualization of G protein-coupled receptor (GPCR) interactions in living cells using bimolecular fluorescence complementation (BiFC). *Current Protocols in Neuroscience*, Chapter 5, Unit 5 29.
Vidi, P. A., & Watts, V. J. (2009). Fluorescent and bioluminescent protein-fragment complementation assays in the study of G protein-coupled receptor oligomerization and signaling. *Molecular Pharmacology*, *75*, 733–739.
Wilson, C. G., Magliery, T. J., & Regan, L. (2004). Detecting protein–protein interactions with GFP-fragment reassembly. *Nature Methods*, *1*, 255–262.

Yost, E. A., Mervine, S. M., Sabo, J. L., Hynes, T. R., & Berlot, C. H. (2007). Live cell analysis of G protein beta5 complex formation, function, and targeting. *Molecular Pharmacology*, *72*, 812–825.

Zhou, J., Lin, J., Zhou, C., Deng, X., & Xia, B. (2011). An improved bimolecular fluorescence complementation tool based on superfolder green fluorescent protein. *Acta Biochimica et Biophysica Sinica*, *43*, 239–244.

CHAPTER FIFTEEN

G Protein–Coupled Receptor Heterodimerization in the Brain

Dasiel O. Borroto-Escuela*, Wilber Romero-Fernandez*, Pere Garriga†, Francisco Ciruela‡, Manuel Narvaez§, Alexander O. Tarakanov¶, Miklós Palkovits||, Luigi F. Agnati#, Kjell Fuxe*,[1]

*Department of Neuroscience, Karolinska Institutet, Stockholm, Sweden
†Departament d'Enginyeria Química, Universitat Politècnica de Catalunya, Barcelona, Spain
‡Unitat de Farmacologia, Departament Patologia i Terapèutica Experimental, Universitat de Barcelona, Barcelona, Spain
§Department of Physiology, School of Medicine, University of Málaga, Málaga, Spain
¶Russian Academy of Sciences, St. Petersburg Institute for Informatics and Automation, Saint Petersburg, Russia
||Human Brain Tissue Bank, Semmelweis University, Budapest, Hungary
#IRCCS Lido, Venice, Italy
[1]Corresponding author: e-mail address: Kjell.Fuxe@ki.se

Contents

1. Introduction 282
2. *In Situ* PLA for Demonstrating Receptor Heteromers and Their Receptor–Receptor Interactions in Brain Tissue 283
3. Brain Tissue Preparation 283
4. Proximity Probes: Conjugation of Oligonucleotides to Antibodies 287
5. PLA Reactions, Reagents, and Solutions 287
6. Quantitative PLA Image Analysis 289
7. Advantages and Disadvantages of the PLA Method 289
8. Application 291
Acknowledgments 293
References 293

Abstract

G protein–coupled receptors (GPCRs) play critical roles in cellular processes and signaling and have been shown to form heteromers with diverge biochemical and/or pharmacological activities that are different from those of the corresponding monomers or homomers. However, despite extensive experimental results supporting the formation of GPCR heteromers in heterologous systems, the existence of such receptor heterocomplexes in the brain remains largely unknown, mostly because of the lack of appropriate methodology. Herein, we describe the *in situ* proximity ligation assay procedure underlining its high selectivity and sensitivity to image GPCR heteromers with confocal

Methods in Enzymology, Volume 521
ISSN 0076-6879
http://dx.doi.org/10.1016/B978-0-12-391862-8.00015-6

microscopy in brain sections. We describe here how the assay is performed and discuss advantages and disadvantages of this method compared with other available techniques.

1. INTRODUCTION

G protein–coupled receptors (GPCRs) play critical roles in cellular processes and signaling and have been shown to form heteromers with diverge biochemical and/or pharmacological activities that are different from those of the corresponding monomers or homomers. The idea of the existence of direct interactions between two different GPCRs at the level of the plasma membrane has its origin in 1980/1981 on the basis of the discovery that peptides like cholecystokinin-8 (CCK-8) and substance P could modulate the density, and especially the affinity, of distinct monoamine receptors in membrane preparations from the CNS with *in vivo* functional correlates (Agnati, Fuxe, Zini, Lenzi, & Hokfelt, 1980; Fuxe et al., 1981). These initial findings were in line with the previous discovery of negative cooperativity between β adrenergic receptors in 1974/1975 by Lefkowitz and colleagues, indicating the possible existence of homodimers of β adrenergic receptors leading to site–site interactions in recognition (Limbird, Meyts, & Lefkowitz, 1975).

Nevertheless, despite extensive experimental results supporting the formation of GPCR heteromers in heterologous systems, the existence of such receptor heterocomplexes in their native environment remains largely unknown, mostly because of the lack of appropriate methodology. For instance, until recent years, the methods that have been developed to study receptor–receptor interactions in heteromers require that genetic constructs be expressed in the cells to enable detection of the receptor interactions, thus excluding the use of tissue samples (Borroto-Escuela, Garcia-Negredo, Garriga, Fuxe, & Ciruela, 2010; Ferre et al., 2009; Fuxe et al., 2012).

In order to demonstrate in native tissue the existence of GPCR heteromers, especially in a manner that can be generally applicable to different receptor pairs, a well-characterized *in situ* proximity ligation assay (*in situ* PLA) has been adapted to confirm the existence of GPCR heteromers in brain slices *ex vivo*.

In situ PLA is based on a pair of antibodies that can bind to target proteins and to which oligonucleotides have been attached. When the so-called proximity probes recognize a target, for example, if the two target receptors interact, the attached oligonucleotides are brought into a sufficiently close spatial proximity to allow them to join followed by ligation of the two linear oligonucleotides into a circular DNA molecule. This newly formed DNA circle strand can serve as a template for rolling circle amplification (RCA), resulting in a long single-stranded rolling circle product (RCP) attached to one of the proximity probes. As the RCP is linked to the proximity probe, it is attached at the site where the proximity probe bound, which means that it can be used to reveal the location of the receptor complex (Soderberg et al., 2006, 2007). The RCPs can then be detected and quantified by hybridizing fluorescent oligonucleotides to the repeated sequences of the RCPs, rendering them visible by fluorescence microscopy (Fig. 15.1).

Herein, we describe the *in situ* PLA procedure as a high selectivity and sensitivity assay to image GPCR heteromers in brain sections by confocal microscopy. We describe how the assay is performed and discuss advantages and disadvantages of this method compared with other available techniques.

2. *In Situ* PLA FOR DEMONSTRATING RECEPTOR HETEROMERS AND THEIR RECEPTOR–RECEPTOR INTERACTIONS IN BRAIN TISSUE

In situ PLA has previously been performed to confirm the existence of striatal $A_{2A}R$–D_2R heteromers (Trifilieff et al., 2011). The PLA technique involved the use of two primary antibodies of different species directed to either D_2R or to $A_{2A}R$ (Fig. 15.2). We recommend to use *in situ* PLA also to indicate the ratio between heteromers versus total number of the two participating receptor populations, using in addition to Western blots, receptor autoradiography, and biochemical binding methods, the two latter methods showing the densities and affinities of the two functional receptor populations. This will also help to normalize the heteromer values for comparison between groups in addition to evaluating the potential changes in the total number of the two receptor populations. The person doing the PLA measurements should be blind to the code of the experimental groups studied.

3. BRAIN TISSUE PREPARATION

As for all antibody-based staining methods, the samples should be sufficiently pretreated to fit the primary antibodies with respect to fixation, permeabilization, and antigen retrieval of the tissue to be investigated. As the protocols provided in this section are general, it is highly suggested to apply the same protocol that has been working previously for your receptor

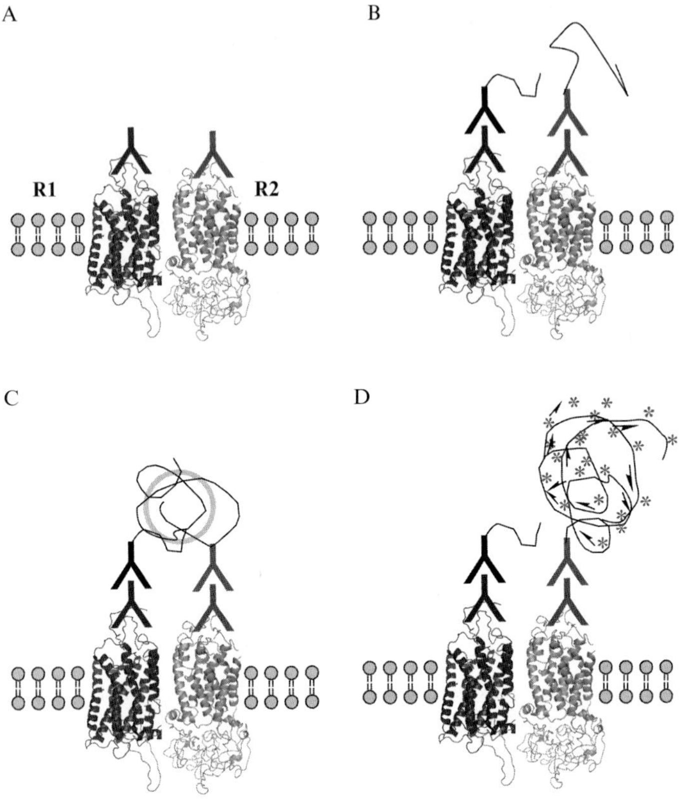

Figure 15.1 Schematic presentation of *in situ* PLA for detection of GPCR heteromers in the brain. (A) Two primary antibodies from different species are used to detect each receptor heteromer protomers (R1 and R2). (B) Each species-specific secondary antibody with attached oligonucleotide DNA probes (proximity probes) is targeted to the corresponding primary antibody, and when they are in a close proximity (the distance between the two secondary antibodies is a maximum of 10–20 nm as calculated from a DNA arm of 25–35 bp), the proximity probes serves as a template for the hybridization of circularization oligonucleotides, which are then joined by ligation into a circular DNA molecule (C). The circular DNA molecule is then amplified by rolling circle amplification (RCA) primed by one of the proximity probes and using a polymerase to yield a long concatemeric copy of the rolling circle that remains covalently attached to the proximity probe (D). The RCA product can subsequently be identified by hybridization of fluorophore-labeled complementary oligonucleotide probes added to highlight the product. (For color version of this figure, the reader is referred to the online version of this chapter.)

pairs in immunohistochemistry. Thus, similar conditions as employed for immunohistochemistry can be used for *in situ* PLA reactions.

For *in situ* PLA, the common options are fixed (paraffin-embedded or cryostat sections and vibratome sections) or unfixed cryostat (frozen) sections. The choice of section is determined by a number of conditions,

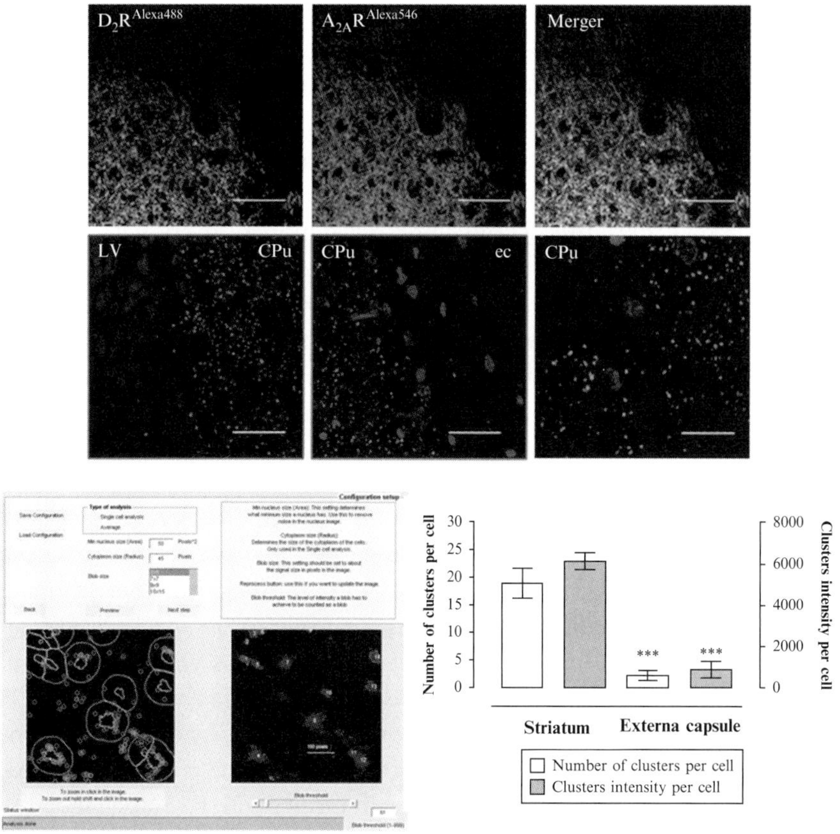

Figure 15.2 Upper-upper panel: Specific D_2R (green) and $A_{2A}R$ (red) immunoreactivities and colocalization (yellow) in striatal sections. D_2R immunoreactivity was high in the striatum of rat surrounded by the external capsule (ec) using fluorescence immunohistochemistry (left). $A_{2A}R$ immunoreactivity (middle) showed a high level of colocalization with D_2R (right) in the striatum of wild-type rat. Scale bars, 75 μm. Upper-down panel: PLA-positive $A_{2A}R$–D_2R heteromers in striatal sections adjacent to the sections with immunoreactivity $A_{2A}R$–D_2R heteromers were visualized as red clusters (blobs, dots) within the striatum which were almost absent within the lateral ventricle (LV, left) and the external capsule (ec, middle panel). Higher magnification image revealed a large number of PLA-positive red clusters within the caudate putamen (CPu, right). Each cluster represents a high concentration of fluorescence from the single-molecule amplification resulting from several hundred-fold replication of the DNA circle formed as a result of the probe proximity; the cluster/dot number can be quantified independently of the intensity. Nuclei are shown in blue (DAPI). Scale bars, 50 μm for left and middle panels; 20 μm for the right panels. Lower-left panel: Screendump from the corresponding BlobFinder analysis. The left pictures show how the software has identified the PLA signals, the nuclear limit, and the approximate limit of the cytoplasm based on a user-defined radius. The right picture shows the raw image based on 14 *Z*-planes with the nuclei enumerated. Lower-right panel: PLA-positive red clusters in striatum were quantified per cell using BlobFinder and the results are presented.

including the time and skill of the investigator. However, careful consideration of the fixation protocol is especially necessary to ensure the optimal preservation of the morphology of the specimen and target antigen (receptors). Incorrect specimen preparation can block or impede antigen labeling in the tissue. Unfortunately, the methods that are best for the preservation of tissue structure do so by modifying proteins, thereby reducing the efficiency of antigen detection. In cases of failure, it is important to try with multiple different conditions before you give up the *in situ* PLA.

1. Because of the ease of use, fixed frozen free-floating sections are often employed in most of the *in situ* PLA experiments. First, animals are anesthetized by an intraperitoneal injection of, for example, pentobarbital (60 mg/ml, [0.1 ml/100 g]) and perfused intracardially with 30–50 ml of ice-cold 4% paraformaldehyde (PFA) in 0.1 *M* phosphate-buffered saline (PBS), pH 7.4, solution. After perfusion, brains are collected and transferred into well-labeled glass vials filled with 4% PFA fixative solution for 6 h. Then, the brain pieces are placed in sucrose 20% in PBS and incubated for 24 h until sections (10–30 μm thick) are generated and serially collected using a cryostat. Alternatively to the use of fixed free-floating sections, we can use tissue fixed frozen sections attached to microscopy slides. Mounted sections on slides must be kept at −20 °C until use. Encircle the tissue section on the glass slide by creating a hydrophobic barrier using a grease pen or a silicon mask and proceed as follows.
2. Wash the fixed free-floating sections four times with PBS, then incubate with the blocking solution (10% fetal bovine serum [FBS] and 0.5% Triton X-100 or Tween-20 in Tris buffer saline [TBS], pH 7.4) for 2 h at room temperature or 1 h at 37 °C and then follow the Protocol step 5.

To reduce the likelihood of unspecific binding of the antibodies to the tissue, the tissue needs to be blocked by a blocking agent, such as bovine serum albumin (BSA; by adding 1 μl BSA (10 mg/ml) and 1 μl sonicated salmon sperm DNA (0.1 mg/ml) to 38 μl of 0.5% Triton X-100 or Tween-20 in TBS, pH 7.4; Leuchowius, Weibrecht, & Soderberg, 2011) or animal serum like 10% FBS (if animal serum is used, make sure that it is sterile filtered, as unfiltered serum may increase the amount of background signals). Use the blocking agent best suited for the antibodies used. Each time must be

Quantification of $A_{2A}R$–D_2R heteromers demonstrates highly significant differences in PLA clusters per cell between caudate putamen and external capsule (***$P < 0.001$ by Student's *t*-test). (See Color Insert.)

checked that the reaction should never become dry as this will cause high background.

4. PROXIMITY PROBES: CONJUGATION OF OLIGONUCLEOTIDES TO ANTIBODIES

Proximity probes are created through the attachment of oligonucleotides to antibodies. The oligonucleotide component of the proximity probes can be covalently coupled to an antigen-binding component or attached to secondary antibodies specific for antibodies raised in different species. This approach avoids the need to conjugate the oligonucleotide components to each primary antibody pair.

Several different types of chemistry can be used for the conjugation of oligonucleotides to antibodies. Mainly three methods have been used extensively in recent years: the maleimide/NHS-esther chemistry (SMCC; Soderberg et al., 2006), the succinimidyl 4-hydrazinonicotinate acetone hydrazone (SANH; Leuchowius et al., 2011), or the commercially available Antibody-Oligonucleotide All-in-One Conjugation Kit from Solulink company (http://www.solulink.com/), based on two complementary heterobifunctional linkers (Sulfo-S-4FB (formylbenzamide) and S-HyNic (hydrazino-nicotinamide)). Because the act of conjugation can severely affect the ability of some antibodies to bind antigen, it may be necessary to analyze different antibodies, conjugation chemistries, and reaction conditions to obtain suitable proximity probes.

Another possibility, not less useful, is to buy directly proximity probes from specialized companies on antibody-oligonucleotide conjugation, for example, Duolink (Uppsala, Sweden; http://www.olink.com/).

5. PLA REACTIONS, REAGENTS, AND SOLUTIONS

1. If primary antibodies directly labeled with oligonucleotides or primary antibodies are used in combination with secondary proximity probes (see Protocol step 4), the conditions for incubation with the primary antibodies should be chosen according to the manufacturer's recommendations or will have to be identified by the users. For instance, incubate the tissue with the primary antibodies diluted into a suitable concentration in the wash buffer (0.5% Triton X-100 or Tween-20 in TBS, pH 7.4) at 1–2 h at 37 °C or +4 °C overnight.

2. After incubation of the primary antibodies at conditions specified above or determined by the user, excess antibody should be removed. Wash the slides four times, 5 min each time, with wash buffer.
3. In the mean time, if primary antibodies are used in combination with secondary proximity probes, dilute the proximity probes to a suitable concentration in the wash buffer. It is important to use the same buffer as those for the primary antibody to avoid background staining. Apply the proximity probe mixture to the sample and incubate for 1 h at 37 °C in a humidity chamber. Do not allow the samples to dry, as this will cause also artifacts. To remove unbound proximity probes, wash the slides four times, 5 min each time, with wash buffer.
4. Prepare the hybridization–ligation solution. To ensure optimal conditions for the enzymatic reactions, the sections should be soaked for 1 min in 1 × ligation buffer (10 m*M* Tris–acetate, 10 m*M* magnesium acetate, 50 m*M* potassium acetate, pH 7.5; Soderberg et al., 2008), prior to addition of the hybridization–ligation solution (final concentration: BSA (250 μg/ml), 1 × T4 DNA ligase buffer, Tween-20 (0.05%), NaCl 250 m*M*, ATP 1 m*M*, and the circularization or connector oligonucleotides 125–250 n*M*). Circularization or connector oligonucleotides can be designed and synthesized as described previously (Soderberg et al., 2008). Remove the soaking solution (ligation buffer) and add T4 DNA ligase at a final concentration of 0.05 U/μl to the hybridization–ligation solution. Vortex briefly to mix the ligase with the solution. Apply the mixture immediately to the sections and slides in a humidity chamber for 30 min at 37 °C.
5. Wash the sections three times with wash buffer in a washing jar for 5 min to remove excess connector oligonucleotides.
6. Prepare the RCA mixture. Soak the sections in 1 × RCA buffer (50 m*M* Tris–HCl, 10 m*M* $MgCl_2$, 10 m*M* $(NH_4)_2SO_4$, pH 7.5 adjusted with HCl) for 1 min. Remove the soaking solution and add the RCA solution (final concentration: phi-29 polymerase 0.125–0.200 U/μl, BSA (250 μg/ml), 1 × RCA buffer, Tween-20 (0.05%), and dNTP (250 μ*M* for each)). Vortex briefly the RCA solution and incubate in a humidity chamber for 100 min at 37 °C.
7. Wash the sections three times with wash buffer in a washing jar for 5 min. Prepare the detection solution (final concentration: BSA (250 μg/ml), 2 × sodium citrate, sodium chloride buffer, and the fluorescence detection (e.g., Texas Red or Alexa 555)-oligonucleotide strand (6.25 n*M*)), see Soderberg et al., 2008) and incubate the sections in a humidity chamber for 30 min at 37 °C. Keep the detection

solution in the dark to prevent fluorophore bleaching. From now on, all reactions and wash steps should be performed in the dark.

8. Wash the sections twice, each time with wash buffer in a washing jar for 5 min in the dark.
9. Wash the sections once with TBS in a washing jar for 5 min in the dark.
10. Dry and mount sections with the appropriate mounting media (e.g., VectaShield or Dako). The sections should be protected against light and can be stored for several days at 4 °C or for several months at −20 °C.

6. QUANTITATIVE PLA IMAGE ANALYSIS

1. Visualize the sections with fluorescence microscopy equipped with excitation/emission filters compatible with the fluorophores used. The *in situ* PLA signals have a very characteristic appearance that is easily recognized once you know what to look for. The PLA detection reaction products are seen as bright fluorescent puncta of submicrometer size (see Fig. 15.2). By moving the focus up and down in your sample tissue, you should note appearance and disappearance of PLA signals. Up to a certain density of PLA signals, they appear as discrete dots (puncta, blobs) that can be easily enumerated using image analysis software.
2. Analyze the captured images by image techniques to quantify the number of dots (Fig. 15.2, lower panel). Many commercial image analysis software packages can also be used in addition to free software packages, such as BlobFinder (Allalou & Wahlby, 2009) or Cellprofiler (Carpenter et al., 2006). The BlobFinder is a free software tool developed by the Centre for Image Analysis-Uppsala University for such objectives (the freeware is available for download from http://www.cb.uu.se/~amin/BlobFinder/; Fig. 15.2, lower panel). At higher densities per number of nuclei, the dots start to coalesce, thus making it more difficult to resolve and enumerate individual signals. It is important to use the same settings for image acquisition for all images in a series.

7. ADVANTAGES AND DISADVANTAGES OF THE PLA METHOD

Comparison with other methods to study receptor–receptor interactions in heteromers, such as FRET, BRET, and bimolecular fluorescence complementation: In situ PLA can offer advantages by permitting analyses of interactions among any receptors for which suitable antibodies are available without using genetic constructs. Such constructs can perturb cellular function and cannot always be used, for example, in cells and tissues from subjects. *In situ* PLA can be performed in all samples of cells and tissues, and the method is highly suited to investigate human specimens collected from biobanks, in order to investigate patterns of changes in heteromers that could provide insights on the role of basic heteromer mechanisms or have a diagnostic value (Nilsson et al., 2010). The method has also proven useful to monitor the effects of different compounds like agonists and antagonists or their combined treatment on the receptor heteromers in cells and tissue (Borroto-Escuela et al., 2011). The information is obtained at a resolution of individual cells or even of subcellular compartments, providing profound insights into cellular heterogeneity in tissues. The method also provides an enhanced sensitivity and selectivity compared with many other methods as powerful RCA and dual target recognition are used (Clausson et al., 2011).

As with any method there are limitations, for instance, *in situ* PLA cannot be used with live cells, as it requires cell fixation and, in some cases, permeabilized cells. When studying receptor–receptor interactions, it is important to remember that the method, like many other methods for studying protein–protein interactions, can show that two proteins are in close proximity and likely directly interact. Proteins can also interact indirectly through an adapter protein. The maximal distance between two epitopes to give rise to a signal with *in situ* PLA is 10–30 nm with direct-conjugated proximity probes, and slightly longer when secondary proximity probes are used. By changing the length of the oligonucleotides, the maximal distance limits can be reduced or increased.

Other critical parameters for achieving good results is the use of excellent antibodies. The antibodies must also be used under optimal conditions taking into consideration parameters such as antibody concentration, epitopes targeted by the antibodies, fixation, antigen retrieval, blocking conditions, etc. A range of controls both positive and negative ones should be used to guarantee the specificity of the PLA signal. Positive controls can include cells where the protein is known to be expressed, such as in certain cells or tissues or in cells transfected to express the protein. Negative controls include cells or tissues that do not express the protein or where the protein has been knocked out or downregulated by, for example, siRNA.

8. APPLICATION

In situ PLA has been used to study proteins and protein–protein interactions in a range of applications (Leuchowius et al., 2010; Nilsson et al., 2010). In 2011, the method was employed to study GPCRs heteromers, mainly adenosine A_{2A} and dopamine D_2 receptor heteromers in striatal sections (Trifilieff et al., 2011) and dopamine D_2R and D_4R in transiently transfected HEK293T cells (Borroto-Escuela et al., 2011). In addition, the methods were employed to demonstrate for the first time the existence of FGFR1 and 5-HT1A receptor heterocomplexes in the rat hippocampus and dorsal and median raphe in the midbrain (Fig. 15.3A; Borroto-Escuela et al., 2012).

In the analysis using *in situ* PLA, it is also important to determine the ratio between heteromers versus total number of the two participating receptor populations, using in addition to Western blots, receptor autoradiography, and biochemical binding methods, the two latter methods showing the densities and affinities of the two functional receptor populations. The relationship between these parameters will help to normalize the heteromer values for comparison between groups in addition to evaluating the potential

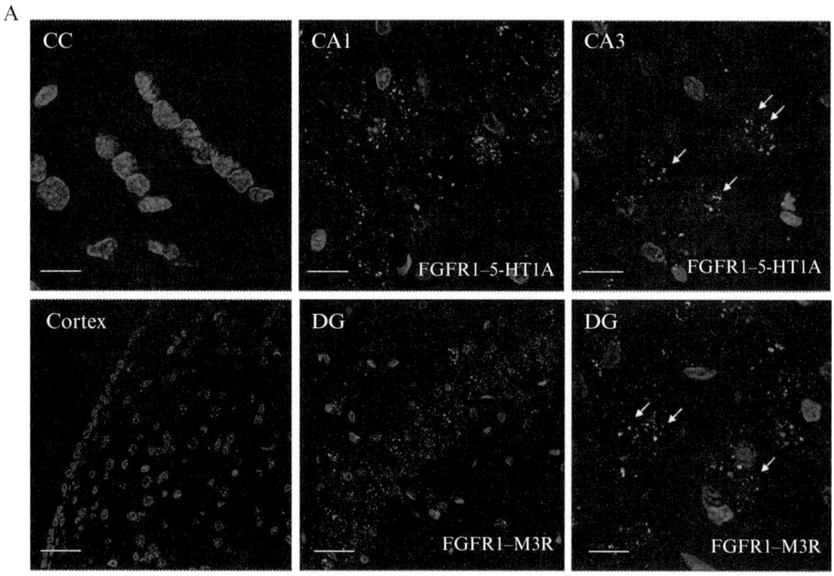

(Continued)

B

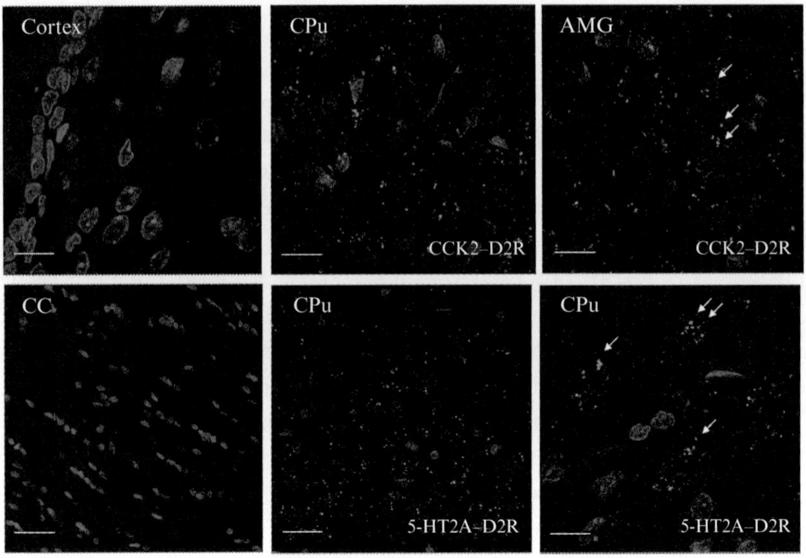

Figure 15.3 Detection of different GPCR–GPCR heteromers and GPCR–RTK heterocomplexes in dorsal rat hippocampal and striatal sections by *in situ* PLA. (A-upper panel) Constitutive FGFR1–5-HT1A heteroreceptor complexes are detected by *in situ* PLA (red clusters) in dorsal rat hippocampus (Ammon's horn 1 and 3 (CA1, CA3) but not, as an example, in the corpus callosum (cc). Scale bars, 20 μm. (A-lower panel) Constitutive FGFR1–M3R heteroreceptor complexes are detected by *in situ* PLA (red clusters) in dorsal rat hippocampus (granular layer of the dentate gyrus [DG]) but not in the cortex cerebri. Scale bars from the left to the right, 50, 50, and 20 μm. (B-upper panel) Constitutive CCK2–D_2R heteromers are detected by *in situ* PLA (red clusters) in striatal sections (caudate putamen: CPu; amygdaloid cortex: AMG) but not, for example, in the cortex cerebri. Scale bars, 20 μm. (B-upper panel) Constitutive 5-HT2A–D_2R heteromers are detected by *in situ* PLA (red clusters) in striatal sections (caudate putamen: CPu; amygdaloid cortex: AMG) but not, for example, in the corpus callosum (cc). Scale bars from the left to the right, 75, 50, and 20 μm. Nuclei appear as a blue color in all panels and the white arrows indicate the red cluster formation (PLA signal). (See Color Insert.)

changes in the total number of the two receptor populations. Of increasing importance will be to determine the agonist/antagonist regulation of these receptor heteromers in order to understand their potential roles as targets for drugs used in neuropsychopharmacology for treatment of psychiatric and neurological diseases. Analysis of human brain material with *in situ* PLA can also reveal if the relative abundance of specific receptor heteromers in discrete brain regions is altered in brain diseases (Fig. 15.3A and B).

ACKNOWLEDGMENTS

This work has been supported by the Swedish Medical Research Council (04X-715) Torsten and Ragnar Söderberg Foundation (MN 65/08), Telethon TV3's La Marató Foundation 2008, and M.M. Wallenberg Foundation to K. F., and Karolinska Institutets Forskningsstiftelser 2010 and 2011 to D. O. B.- E. Telethon TV3's La Marató Foundation 2008 to P.G. and K.F. Also by grants SAF2008-01462 and Consolider-Ingenio CSD2008-00005 from Ministerio de Ciencia e Innovación to F. C. A. O. T. has not received any support for this work.

REFERENCES

Agnati, L. F., Fuxe, K., Zini, I., Lenzi, P., & Hokfelt, T. (1980). Aspects on receptor regulation and isoreceptor identification. *Medical Biology*, *58*, 182–187.

Allalou, A., & Wahlby, C. (2009). BlobFinder, a tool for fluorescence microscopy image cytometry. *Computer Methods and Programs in Biomedicine*, *94*, 58–65.

Borroto-Escuela, D. O., Garcia-Negredo, G., Garriga, P., Fuxe, K., & Ciruela, F. (2010). The M(5) muscarinic acetylcholine receptor third intracellular loop regulates receptor function and oligomerization. *Biochimica et Biophysica Acta*, *1803*, 813–825.

Borroto-Escuela, D. O., Romero-Fernandez, W., Mudo, G., Perez-Alea, M., Ciruela, F., Tarakanov, A. O., et al. (2012). Fibroblast growth factor receptor 1–5-hydroxytryptamine 1A heteroreceptor complexes and their enhancement of hippocampal plasticity. *Biological Psychiatry*, *71*, 84–91.

Borroto-Escuela, D. O., Van Craenenbroeck, K., Romero-Fernandez, W., Guidolin, D., Woods, A. S., Rivera, A., et al. (2011). Dopamine D2 and D4 receptor heteromerization and its allosteric receptor-receptor interactions. *Biochemical and Biophysical Research Communications*, *404*, 928–934.

Carpenter, A. E., Jones, T. R., Lamprecht, M. R., Clarke, C., Kang, I. H., Friman, O., et al. (2006). Cell Profiler: Image analysis software for identifying and quantifying cell phenotypes. *Genome Biology*, *7*, R100.

Clausson, C. M., Allalou, A., Weibrecht, I., Mahmoudi, S., Farnebo, M., Landegren, U., et al. (2011). Increasing the dynamic range of in situ PLA. *Nature Methods*, *8*, 892–893.

Ferre, S., Baler, R., Bouvier, M., Caron, M. G., Devi, L. A., Durroux, T., et al. (2009). Building a new conceptual framework for receptor heteromers. *Nature Chemical Biology*, *5*, 131–134.

Fuxe, K., Agnati, L. F., Benfenati, F., Cimmino, M., Algeri, S., Hokfelt, T., et al. (1981). Modulation by cholecystokinins of 3H-spiroperidol binding in rat striatum: Evidence for increased affinity and reduction in the number of binding sites. *Acta Physiologica Scandinavica*, *113*, 567–569.

Fuxe, K., Borroto-Escuela, D. O., Marcellino, D., Romero-Fernandez, W., Frankowska, M., Guidolin, D., et al. (2012). GPCR heteromers and their allosteric receptor-receptor interactions. *Current Medicinal Chemistry*, *19*, 356–363.

Leuchowius, K. J., Jarvius, M., Wickstrom, M., Rickardson, L., Landegren, U., Larsson, R., et al. (2010). High content screening for inhibitors of protein interactions and post-translational modifications in primary cells by proximity ligation. *Molecular & Cellular Proteomics*, *9*, 178–183.

Leuchowius, K. J., Weibrecht, I., & Soderberg, O. (2011). In situ proximity ligation assay for microscopy and flow cytometry. *Current Protocols in Cytometry*, 36 chapter 9, unit 9.

Limbird, L. E., Meyts, P. D., & Lefkowitz, R. J. (1975). Beta-adrenergic receptors: Evidence for negative cooperativity. *Biochemical and Biophysical Research Communications*, *64*, 1160–1168.

Nilsson, I., Bahram, F., Li, X., Gualandi, L., Koch, S., Jarvius, M., et al. (2010). VEGF receptor 2/-3 heterodimers detected in situ by proximity ligation on angiogenic sprouts. *The EMBO Journal, 29*, 1377–1388.

Soderberg, O., Gullberg, M., Jarvius, M., Ridderstrale, K., Leuchowius, K. J., Jarvius, J., et al. (2006). Direct observation of individual endogenous protein complexes in situ by proximity ligation. *Nature Methods, 3*, 995–1000.

Soderberg, O., Leuchowius, K. J., Gullberg, M., Jarvius, M., Weibrecht, I., Larsson, L. G., et al. (2008). Characterizing proteins and their interactions in cells and tissues using the in situ proximity ligation assay. *Methods, 45*, 227–232.

Soderberg, O., Leuchowius, K. J., Kamali-Moghaddam, M., Jarvius, M., Gustafsdottir, S., Schallmeiner, E., et al. (2007). Proximity ligation: A specific and versatile tool for the proteomic era. *Genetic Engineering, 28*, 85–93.

Trifilieff, P., Rives, M. L., Urizar, E., Piskorowski, R. A., Vishwasrao, H. D., Castrillon, J., et al. (2011). Detection of antigen interactions ex vivo by proximity ligation assay: Endogenous dopamine D2-adenosine A2A receptor complexes in the striatum. *BioTechniques, 51*, 111–118.

CHAPTER SIXTEEN

Experimental Strategies for Studying G Protein-Coupled Receptor Homo- and Heteromerization with Radioligand Binding and Signal Transduction Methods

Roberto Maggio[*,1], **Cristina Rocchi**[*], **Marco Scarselli**[†]

[*]Department of Biotechnological and Applied Clinical Sciences, University of L'Aquila, L'Aquila, Italy
[†]Department of Translational Research and of New Technologies, University of Pisa, Pisa, Italy
[1]Corresponding author: e-mail address: roberto.maggio@univaq.it

Contents

Abstract

Before the molecular biology era, functional experiments on isolated organs and radioligand binding and biochemical experiments on animal tissues were widely used to characterize G protein-coupled receptors (GPCRs). The introduction of recombinant cell lines expressing a single GPCR type has been a big step forward for studying both drug–receptor interactions and signal transduction. Before the introduction of the concept of receptor oligomerization, all data generated were attributed to the interaction of drugs with receptor monomers. Now, considerable data must be reinterpreted in light of receptor homo- and heteromerization. In this chapter, we will review some of the methods used to study radioligand binding and signal transduction modifications induced by GPCR homo- and heteromerization.

Methods in Enzymology, Volume 521
ISSN 0076-6879
http://dx.doi.org/10.1016/B978-0-12-391862-8.00016-8

1. INTRODUCTION

Methods to study G protein-coupled receptors (GPCRs) homo- and heteromerization are in constant evolution and modern techniques today attempt to study the mechanisms of receptor interaction at the level of single molecules (Kasai et al., 2011; Scarselli, Annibale, & Radenovic, 2012). Nevertheless, a few laboratories can take advantage of such sophisticated approaches, so radioligand binding and biochemical assays still remain the most commonly utilized techniques worldwide to study GPCR homo- and heteromerization. In the following paragraphs, we describe and discuss some of the radioligand binding and signaling methods currently used to study receptor homo- and heteromerization. For fundamental aspects of the binding methods, we redirect the reader to McKinney (1998).

2. RADIOLIGAND BINDING IN RECEPTOR OLIGOMERIZATION

Radioligand binding studies might, in principle, be used to obtain evidence on the oligomeric state of GPCRs. If GPCRs exist as monomers with single binding sites, drug–receptor interactions should be governed by the law of mass action with a drug that combines with a receptor at a rate dependent on the concentration of the drug and the receptor. Therefore, at equilibrium, in saturation-binding studies, the maximal number of receptors expressed by the cell (B_{max}) should be the same for all radioligands and should be independent of assay conditions. In competition binding assays, all ligands should fully compete with each other, and dissociation constants determined in saturation and competition studies should be the same. If, however, oligomerization is taking place, it would be expected that this will alter the properties of the receptor, with equilibrium and kinetic binding exhibiting cooperativity (Strange, 2005).

2.1. Equilibrium and kinetic binding experiments in receptor homomerization

Early evidence of receptor homomerization modifying ligand binding was provided by Wreggett and Wells (1995) on muscarinic receptors solubilized from porcine atria. In their work, they demonstrated that solubilized purified receptors bound muscarinic ligands in an apparently cooperative manner. This was deduced from the fact that, for binding at equilibrium, the apparent capacity for the antagonist [^{3}H]quinuclidinylbenzilate exceeded that for the antagonists [^{3}H]AF-DX 384 and *N*-[^{3}H]methylscopolamine. Binding of a high concentration of [^{3}H]quinuclidinylbenzilate to muscarinic M_2 receptors was fully inhibited by unlabeled methylscopolamine, which therefore affected sites not labeled at similar concentrations of *N*-[^{3}H]methylscopolamine.

Similar results were found by Vivo, Lin, and Strange (2006) for dopamine D_2 receptors. In saturation-binding experiments, [^{3}H]spiperone labeled twice the number of sites as [^{3}H]raclopride. This nonstandard behavior of the two radioligands could have been attributed to labeling of different populations ofbinding sites, but unlabeled raclopride was able to fully inhibit [^{3}H]spiperone binding, with behavior inconsistent with different populations of sites. In Fig. 16.1A and B, we reproduced these data in stably transfected CHO-D_2 cells. The basic model of the receptor emerging from these studies is of an oligomer, in which the binding of ligands to some of the monomers in the array exerts homotropic negative cooperativity on the binding of subsequent molecules.

The critical parameter considered in these experiments is the B_{max} of the different radioligands. The extent of the negative cooperativity of ligands binding to different monomers in the array determines the differences in B_{max} between the radioligands, hence the possibility of discerning these differences. In order to measure the difference in B_{max} among radioligands, we suggest using a reproducible system that allows for constant receptor expression. It can be a tissue homogenate or a solubilized protein preparation, or a cell line

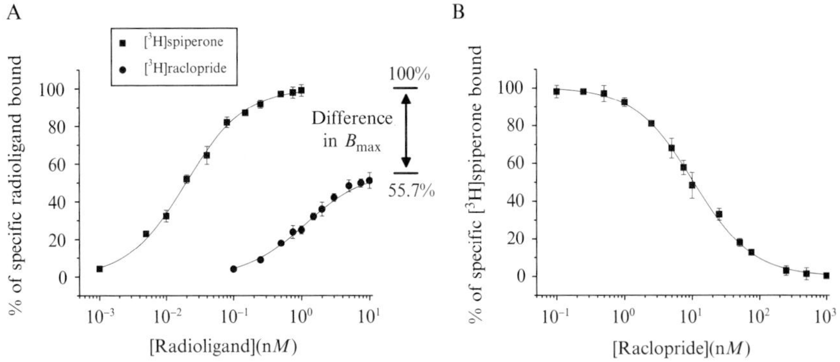

Figure 16.1 Binding of [^{3}H]spiperone and [^{3}H]raclopride to D_2 dopamine receptors expressed in CHO cells. (A) [^{3}H]Spiperone and [^{3}H]raclopride saturation assays were performed in the presence of 100 m*M* NaCl. B_{max} determinations were performed on the same preparation of membranes in order to allow comparisons. (B) Raclopride/[^{3}H]spiperone competition experiments were performed with 200 p*M* [^{3}H]spiperone. Specific [^{3}H]spiperone binding was determined in the presence of 3 m*M* dopamine. The curves shown are representative curves from single experiments with data points determined in triplicate and are best described by a one-binding site model.

(CHO, HeLa, HEK293, and Sf9 among others) stably transfected with the receptor of interest. When using stable cell lines, we recommend preparing membrane stocks to be used for several experiments in order to compare B_{max} across experiments. Many GPCRs have specific ionic requirements for the optimal binding of ligands (Ca^{2+}, Na^{+}) and differences between ligands can be unveiled in the presence or absence of ions. As shown by Armstrong and Strange (2001), we suggest attempting binding in the absence and in the presence of different ion concentrations. In addition, the influence of G protein-coupling should be considered in these experiments, especially when one or both radioligands are agonists. G protein-coupling can be reduced by preincubation of the receptors with GTP or nonhydrolyzable GTP analogues, such as Gpp(NH)p or GTPγS (guanosine gamma thiophosphate). Alternatively, GPCRs mutants unable to couple G proteins can be used.

In the case where two competitive radioligands show a difference in B_{max}, competition binding assays should be performed in order to see if the radioligand with the lower B_{max} inhibits in full the binding of the radioligand with the higher B_{max}. This will confirm that the two radioligands are labeling the same population of binding sites and the difference in B_{max} is due to negative cooperativity between sites.

While for most class A (rhodopsin-like) receptors there are several commercially available radioligands, for the other GPCRs, there are few if any commercially available radioligands; consequently, this approach might be impracticable. A more sensitive method to detect cooperative interactions between two topographically distinct binding sites is an investigation into the dissociation kinetics of a tracer ligand in the absence and presence of a second ligand.

Drugs bound to independent orthosteric sites should come off the receptors according to a simple "exponential decay" curve. Furthermore, radioligand dissociation from independent sites should be independent of the method used to initiate dissociation (dilution or addition of an excess of an unlabeled drug).

Data showing that complex dissociation kinetics could be attributed to receptor oligomerization was first provided by Franco et al. (1996) with adenosine A_1 receptors. In their work, they showed that there were differences in the dissociation kinetics of [^{3}H]-(*R*)-(phenylisopropyl)adenosine ([^{3}H]R-PIA) from the receptor for the various ligand concentrations when dissociation was performed in the presence of a 300-fold excess of unlabeled R-PIA. Similar results were obtained with dopamine D_2 receptors, when the agonist [^{3}H]*N*-propyl-norapomorphine dissociation was initiated by dilution, the dissociation rate in the absence of sodium ions was unaffected by the addition of the antagonist/inverse agonist (+)-butaclamol, but was accelerated by the addition of agonists, for example, dopamine, suggesting that the receptor was not behaving as a monomer with a single binding site (Kara et al., 2010).

Investigations into dissociation binding kinetics attempt to test a series of unlabeled drugs for each single radioligand. Modification in the off rate of a radioligand indicates cooperativity between the radioligand and the unlabeled drug. Still, this is not unambiguous evidence of receptor oligomerization, as cooperativity between ligands can occur either across receptors (off-target allosterism) or intrareceptor (in-target allosterism) if the two drugs bind to two different sites in the same monomer. An attempt to distinguish between these two alternatives could be to initiate dissociation kinetics with saturated and unsaturated receptors. If we consider an allosteric site in a single monomer, this is topographically and morphologically distinct from the orthosteric site; therefore, it will very likely have very low affinity for a radioligand designed for the orthosteric site. Given that, if modification of the off rate depends on intrareceptor allosterism, using a large excess of unlabeled drug should show no difference in the off rate of the radioligand bound to the orthosteric site under conditions of saturated and unsaturated receptors, because the allosteric sites will be always fully available to the unlabeled drug (not occupied by the radioligand). Conversely, receptor occupancy should modify the off rate of the radioligand if allosterism is across the orthosteric sites of the two receptors: (a) when dissociation is initiated by dilution, the extent of cooperativity between two molecules of radioligand binding contemporary to the orthosteric sites of the dimer is proportional to the number of receptors occupied; conversely, (b) when dissociation is initiated by a large excess of unlabeled drug, the extent of radioligand/unlabeled drug cooperativity depends on the number of orthosteric sites available for binding of the unlabeled drug, which diminishes as the radioligand tends to saturate all the receptors (Fig. 16.2). As illustrated above for equilibrium binding assays, the influence of G protein-coupling should be analyzed especially when using agonists.

2.2. Equilibrium and kinetic binding experiments in receptor heteromerization

While there is no hint for knowing which protomer a radioligand binds in a homomeric assembly, this is possible in a heteromeric assembly where different ligands bind to different protomers. Although we do not know how receptors forming heteromers assemble in the plasma membrane, whether as true 1:1 heterodimers or distinct homodimers assembled into

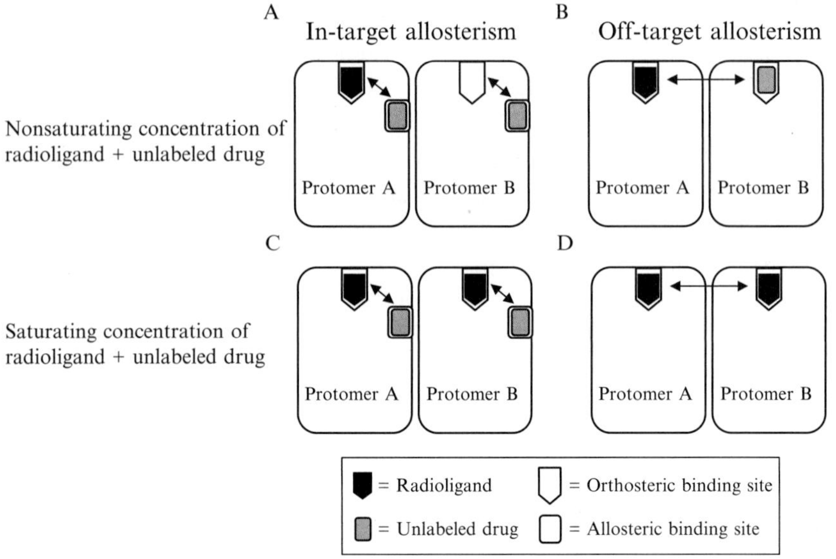

Figure 16.2 Allosterism at a GPCR can be in-target when both allosteric and orthosteric sites are present in the same protomer (A, C) or off-target when they are found on different protomers (B, D). In principle, when considering in-target allosterism, there should not be a difference in the off rate of a radioligand when dissociation is initiated in the presence of a nonsaturating (A) or saturating (C) concentration of the radioligand itself, as the allosteric site will always be fully available to the unlabeled drug. The only difference should be seen between dissociation initiated by dilution alone and dissociation initiated in the presence of an excess of the unlabeled drug. In contrast, when considering allosterism across the orthosteric site (off-target allosterism), the concentration of the radioligand should influence the off rate. In the nonsaturating condition (B), a fraction of the protomers in the dimers will be available to the cold ligand to modify the off rate of the radioligand; conversely, in the saturating condition (D), all the protomers will be occupied by the radioligand and, at least at the beginning, the cold ligand will not modify radioligand dissociation. Likewise, differences in the off rate should be seen in off-target allosterism, when there is cooperativity among the binding of subsequent molecules of the radioligand to the protomers in the dimer. Dissociation initiated by dilution only, in the presence of a saturating or nonsaturating concentration of the radioligand, should give different off rates. For an extensive review of allostery at GPCR homo- and heteromers, see Smith and Milligan (2010).

a heterooligomer, several studies have clearly demonstrated modulation in binding affinities across receptor heteromers. An early work showing changes in pharmacology when two functional receptors were coexpressed together comes from Devi group (Jordan & Devi, 1999). They provided pharmacological evidence that coexpression of kappa and delta opioid receptors results in a new heteromeric receptor that exhibits ligand binding characteristics distinct from those of either receptor.

While there is broad consensus that GPCR oligomerization is not necessary for efficient activation of heterotrimeric G proteins, as monomeric receptors are sufficient to ascertain this basic function, GPCR oligomerization appears to be important for allosteric regulation and fine-tuning of signaling and receptor trafficking. Given, it is not always easy to detect modifications in radioligand binding due to receptor heteromerization, as they could be small and restricted to the fraction of the receptors that form heteromers. For this reason, in studying heteromerization in recombinant cell lines, we usually transfect different amounts of receptor cDNAs in order to have unbalanced expression of receptors, such that one receptor (Receptor-A) will be overexpressed with respect to the other (Receptor-B). In this way, we increase the fraction of Receptor-B engaged in heteromerization with Receptor-A. Then, binding is performed using a radioligand specific for Receptor-B. Unlabeled drugs specific for Receptor-A and -B are then tested against the radioligand in order to reveal modifications in the profile of saturation and inhibition curves. Curve profiles should be compared with data obtained from cells separately transfected with the same amount of Receptor-A and -B and then mixed. This control provides a similar amount of receptors A and B, but no heteromers.

The modifications described so far are either an increase or decrease in binding affinity for specific agonists and antagonists, and this is the consequence of an alteration in the binding pockets, revealing that some ligands may prefer receptors in their homomeric or heteromeric conformation. For example, a decrease in the affinity of receptor-selective agonists was observed in the case of delta–kappa opioid receptors (DOR–KOR) (Jordan & Devi, 1999) and delta–mu opioid receptors (DOR–MOR) (George et al., 2000). In contrast, an increase in the affinity of β_2-adrenergic receptor ligands was observed for β_1–β_2-adrenergic receptor pairs (Zhu et al., 2005).

As mentioned above for receptor homomerization, kinetic experiments are more sensitive to cooperative interactions between two topographically distinct binding sites. We suggest using the same approach as in the equilibrium binding experiment overexpressing one receptor (Receptor-A) with respect to the other (Receptor-B) and using a radioligand specific for the less expressed Receptor-B. In order to study the modification in the off rate of a Receptor-B-specific radioligand, we suggest comparing the profile of dissociation curves initiated by dilution plus an excess of a cold Receptor-B-specific ligand with the profile of dissociation curves in which we further add different concentrations of a ligand specific for Receptor-A. In this way, both decreases and increases in the apparent dissociation rate constant have been described. The rate of dissociation of DOR and MOR radioligands at DOR/MOR heteromers was decreased when dissociation was performed in the presence of cold ligands for MOR and DOR receptors, respectively (Gomes et al., 2000). Conversely, an increase in the apparent dissociation rate constant has been observed for chemokine CCR2/CCR5 heteromers, in which the rate of radioligand dissociation from one unit of the heteromer was strongly increased in the presence of an unlabeled chemokine ligand of the other unit (Springael et al., 2006). These results indicate that changes in receptor pharmacology do not seem to follow a common pattern, but depend on the receptor pairs that are engaged in the formation of heteromers and from the ligands tested.

3. SIGNAL TRANSDUCTION IN RECEPTOR OLIGOMERIZATION

Qualitative and quantitative changes in signal transduction have been clearly demonstrated in receptor homo- and heteromerization (Maggio, Innamorati, & Parenti, 2007). Strategies to investigate modifications in signal transduction induced by receptor oligomerization are essentially based on the use of receptor mutants and cell systems engineered to dissect the activity of a single protomer from that of both protomers in a homo- or heteromer. What is emerging from these studies is that the way how oligomerization influences signal transduction is not univocal, but depends on the GPCR homo- and heteromers that are investigated.

3.1. Signal transduction in GPCR homomerization

While there are no supporting data that the signal transduction of homomers or homooligomers can be qualitatively different from that of monomers, there is some evidence that synergistic activation of the two protomers in a homomer can exhibit stronger activation of G protein or recruitment of β-arrestin. In a recent study, by using wild-type and mutant serotonin type 4 (5-HT_4) receptors expressed in COS-7 cells as models of class A GPCRs, Pellissier et al. (2011) showed that activation of one protomer in a homomer was sufficient to stimulate G proteins. However, coupling efficiency was two times higher when both protomers were activated. Similar results were provided by Novi, Scarselli, Corsini, and Maggio (2004) and Novi et al. (2005) for mitogen-activated protein kinase (ERK1/2) phosphorylation and the recruitment of β-arrestin, respectively. These studies require cotransfection of the wild-type receptor together with nonfunctional mutant receptors or receptors activated solely by synthetic ligands (RASSLs). There are still few RASSLs that can be utilized for these types of experiments, but it is expected that they will increase in the future as they have a huge potential to define the physiological roles of GPCRs and to validate receptors in animal models as therapeutic targets to treat human disease (Alvarez-Curto et al., 2011).

In order to demonstrate an increase in signaling efficacy by the synergic activation of both protomers in a homomer, Pellissier et al. (2011) cotransfected the wild-type 5-HT_4 receptor together with a series of 5-HT_4 mutants. In particular, they used a mutant 5-HT_4-(D^{100}A) which, in contrast with the wild-type receptor, was able to bind the synthetic agonist ML10375 but not the physiological neurotransmitter 5-hydroxytriptamine. Despite its difference in agonist recognition, 5-HT_4-(D^{100}A) maintained the same efficacy in signal transduction as the wild-type receptor as shown by a similar increase in cAMP accumulation when stimulated by the reciprocal agonist BIMU8. By using alternatively 5-HT or ML10375, it was possible to separately stimulate the two protomers of the functional 5-HT_4/5-HT_4-(D^{100}A) heteromer. In contrast, using BIMU8, they could simultaneously activate the two protomers. After titrating the 5-HT_4 and 5-HT_4-(D^{100}A) cDNAs in order to obtain an equal amount of receptors expressed on the membrane, they demonstrated that BIMU8 was twice as effective as 5-HT or ML10375, indicating that signaling is more efficient when two protomers are activated.

In order for these experiments to be interpreted correctly, a strict titration of the cotransfected receptor must be performed. Furthermore, combinations of the relative amount of the wild-type and mutant receptor should be measured. The easiest method to measure this amount is to coimmunoprecipitate wild-type and mutant receptors with antibodies directed to tags previously added at their N-terminal, or in a region of the receptor that does not influence its function and is readily accessible to antibodies. In this way, it should be possible to measure the relative amount of wild type/mutant receptor complexes. If we assume that the mutation does not change the relative affinity between the wild-type and mutant receptors, this amount should be half of the entire pool of receptor complexes. It should be kept in mind, though, that artifacts like the dissociation of complexes during extraction could eventually alter the results. The use of energy transfer-based techniques (bioluminescence resonance energy transfer and fluorescence resonance energy transfer) could help to assess whether or not the relative affinity for the wild-type and the mutant receptor has changed (for a review, see Alvarez-Curto, Pediani, & Milligan, 2010).

More complex experiments can be engineered in which the only functional receptor complex is the heteromer formed by the complementation of two nonfunctional receptors. Here, we will describe the brilliant strategy used by Javitch's group (Han, Moreira, Urizar, Weinstein, & Javitch, 2009) to study changes in the efficacy of dopamine receptor homomers. They stably expressed aequorin in Flp-In T-REx-293 cells and used the luminescence produced by this protein in the presence of the substrate coelenterazine as a readout of $G_{\alpha q}$-induced intracellular calcium redistribution. Activation of $G_{\alpha q}$-coupled receptors in these cells induced a robust luminescence signal, while activation of $G_{\alpha i}$ coupled to the dopamine D_2 receptor did not lead to luminescence. To couple D_2 activation to the luminescence readout in these cells, they expressed a chimeric, pertussis toxin (PTX)-resistant $G_{\alpha q}$ ($G_{\alpha qi5}$) that could signal from $G_{\alpha i}$-coupled receptors (Conklin, Farfel, Lustig, Julius, & Bourne, 1993). Then, they constructed a fusion D_2-$G_{\alpha qi5}$ protein in which the short-tethered $G_{\alpha qi5}$ protein was not able to couple to the fused D_2 receptor and prevented the coupling of heterologous $G_{\alpha qi5}$ proteins. Conversely, coexpression of wild-type D_2 receptor (termed "Protomer A") and D_2-$G_{\alpha qi5}$ (Protomer B), each of which was incapable of signaling when expressed alone in aequorin-transfected Flp-In T-REx-293 cells, led to robust agonist-mediated receptor activation. Further modifying the wild-type D_2 or the mutant D_2-$G_{\alpha qi5}$ receptor, they could selectively activate one of the two protomers of the D_2/D_2-$G_{\alpha qi5}$ complex. A further addition to this strategy that we could suggest is the use of RASSLs in order to stimulate the A or B protomers alone or together when desired.

A similar strategy can be devised to study how receptor homomers recruit β-arrestin or other scaffolding proteins. It should be kept in mind, though, that mutant receptors could eventually modify the dimerization interface and signaling ability, which is why the interpretation of results should be based on multiple approaches and not just on a single mutation.

3.2. Signal transduction in GPCR heteromerization

Receptor heteromerization has been shown to result in modifications in coupling efficacy or changes in coupling specificity (Rozenfeld & Devi, 2010). Unless we are working with receptor subtypes, for which selective ligands may not be available, the pharmacology of the interacting protomers is usually different enough to specifically target the two protomers in the heteromer. Few studies, at the moment, have reported changes in coupling specificity: opioid MOR and DOR switched from $G_{\alpha i}$ coupling, when individually expressed, to $G_{\alpha z}$ coupling when expressed together; dopamine D_1 and D_2 switched from $G_{\alpha s/olf}$ (D_1) and $G_{\alpha i}$ (D_2), when individually expressed (or stimulated), to a $G_{\alpha q/11}$-mediated response when coexpressed and stimulated together; finally, heteromerization between cannabinoid CB_1 and dopamine D_2 apparently involves a switch in CB_1 coupling from $G_{\alpha i}$ to $G_{\alpha s}$ upon coactivation of D_2-coupled $G_{\alpha i}$. More common are reports that show changes in coupling efficacy when one receptor is stimulated in the presence of agonists or antagonists for the cognate receptor.

In the first case, experiments should be straightforward as only when the two interacting receptors are coexpressed should the coupling activity pertinent to the stimulation of the heteromers become evident. Usually, this occurs when both protomers of the heteromer are stimulated. Interpretation of these data is not always simple, and alternative possibilities must be considered and excluded before definitively attributing the coupling shift to receptor heteromerization. For instance, many GPCRs couple to multiple G proteins, but with different affinity. Stimulation of one of the coexpressed receptors can lead to the dilution of a specific G protein and favor the activation of other G proteins by a promiscuous cognate receptor. This can easily occur when the two interacting receptors couple to the same G protein. Strategies like those illustrated to study receptor homomerization can eventually be used to differentiate between a heteromer-induced switch in coupling selectivity and G protein dilution. For instance, a fusion construct receptor-G protein that canalizes the signaling toward the G protein responsible for the new activity could help to solve the problem. Furthermore, G protein dilution could eventually be prevented by cotransfecting the G protein suspected to be diluted together with the two interacting receptors.

More often, changes in coupling efficiency have been reported when two receptors are coexpressed together. As a model system, we will report the results we have obtained with the cotransfection of dopamine D_2 and D_3 receptors (Scarselli et al., 2001). Dopamine D_2 and D_3 receptors share a high degree of sequence homology, but they show substantial differences in their binding profiles and in their coupling to cellular transduction mechanisms, with the D_2 receptor strongly coupled to $G_{\alpha i/o}$ protein and the D_3 receptor weakly coupled to this G protein.

In our system, we completely knocked out the activity of the D_3 receptor by cotransfecting a chimeric adenylyl cyclase ACV/VI together with this receptor in COS-7 cells. In cells cotransfected with D_3 and ACV/VI, agonists did not inhibit the forskolin-stimulated cAMP increase; conversely, in cells cotransfected with D_2 and ACV/VI, all agonists strongly inhibited forskolin-induced cAMP increases. In the absence of any alteration in efficacy (maximal effect), the potency of some agonists was markedly amplified in cells cotransfected with D_3, D_2 receptors and ACV/VI as compared to those transfected with D_2 and ACV/VI. These results support the concept that in cells cotransfected with D_3 and D_2 receptors, the agonists recognize D_3/D_2 heterodimers with higher potency.

Several other studies have reported that heteromerization alters receptor activity and G protein activation. For instance, MOR association with DOR decreases MOR activity in response to selective agonists (Gomes et al., 2000), while MOR association with the α_2 adrenergic receptor increases its activity (Rios, Gomes, & Devi, 2006). Furthermore, heteromerization between the angiotensin II (AngII) type 1 (AT_1) receptor and the bradykinin B_2 receptor increases the G protein-coupling of AT_1 in response to its ligand AngII (AbdAlla, Lother, & Quitterer, 2000).

Two possibilities can explain these results: (a) allosteric modification in the affinity of agonists at receptor A when receptor B is present either in the bound or unbound state or (b) a conformational change in the heteromer complex G protein interface with increased or decreased G protein-coupling.

By knocking down the function or the binding activity of one of the two protomers in heteromers, these two alternatives can be analyzed in detail. For instance, modifications in agonist affinity at receptor A when receptor B is stimulated should be prevented using a binding-deficient mutant receptor B. Furthermore, agonists and antagonists for receptor B should have an opposite influence on the activity stimulated by receptor A. Conversely, if these modifications are independent of the stimulation of receptor B, they should persist. Nonfunctional receptors will help with understanding whether both protomers are required to provide the increase (or decrease) in coupling efficacy in the heteromer or whether the G protein couples preferentially to one of the two protomers and heteromerization modifies the coupling efficacy of the coupled protomer.

4. DOMAIN SWAPPING IN RECEPTOR OLIGOMERIZATION

To complete our discussion on GPCR oligomerization and signaling, we will mention the phenomenon of domain swapping and how this can be studied. This phenomenon was originally described for GPCRs by Maggio, Vogel, and Wess (1993) using two reciprocally inactive chimeric α_2 adrenergic/muscarinic M_3 receptors in which the first five transmembrane regions of one receptor were joined to the last two transmembrane regions of the other receptor. The cotransfection of the two chimeric receptors resulted in the rescue of muscarinic and adrenergic binding and functional activity. Other studies have reported the rescue of nonfunctional receptor mutants by domain swapping in recombinant cell lines and in transgenic animals *in vivo* (Conn, Ulloa-Aguirre, Ito, & Janovick, 2007), but the most prominent effect that probably occurs under physiological conditions and can be explained by domain swapping is transactivation. This effect has been described for glycoprotein hormone receptors: luteinizing hormone (LH) and follicle-stimulating hormone (FSH) receptors (Jeoung, Lee, Ji, & Ji, 2007). These two receptors, together with the thyrotrophin-stimulating hormone receptor, the other component of the glycoprotein hormone receptor family, have a long N-terminal extracellular extension responsible for high-affinity hormone binding and a seven transmembrane domain responsible for signal transduction. Jeoung et al. (2007) have demonstrated that the N-terminal domain tethered to one receptor can bind the transmembrane domain of another receptor and activate it. This was achieved by constructing binding and functionally deficient receptor mutants, in which modifications were introduced respectively in the N-terminal segment of an LH or FSH receptor and a transmembrane domain of another LH or FSH receptor. When these nonfunctional receptors were cotransfected together, they regained the ability to signal in presence of the agonists. The only possible explanation of these data is that the wild-type N-terminal extracellular domain of receptor A transactivated the wild-type transmembrane domain of receptor B.

The design of these types of experiments requires knowledge of the overall structure of the receptor to be modified in order to prepare binding and functionally deficient mutants that maintain correct folding and are still able to reach the plasma membrane. An exhaustive list of receptor mutants with their binding and functional characteristics that may be helpful in designing the right strategy for these experiments can be found in the G protein-coupled receptor database at the Web site http://www.gpcr.org/7tm/.

5. CONCLUDING REMARKS

A major limitation of the experiments described above is that, at least in routine binding experiments, we do not know the ratio of receptors in the monomeric, homomeric, or oligomeric forms. Furthermore, we do not have a means of altering this ratio, having preparations with only monomers or homooligomers. If we assume the existence of a dynamic equilibrium between monomers and dimers (Kasai et al., 2011), cells expressing different levels of receptors should probably behave differently with regard to the binding experiments illustrated above. Cells expressing a low number of receptors should have a relatively low number of homooligomers, as receptors would have to travel longer to meet their partners; on the contrary, the ratio of homooligomers should increase in cells expressing a high number of receptors. Furthermore, the relative affinity of the interacting protomers will affect the level of homo/heteromers expressed.

This limitation still arouses criticism about the real interpretation of binding data. For example, potential allosteric interaction of ligands at different sites of the same receptor monomer could explain some of the data published in the literature. Furthermore, additional GPCR-interacting proteins could eventually explain the data. This is why multiple experimental approaches are required if equal, two-way, cross-receptor interactions within a GPCR homomer or heteromer, at the level of binding or function, are to be unequivocally demonstrated (Birdsall, 2010).

REFERENCES

AbdAlla, S., Lother, H., & Quitterer, U. (2000). AT1-receptor heterodimers show enhanced G protein activation and altered receptor sequestration. *Nature*, *407*, 94–98.

Alvarez-Curto, E., Pediani, J. D., & Milligan, G. (2010). Applications of fluorescence and bioluminescence resonance energy transfer to drug discovery at G protein coupled receptors. *Analytical and Bioanalytical Chemistry*, *398*, 167–180.

Alvarez-Curto, E., Prihandoko, R., Tautermann, C. S., Zwier, J. M., Pediani, J. D., Lohse, M. J., et al. (2011). Developing chemical genetic approaches to explore G protein-coupled receptor function—Validation of the use of a receptor activated solely by synthetic ligand (RASSL). *Molecular Pharmacology*, *80*, 1033–1046.

Armstrong, D., & Strange, P. G. (2001). Dopamine D_2 receptor dimer formation: Evidence from ligand binding. *The Journal of Biological Chemistry*, *276*, 22621–22629.

Birdsall, N. J. M. (2010). Class A GPCR heterodimers: Evidence from binding studies. *Trends in Pharmacological Sciences*, *31*, 499–508.

Conklin, B. R., Farfel, Z., Lustig, K. D., Julius, D., & Bourne, H. R. (1993). Substitution of three amino acids switches receptor specificity of $G_q\alpha$ to that of $G_i\alpha$. *Nature*, *363*, 274–276.

Conn, P. M., Ulloa-Aguirre, A., Ito, J., & Janovick, J. A. (2007). G protein-coupled receptor trafficking in health and disease: Lessons learned to prepare for therapeutic mutant rescue in vivo. *Pharmacological Reviews*, *59*, 225–250.

Franco, R., Casadó, V., Ciruela, F., Mallol, J., Lluis, C., & Canela, E. I. (1996). The cluster-arranged cooperative model: A model that accounts for the kinetics of binding to A_1 adenosine receptors. *Biochemistry*, *35*, 3007–3015.

George, S. R., Fan, T., Xie, Z., Tse, R., Tam, V., Varghese, G., et al. (2000). Oligomerization of mu- and delta-opioid receptors. Generation of novel functional properties. *The Journal of Biological Chemistry*, *275*, 26128–26135.

Gomes, I., Jordan, B. A., Gupta, A., Trapaidze, N., Nagy, V., & Devi, L. A. (2000). Heterodimerization of μ and δ opioid receptors: A role in opiate synergy. *The Journal of Neuroscience*, *20*, RC110.

Han, Y., Moreira, I. S., Urizar, E., Weinstein, H., & Javitch, J. A. (2009). Allosteric communication between protomers of dopamine class A GPCR dimers modulates activation. *Nature Chemical Biology*, *5*, 688–695.

Jeoung, M., Lee, C., Ji, I., & Ji, T. H. (2007). Trans-activation, cis-activation and signal selection of gonadotropin receptors. *Molecular and Cellular Endocrinology*, *260*, 137–143.

Jordan, B. A., & Devi, L. A. (1999). G protein-coupled receptor heterodimerization modulates receptor function. *Nature*, *399*, 697–700.

Kara, E., Lin, H., & Strange, P. G. (2010). Co-operativity in agonist binding at the D_2 dopamine receptor: evidence from agonist dissociation kinetics. *Journal of Neurochemistry*, *112*(6), 1442–1453.

Kasai, R. S., Suzuki, K. G., Prossnitz, E. R., Koyama-Honda, I., Nakada, C., Fujiwara, T. K., et al. (2011). Full characterization of GPCR monomer-dimer dynamic equilibrium by single molecule imaging. *Biological Chemistry*, *192*, 463–480.

Maggio, R., Innamorati, G., & Parenti, M. (2007). G protein-coupled receptor oligomerization provides the framework for signal discrimination. *Journal of Neurochemistry*, *103*, 1741–1752.

Maggio, R., Vogel, Z., & Wess, J. (1993). Coexpression studies with mutant muscarinic/adrenergic receptors provide evidence for intermolecular 'cross-talk' between G protein-linked receptors. *Proceedings of the National Academy of Sciences of the United States of America*, *90*, 3103–3107.

McKinney, M. (1998). Practical aspects of radioligand binding. *Current Protocols in Pharmacology*, Chapter 1:Unit1.3.

Novi, F., Scarselli, M., Corsini, G. U., & Maggio, R. (2004). The paired activation of the two components of the muscarinic M_3 receptor dimer is required for induction of ERK1/2 phosphorylation. *The Journal of Biological Chemistry*, *279*, 7476–7486.

Novi, F., Stanasila, L., Giorgi, F., Corsini, G. U., Cotecchia, S., & Maggio, R. (2005). Paired activation of two components within muscarinic M_3 receptor dimers is required for recruitment of β-arrestin-1 to the plasma membrane. *The Journal of Biological Chemistry*, *280*, 19768–19776.

Pellissier, L. P., Barthet, G., Gaven, F., Cassier, E., Trinquet, E., Pin, J. P., et al. (2011). G protein activation by serotonin type 4 receptor dimers: Evidence that turning on two protomers is more efficient. *The Journal of Biological Chemistry*, *286*, 9985–9997.

Rios, C., Gomes, I., & Devi, L. A. (2006). μ-Opioid and CB_1 cannabinoid receptor interactions: Reciprocal inhibition of receptor signaling and neuritogenesis. *British Journal of Pharmacology*, *148*, 387–395.

Rozenfeld, R., & Devi, L. A. (2010). Exploring a role for heteromerization in GPCR signalling specificity. *The Biochemical Journal*, *433*, 11–18.

Scarselli, M., Annibale, P., & Radenovic, A. (2012). Cell type-specific β_2-adrenergic receptor clusters identified using photoactivated localization microscopy are not lipid raft related, but depend on actin cytoskeleton integrity. *The Journal of Biological Chemistry*, *287*, 16768–16780.

Scarselli, M., Novi, F., Schallmach, E., Lin, R., Baragli, A., Colzi, A., et al. (2001). D_2/D_3 dopamine receptor heterodimers exhibit unique functional properties. *The Journal of Biological Chemistry*, *276*, 30308–30314.

Smith, N. J., & Milligan, G. (2010). Allostery at G protein-coupled receptor homo- and heteromers: Uncharted pharmacological landscapes. *Pharmacological Reviews*, *62*, 701–725.

Springael, J. Y., Le Minh, P. N., Urizar, E., Costagliola, S., Vassart, G., & Parmentier, M. (2006). Allosteric modulation of binding properties between units of chemokine receptor homo- and hetero-oligomers. *Molecular Pharmacology*, *69*, 1652–1661.

Strange, P. G. (2005). Oligomers of D_2 dopamine receptors. Evidence from ligand binding. *Journal of Molecular Neuroscience*, *26*, 155–160.

Vivo, M., Lin, H., & Strange, P. G. (2006). Investigation of cooperativity in the binding of ligands to the D_2 dopamine receptor. *Molecular Pharmacology*, *69*, 226–235.

Wreggett, K. A., & Wells, J. W. (1995). Cooperativity manifest in the binding properties of purified cardiac muscarinic receptors. *The Journal of Biological Chemistry*, *270*, 22488–22499.

Zhu, W. Z., Chakir, K., Zhang, S., Yang, D., Lavoie, C., Bouvier, M., et al. (2005). Heterodimerization of β_1- and β_2-adrenergic receptor subtypes optimizes β-adrenergic modulation of cardiac contractility. *Circulation Research*, *97*, 244–251.

CHAPTER SEVENTEEN

Analysis of GPCR Dimerization Using Acceptor Photobleaching Resonance Energy Transfer Techniques

Marta Busnelli[*,†,1], **Mario Mauri**[‡,1], **Marco Parenti**[‡], **Bice Chini**[*,2]
[*]CNR Institute of Neuroscience, University of Milan, Milan, Italy
[†]Department of Medical Biotechnology and Translational Medicine, University of Milan, Milan, Italy
[‡]Department of Experimental Medicine, University of Milan-Bicocca, Monza, Italy
[1]These two authors equally contributed to the chapter.
[2]Corresponding author: e-mail address: b.chini@in.cnr.it

Contents

1. Introduction 312
2. Resonance Energy Transfer 313
3. Fluorescent Resonance Energy Transfer 313
   3.1 Methods 316
   3.2 Generation of FRET fusion constructs 317
   3.3 Coexpression of FRET fusion constructs in mammalian cells 318
   3.4 Immunostaining 318
   3.5 FRET assay 318
   3.6 FRET analysis 319
4. BRET 321
   4.1 Methods 322
   4.2 Generation of BRET fusion constructs 323
   4.3 Coexpression of BRET fusion constructs in mammalian cells 323
   4.4 BRET assay 324
   4.5 BRET measurement 324
   4.6 BRET result analysis 325
5. Conclusions 326
Acknowledgments 327
References 327

Abstract

The ability of GPCRs to assemble into multimeric complexes is one of the most recently studied and discussed topics for many reasons, including the possibility that GPCR assemblies show a distinct pharmacological profile offering an innovative avenue for the drug synthesis. In addition, the possible differential coupling of monomeric versus

Methods in Enzymology, Volume 521
ISSN 0076-6879
http://dx.doi.org/10.1016/B978-0-12-391862-8.00017-X

multimeric GPCRs to G proteins and other downstream partners, as well as the signaling, the regulation through desensitization and internalization, and the subcellular localization can well represent additional factors that contribute to GPCR-mediated physiopathological states.

The standard biochemical techniques used to identify GPCR interactions, such as coimmunoprecipitation, have obvious limitations owing to the use of nonphysiological buffers and detergents that disrupt the natural cell environment and biological interactions and preclude the analysis of subcellular localization and compartmentalization.

In the past decade, new biophysical proximity assays based on the resonance energy transfer (RET) between two chromophores allow the study of dimerization in intact living cells, thus proving more information on GPCR physiological roles.

In this chapter, we detail the application of two RET techniques based on fluorescence (FRET) and bioluminescence (BRET) to the study of GPCR dimerization and describe the results that can be obtained.

1. INTRODUCTION

G protein-coupled receptors (GPCRs), also known as seven-transmembrane receptors (7TMRs), constitute the largest family of signal-transducing proteins that regulate virtually all known physiological processes in mammalian cells (Lefkowitz, 2000). A substantial body of evidence suggests that GPCRs can form homo- and/or heterodimeric and/or oligomeric complexes (Terrillon & Bouvier, 2004). However, detailed knowledge on the structural mechanism(s), timing, and physiological relevance of complex formation is far from being clear for most GPCRs. This is mainly due to the limited stringency of existing experimental approaches, thus requiring the combination of multiple strategies to overcome the rather high probability of artifactual findings.

Coimmunoprecipitation was considered for years a valid assay to demonstrate the occurrence of a physical interaction between proteins, but the need for prior solubilization of cell and tissue samples using nonphysiological buffers and detergents can easily give rise to artifacts (Milligan, 2006) and does not contribute any information about cellular and/or subcellular localization of interactions. In addition, immunoprecipitating antibodies are not available for most GPCRs.

During the past decade, there has been an exponential growth of imaging techniques allowing the detection of protein–protein interactions based on the resonance energy transfer (RET) process, offering the unique advantage over biochemical approaches, to identify the occurrence of protein interactions in living cells without perturbing the physiological environment, both in terms of cell compartments and protein–lipid composition and architecture (Marullo & Bouvier, 2007).

2. RESONANCE ENERGY TRANSFER

RET occurs between an excited chromophore, known as the "donor," and a compatible "acceptor" molecule in a nonradiative manner as a result of a dipole–dipole coupling.

A variety of RET-based approaches have been developed, including fluorescent resonance energy transfer (FRET), in which both donor and acceptor are fluorescence emitter molecules, and bioluminescent resonance energy transfer (BRET), in which the donor is an enzyme catalyzing the conversion of a substrate to a bioluminescent molecule that in turn excites a fluorescent acceptor. In both cases, the RET leads to a reduced emission energy of the donor and an increased emission energy of the acceptor. Noteworthy, the efficiency (E) of RET is inversely proportional to the sixth power of the distance (R) between donor and acceptor molecules (Eq. 17.1; Angers et al., 2000; Marullo & Bouvier, 2007):

$$E = 1/\left(1 + R^6/\mathrm{RO}^6\right) \qquad [17.1]$$

where RO is the intermolecular distance producing the 50% of RET from donor to acceptor. Typically, RO is 2–6 nm and RET is effective within 10 nm distance. Interestingly, this allows the study of GPCR interactions as the diameter of the seven-helical bundle is estimated to be approximately 5 nm (Mercier, Salahpour, Angers, Breit, & Bouvier, 2002).

The efficiency of RET also depends on the overlapping between the emission spectrum of the donor and the excitation spectrum of the acceptor, the donor life span, and the relative orientations of donor and acceptor molecules. As a result, changes of RET can derive from variable distances between donor and acceptor molecules, their different relative orientations, and the sample preparation (Rodighiero et al., 2008).

3. FLUORESCENT RESONANCE ENERGY TRANSFER

The advantage of FRET is that it allows to determine not only the proximity between two chromophores but also the spatial localization where the interaction occurs within a single cell.

The most common FRET donor and acceptor couples of chromophores are, respectively, the enhanced cyan fluorescent protein (ECFP) and the enhanced yellow fluorescent protein (EYFP) variants of enhanced green fluorescent protein (EGFP). These two mutants contain a few amino acid substitutions that affect the fluorescence emission spectra from green to cyan and yellow, thus ensuring the optimal compromise required for RET experiments as mentioned earlier and shown in Fig. 17.1.

In the "classic" FRET experimental setup, called "sensitized emission," all the analysis have to be carefully corrected for a high background signal contribution due to the spectral "bleed-through" and "cross talk."

The "bleed-through" represents the fluorescence contributed by the excited donor into the acceptor detection channel. This interference depends on the overlapping emission spectra of donor and acceptor molecules. Thus, it cannot be eliminated through the adjustments of instrumental settings, but it must be experimentally calculated and then subtracted as a background signal noise.

The "cross talk," that represents the partial excitation of the acceptor directly exerted from the donor excitation wavelength, highly contributes to the background noise in the FRET setup that employs wide-field microscopes and lamps as excitation sources. Using a confocal microscope with laser excitation, this inconvenient is overcome because the experimenter can choose a single-point donor excitation wavelength that does not interfere with the acceptor excitation spectrum.

The FRET setup must include the following controls to define the background noise and the unspecific signal:

- sample with only donor chromophore, to define the bleed-through levels
- sample with only acceptor chromophore, to calculate the cross talk levels
- sample with both chromophores having no proximity, such as a not interacting GPCR receptor couple, to obtain the unspecific FRET signal given by random chromophore interactions.

The "sensitized emission" remains the common FRET technique for the analysis of interactions in signaling processes and cellular reactions (calcium levels, enzymatic reactions, etc.), although it requires appropriate controls and the check of the levels of expression of the two chromophores that have to be kept constant in all samples to ensure a reliable estimate of the physical background parameters obtained by bleed-through and cross talk analyses.

A second way to detect a RET signal between two chromophores is by means of the "acceptor photobleaching" method based on the evaluation of

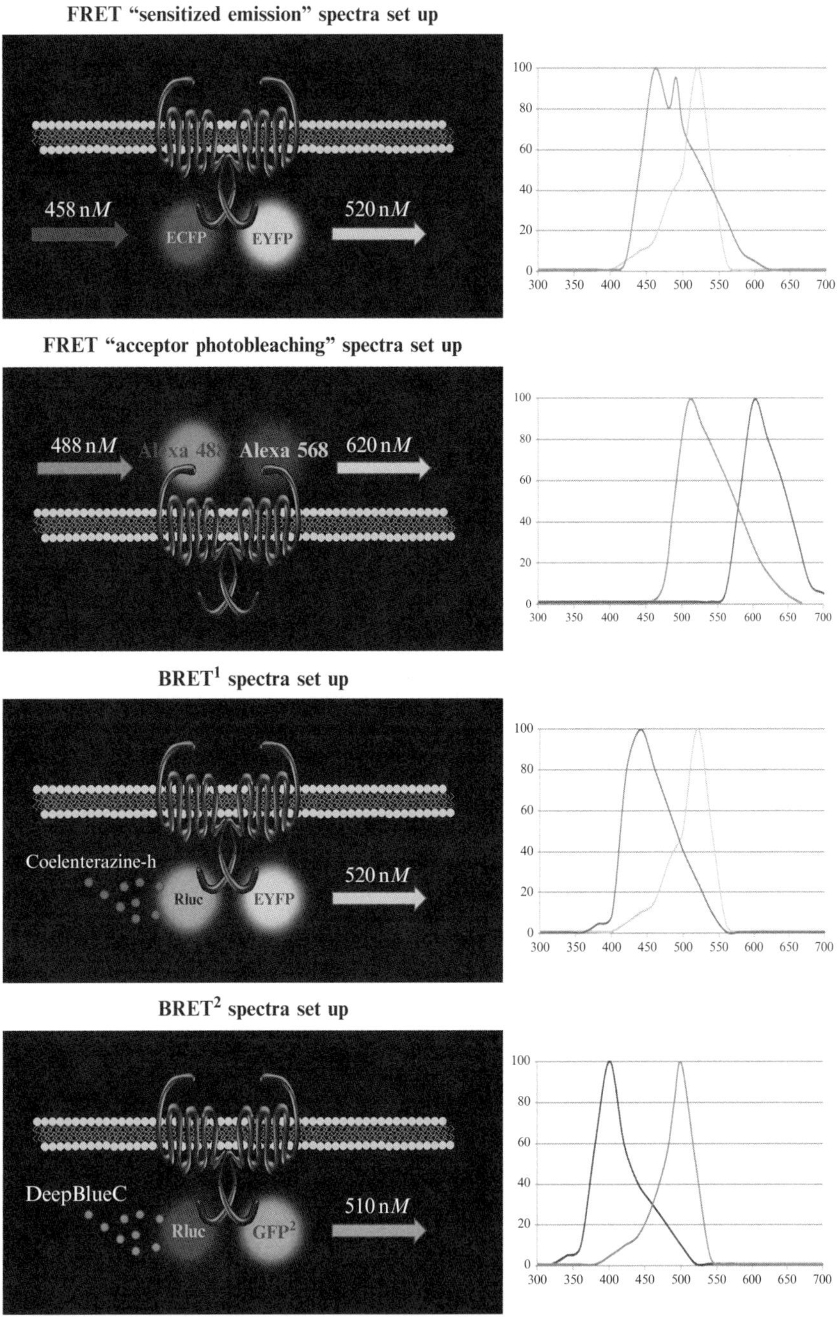

Figure 17.1 The most common couples of chromophores used for different RET techniques. (For color version of this figure, the reader is referred to the online version of this chapter.)

intensity of the signal emitted by the donor chromophore before and after the acceptor disruption by photobleaching (Konig et al., 2006). If the two chromophores were in proximity at steady state, the acceptor photobleaching would lead to an increased donor fluorescence, whereas if they were not, the acceptor photobleaching does not alter the donor fluorescence because of the absence of any previous RET. This method requires less control than the previous, but it is almost impossible to be performed on living cells because of the delay between the recordings before and after photobleaching that could lead to a loss of information due to cell motility.

For the analysis of GPCR homo-/heterodimerization, the "acceptor photobleaching" analysis offers certain advantages over the "sensitized emission" assay, as (1) it avoids the need of running the above mentioned controls and (2) the results are expressed as ratios, thus limiting the variability due to erratic measurements and analysis.

To highlight the potential advantages and limitations of the FRET technique, we describe here an experiment dealing with the thromboxane A2 receptor (TP), a rhodopsin-like, class A GPCR, which exists in two isoforms, termed α and β, whose primary sequences are identical for the first N-terminal 328 amino acids and differ in their C-terminal tails of 15 and 79 residues, respectively. It has been previously demonstrated using biochemical techniques that TPα and TPβ form homo- and heterodimers in transfected cells (Laroche et al., 2005).

The aim of the experiment is to assess whether a set of mutations in the transmembrane helix 1 (TM1) of TPs compromises dimerization. If verified, this would allow to study the cellular behavior of a monomeric TM1 variant of TP as compared to dimerizing wild-type TPs.

3.1. Methods

The HEK 293 cell line was employed to transiently coexpress the various dual combinations of wild type, and TM1 mutant TPα and TPβ were N-terminally tagged with c-myc and HA epitopes to allow the extracellular recognition by epitope-specific antibodies. This avoids the prior cell permeabilization step, and hence, it allows to analyze only the interactions occurring at the cell surface.
Materials

- pGW1 vector (British Biotechnology, Oxford, UK), an eukaryotic expression vector under the control of CMV promoter available in two variants that encode at the 5′ end of the multiple cloning sites for c-myc (*DLDYDSVQPY*) or HA (YPYDVPDVA)

- HEK293 cells (American Type Culture Collection, ATCC)
- DMEM culture medium supplemented with 10% FCS, 2 m*M* L-glutamine, 100 U/ml penicillin, and 100 μg/ml streptomycin (Sigma-Aldrich)
- Primary antibodies against c-myc and HA epitopes (Santa Cruz, cat n° sc-40 and sc-805) and Alexa Fluor® 488- and 568-conjugated secondary fluorescent antibodies (Invitrogen, cat n° 11008 and 11031)
- FuGene® transfection reagent (Roche, cat n° 11814443001)
- ImageJ software for image analysis (http://rsbweb.nih.gov/ij/) and GraphPad Prism® software for statistical analysis

Disposables

- 12-well tissue culture plates (Cellstar® Greiner Bio-One, cat n° 665-180)
- Poly-D-Lysine (Sigma-Aldrich-Aldrich, cat n° P-2636) for coating coverslips
- 19-mm-diameter coverslips suitable for laser scanning microscopy (VWR, cat n° 631-0155)

Instruments

- Zeiss LSM 710 confocal laser scanning microscope equipped with 488 and 565 nm laser lines, bleaching and time-lapse modules, and regions of interest (ROI) manager

3.2. Generation of FRET fusion constructs

Two suitable restriction sites close to the starting ATG and the stop codon have to be introduced to subclone the cDNAs into the plasmid vector containing the epitope tag sequences.

The cDNAs of TPs were subcloned from the original pCDNA3 expression vectors into the pGW1 vector containing the HA (YPYDVPDYA) or the c-myc epitope (DLDYDSVQPY, corresponding to amino acids 12–22 of human c-myc) for the recognition by anti-tag-specific antibodies. Briefly, the TP cDNAs were amplified by PCR introducing the *Asc*I and *Eco*RI recognition sites before and after the TP open reading frames. The specific oligonucleotide primers were as follows: forward primer (for both TPα and TPβ): 5′-GTTGGCGCGCCATGTGGCCCAACG-3′; reverse primer for TPα: 5′-CCCGAATTCCTACTGAGCCCGGAGC-3′; reverse primer for TPβ: 5′-CGCCGAATTCTCAATCCTTTCTGGACA-3′. The PCR reaction products were digested with *Asc*I and *Eco*RI restriction enzymes and inserted into the purified expression vector that had been opened between the *Asc*I/*Eco*RI poly-linker sites using the corresponding enzymes. The resulting constructs were verified by sequencing.

3.3. Coexpression of FRET fusion constructs in mammalian cells

Approximately 170,000 HEK 293 cells were seeded onto 19-mm-diameter poly-D-lysine-coated coverslips, placed into 12-well tissue culture plates and incubated overnight at 37 °C in a humidified atmosphere and 5% CO_2. Twenty-four hours later, cells were cotransfected with a receptor couple using FuGene® reagent following manufacturer's instructions. The expression levels of receptors were previously verified by binding studies to set the optimal cDNA concentrations that ensure a detectable signal at the cell surface maintaining the physiological levels of expression.

In our study, we transfected 100 ng of each receptor cDNA for all FRET pairs and 200 ng of cDNA for the single control transfection.

3.4. Immunostaining

Forty-eight hours after transfection, coverslips were incubated for 1 h with anti-myc or anti-HA antibodies under nonpermeabilizing conditions in DMEM plus 10% FBS at 37 °C. After treatment, cells were washed twice with PBS/Ca^{2+}/Mg^{2+} and fixed for 10 min at 25 °C with 4% (w/v) *p*-formaldehyde in 0.12 *M* sodium phosphate buffer, pH 7.4. Fixed cells were rinsed with PBS/Ca^{2+}/Mg^{2+} and incubated for 1 h with AlexaFluor® 488- or 568-conjugated secondary anti-mouse or anti-rabbit IgG antibodies in GDB buffer (0.02 *M* sodium phosphate buffer, pH 7.4, containing 0.45 *M* NaCl, 0.2% (w/v) bovine gelatine), washed with PBS/Ca^{2+}/Mg^{2+} and mounted on glass slides with a 90% (v/v) glycerol/PBS solution. Samples were analyzed using Zeiss LSM 710 confocal laser scanning microscope.

3.5. FRET assay

To perform the FRET measurements according to the laser-induced acceptor photobleaching method, the following steps were undertaken:

- Carefully define the sample cells to analyze avoiding exceedingly high or low protein expression leading to artifacts.
- Set laser and gain parameters at least 25% below the saturation limits of images in acquisition to ensure a precise detection of all the light increments due to the donor recovery after photobleaching.
- Define the two ROIs on the plasmalemma, one to be bleached (analysis ROI) and another to be used as a "sentinel ROI" during data processing.
- Acquire at least three control images of the selected cell using a line-by-line sequential mode without any averaging steps.

- Bleach the selected "analysis ROI" with an appropriate number of pulses with full power acceptor excitation laser line; the pulse number depends on the used chromophore, the laser power, and the confocal pixel dwell, defined as the period of time during which excitation and detection occur for each pixel that needs to be as fast as possible to avoid any sample overheating.
- Acquire at least seven processed images of the selected cell using a line-by-line sequential mode without any averaging steps.

In our study, cells cotransfected with wild-type or TM1 HA- and myc-tagged TP pairs were incubated for 1 h with the antibodies against HA and c-myc epitopes, fixed, and exposed to fluorochrome-conjugated secondary antibodies. All images were acquired with the laser scanning confocal microscope using a 63× oil-immersion objective and applying an additional 1.5–2× zoom. Before bleaching, three images were captured at 488 (donor wavelength) and 561 nm (acceptor wavelength) channels using the line-by-line sequential mode without any averaging steps to reduce the basal bleaching. Then, the defined ROI was bleached by 20 pulses of full power 561 nm laser line (each pulse 1.28 μs/pixel). After bleaching, seven images were acquired to obtain a full curve for analysis. The number of bleaching steps was held constant throughout each experiment.

3.6. FRET analysis

The FRET signal was quantified by measuring the average intensities of ROIs in the donor and acceptor fluorochrome channels before and after bleaching using the ImageJ software. In parallel, the distinct membrane "sentinel" ROI of approximately the same size of the bleached ROI was measured to determine any change of fluorescence intensity not due to the FRET occurring during the measurements (Fanelli et al., 2011). All the results were normalized to the background bleaching recorded within the sentinel ROI as follows:

1. calculate frame by frame the average fluorescence intensity (AI) of the bleached and sentinel ROIs;
2. calculate the percent fluorescence for each channel corresponding to donor and acceptor signals and for each frame (Eq. 17.2) as follows:

$$\text{Percent fluorescence} = (\text{AI of bleached ROI})/(\text{AI of sentinel ROI}) \times 100 \quad [17.2]$$

For each experimental condition, 20 measurements were performed from two independent transfections. Individually transfected samples were used as controls to verify that no artifacts were generated in the emission spectra throughout the experiment due to sample overheating.

The obtained normalized values represent the change of donor signal after acceptor photobleaching corrected for the chromophore basal bleaching during acquisition, and these values were plotted on a graph where the *Y*-axis is the percent donor recovery and the *X*-axis the frame number. Higher percent recovery means higher proximity between the two chromophores.

The data analysis was performed as described. The results from wild type and TM1 are shown in Fig. 17.2. Statistics was performed using the one-way ANOVA repeated test with one grouping factor.

As shown in Fig. 17.2, we observed a significant donor fluorescence recovery after acceptor photobleaching in the membrane ROI of cells expressing the wild-type TP couple, whereas ROIs of cells expressing the

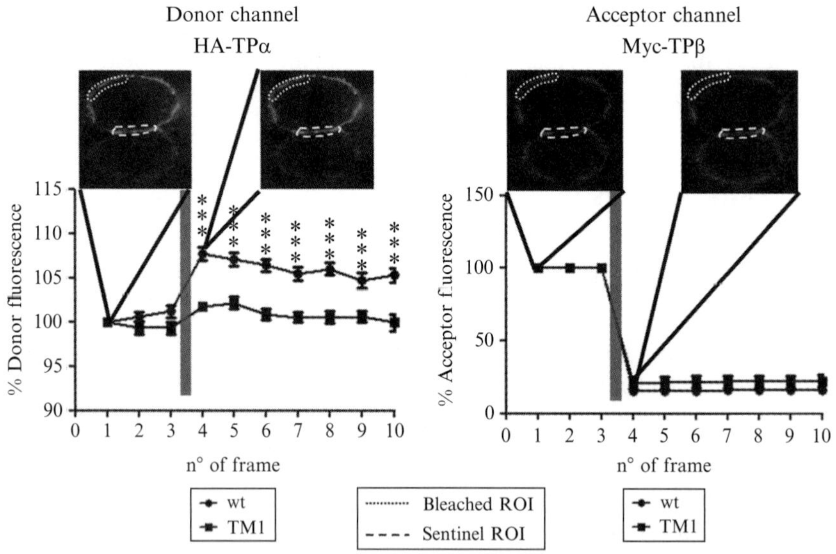

Figure 17.2 Representative results from an acceptor photobleaching experiment on transfected HEK 293 cells expressing wild-type or TM1 TP receptors. (A) Percent donor fluorescences recorded frame by frame after laser-induced acceptor photobleaching (red line) of membrane ROIs (dotted areas) of wild-type and TM1-transfected cells. (B) Acceptor fluorescence recorded before and after photobleaching of membrane ROIs (dotted areas) in cells expressing wild-type or TM1 receptors. The results were normalized to the sentinel ROIs (broken lines). ***$P<0.001$, $n=30$. (For interpretation of the references to color in this figure legend, the reader is referred to the online version of this chapter.)

TM1 couple showed no modification of donor fluorescence after acceptor photobleaching. This indicates that the donor/acceptor chromophores lie in proximity in the case of wild-type TP couple, thus suggesting dimerization, whereas no proximity of TM1-associated chromophore couple is detected, hence suggesting that mutations in helix 1 impairs dimerization. It is worth noting that detection requires the acceptor fluorescence be significantly reduced after photobleaching (Fig. 17.2, right panel).

4. BRET

BRET is a natural biophysical phenomenon occurring in marine organisms, such as the sea pansy *Renilla reniformis* and the jellyfish *Aequorea victoria*, involving the nonradiative energy transfer between donor and acceptor molecules. The energy donor is an enzyme, typically *Renilla* luciferase (Rluc), that generates bioluminescence upon oxidation of the membrane-permeable coelenterazine substrate. The energy acceptor is a fluorophore, typically a GFP variant, that absorbs light at lower wavelength and reemits it at a longer wavelength (Wu & Brand, 1994).

Initial BRET studies ($BRET^1$) employed the Rluc and EYFP energy transfer pair and benzyl-coelenterazine (coelenterazine-h) as the luciferase substrate. In this configuration, Rluc emits at 475–480 nm wavelength and EYFP at 525–530 nm. The small separation between the two spectra results in a relatively high background signal.

The second-generation BRET ($BRET^2$) was improved by using a new Rluc substrate, the coelenterazine derivative bis-deoxy-coelenterazine, called DeepBlueC, and the GFP^2 variant. In this setup, the Rluc oxidation of DeepBlueC emits at a wavelength that is substantially shifted toward blue as compared to coelenterazine-h. In addition, the donor and acceptor emissions peak at 395 and 510 nm, respectively, thus resulting in a greater spectral separation and an improved signal-to-noise ratio (Ramsay, Kellett, McVey, Rees, & Milligan, 2002).

BRET has various advantages over FRET; most notably, it does not require a light source to excite the donor molecule that causes damage to living cells, induces fluorophore photobleaching, and generates autofluorescence. However, the limitation of BRET is that the low amount of light emitted in the reaction is below the sensitivity of most current imaging equipments, hence not allowing a high-resolution analysis and single cell imaging and not providing any information about the subcellular localization of probes.

4.1. Methods

The use of BRET to study the homo-/heterodimerization of GPCRs generally involves multiple steps: (1) generation of the two receptor protomers fused to Rluc energy donor and to GFP^2 or YFP acceptor at their N- or C-terminal ends, (2) coexpression of the two fusion receptors in a suitable mammalian cell vector, (3) measurement of BRET signal at steady state and/or following receptor activation.

As an example, we report here the use of BRET to monitor the homodimerization of oxytocin receptor (OTR) in living cells.

Materials

- Receptor template DNA (entire coding sequence)
- Codon humanized pRluc vector (Perkin Elmer, cat n° 6310220) containing the gene coding for Rluc
- Codon humanized $pGFP^2$ vector (Perkin Elmer, cat n° 6310220) containing the gene encoding the blue shift variant of *Aequorea victoria* GFP
- HEK 293 cells obtained from ATCC
- HEK 293 culture medium: DMEM supplemented with 10% FCS, 2 m*M* L-glutamine, 100 U/ml penicillin, and 100 μg/ml streptomycin (Sigma-Aldrich)
- FuGene® transfection reagent (Roche, cat n° 11814443001)
- BRET assay buffer: phosphate-buffered saline (PBS), pH 7.4, supplemented with 0.5 m*M* $MgCl_2$ and 0.1% (w/v) glucose
- Coelenterazine-h (Invitrogen, cat n° c-6780) and DeepBlueC® (Perkin Elmer), 1 m*M* stock in ethanol, diluted to 50 μ*M* in PBS just prior to use
- Receptor agonist and antagonist
- GraphPad Prism® or Excel® softwares for data analysis

Disposables

- 6-well tissue culture plates (Cellstar® Greiner Bio-One, cat n° 657-160)
- black polystyrene 96-well microplates, with white wells (for BRET measurements, Black/White Isoplate-96 Black Frame White Well; Perkin Elmer, cat n° 6005030)
- white clear-bottom polystyrene 96-well microplates (for fluorescence measurements, View-plate 96, Perkin Elmer, cat n° 6005181) with white opaque-backed adhesive tape for 96-well plates (for luminescence measurement, Perkin Elmer, cat n° 651-99)

Equipment

- plate reader for luminescence, fluorescence, and BRET detection (Infinite F500, Tecan)

4.2. Generation of BRET fusion constructs

Amplify by PCR the entire coding sequence of the receptor of interest without the stop codon, and after digestion with the appropriate enzymes, subclone it in frame with GFP^2 or EYFP and Rluc to generate a fusion protein.

In our study, we generated the OTR–GFP^2 fusion construct from the cDNA of the human OTR cloned in the pEGFP-N3 vector (Clontech) using the proofreading Platinum® Pfx DNA polymerase (Invitrogen) and specific sense (5′-CAAAAGCTTATGGAGGGCGCGCTCGCAG-3′) and antisense (5′ GACGAGGGTCGGTAGGTGCCCTAGGTTTG-3′) primers harboring unique *Hin*dIII and *Bam*HI restriction sites. The resulting PCR fragment was cloned in frame into the *Hin*dIII/*Bam*HI sites of pGFP^2-N2 vector (cat no. 6310240, Perkin Elmer). In the case of OTR–Rluc, we used the same construct described by (Terrillon et al., 2003) in which the *Age*I site was created at the beginning of the Rluc coding region, and the *Age*I/*Xba*I fragment was inserted in frame into pRK5-OTR vector.

4.3. Coexpression of BRET fusion constructs in mammalian cells

Seed approximately 300,000 HEK293 cells into each of a six-well tissue culture plate and incubate overnight at 37 °C in a humidified atmosphere with 5% CO_2.

Twenty-four hours later, perform cell transfection using the FuGene® reagent according to the manufacturer's instructions to express the Rluc-fused receptor alone, or constant amounts of Rluc-fused receptor and increasing concentrations of GFP^2-fused receptor cDNAs for saturation curves. The expression level of Rluc used in saturation curves should correspond to the lowest amount of protein needed to get a detectable BRET signal. To verify the specificity of the interaction between the donor and acceptor molecules, cotransfect the donor receptor–Rluc fusion together with increasing concentrations of the soluble GFP^2 or with increasing concentrations of a noninteracting GPCR known to have the same localization of your receptor of interest.

In our study, we transfected 0.25 μg OTR–Rluc alone, and for titration experiments, we performed several independent transfections using a constant amount of OTR–Rluc (500 ng) and increasing concentrations (2.5, 5, 10, 20, 40, 80, 160, 320, 640, 1280 ng/well in six-well plate) of OTR–GFP^2 cDNAs or the soluble GFP^2 used as a negative control.

4.4. BRET assay

4.4.1 Sample preparation

Forty-eight hours posttransfection, detach cells, wash twice with PBS, and centrifuge at 800 × *g* for 5 min at room temperature, and then resuspend the cell pellets in 100 μl of BRET buffer. For each transfection condition, (1) determine the number of cells by measuring the protein concentration of an aliquot of cell suspension and calculating the number of cells/ml from a preestablished standard curve of protein versus cell number and (2) adjust the cell concentration to 1×10^6 cell/ml.

4.4.2 Measurement of total fluorescence and luminescence to assess protein expression levels

Distribute 90 μl cell suspension into each well of a white-wall, clear-bottomed 96-well plate. Measure total fluorescence levels (GFP^2_{Total}) by exciting at 400 nm and recording the GFP^2 emission at 510 nm. Calculate the specific GFP^2 signal by subtracting the background fluorescence signal ($GFP^2_{Background}$) measured in cells transfected with the receptor–Rluc fusion construct alone as follows (Eq. 17.3):

$$GFP^2 = GFP^2{}_{Total} - GFP^2{}_{Background} \qquad [17.3]$$

Use a white opaque-backed adhesive tape to cover the bottom of the same plate used to measure fluorescence, and add 10 μl of freshly diluted 50 μ*M* coelenterazine-h, wait for 8 min, and measure total luminescence.

Calculate the relative expression of fusion proteins for each data point as GFP^2/Rluc ratio.

4.5. BRET measurement

Add 80 μl of the cell suspension to each well of a black-wall, white-bottom, 96-well plate (Perkin Elmer), and if required, add the test compounds (such as agonist or antagonist) for the desired period (usually 1–30 min).

In our study, we investigated the effect of the ligand binding and the receptor activation on dimer formation. We stimulated the cells with the OTR agonist dLVT (Chini et al., 2003) for 2 min at RT.

After the cell stimulation, add 10 μl of freshly prepared 50 μ*M* DeepBlueC (Perkin Elmer) and immediately read BRET signal using a multiplate reader that allows the sequential integration of signals detected with Rluc filter (370–450 nm) and GFP^2 filter (510–540 nm).

Calculate BRET ratio as follows (Eq. 17.4):

$$\mathrm{BRET}^2 = (\text{emission of } \mathrm{GFP}^2 \text{ at } 510\,\mathrm{nm})/(\text{emission Rluc at } 400\,\mathrm{nm}) \quad [17.4]$$

Likewise, calculate the background BRET signal in cells transfected with the receptor–Rluc fusion construct alone (Eq. 17.5):

$$\mathrm{BRET}^2{}_{\mathrm{Background}} = (\text{emission of } \mathrm{GFP}^2 \text{ at } 510\,\mathrm{nm})/(\text{emission Rluc at } 400\,\mathrm{nm}) \quad [17.5]$$

Correct the values by subtracting the background signal (Eq. 17.6):

$$\text{Net BRET} = \mathrm{BRET}^2 - \mathrm{BRET}^2{}_{\mathrm{Background}} \quad [17.6]$$

4.6. BRET result analysis

Plot net BRET values as a function of the GFP^2/Rluc fusion protein ratio, and fit the data using a nonlinear regression equation by assuming a single binding site (GraphPad Prism®).

In the case of a specific interaction, the BRET signal rapidly rises with increasing expression of the receptor–GFP^2 fusion protein and reaches a plateau when all the available receptor–Rluc fusion proteins are saturated with their receptor–GFP^2 counterparts. The maximal level (BRET_{max}) should be a function of the total number of dimers formed and the distance between donor and acceptor within the dimer.

A BRET_{50} value can be calculated from a BRET saturation curve and represents the acceptor/donor ratio required to obtain 50% of the BRET_{max}. The BRET_{50} should provide the relative affinity of the acceptor and donor fusion proteins. A decrease in the affinity between the test proteins should result in a right shifted curve and in an increased BRET_{50} value, which may or may not be accompanied by a change in the BRET_{max} signal. Changes in the conformation of the two proteins that do not alter the affinity of their association may result in changes of BRET_{max} without affecting BRET_{50}.

If the obtained BRET signal results from random collision between the energy donor and acceptor, "the by-stander BRET" signal should increase almost linearly with the increase of the acceptor fusion protein.

In our study, as shown in Fig. 17.3, we observed a significant BRET signal for the OTR–Rluc/OTR–GFP^2 pair and the BRET saturation curves generated behaved as hyperbolic function reaching a saturation level and

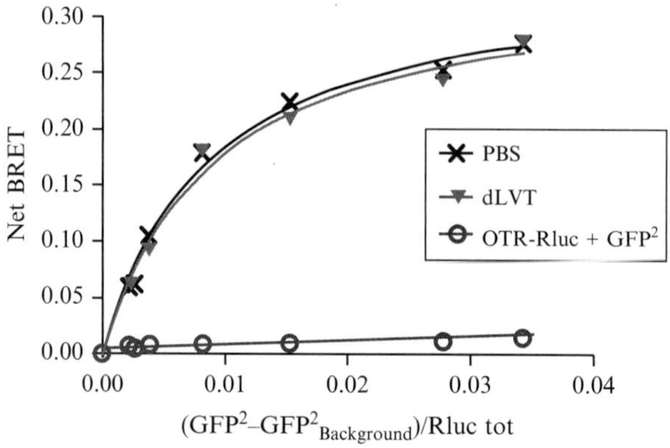

Figure 17.3 Effect of ligand binding and receptor activation on oxytocin receptor homodimerization. $BRET^2$ saturation curves were performed in HEK 293 cells coexpressing increasing concentrations of OTR–GFP^2 or soluble GFP^2 (OTR–Rluc + GFP^2 used as negative control) with constant amounts of OTR–Rluc. The specific formation of OTR–Rluc/OTR–GFP^2 dimers in basal condition (PBS) is confirmed by the NET BRET signal that rapidly rises upon increasing expression of the OTR–GFP^2 fusion protein. Treatment with dLVT (100 n*M*) for 2 min at RT does not change OTR homo-dimer formation. The random collisions between OTR–Rluc and the soluble GFP^2 (OTR–Rluc + GFP^2) in basal conditions lead to a linear relationship between the NET BRET and the GFP^2 amount. (For color version of this figure, the reader is referred to the online version of this chapter.)

confirming the previous finding that OTRs can indeed form homodimers (Devost & Zingg, 2003; Terrillon et al., 2003). Coexpression of OTR–Rluc with soluble GFP^2 led to a marginal signal that increased linearly with an increasing amount of GFP^2 added. The stimulation with dLVT (100 n*M*) did not promote any consistent change in the BRET saturation curve, indicating that the ligand binding and the change in conformation due to receptor activation do not affect OTR oligomerization state.

5. CONCLUSIONS

Despite the potentialities of the techniques described above, some caution is warranted in the interpretation of results from FRET and BRET assays for a variety of reasons. All of these approaches are proximity indicators and do not provide a direct proof of physical interaction between proteins. Finally, such methods do not define the total number of interacting proteins, leaving open the possibility of higher order oligomers.

ACKNOWLEDGMENTS

This study was supported by the Cariplo Foundation (Grant 2008.2314 to B. C. and M. P.), the Regione Lombardia (Progetto TerDisMental, ID 16983—Rif. SAL-50 to B. C.), and the Italian Ministry for University and Research (MiUR-FIRB Grant no. RBIN04CKYN to M. P.).

REFERENCES

Angers, S., Salahpour, A., Joly, E., Hilairet, S., Chelsky, D., Dennis, M., et al. (2000). Detection of beta 2-adrenergic receptor dimerization in living cells using bioluminescence resonance energy transfer (BRET). *Proceedings of the National Academy of Sciences of the United States of America*, *97*, 3684–3689.

Chini, B., Chinol, M., Cassoni, P., Papi, S., Reversi, A., Areces, L., et al. (2003). Improved radiotracing of oxytocin receptor-expressing tumours using the new [111In]-DOTA-Lys8-deamino-vasotocin analogue. *British Journal of Cancer*, *89*, 930–936.

Devost, D., & Zingg, H. H. (2003). Identification of dimeric and oligomeric complexes of the human oxytocin receptor by co-immunoprecipitation and bioluminescence resonance energy transfer. *Journal of Molecular Endocrinology*, *31*, 461–471.

Fanelli, F., Mauri, M., Capra, V., Raimondi, F., Guzzi, F., Ambrosio, M., et al. (2011). Light on the structure of thromboxane Areceptor heterodimers. *Cellular and Molecular Life Sciences*, *68*, 3109–3120.

Konig, P., Krasteva, G., Tag, C., Konig, I. R., Arens, C., & Kummer, W. (2006). FRET-CLSM and double-labeling indirect immunofluorescence to detect close association of proteins in tissue sections. *Laboratory Investigation*, *86*, 853–864.

Laroche, G., Lepine, M. C., Theriault, C., Giguere, P., Giguere, V., Gallant, M. A., et al. (2005). Oligomerization of the alpha and beta isoforms of the thromboxane A2 receptor: Relevance to receptor signaling and endocytosis. *Cellular Signalling*, *17*, 1373–1383.

Lefkowitz, R. J. (2000). The superfamily of heptahelical receptors. *Nature Cell Biology*, *2*, E133–E136.

Marullo, S., & Bouvier, M. (2007). Resonance energy transfer approaches in molecular pharmacology and beyond. *Trends in Pharmacological Sciences*, *28*, 362–365.

Mercier, J. F., Salahpour, A., Angers, S., Breit, A., & Bouvier, M. (2002). Quantitative assessment of beta 1- and beta 2-adrenergic receptor homo- and heterodimerization by bioluminescence resonance energy transfer. *The Journal of Biological Chemistry*, *277*, 44925–44931.

Milligan, G. (2006). G-protein-coupled receptor heterodimers: Pharmacology, function and relevance to drug discovery. *Drug Discovery Today*, *11*, 541–549.

Ramsay, D., Kellett, E., McVey, M., Rees, S., & Milligan, G. (2002). Homo- and hetero-oligomeric interactions between G-protein-coupled receptors in living cells monitored by two variants of bioluminescence resonance energy transfer (BRET): Hetero-oligomers between receptor subtypes form more efficiently than between less closely related sequences. *The Biochemical Journal*, *365*, 429–440.

Rodighiero, S., Bazzini, C., Ritter, M., Furst, J., Botta, G., Meyer, G., et al. (2008). Fixation, mounting and sealing with nail polish of cell specimens lead to incorrect FRET measurements using acceptor photobleaching. *Cellular Physiology and Biochemistry*, *21*, 489–498.

Terrillon, S., & Bouvier, M. (2004). Roles of G-protein-coupled receptor dimerization. *EMBO Reports*, *5*, 30–34.

Terrillon, S., Durroux, T., Mouillac, B., Breit, A., Ayoub, M. A., Taulan, M., et al. (2003). Oxytocin and vasopressin V1a and V2 receptors form constitutive homo- and heterodimers during biosynthesis. *Molecular Endocrinology*, *17*, 677–691.

Wu, P., & Brand, L. (1994). Resonance energy transfer: Methods and applications. *Analytical Biochemistry*, *218*, 1–13.

CHAPTER EIGHTEEN

Techniques for the Discovery of GPCR-Associated Protein Complexes

Avais Daulat[*,†,‡], **Pascal Maurice**[*,†,‡,1], **Ralf Jockers**[*,†,‡,2]
[*]INSERM, U1016, Institut Cochin, Paris, France
[†]CNRS UMR 8104, Paris, France
[‡]Université Paris Descartes, Paris, France
[1]Present address: FRE CNRS/URCA 3481, UFR Sciences Exactes et Naturelles, Reims, France
[2]Corresponding author: e-mail address: ralf.jockers@inserm.fr

Contents

1. Introduction 330
2. TAP of GPCR and Its Associated Protein Complexes 331
   2.1 Required materials 331
   2.2 Generation of constructs, stable cell line, and amplification of cells for large-scale purification 333
   2.3 Preparation of crude membranes and receptor solubilization 334
   2.4 TAP of IgG BD/CBP-tagged GPCRs and associated protein complexes 334
   2.5 Sample preparation for MS analysis 335
3. Alternative Methodology to Perform TAP of GPCR-Associated Protein Complexes 337
   3.1 Reagents 337
   3.2 Generation of constructs, establishment of stable cell line, and amplification of cells for large-scale purification 339
   3.3 TAP of the SBP/CBP-tagged GPCRs and associated protein complexes 339
4. Purification of GPCR-Associated Protein Complexes by Peptide Affinity Chromatography 339
   4.1 Required materials 340
   4.2 Preparation of the cell/tissue lysate 340
   4.3 Preparation of peptide columns 341
   4.4 Peptide affinity purification 342
5. Conclusion 343
Acknowledgments 344
References 344

Abstract

Biosynthesis and function of G protein-coupled receptors (GPCR) are accompanied by multiple GPCR-associated protein complexes. Despite considerable sequence diversity, all GPCRs are assumed to share a common 7-transmembrane-spanning architecture

Methods in Enzymology, Volume 521
ISSN 0076-6879
http://dx.doi.org/10.1016/B978-0-12-391862-8.00018-1

giving rise to an extracellular, intracellular, and transmembrane interface for the interaction with protein partners recognizing either linear or structural receptor epitopes. Different purification techniques have been developed in the past to identify GPCR-associated proteins other than classically known interacting proteins like heterotrimeric G proteins and β-arrestins. These techniques use either entire receptors or receptor subdomains as baits. We are presenting here two proteomic approaches developed in our laboratory to purify protein complexes interacting either with receptor subdomains from cell or tissue lysates or with entire receptors from intact cells.

1. INTRODUCTION

Membrane proteins represent a significant proportion (15–30%) of the human proteome (Almen, Nordstrom, Fredriksson, & Schioth, 2009). Many of them are involved in sensing the extracellular environment as the superfamily of G protein-coupled receptors (GPCRs), composed of approximately 900 members in humans. The recent structural resolution of several GPCRs confirmed the high degree of conservation of their overall structure (Rosenbaum, Rasmussen, & Kobilka, 2009). GPCRs are characterized by 7-transmembrane (TM) alpha-helices that are connected by three extracellular and three intracellular loops and an amino-terminal extracellular and a carboxyl-terminal intracellular tail (C-tail). These structural elements define three potential interfaces for protein–protein interactions, extracellular, intracellular, and TM, each composed of several receptor subdomains. Purification methods of GPCR-associated complexes (GAPCs) can be divided into two groups: those that are based on receptor subdomains, typically the C-tail, and those based on the entire receptor as bait. Due to the hydrophobic nature of the 7TM domain, receptors need to be extracted from their membrane environment with detergents, ideally without disturbing the integrity of the receptor and associated proteins. For a detailed comparison of different proteomic and genetic methods dedicated to the identification of GAPCs, please see Daulat, Maurice, and Jockers (2009).

We developed two complementary purification methods that will be described in detail in this chapter. The first is based on the expression of the entire GPCR tagged with a tandem affinity purification (TAP) tag (Daulat et al., 2007). The second method is an optimized peptide affinity purification protocol (Maurice et al., 2008).

2. TAP OF GPCR AND ITS ASSOCIATED PROTEIN COMPLEXES

In this section, we provide the optimal condition and procedures developed in our laboratory to purify protein complexes associated with GPCRs. TAP is a powerful method to purify, from a cellular context, intact protein complexes associated to a protein of interest. As the topology of GPCRs is complex, the TAP method is appropriate for the identification of proteins interacting with multiple intracellular and TM subdomains. The TAP protocol has been extensively used in mammalian cells since the pioneering development in yeast by Dr. Seraphin (Rigaut et al., 1999). The method consists of fusing two affinity tags to a bait protein and to perform a two-step purification under mild condition in order to preserve the integrity of protein complexes and to reduce the contamination by nonspecific proteins. Originally, the TAP method has been extensively used for large-scale purification of soluble protein complexes. In our laboratory, we adapted the TAP methodology to the specific needs of GPCRs. We were able to isolate GPCR-associated proteins for MT_1 and MT_2 melatonin receptors from HEK293 cells stably expressing C-terminally TAP-tagged receptors (Daulat et al., 2007). In this context, the TAP tag was composed of two immunoglobulin (IgG) binding units of protein A derived from *Staphylococcus aureus*, a cleavage sequence for the tobacco etch virus (TEV) protease and a calmodulin binding peptide (CBP) (Fig. 18.1A). The protein of interest and its associated protein complexes were first purified using IgG-coated agarose beads and eluted by the TEV protease that targets the Glu-X-X-Tyr-X-Gln/Ser consensus sequence. The eluted material is then immobilized on calmodulin-coated sepharose beads in the presence of calcium and finally eluted with EGTA.

2.1. Required materials

- Rabbit IgG-agarose beads
- TEV protease
- Calmodulin-sepharose beads
- Costar® Spin-X®
- Coelenterazine H
- Ultraturax T25; spectrophotometer; high-speed, refrigerated centrifuge; shaker/roller; protein quantification kit; luminometer

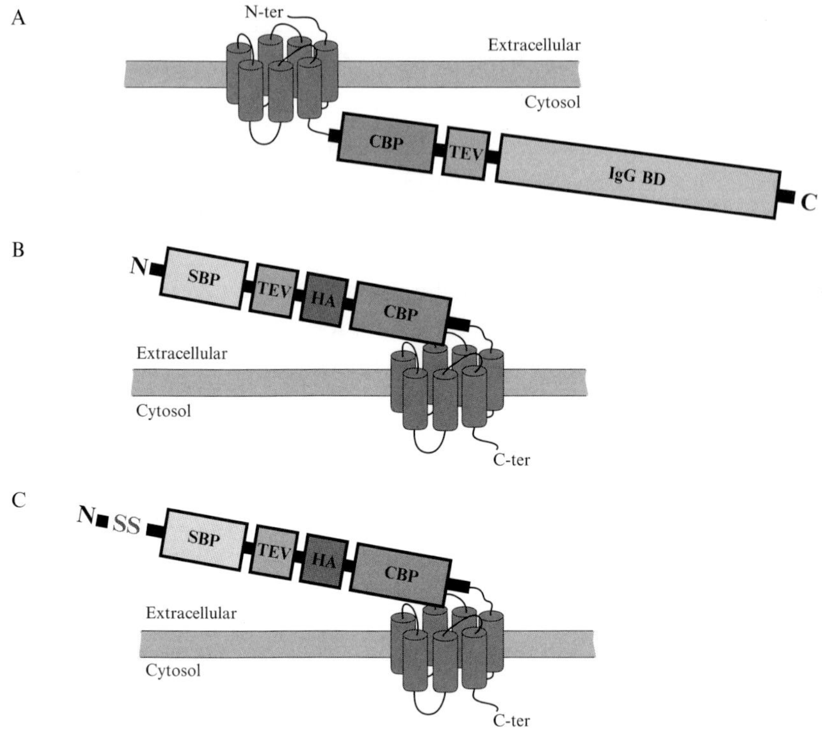

CBP tag = KRRWKKNFIAVSAANRFKKI (≈ 3.2 kDa)

SBP tag = AGELEQLRARLEHHPQGQRE (≈ 2.5 kDa)

IgG BD tag = KTAALAQHDEAVDNKFNKEQQNAFYEILHLPNLNEEQRNAFIQSLKDDPSQSANLLAE AKKLNDAQAPKVDNKFNKEQQNAFYEILHLPNLNEEQRNAFIQSLKDDPSQSANLLAEAKKLNGAQ APKVDANSAGKST (≈ 15.3 kDa)

SS = MRCSPGGVWLGLAASLLHVSLHGEGT

TEV site = ENLYFQG

HA epitope = YPYDVPDYA

Figure 18.1 Different TAP methods. Schematic representation of TAP tags used to purify GPCR-associated protein complexes. (A) IgG BD/CBP tag fused to the receptor C-tail. (B) SBP/CBP tag fused to the N-ter of the receptor. (C) Same as (B) but with the addition of an SS sequence to facilitate cell surface expression. CBP, calmodulin binding domain; HA, hemagglutinin tag; IgG BD, IgG binding domain; SBP, streptavidin binding domain; SS, signal peptide sequence; TEV, tobacco etch virus protease cleavage site. (See Color Insert.)

Buffer composition

- Lysis buffer: 5 m*M* Tris/HCl, 2 m*M* EDTA, pH 8.0
- Solubilization buffer: 10% (v/v) glycerol, 75 m*M* Tris/HCl, 2 m*M* EDTA, 5 m*M* $MgCl_2$, protease inhibitor EDTA-free cocktail, 1 m*M* orthovanadate, 2 m*M* NaF, detergent of choice, pH 8.0
- Calmodulin binding buffer (CBB): 75 m*M* Tris/HCl, 5 m*M* $MgCl_2$, 2 m*M* $CaCl_2$, detergent of choice, pH 8.0
- Calmodulin rinsing buffer: 50 m*M* NH_4HCO_3, 2 m*M* $CaCl_2$, detergent of choice, pH 8.0
- Calmodulin elution buffer: 50 m*M* NH_4HCO_3, 100 m*M* EGTA, detergent of choice, pH 8.0

2.2. Generation of constructs, stable cell line, and amplification of cells for large-scale purification

2.2.1 Generation of TAP-tagged constructs

We generated our construct by using the original tag composed of two IgG binding domains and a calmodulin binding domain separated by a proteolysis site for the TEV protease (Fig. 18.1A). The tag has been cloned either upstream or downstream of the gene of interest under the control of the CMV promoter in a pcDNA3.1 plasmid. Of note, if the tag is fused upstream of the gene of interest, it is highly recommended to add a signal peptide sequence for proper folding in the endoplasmic reticulum of the neo-synthesized GPCR and its surface expression. In our case, we used the TAP tag fused at the carboxyl-terminus of the melatonin MT_1 and MT_2 receptor. All constructs should be verified by sequencing.

2.2.2 Generation of stable cell lines expressing TAP-tagged GPCRs

We typically generate stable cell lines in HEK293 cells, a validated cell type for expression and signaling studies of GPCRs (growth medium: DMEM supplemented with 10% FBS, 4.5 g/L glucose, 100 U/mL penicillin, 0.1 mg/mL streptomycin, 1 m*M* glutamine). We usually transfect our cells with FuGENE®6 or calcium phosphate. As the pcDNA3.1 vector confers neomycin resistance to cells having successfully incorporated the plasmid into their genome, we select our stable cell lines using Geneticin® (G418) at 0.4 mg/mL. Resistant clones are isolated using serial dilutions, and the expression of the protein of interest in stable clones is checked using radioligand binding assays or Western blot analysis. Preferably, clones with expression levels close to physiological levels of the respective GPCR should be selected for further experiments. Overexpression of receptors might promote their mislocalization and favor the interaction with abundant low-affinity binding partners, which are of limited relevance under physiological conditions.

2.2.3 Amplification of cells for large-scale purification

The quantity of required cells will depend on the expression level of the TAP-tagged GPCR, the purification yield, and the amount of protein required for sample processing and mass spectrometry (MS) identification. We suggest to expand the cells from one 10-cm Petri dish to two 15-cm Petri dishes followed by ten and, if necessary, forty 15-cm Petri dishes. In parallel, the same quantity of nontransfected control HEK293 cells should be prepared and processed in a similar way throughout the purification procedure.

2.3. Preparation of crude membranes and receptor solubilization

Cells are lysed and crude membranes prepared to enrich the fraction containing the GPCRs and to remove, at the same time, nuclear and cytosolic proteins. HEK293 cells expressing the TAP-tagged GPCR of interest in 15-cm Petri dishes are washed once with phosphate-buffered saline (PBS) and detached with 10 mL of Ca^{++}- and Mg^{++}-free PBS supplemented with 2 m*M* EDTA. Cells from five 15-cm Petri dishes are gently lifted by pipetting up and down, transferred into 50-mL conical tubes, and spun down by centrifugation at $500 \times g$ for 5 min at 4 °C. The pellet is resuspended with 35 mL (5 × 7 mL) of lysis buffer supplemented with protease and phosphatase inhibitors. The lysate is homogenized with a polytron at maximal speed at 4 °C for 2 × 15 s and the homogenate is centrifuged at $48{,}000 \times g$ for 45 min at 4 °C.

The pellet (crude membranes) is then resuspended in solubilization buffer. The concentration of proteins is determined and adjusted to 2 mg/mL. The detergent of choice is then added. Typically, for melatonin receptors, we use either 0.5% digitonin or 0.25% Brij96V for solubilization for 16 h at 4 °C with gentle rocking (Daulat et al., 2007). Then, the solubilized fraction (supernatant) is separated by centrifugation at $48{,}000 \times g$ for 45 min at 4 °C.

2.4. TAP of IgG BD/CBP-tagged GPCRs and associated protein complexes

The supernatant is collected and incubated with 200 μL IgG-coated agarose beads, pre-equilibrated in solubilization buffer for 6–8 h at 4 °C with gentle rocking. Beads are spun down by centrifugation at $800 \times g$ for 1 min, transferred into a 1.5-mL microcentrifuge tube, and washed three times with

1 mL of CBB by successive spin down and resuspension. The protein complex bound to IgG beads are then eluted using 400 μL of CBB supplemented with 100 U of TEV protease with gentle rocking for 16 h at 4 °C. The eluted fraction is recovered by spinning down the beads, which are then washed twice with 400 μL of CBB. The eluted fraction and the two washes are combined and passed through Costar® Spin-X® column by centrifugation to remove any traces of IgG-coated beads. The second step of purification is performed by incubating the combined TEV eluates (~1200 μL) with 50 μL of calmodulin-sepharose beads, pre-equilibrated in CBB, in the presence of 50 m*M* $CaCl_2$ for 4 h at 4 °C with gentle rocking. Finally, the beads are washed three times with 1 mL of CBB supplemented with 50 m*M* of $CaCl_2$.

Note: Before performing large-scale experiments, it is recommended to perform small pilot experiments in order to optimize solubilization conditions and the TAP procedure. If a radioligand for the GPCR of interest is available and radioligand binding maintained after receptor solubilization, the purification yield can be determined at every step of the TAP procedure. Alternatively, specific antibodies can be used to detect the GPCR of interest by Western blot at the different steps of the purification. To assess the successful copurification of GAPCs, it is recommended to follow by Western blot analysis, known interacting proteins such as G proteins throughout the TAP procedure. Furthermore, we developed an original methodology to screen for the detergent of choice and optimize the solubilization procedure. Briefly, we fused the cDNA of *Renilla* luciferase at the C-terminus of our GPCR of interest. Forty-eight hours post-transfection, crude membranes are prepared and the concentration of membrane proteins is measured and adjusted at 2 mg/mL. Different detergents at various concentrations are added to 1 mL of crude membranes incubated for 8 h at 4 °C under gentle end-over-end mixing. The solubilized fraction (supernatant) is then separated from the insoluble fraction (pellet) as described above (Section 2.3). The insoluble fraction is resuspended in 1 mL of solubilization buffer, and luciferase activity measured in both fractions by adding coelenterazine H (5 μ*M*) according to the manufacturer's instruction (Promega). The ratio of the luciferase activity between the supernatant and the pellet fraction will provide the solubilization yield.

2.5. Sample preparation for MS analysis

As the presence of detergents is not compatible with MS analysis of peptides and the GPCR of interest has, however, to be kept in its soluble form, peptides for MS analysis are generated by in-gel trypsin digestion. Proteins are eluted from calmodulin beads using NuPAGE® LDS loading buffer followed by heating at 95 °C for 10 min. The eluate is then loaded on a 1D electrophoresis NuPAGE® precast gel system (4–12%). After migration, gels are silver stained using MS-compatible protocols (Rabilloud, 1999). The proteins are fixed in-gel by 30% ethanol and 5% acetic acid (3 × 30 min). Then, the gel is washed with bi-distilled water (4 × 10 min) and sensitized by 0.02% $Na_2S_2O_3$ (1 min). After two washes with bi-distilled water (2 × 1 min), the gel is stained with 0.2% $AgNO_3$ (60 min), rinsed with bi-distilled water (1 min), and developed by 2.5% Na_2CO_3, 0.00925% formaldehyde, and 0.00125% $Na_2S_2O_3$. The staining is finally stopped by 4% Tris and 2% acetic acid (60 min), and the gel is stored in bi-distilled water containing 1% acetic acid.

Bands or spots of interest are excised from the gel, washed with 200 m*M* ammonium bicarbonate (ABC), and immediately destained according to Gharahdaghi, Weinberg, Meagher, Imai, and Mische (1999). In-gel trypsin digestion is carried out as described in a protocol based on Zip-Tip Plate (Millipore) with minor modifications. After destaining, spots are rinsed three times with bi-distilled water and shrunk with 50 m*M* ABC/50% acetonitrile (ACN) for 20 min at room temperature. Gel pieces are dried using 100% ACN for 15 min and then incubated in 50 m*M* ABC containing 10 m*M* DTT for 1 h at 56 °C. The solution is then replaced by 55 m*M* iodoacetamide in 50 m*M* ABC for 30 min in the dark at room temperature. The gel pieces are washed twice with 50 m*M* ABC and finally shrunk with 25 m*M* ABC/50% ACN for 30 min and dried using 100% ACN for 10 min. Gel pieces are rehydrated in 20 μL of 40 m*M* ABC/10% ACN (pH 8.0), containing 12.5 μg/mL Sequencing Grade Modified Trypsin (Promega). Proteins are digested overnight at 37 °C. After digestion, the gel pieces are shrunk with 100% ACN and peptides are extracted with 0.2% TFA. Peptides are then desalted using C18 phase on ZipPlate. Two elutions are performed successively to recover products from C18 phase, first with 50% ACN/0.1% trifluoroacetic acid (TFA) and then using 90% ACN/0.1% TFA. Pooled elutions are concentrated using a vacuum centrifuge and generated peptides are redissolved in 3 μL of 1% formic acid before MS analysis.

Alternatively, GAPCs can be directly trypsin digested from calmodulin beads. Although this protocol is simpler and faster, the confidence score and the amount of identified proteins are often smaller compared to results obtained with electrophoretically separated samples. After washing calmodulin beads with CBB, we rinse the beads two times with 1 mL of calmodulin rinsing buffer without detergent and resuspend the beads in 100 μL of 50 m*M* ABC, pH 8.0. Prior to trypsin digestion, we first reduce the sample by adding 25 m*M* DTT for 20 min at 50 °C and then alkylate the sample by adding 100 m*M* of iodoacetamide for 40 min at RT. Then 1 μg of sequence-grade trypsin is added and the protein mixture incubated at 37 °C for 4 h to overnight.

3. ALTERNATIVE METHODOLOGY TO PERFORM TAP OF GPCR-ASSOCIATED PROTEIN COMPLEXES

Since its first description, several variants of the TAP approach have been reported introducing new tag combinations (Drakas, Prisco, & Baserga, 2005; Gloeckner, Boldt, Schumacher, & Ueffing, 2009; Honey, Schneider, Schieltz, Yates, & Futcher, 2001; Nakatani & Ogryzko, 2003). Recently, we applied a new TAP tag developed by Dr. Angers (Angers, Li, et al., 2006; Angers, Thorpe, et al., 2006) to the purification of GAPCs. This tag consists of a combination of a streptavidin binding peptide (SBP) and a CBP (Fig. 18.1B and C). Attractive features of this tag are its small size (~10 kDa as compared to ~25 kDa for the original tag described above) and the high affinity of the SBP for streptavidin-coated beads. This tag offers two options for elution, by adding biotin or TEV protease. We suggest to first eluate with biotin, which is fast, efficient, and cheap. Eluted proteins are then further purified using calmodulin beads through the classical CBP tag.

We recently fused this tag to the amino-terminus of the MT_1 melatonin receptor. Whereas previous studies with the classical IgG BD/CBP tag showed that fusion of this large tag completely abolished cell surface expression of the receptor (Avais M Daulat and Ralf Jockers, unpublished data), we observed that the small SBP/CBP tag was compatible with cell surface expression of the receptor (Fig. 18.2). To further enhance the cell surface expression of the receptor, a signal peptide has been introduced (Fig. 18.2C). The same SBP/CBP tag has been successfully used by another group to identify interacting proteins of GPR54 (Wacker et al., 2008). In this section, we present the optimized methodology for the purification of GAPCs.

3.1. Reagents

- Fast-Flow Streptavidin Sepharose
- Calmodulin-sepharose 4B

Buffer composition

- TAP lysis buffer: 10% (v/v) glycerol, 50 m*M* HEPES–NaOH, 150 m*M* NaCl, 2 m*M* EDTA, 2 m*M* dithiothreitol, protease inhibitor

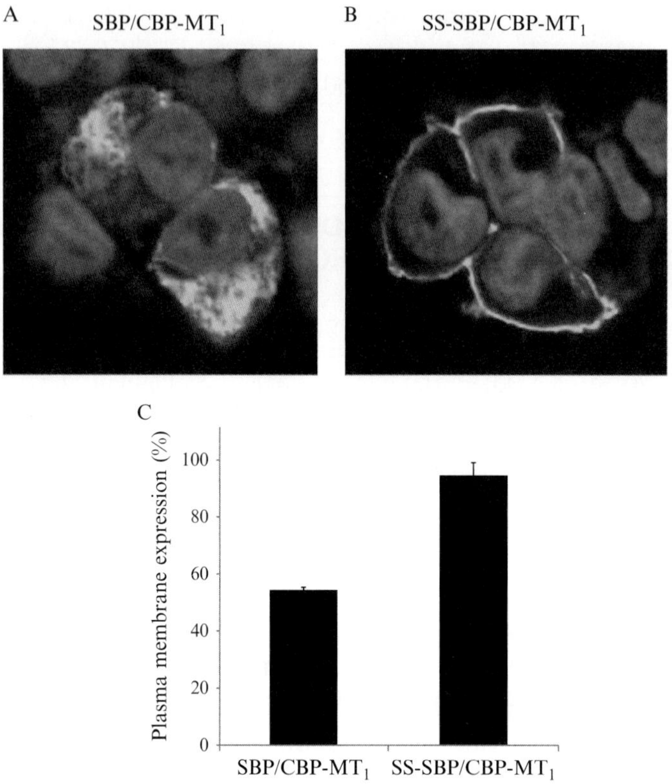

Figure 18.2 Surface expression of SBP/CBP-tagged melatonin MT_1 receptor. Expression profile of SBP/CBP-tagged MT_1 receptor without (A) or with (B) a signal peptide sequence (SS) determined by immunofluorescence in permeabilized HEK293 cells. Receptors are labeled with anti-HA antibodies and cell nuclei with DAPI. Cell surface expression of these constructs is quantified by Western in-cell nonpermeabilized cells using anti-HA antibodies. Total expression is determined in permeabilized cells and the % of surface expression calculated. (See Color Insert.)

EDTA-free cocktail, 1 m*M* orthovanadate, 2 m*M* NaF, detergent of choice, pH 8.0

- CBB: 75 m*M* Tris/HCl, 5 m*M* $MgCl_2$, 2 m*M* $CaCl_2$, detergent of choice, pH 8.0
- Streptavidin elution buffer: CBB supplemented with 50 m*M* of D-biotin
- Calmodulin rinsing buffer: 50 m*M* NH_4HCO_3, 2 m*M* $CaCl_2$, detergent of choice, pH 8.0
- Calmodulin elution buffer: 50 m*M* NH_4HCO_3, 100 m*M* EGTA, detergent of choice, pH 8.0

3.2. Generation of constructs, establishment of stable cell line, and amplification of cells for large-scale purification

The cDNA sequence of the SBP/CBP tag preceded or not by a signal sequence peptide from the α7-nicotinic acetylcholine receptor is fused upstream of the cDNA coding sequence of our GPCR of interest, that is, melatonin MT_1 receptor (Fig. 18.1B and C). The pGLUE vector (http://www.addgene.org/) is composed of an internal ribosome entry site, which allows the expression of the puromycin-resistance gene required for clonal selection of stable cell line expressing the tagged GPCR. Transfection of HEK293 cells and generation of clonal cell lines can be performed as described in the Section 2.2 by using puromycin (2 μg/mL) for the selection of positive clones.

3.3. TAP of the SBP/CBP-tagged GPCRs and associated protein complexes

Crude membranes containing SBP–CBP-tagged receptors and receptor solubilization can be performed as described in Section 2.3. The supernatant containing solubilized receptors and interacting proteins is incubated with 50 μL of streptavidin-coated sepharose beads, pre-equilibrated with TAP lysis buffer, for 4–6 h at 4 °C with gentle rocking. Beads are spun down by centrifugation at 800 × *g* for 1 min, transferred into a 1.5-mL microcentrifuge tube, and subsequently washed twice with 1 mL of TAP lysis buffer and three times with CBB by successive spin down and resuspension. The streptavidin-bound proteins are eluted with three consecutive elutions with 200 μL of CBB supplemented with 50 m*M* of D-biotin for 5 min at 4 °C, while resuspending the beads regularly. The eluted material (600 μL) is supplemented with 400 μL of CBB. The eluted fraction is then immobilized on 50 μL calmodulin-sepharose beads, pre-equilibrated with CBB, supplemented with 50 m*M* of $CaCl_2$ for 4 h at 4 °C with gentle rocking. Finally, beads are washed three times with 1 mL of CBB and processed for the MS analysis as described above (Section 2.5).

4. PURIFICATION OF GPCR-ASSOCIATED PROTEIN COMPLEXES BY PEPTIDE AFFINITY CHROMATOGRAPHY

In this section, we provide a detailed description of the procedure that uses isolated GPCR subdomains as bait to purify GAPCs from cell or tissue lysates. This peptide affinity chromatography combines chemically synthesized 6 × His-tagged baits with metal affinity immobilization on a Ni–NTA matrix. It is important to note that in contrast to the previously mentioned

TAP approach (see Sections 2 and 3), interacting partners identified with this protocol need to be confirmed in intact cells using entire GPCR to eliminate false positives.

4.1. Required materials

- Ni–NTA agarose beads
- Ultra-Turrax T25
- Spectrophotometer
- High-speed, refrigerated centrifuge
- Shaker/roller
- Protein quantification kit

Reagents

- Crushing buffer: 20 m*M* NaH_2PO_4, 2 m*M* Na_3VO_4, 10 m*M* NaF, protease inhibitor EDTA-free cocktail, pH 8.0
- Homogenization buffer: crushing buffer + 150 m*M* NaCl, detergent of choice, pH 8.0
- Washing buffer: 20 m*M* NaH_2PO_4, 2 m*M* Na_3VO_4, 10 m*M* NaF, protease inhibitor EDTA-free cocktail, 150 m*M* NaCl, detergent of choice, 20–30 m*M* imidazole, pH 8.0

4.2. Preparation of the cell/tissue lysate

This approach was initially designed to isolate GAPCs of MT_1 and MT_2 C-tails from mouse brain lysates (Maurice et al., 2008), but any other tissue or cell lysate can be used. Of note, the choice of the cell/tissue lysate is important as it determines the repertoire of potential interacting partners to be identified.

After rinsing in PBS, cells or tissues are incubated in crushing buffer and mechanically disrupted by using an Ultra-Turrax T25 at maximum speed (4 °C, 30–60 s). As cell lysates are typically acidic, the pH has to be adjusted to pH 8.0. Subsequently, the detergent of choice (in our study, 10 m*M* CHAPS) and 150 m*M* NaCl (homogenization buffer) are then added and the homogenate is solubilized for 3–5 h at 4 °C under gentle end-over-end mixing. Note that the detergent is added at last to avoid detergent foaming and poor mechanical disruption of the sample. Following solubilization, the homogenate is centrifuged at 10,000 × *g* for 1 h at 4 °C. The protein concentration of the supernatant (=cell/tissue lysate) is determined.

4.3. Preparation of peptide columns

During solubilization of the homogenate, prepare your peptide columns. Dissolve your lyophilized synthetic 6 × His-peptide at 1 mg/mL, ideally in 20 m*M* NaH_2PO_4, pH 8.0. Please note that it is important to adjust all the buffers to pH 8.0 as histidine residues have a pK_a of ≈6.0 and will become protonated at lower pH (4.5–5.3). Under these conditions, 6 × His-tagged baits will not bind to nickel ions of the Ni–NTA matrix.

If your peptide is insoluble in phosphate buffer, denaturing agents such as urea (<8 *M*) or guanidine hydrochloride (<6 *M*) can be used according to the manufacturer. Note that we successfully used 6 *M* urea in 20 m*M* NaH_2PO_4, pH 8.0 without any interference with the binding of 6 × His-tagged bait to Ni–NTA matrix.

Prepare Ni–NTA beads by performing three fast washes with the same buffer used to dissolve the 6 × His-tagged peptide (20 m*M* NaH_2PO_4, pH 8.0 ± urea or guanidine hydrochloride) to eliminate any preservatives present in the bead stock solution. Equilibrate then the Ni–NTA agarose beads (typically 20 μL/condition) with 1 mL of the same buffer as above for 15 min at 4 °C under gentle end-over-end mixing. A control condition must be added (uncoated beads) to determine the nonspecific binding of proteins present in cell/tissue lysates to the beads. Then, incubate the dissolved 6 × His-peptide with the beads for 30–90 min at 4 °C under gentle end-over-end mixing. After centrifugation (6500 × *g*, 30 s, 4 °C), take an aliquot of the supernatant and measure the absorbance to determine the amount of immobilized 6 × His-peptide. Absorbance can be measured at 280 nm if aromatic residues are present. If your peptide does not contain aromatic residues, absorbance can be measured at 214 nm, which detects absorbance by the peptide backbone. According to the manufacturer, the binding capacity of the Ni–NTA agarose beads is up to 50 mg His-tagged peptide/mL of resin. In our experiments, we used bait-saturated beads and incubated 500 μg of 6 × His-tagged peptide at 1 mg/mL with 20 μL of beads. After 90 min incubation, 300–350 μg was immobilized. If the immobilization rate is not satisfying, verify that the binding conditions are correct by checking the pH of the peptide reconstitution buffer and ensuring that there is no chelating or reducing agent present in the buffer. Peptide-coated Ni–NTA agarose beads are then washed several times with the same buffer used to dissolve the 6 × His-tagged peptide until no free peptide is detected anymore in the washing solution. If the immobilization buffer contained denaturing agents, they have to be eliminated by progressively decreasing its concentration during the washing steps.

4.4. Peptide affinity purification

Before performing a large-scale purification for MS analysis of GAPCs, it is important to optimize peptide affinity chromatography conditions (amount of protein lysate and imidazole concentrations to be added) using noncoated beads. Typically, 2–10 mg of cell/tissue lysate at 0.5–1 mg of protein/mL and 20–30 m*M* imidazole is used. Note that it is important to add low concentrations of imidazole during the incubation of the beads with the cell/tissue lysate to prevent nonspecific binding of proteins to the Ni–NTA beads. The imidazole ring is part of the structure of histidine. It binds to the nickel ions and disrupts the binding of dispersed histidine residues in nontagged background proteins. At low imidazole concentrations, nonspecific, low-affinity binding of background proteins will be prevented, while 6 × His-tagged baits remain bound to the Ni–NTA matrix.

Incubate noncoated beads (20 μL/condition) with increasing concentrations of cell/tissue lysate in the presence of increasing concentrations of imidazole overnight at 4 °C under gentle end-over-end mixing. Wash the beads five times with 500 μL of washing buffer containing the same concentration of imidazole and elute the nonspecifically retained proteins with 50 μL of 2% SDS in PBS (95 °C, 10 min) to determine the amount of nonspecifically bound protein. Two percent SDS is typically used because of its compatibility with the protein quantification kits, but other elution buffers can be used such as imidazole, acidic pH, EDTA, and EGTA according to the manufacturer's instructions.

Once optimized, you have to validate, if possible, the functionality and specificity of the peptide column by verifying in a small-scale purification with bait-coated beads, the binding of known interacting proteins by conventional Western blotting protocols. For instance, G proteins or G protein-coupled receptor kinases can be monitored when using the C-tail of MT_1 or MT_2 receptors as bait (Maurice et al., 2008). If the 6 × His-tagged bait contains a PSD-95/Disc-large/ZO-1 (PDZ) domain-binding motif, you can also check for recruitment of known PDZ domain-containing proteins. To perform these verifications, incubate the optimized amount of cell/tissue lysate with uncoated (control) and bait-coated beads in the presence of the optimized imidazole concentration overnight at 4 °C under gentle end-over-end mixing. Then, wash the beads five times with 500 μL of washing buffer containing the optimized imidazole concentration and elute retained proteins with 50 μL of 2% SDS in PBS (95 °C for 10 min) if you want to determine the quantity of eluted proteins or directly with Laemmli buffer for 1D electrophoresis and Western blotting.

To scale up the purification procedure for MS analysis, we suggest performing multiple small-scale experiments in parallel and pool protein eluates at the end. Retained proteins are eluted with standard elution buffers for 1D and/or 2D electrophoresis and gels are stained with Coomassie Blue or silver nitrate (see Section 2.5) using conventional MS-compatible protocols. For Coomassie Blue staining, the gel is fixed and stained with 0.1% Coomassie Blue R-250 (or G-250) in 40% methanol and 10% acetic acid (60 min) and then rinsed with 25% methanol and 10% acetic acid. The gel is finally stored in bi-distilled water containing 10% acetic acid. Then excise the protein spots or bands from the gel for trypsin digestion and MS analysis.

5. CONCLUSION

Over the past two decades, proteomic approaches combined with MS-based analysis became powerful tools to identify members of protein complexes associated to a protein of interest. In early 2000, the pioneering work of Dr. Marin and colleagues paved the way for the identification of GAPCs (Becamel et al., 2002; Bockaert, Perroy, Becamel, Marin, & Fagni, 2010). Using the C-tail of serotonin 5-HT_{2C} and 5-HT_{2A} receptors, the authors identified several novel interacting proteins, mostly interacting through their PDZ domain with the last four amino acids of these receptors. Subsequently, we further invested in the development of purification approaches dedicated to the purification of GAPCs by developing two new complementary approaches, a TAP-based and a peptide affinity-based method (Daulat et al., 2007; Maurice et al., 2008).

Before deciding which technique applies best to your GPCR of interest, it is helpful to first identify the scientific question to be answered as precisely as possible, as both approaches have their advantages and limitations. The TAP approach is a powerful methodology because it allows purification of protein complexes associated with full-length GPCRs in intact cells under natural conditions (Daulat et al., 2009). However, the presence of the TAP tag might also interfere with some interactions. For example, its presence at the carboxyl-terminus of GPCRs displaying a PDZ ligand motif is likely to abolish the interaction with PDZ domain-containing proteins. Moreover, the presence of the TAP tag at the amino-terminus of the GPCR might interfere with proper receptor cell surface expression and trafficking, leading to receptor mislocalization and to interaction of the bait with non-physiological protein complexes. Furthermore, the establishment of stable cell lines in a given cellular context is time-consuming.

The major limitation of the peptide affinity chromatography approach as compared to the TAP approach is its restriction to receptor subdomains as baits. However, the protocol is quite easy to establish and extremely versatile in terms of protein identification from different cell types and tissues. Protein complexes associated to your GPCR of interest can be compared between different tissues and cell lysates at different developmental stages or after drug treatment to obtain comparative and functionally relevant data sets. Based on our experience, we highly recommend developing both methodologies in parallel to increase your chance to discover new proteins and new functions for your GPCR of interest.

ACKNOWLEDGMENTS

We thank Cédric Broussard (Plate-forme Protéomique 3P5, Université Paris Descartes, Sorbonne-Paris-Cité) for advice in MS sample pretreatment. This work was supported by grants from the Association pour la Recherche sur le Cancer (ARC, n° 5051), FP7-HEALTH-2009-241592 (European Union) EurOCHIP, Institut National de la Santé et de la Recherche Médicale (INSERM), Centre National de la Recherche Scientifique (CNRS), and Servier.

REFERENCES

Almen, M. S., Nordstrom, K. J., Fredriksson, R., & Schioth, H. B. (2009). Mapping the human membrane proteome: A majority of the human membrane proteins can be classified according to function and evolutionary origin. *BMC Biology*, 7, 50.

Angers, S., Li, T., Yi, X., MacCoss, M. J., Moon, R. T., & Zheng, N. (2006). Molecular architecture and assembly of the DDB1-CUL4A ubiquitin ligase machinery. *Nature*, *443*, 590–593.

Angers, S., Thorpe, C. J., Biechele, T. L., Goldenberg, S. J., Zheng, N., MacCoss, M. J., et al. (2006). The KLHL12-Cullin-3 ubiquitin ligase negatively regulates the Wnt-beta-catenin pathway by targeting Dishevelled for degradation. *Nature Cell Biology*, *8*, 348–357.

Becamel, C., Alonso, G., Geleotti, N., Demey, E., Jouin, P., Ullmer, C., et al. (2002). Synaptic multiprotein complexes associated with 5-HT2C receptors: A proteomic approach. *The EMBO Journal*, *21*, 2332–2342.

Bockaert, J., Perroy, J., Becamel, C., Marin, P., & Fagni, L. (2010). GPCR interacting proteins (GIPs) in the nervous system: Roles in physiology and pathologies. *Annual Review of Pharmacology and Toxicology*, *50*, 89–109.

Daulat, A. M., Maurice, P., Froment, C., Guillaume, J. L., Broussard, C., Monsarrat, B., et al. (2007). Purification and identification of G protein-coupled receptor protein complexes under native conditions. *Molecular & Cellular Proteomics*, *6*, 835–844.

Daulat, A. M., Maurice, P., & Jockers, R. (2009). Recent methodological advances in the discovery of GPCR-associated protein complexes. *Trends in Pharmacological Sciences*, *30*, 72–78.
Drakas, R., Prisco, M., & Baserga, R. (2005). A modified tandem affinity purification tag technique for the purification of protein complexes in mammalian cells. *Proteomics*, *5*, 132–137.
Gharahdaghi, F., Weinberg, C. R., Meagher, D. A., Imai, B. S., & Mische, S. M. (1999). Mass spectrometric identification of proteins from silver-stained polyacrylamide gel: A method for the removal of silver ions to enhance sensitivity. *Electrophoresis*, *20*, 601–605.
Gloeckner, C. J., Boldt, K., Schumacher, A., & Ueffing, M. (2009). Tandem affinity purification of protein complexes from mammalian cells by the Strep/FLAG (SF)-TAP tag. *Methods in Molecular Biology*, *564*, 359–372.
Honey, S., Schneider, B. L., Schieltz, D. M., Yates, J. R., & Futcher, B. (2001). A novel multiple affinity purification tag and its use in identification of proteins associated with a cyclin-CDK complex. *Nucleic Acids Research*, *29*, E24.
Maurice, P., Daulat, A. M., Broussard, C., Mozo, J., Clary, G., Hotellier, F., et al. (2008). A generic approach for the purification of signaling complexes that specifically interact with the carboxy-terminal domain of G protein-coupled receptors. *Molecular & Cellular Proteomics*, 7, 1556–1569.
Nakatani, Y., & Ogryzko, V. (2003). Immunoaffinity purification of mammalian protein complexes. *Methods in Enzymology*, *370*, 430–444.
Rabilloud, T. (1999). Silver staining of 2-D electrophoresis gels. *Methods in Molecular Biology*, *112*, 297–305.
Rigaut, G., Shevchenko, A., Rutz, B., Wilm, M., Mann, M., & Seraphin, B. (1999). A generic protein purification method for protein complex characterization and proteome exploration. *Nature Biotechnology*, *17*, 1030–1032.
Rosenbaum, D. M., Rasmussen, S. G., & Kobilka, B. K. (2009). The structure and function of G-protein-coupled receptors. *Nature*, *459*, 356–363.
Wacker, J. L., Feller, D. B., Tang, X. B., Defino, M. C., Namkung, Y., Lyssand, J. S., et al. (2008). Disease causing mutation in GPR54 reveals importance of second intracellular loop for class A GPCR function. *The Journal of Biological Chemistry*, *283*, 31068–31078.

CHAPTER NINETEEN

Expression, Purification, and Analysis of G-Protein-Coupled Receptor Kinases

Rachel Sterne-Marr[*], Alison I. Baillargeon[†], Kevin R. Michalski[†], John J.G. Tesmer[‡,1]

[*]Biology Department, Siena College, Morrell Science Center, Loudonville, New York, USA
[†]Department of Chemistry and Biochemistry, Siena College, Morrell Science Center, Loudonville, New York, USA
[‡]Life Sciences Institute and the Department of Pharmacology, University of Michigan, Ann Arbor, Michigan, USA
[1]Corresponding author: e-mail address: tesmerjj@umich.edu

Contents

1. Introduction 348
2. Expression and Purification of GRKs 349
   2.1 Previous recombinant GRK expression systems 349
   2.2 Baculovirus-mediated expression of C-terminal hexahistidine-tagged GRKs 350
   2.3 GRK2-H_6 purification 353
3. GRK Functional Assays 357
   3.1 Rho* phosphorylation assay 357
   3.2 Peptide C phosphorylation assay 358
   3.3 ^{329}G-Rho preparation and activation assay 358
   3.4 Cell-based β_2-adrenergic receptor (β_2AR) phosphorylation assay 359
   3.5 Days 1–3: DNA transfection 359
   3.6 Day 4: agonist (isoproterenol)/inverse agonist (alprenolol) treatment, cell harvest 360
   3.7 Days 4 and 5: Peptide *N*-glycosidase F (PNGase) treatment and immunoblotting 361
Acknowledgments 364
References 364

Abstract

G-protein-coupled receptor (GPCR) kinases (GRKs) were first identified based on their ability to specifically phosphorylate activated GPCRs. Although many soluble substrates have since been identified, the chief physiological role of GRKs still remains the uncoupling of GPCRs from heterotrimeric G-proteins by promoting β-arrestin binding through the phosphorylation of the receptor. It is expected that GRKs recognize activated GPCRs through a docking site that not only recognizes the active conformation

Methods in Enzymology, Volume 521
ISSN 0076-6879
http://dx.doi.org/10.1016/B978-0-12-391862-8.00019-3

of the transmembrane domain of the receptor but also stabilizes a more catalytically competent state of the kinase domain. Many of the recent gains in understanding GRK-receptor interactions have been gleaned through biochemical and structural analysis of recombinantly expressed GRKs. Described herein are current techniques and procedures being used to express, purify, and assay GRKs in both *in vitro* and living cells.

1. INTRODUCTION

Most activated G-protein-coupled receptors (GPCRs) are subject to homologous desensitization, wherein GPCR kinases (GRKs) phosphorylate serine and threonine residues in the cytoplasmic tails and loops of the receptor, which in turn recruits arrestin and blocks the binding of heterotrimeric G-proteins (Gurevich, Tesmer, Mushegian, & Gurevich, 2012). There are seven GRKs in man (GRK1–7), constituting three subfamilies. The GRK1 subfamily consists of GRK1 (rhodopsin kinase) and 7, which are expressed in the rod and cone cells of the retina. The GRK2 subfamily consists of GRK2 (βARK1) and GRK3 (βARK2), which are ubiquitously expressed. The GRK4 subfamily consists of GRK4–6. GRK5 and GRK6 are ubiquitously expressed, while GRK4 is found primarily in testes and kidneys.

All GRKs have a common catalytic core composed of what is best thought of as a protein kinase domain inserted into the loop of a regulator of G-protein signaling homology domain (Tesmer, 2009). The C-terminal regions of GRKs are their most distinguishing characteristic, but in each GRK, they contain motifs that help target the enzyme to the plasma membrane. The GRK kinase domain belongs to the PKA, PKG, and PKC (AGC) kinase family. As in other AGC kinase domains, it includes a C-terminal extension that contributes residues to the active site and is a locus for posttranslational regulation (Kannan, Haste, Taylor, & Neuwald, 2007). Unlike other AGC kinases, GRKs are not regulated via phosphorylation of the activation loop of the kinase domain. Instead, GRKs have evolved an activation mechanism that relies on docking with the cytoplasmic surface of activated GPCRs (Tesmer, 2011).

The characteristic feature of GRKs is their ability to specifically recognize and phosphorylate activated GPCRs. GRKs phosphorylate activated receptors up to 1000-fold better than the peptide substrates derived from the same receptor, indicating the existence of an allosteric docking site on GRKs (Palczewski & Benovic, 1991). When the C-terminal tail of rhodopsin is proteolytically removed, light-activated rhodopsin (Rho*) enhances the phosphorylation of soluble peptide substrates over 100-fold (Palczewski, Buczylko, Kaplan, Polans, & Crabb, 1991), indicating that GRKs interact with both the C-terminal tail and the transmembrane domain of GPCRs. The recent crystal structure of GRK6 in a closed conformation suggests that the N-terminal helix (αN), which is critical for efficient receptor phosphorylation, and the AGC kinase C-terminal extension coalesce to form a receptor-docking domain that is allosterically coupled to a closed, more active conformation of the kinase domain (Boguth, Singh, Huang, & Tesmer, 2010), consistent with biochemical data from functional analysis of GRK1 (Huang & Tesmer, 2011; Huang, Yoshino-Koh, & Tesmer, 2009).

2. EXPRESSION AND PURIFICATION OF GRKs

Structural and functional analysis requires homogenous preparations of GRKs with high-specific activity for structure determination by X-ray crystallography and for unambiguous interpretation of enzymatic data.

2.1. Previous recombinant GRK expression systems

GRK2 and GRK3 were first successfully expressed in *Spodoptera frugiperda* (Sf9) cells using *Autographa californica* nuclear polyhedrosis virus as the vector (Kim, Dion, Onorato, & Benovic, 1993). GRK2 and GRK3 were sequentially purified using S-Sepharose, heparin-Sepharose, and Mono S chromatography, generating 5–7 mg l^{-1} culture of pure GRK. Assuming Michaelis–Menten kinetics, a V_{max} of ~3000 nmol min^{-1} mg^{-1} and a K_M of 14.5 μ*M* for rhodopsin were achieved.

GRK5 and GRK6 were expressed in Sf9 cells using the BacPAK system (Clontech) for the generation of baculoviruses (Benovic & Gomez, 1993; Kunapuli & Benovic, 1993). GRK5 was purified to homogeneity using S-Sepharose and Mono S chromatography, exhibiting a V_{max} of ~1000 nmol min^{-1} mg^{-1} and a K_M of 14.5 μ*M* for rhodopsin (Kunapuli, Onorato, Hosey, & Benovic, 1994; Pronin, Loudon, & Benovic, 2002). GRK6 was purified to homogeneity using SP-Sepharose and heparin-Sepharose columns, exhibiting a V_{max} of ~51 nmol min^{-1} mg^{-1} and a K_M of ~10 μ*M* for rhodopsin (Loudon & Benovic, 1994). Note that GRK2–5 were all purified in the presence of 0.02% (v/v) Triton X-100, which is known to stabilize GRK2 (Benovic, 1991) but is also thought to be detrimental to crystallization. Thus, a purification protocol in the absence of detergent was used to purify GRK2 for the purpose of structural analysis (Lodowski, Pitcher, Capel, Lefkowitz, & Tesmer, 2003).

Recombinant GRK1 and GRK4 were generated using the pVL1393 transfer vector and BaculoGold transfection (Pharmingen) (Cha, Bruel, Inglese, & Khorana, 1997; Ohguro et al., 1996; Premont et al., 1996). GRK1, which contains a farnesylation site, was purified by various combinations of heparin-Sepharose and anion and cation exchangers in the presence of detergent from insect cells lysed in either Tween 80 or dodecylmaltoside (Cha et al., 1997; Ohguro et al., 1996). A relatively low-specific activity of 94 nmol min^{-1} mg^{-1} was reported using 20 μ*M* rhodopsin. GRK1 undergoes both heterogeneous autophosphorylation and prenylation (Cha et al., 1997). More recently, a higher-specific activity, C-terminally truncated GRK1 was expressed and purified from High Five cells, yielding an enzyme with a V_{max} of 2300 nmol min^{-1} mg^{-1} and a K_M of 2.1 μ*M* (Singh, Wang, Maeda, Palczewski, & Tesmer, 2008). GRK4α, which contains C-terminal palmitoylation sites, was extracted from homogenates with 1% Lubrol PX and partially purified by Mono Q anion exchange chromatography (Premont et al., 1996).

GRK7, which contains a geranylgeranylation site, was expressed in the presence of supplemental mevalonolactone in High Five insect cells using the Bac-to-Bac Baculovirus Expression System (Invitrogen) but was not purified (Chen et al., 2001).

GRK1 and GRK7 have also been N-terminally FLAG tagged, expressed in HEK293 cells and used in Michaelis–Menten kinetic studies after extraction in 0.5% dodecylmaltoside and purification by anti-FLAG affinity chromatography (Horner, Osawa, Schaller, & Weiss, 2005). V_{max} values for FLAG-GRK1 and FLAG-GRK7 were ~1130 and ~920 nmol min^{-1} mg^{-1}, respectively, and K_M values were ~3.5 and ~2 μ*M*, respectively.

Whereas all GRKs can be expressed in insect cells, at least some active GRK2 reportedly can be expressed in *E. coli* BL21(DE3) cells using the Studier system (Studier & Moffatt, 1986). The reported V_{max} and K_M values for rhodopsin were ~450 nmol min^{-1} mg^{-1} and ~8 μ*M*, respectively (Gan et al., 2000).

2.2. Baculovirus-mediated expression of C-terminal hexahistidine-tagged GRKs

In our labs, Sf9 cells are used to generate baculoviruses, whereas High Five cells, which in our hands and have a faster doubling time, are used for baculovirus infection and GRK purification. Our preferred baculovirus expression system for the past decade has been Invitrogen's Bac-to-Bac. The use of hexahistidine-tagged variants now greatly facilitates GRK purification

```
             BamHI(1805-10)                        SalI(1843-48)                SpeI(1855-60)
pFBDual      GGATCCCGGTCCGAAGCGCGCGGAATTCAAAGGCCTACGTCGACGAGCTC---------------ACTAGTCGCGGCCG
pFBDual-H6   GGATCCCGGTCCGAAGCGCGCGGAATTCAAAGGCCTACGTCGACCACCATCACCATCACCATTAAACTAGTCGCGGCCG
                                                   V  D  H  H  H  H  H  H stop
```

Figure 19.1 Modification of Invitrogen's pFastBac Dual vector to allow expression of carboxyl terminal hexahistidine-tagged proteins (see text).

to levels suitable for most purposes even after one chromatography step, provided sufficient levels of expression. Because the N-termini of GRKs are known to be critical to their function, the pFastBac Dual vector (Invitrogen) was modified to instead express the open reading frames that encode a C-terminal hexahistidine tag. Two complementary primers encoding the hexahistidine tag and a TAA stop codon and generating 5′ *Sal*I and 3′ *Spe*I overhangs were annealed and ligated into *Sal*I/*Spe*I-cut pFBDual vector. This modification leaves most of the polylinker downstream of the polyhedrin promoter intact (Fig. 19.1). cDNAs from full-length or truncated GRKs can therefore be amplified with primers containing *Bam*HI and *Sal*I restriction sites and ligated into the pFBDual-H_6 vector. The following GRKs have been expressed using this system: bovine GRK1 (1–535-H_6) (Singh et al., 2008), bovine GRK2 (1–689-H_6) (Huang et al., 2009), and human GRK6 (1–531-H_6) (Boguth et al., 2010). In the case of GRK1 and GRK6, C-terminal lipid modification sites were truncated in order to improve expression and solubility and facilitate crystallization. Consistent with loss of C-terminal farnesylation, bovine GRK1 (1-535-H_6) had a sixfold higher K_M for rhodopsin and a twofold lower V_{max} compared to wild type (Singh et al., 2008).

We refer readers to the 2010 Invitrogen Bac-to-Bac manual for most details of how we produce baculovirus for GRK expression, although we use the less expensive bacmid purification described in the 2002 manual. Below, we only describe the steps that are specific to analysis of GRKs and use GRK2-H_6 as an example GRK.

2.2.1 Checking GRK2-H_6 expression in transfected cells

We harvest the VS_0 baculovirus (or the P1 viral stock) after 3 days, and the expression of GRK-H_6 is surveyed in the soluble fraction of cell lystates.

Solutions, reagents, and equipment

- Microprocessor-based probe sonicator equipped with microprobe (e.g., Fisher Sonic Dismembrator F550 Ultrasonic Homogenizer).
- Cell Wash Buffer (20 m*M* HEPES, pH 7.5, 150 m*M* NaCl).
- Nickel Column Base Buffer (20 m*M* HEPES, pH 8, 300 m*M* NaCl).

- Nickel Column Lysis Buffer (NCLB): base buffer with protease inhibitors and reducing agent added before use (20 m*M* HEPES, pH 8, 300 m*M* NaCl, 1 m*M* DTT, 1 m*M* PMSF, 10 μg ml^{-1} leupeptin, 200 μg ml^{-1} benzamidine).
- Dithiothreitol (DTT, Sigma, 1 *M* aliquots stored at −20 °C).
- Benzamidine (Sigma; 20 mg ml^{-1} stock in water stored at −20 °C).
- Leupeptin (Sigma or ThermoFisher; 10 mg ml^{-1} stock in dimethyl sulfoxide).
- Phenylmethylsulfonyl fluoride (PMSF, Sigma; prepared fresh at 100 m*M* in 100% isopropanol).
- 2-ml conical sonication tube (e.g., Axygen MCT-200-C).
- Trypan Blue (select carcinogen).

Protocol

- NCLB with protease inhibitors and reducing agent (NCLB/PI/DTT; enough for 400 μl/sample) is prepared and stored on ice. Virus-containing media are removed from cells, clarified at 300 × *g* for 5 min and stored in the presence of 2% FBS. Cells are washed with 2 ml cold Cell Wash Buffer on the 6-well plate, resuspended with 1 ml wash buffer, delivered to 1.5-ml microfuge tubes, and spun 500 × *g* for 5 min at 4 °C (cell pellets may be frozen at −20 °C at this stage). Cells are resuspended in 400 μl NCLB/PI/DTT and transferred to 2-ml conical sonication tubes, where they are sonicated (with probe placed 2–3 mm from the top of the lysate) in an ice-water bath for three rounds of 30 cycles consisting of 1-s pulses. Lysis is checked under a microscope with trypan blue, and viscosity is assessed by pipetting. The lysate is transferred to a microfuge tube and spun 14,000 × *g* at 4 °C, and the supernatant fraction is pipetted to a fresh tube. The protein concentration is determined (usually 2–3 mg ml^{-1}), and the presence of the target protein is determined by analyzing 20 μg of lysate by SDS-PAGE and Coomassie staining (or immunoblotting). Clarified lysates are stored at −20 °C. If no convincing signal is observed, the pellet fraction from the 14,000 × *g* spin can be solubilized in SDS sample buffer and analyzed to determine whether the target protein has aggregated. Sufficient viral titers can also be verified (Kitts & Green, 1999).

2.2.2 Optimizing expression kinetics with the VS_1 virus

We have generally infected 20 ml of High Five cells with 2–3 concentrations (0.5–1.5 ml) of VS_1 virus (amplified VS_0) and found that optimal GRK2-H_6 expression in various mutants and baculovirus preparations occurs between

44 and 56 h. Three 125-ml baffled Erlenmeyer flasks closed with BugStoppers (Whatman) containing 20 ml of High Five cells at 1×10^6 cells/ml are infected with 0.5, 1, or 1.5 ml of VS_1 viral supernatant, and the infection is allowed to proceed for 44, 48, 52, and 56 h. At each time point, 1 ml of culture is removed, cells are pelleted at $300 \times g$ for 5 min, the supernatant is poured off, and the cells are washed with 1 ml Cell Wash Buffer. After removal of buffer, the cell pellets are stored at −20 °C. Heterologous expression equal to the major High Five proteins in the soluble fraction is indicative of strong expression. One-quarter of that expression may lead to ~4 mg purified GRK2 protein per liter of culture. If expression is weak, viral titer should be determined and a broader range of virus concentrations (0.1–10 MOI) for infection should be used. From the results above, a VS_1 volume and infection time are determined. If expression is strong, we have found that 200 ml infection is sufficient. For GRKs with weaker expression, we typically carry out 2 × 250 ml infections.

2.3. GRK2-H_6 purification

2.3.1 Day 1: harvest, Ni-NTA agarose chromatography

Reagents, solutions, and equipment

- Rotor–stator homogenizer ideally equipped with external speed control.
- Nickel Column Base Buffer (20 m*M* HEPES, pH 8, 300 m*M* NaCl).
- Peristaltic Pump.
- Fraction Collector and rack for 13 × 100 test tubes.
- Low-speed refrigerated centrifuge and an ultracentrifuge and accompanying rotors (e.g., Fiberlite F21S and Beckman 70Ti) to accommodate ~30 ml of lysate.
- 5 ml Ni-NTA agarose (Qiagen) column (e.g., 1 × 10 cm Bio-Rad glass Econo column (737-1012) fitted with a flow adapter (738-0015) and equilibrated in nickel column base buffer.
- DTT (Sigma, 1 *M* aliquots stored at −20 °C).
- Benzamidine (Sigma; 20 mg ml^{-1} stock in water stored at −20 °C).
- Leupeptin (Sigma or ThermoFisher; 10 mg ml^{-1} stock in dimethyl sulfoxide).
- PMSF (Sigma; prepared fresh at 100 m*M* in isopropanol).
- NCLB base buffer with protease inhibitors and reducing agent added before use (20 m*M* HEPES, pH 8, 300 m*M* NaCl, 1 m*M* DTT, 1 m*M* PMSF, 10 μg ml^{-1} leupeptin, and 200 μg ml^{-1} benzamidine. Fresh protease inhibitors are added immediately before lysis and before the Ni^{2+}-NTA agarose column.)

- Imidazole (Sigma, prepared fresh in NCLB at 150 m*M*; this stock used to make 20 and 40 m*M* column buffers).
- Trypan Blue (a select carcinogen).

Cell pellets from 0.2 to 0.5 l cultures are resuspended in 15 ml of NCLB and lysed in an ice-water bath using a mounted rotor–stator homogenizer (usually three bursts of 30 s at ~20,000 rpm). Lysis is verified under a microscope using Trypan Blue staining. The lysate is clarified first by centrifugation at 17,000 × *g* for 20 min and then by centrifugation at 100,000 × *g* for 45 min. The 100k supernatant fraction is collected and then the protein concentration is determined by Bradford assay and adjusted to 5 mg ml^{-1} with NCLB. Imidazole is added to the lysate to 20 m*M* from a 150 m*M* stock in NCLB. If the GRK is sensitive to proteolytic degradation, 1 m*M* EDTA can be added to the NCLB to inhibit metalloproteases but should be diluted to ~100 μ*M* before loading on to Ni-NTA columns.

The column is equilibrated with three column volumes of NCLB supplemented with imidazole to 20 m*M* at a flow rate of 0.5 ml min^{-1} for this and subsequent steps. In the fraction collector, 13 × 100 test tubes are used to hold 1.5-ml microfuge collection tubes. The clarified lysate is loaded on the column, the column is washed with three column volumes of NCLB plus 20 m*M* imidazole, washed with three more column volumes of NCLB plus 40 m*M* imidazole, and then the hexahistidine-tagged protein is eluted in 1 ml fractions with 40 ml of NCLB plus 150 m*M* imidazole. Fifteen microliters of selected column fractions are combined with 5 μl 4 × SDS sample buffer (8% (w/v) SDS, 250 m*M* Tris, pH 6.8, 20 m*M* EDTA, 40% glycerol, trace bromophenol blue, and 10 m*M* DTT), and 10 μl of this mixture is analyzed by 8% SDS polyacrylamide gels and stained with Coomassie Blue (Fig. 19.2A). Fractions are pooled and stored overnight at 4 °C.

2.3.2 Day 2: cation exchange chromatography

A cation exchange step has proven to be greatly useful not only for achieving higher purity but also as a diagnostic tool. Some GRKs, such as GRK1, express with different N-terminal truncations, and a strong cation exchanger can help to separate these pools (Singh et al., 2008). The cation exchanger can also discriminate between different phosphorylation states. For example, MAP kinase-mediated phosphorylation of GRK2 at Ser670 was first revealed by cation exchange chromatography (Pitcher et al., 1999). For this reason, the GRK2-S670A mutant is often used for GRK2 preparations.

Reagents, solutions, and equipment

- 1 ml High S and High Q columns (Econo-Pac High Capacity, Bio-Rad).

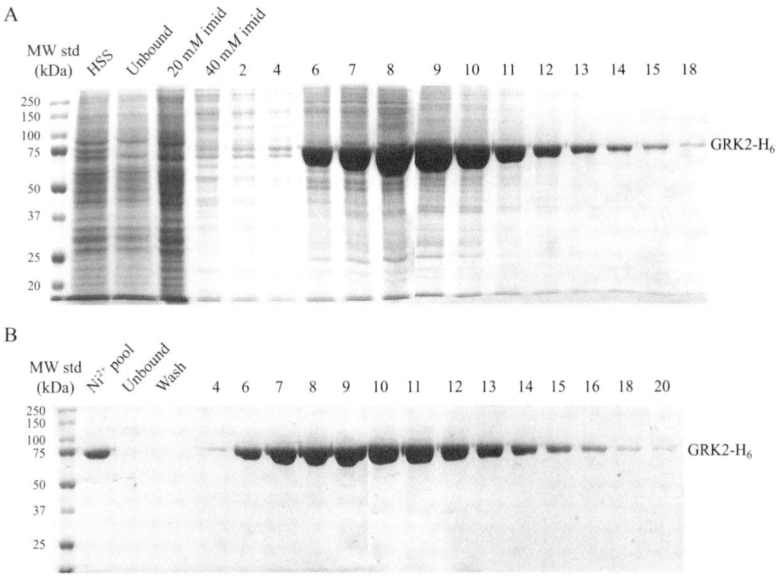

Figure 19.2 Purification of GRK2-H_6 from baculovirus-infected High Five cells. (A) Clarified lysates from a 0.5-l culture were chromatographed on a 5-ml Ni-NTA column. HSS, high-speed supernatant; imid, imidazole. (B) GRK2-H_6-containing fractions (6–14) were pooled and purified over a 1-ml Econo-Pac High Capacity High S column. Column fractions were analyzed by 8% SDS-PAGE and Coomassie staining.

- Gradient maker (e.g., Hoeffer SG 30).
- No salt buffer (20 m*M* HEPES, pH 7.5, 5 m*M* EDTA, 1 m*M* DTT, 0.02% (v/v) Triton X-100, with protease inhibitors; stored at 4 °C without DTT and protease inhibitors).
- Low salt buffer (20 m*M* HEPES, pH 7.5, 50 m*M* NaCl, 5 m*M* EDTA, 1 m*M* DTT, 0.02% (v/v) Triton X-100, with protease inhibitors; stored at 4 °C without DTT and protease inhibitors).
- High salt buffer (20 m*M* HEPES, pH 7.5, 600 m*M* NaCl, 5 m*M* EDTA, 0.02% (v/v) Triton X-100, with protease inhibitors; stored at 4 °C without DTT and protease inhibitors).

For GRK2, the High Q column simply acts as a prefilter for the removal of contaminants, including other protein kinases (Benovic, Mayor, Staniszewski, Lefkowitz, & Caron, 1987; Gan et al., 2004; Sterne-Marr et al., 2009). This property applies to GRK5 and GRK6, but not to GRK1 or GRK4α, which bind to both cation and anion exchangers (Ohguro et al., 1996; Premont et al., 1996; Sallese et al., 1997). The nickel column pool is diluted sixfold in 20 m*M* HEPES, pH 7.5, 5 m*M* EDTA, and

0.02% TX-100, with protease inhibitors to lower the NaCl concentration to 50 m*M*. The High Q and High S columns are hooked up in tandem and equilibrated in low salt buffer at a flow rate of 1 ml min^{-1}. The nickel column pool is loaded on the tandem columns, and then after a one-column volume wash, the High Q column is removed. The High S column is washed with 5–10 column volumes of low salt buffer and then eluted with a 20 ml linear gradient of 50–600 m*M* NaCl. Column fractions are visualized (Fig. 19.2B) as described above, and high GRK2 fractions (e.g., 7–11) and lower GRK fractions (e.g., 6, 12–15) are each pooled and combined with one-third volume of glycerol (25% glycerol stocks), distributed in aliquots, and stored at −20 or −80 °C. We generally obtain 10–20 mg pure GRK2-H_6 per liter of insect cell culture.

2.3.3 Polishing by size exclusion chromatography

Although at this point hexahistidine-tagged GRKs are often >90% pure, sometimes a buffer exchange step is required or additional purification is needed. Size exclusion chromatography achieves both purposes. Typically, two tandem Superdex S200 columns (GE Healthcare) mounted on a high-pressure liquid chromatography system such as a Biologic Duo-Flow (Bio-Rad) are used to separate monodisperse GRKs from aggregates and other minor contaminants. The standard running buffer for the column is 20 m*M* HEPES, pH 8.0, 100–200 m*M* NaCl, and 1–2 m*M* DTT.

2.3.4 Crystallization of GRKs

Currently, there are 23 crystal structures deposited in the Protein Data Bank representing bovine GRK1, human and bovine GRK2 (alone and in complex with heterotrimeric G-proteins), and human GRK6. Ligand-free structures diffract to lower resolution and their electron density maps suggest a high degree of structural heterogeneity (Lodowski et al., 2005; Singh et al., 2008). Crystallization is typically performed at 4 °C using the hanging drop method, and, with only one exception, all GRKs have been crystallized using polyethylene glycol (typically 3350 molecular weight) as the precipitating agent at low ionic strength (typically 100 m*M* NaCl). The exception is the recent structure of GRK6 in complex with the inhibitor sangivamycin, which used ammonium sulfate as the precipitating agent (Boguth et al., 2010). Most GRK crystals were obtained with well solutions adjusted to acidic pH (4.3–6.5).

3. GRK FUNCTIONAL ASSAYS

Three assays are typically used to characterize GRK function *in vitro*: the rhodopsin (Rho) phosphorylation assay, the peptide C phosphorylation assay, and the activation assay. Endogenous Rho isolated from bovine rod outer segments (ROSs) is the most readily available and convenient GPCR substrate for these assays. All phosphorylation sites are found in the carboxyl tail of the receptor. However, the C-terminal Rho peptide, known as peptide C, is a very poor substrate for GRKs. Thus, the peptide C phosphorylation assay is thought to measure the intrinsic ability of the kinase to assume an activated conformation in the absence of activated receptor. In contrast, the activation assays measures the ability of truncated Rho* that lacks the phosphoacceptor sites in the C-terminal tail (G^{329}-Rho*) to enhance peptide C phosphorylation. Residues that when mutated show defects in the activation assay (but not in the peptide C phosphorylation assay, are thought to be involved in direct interactions with the receptor substrate. Residues that show defects in both are thought to impair formation of the receptor-docking site) as this is believed to be allosterically coupled to closure of the kinase domain (Boguth et al., 2010; Huang, Orban, Jastrzebska, Palczewski, & Tesmer, 2011).

3.1. Rho* phosphorylation assay

ROSs are isolated from dark-adapted bovine retinas (W. L. Lawson Co., Omaha, NE). One hundred retinas are processed at a time under dim red light by published procedures (Kuhn & Wilden, 1982; Papermaster, 1982), typically yielding 10–20 mg of Rho. For the standard assay, final assay conditions are ~40 n*M* GRK2, 8 μ*M* Rho, 200 μ*M* ATP, 0.5–1 dpm/fmol γ-^{32}P-ATP, 20 m*M* Tris, pH 7.5, 2 m*M* EDTA, and 7.5 m*M* $MgCl_2$. Stocks of Rho and GRK2 are prepared at 32 μ*M* and 80 n*M*, respectively, in GRK assay buffer. A "4× ATP Hot Mix" is prepared with 800 μ*M* ATP and 1–2 dpm/pmol ^{32}P-ATP (MP Biomedical) in GRK assay buffer. The 4× ATP Hot Mix is diluted 100-fold and 5 μl are counted in a scintillation counter to determine the specific radioactivity. Five microliters of GRK2 and 2.5 μl 4× ATP Hot Mix are combined in 0.5-ml PCR tubes. Samples are taken to a dark room where Rho is resuspended by vortex mixing, and 2.5 μl are delivered to the reaction tube under dim red light. Samples are transferred to a 30 °C incubator for 30 s, and then a desk lamp with a

100 W bulb positioned over the incubator is illuminated. After a 3 min incubation, samples are transferred to an ice-water bath and reactions are quenched with 14 μl 4× SDS/DTT sample buffer (8% (w/v) SDS, 250 m*M* Tris, pH 6.8, 20 m*M* EDTA, 40% glycerol, trace bromophenol blue, and 50 m*M* DTT). Samples are mixed by tapping periodically during a 30 min incubation at 65 °C and briefly spun at 500 × *g* before separation on 10% SDS-PAGE. Afterward, the bottom of the gel with the running dye (and free ATP) is removed, the gel is stained with Coomassie dye, destained, the Rho bands are excised (from the hydrated gel), and the radioactivity is determined in a scintillation counter. For steady state kinetics studies, 2 × concentrations of Rho in GRK assay buffer are prepared, and the reaction contains 5 μl Rho, 2.5 μl 80 n*M* GRK2, and 2.5 μl 4 × ATP Hot Mix.

3.2. Peptide C phosphorylation assay

Peptide C contains residues 330-DDEASTTVSKTETSQVA-346 of bovine Rho. Three arginines are often appended to the peptide sequence shown above to facilitate binding to negatively charged filters (such as P81 paper) in a filter-based kinase assay, but we have found that the triargininyl peptide C can be separated from free ^{32}P-ATP on 18% SDS polyacrylamide gels where the radioactivity can be more confined. Peptide C phosphorylation reactions (~500 n*M* GRK2, 1 m*M* peptide C, 100 μ*M* ATP, 1.5 dpm/fmol ^{32}P-ATP, 20 m*M* Tris, pH 7.5, 2 m*M* EDTA, and 7.5 m*M* $MgCl_2$) are carried out for 50 min at 30 °C before quenching the reaction, loading 10 μl on an 18% gel, and quantifying phosphorylation as described above.

3.3. ^{329}G-Rho preparation and activation assay

Endoprotease Asp-N cleavage of Rho to prepare G^{329}-Rho has been previously described (Palczewski et al., 1991). After inactivation of the protease by the addition of EDTA and DTT to 2 m*M*, 12% SDS-PAGE/Coomassie staining is used to verify reaction completion. The ROS are stripped with 5 *M* urea and washed three times with 50 m*M* Tris, pH 7.5. Following the second wash, an aliquot is removed for protein concentration determination, and the final G^{329}-Rho is resuspended at 12 μ*M* in GRK assay buffer.

The activation assay is carried out in 12 μl reactions with 100 n*M* GRK, 2 μ*M* ^{329}G-Rho, 100 μ*M* peptide C, and 100 μ*M* ATP (1.5 dpm/fmol ^{32}P-ATP). Sixfold concentrated stocks of GRK, ^{329}G-Rho, and peptide C are prepared in GRK assay buffer. Two microliters each of GRK2 and peptide C are combined with 6 μl 200 μ*M* ATP (3 dpm/fmol ^{32}P-ATP) in 0.5-ml PCR tubes. Two microliters of 12 μ*M* ^{329}G-Rho is added under dim red light. Two identical sets of reactions are prepared to allow for reaction in the light and dark for 30 min at 30 °C with periodic tap mixing. Reactions are quenched and analyzed as described for the peptide C assay.

3.4. Cell-based β_2-adrenergic receptor (β_2AR) phosphorylation assay

Cell-based detection of GPCR phosphorylation has emerged as a sensitive and direct assessment of GRK activity in intact cells. GRK2 phosphorylates Ser355, 356, and 364 of the β_2AR in living cells (Seibold et al., 2000). Based upon the success of detecting *in vivo* phosphorylation of chemokine receptor type 5 using monoclonal phosphosite antibodies (Pollok-Kopp, Schwarze, Baradari, & Oppermann, 2003), a similar approach was used to measure GRK2 phosphorylation of the β_2AR (Tran et al., 2004). This methodology generates a robust agonist-induced and dose-dependent phosphorylation signal in HEK293 cells stably expressing hexahistidine-tagged β_2AR.

Unfortunately, using the above protocol we were unable to detect any dependence of phosphorylation on transiently transfected GRK2, as was also reported by others for exogenously supplied GRK2 (Shenoy et al., 2006). We have, therefore, modified the procedures of Tran et al. (2004) to allow detection of exogenous GRK2 promoted phosphorylation of the transiently transfected β_2AR in COS-7 cells. This procedure does not require enrichment of β_2AR by affinity chromatography or immunoprecipitation, and therefore untagged receptors can be used.

3.5. Days 1–3: DNA transfection

Reagents

- FuGENE® HD transfection (Promega).
- Transfection Quality DNA (Qiagen midi preps).

DNAs

- pcDNA3.1-HA-hβ_2AR (HA-tagged human β_2AR from Drs. Marc Caron and Larry Barak).
- pcDNA3-bGRK2 (bovine GRK2 from Dr. Jeffrey Benovic).
- pcDNA3.1-Gα_s (human Gα_s long, UMR cDNA Resource Center).
- pcDNA3.1-Gβ_1 (human Gβ_1, UMR cDNA Resource Center).
- pcDNA3.1-Gγ_2 (human Gγ_2, UMR cDNA Resource Center).

Approximately 3×10^6 COS-7 cells/well are plated in 6-well plates (~25% confluency) and incubated overnight in a humidified atmosphere at 37 °C/ 5% CO_2. In general, DNA:FuGENE® HD complexes are prepared with 2 μg of total DNA and 8 μl of FuGENE® HD. Transfection mixes are prepared by the addition of DNA to 100 μl serum-free and antibiotic-free DMEM in microfuge tubes followed by tap mixing. FuGENE® HD is added, the transfection mix is vortexed for 1 s, and then incubated for 15 min. Transfection mixes are added drop-wise to each well of the 6-well plate, and the cells are incubated for 48 h in a humidified atmosphere at 37 °C/5% CO_2.

3.6. Day 4: agonist (isoproterenol)/inverse agonist (alprenolol) treatment, cell harvest

Reagents and Equipment

- Cell scrapers for 6-well tissue culture plate.
- Microprocessor-based probe sonicator equipped with microprobe (e.g., Fisher Sonic Dismembrator F550 Ultrasonic Homogenizer).
- Cell Wash Buffer (20 m*M* HEPES, pH 7.5, 150 m*M* NaCl), 4 °C.
- Receptor Solubilization Buffer (RSB, 200 μl/well; 20 m*M* HEPES, pH 7.4, 150 m*M* NaCl, 10 mg ml^{-1} *n*-dodecyl-β-D-maltoside (DDM), 10 m*M* DTT (reducing agent), 1 m*M* PMSF, 10 μg ml^{-1} leupeptin, 200 μg ml^{-1} benzamidine (protease inhibitors, PI), 20 m*M* tetrasodium pyrophosphate, 10 m*M* sodium fluoride, and ±100 n*M* okadaic acid (phosphatase inhibitors, PhI)).
- (−)Isoproterenol (ISO, Sigma, prepared fresh as described below).
- Alprenolol (ALP, Sigma, 1 m*M* stock in ethanol, stored at −20 °C).
- DDM (Anatrace Inc., Maumee, OH; prepared fresh at 10 mg ml^{-1} in RSB).
- DTT (Sigma, 1 *M* aliquots stored at −20 °C).
- Benzamidine (Sigma; 20 mg ml^{-1} stock in water stored at −20 °C).
- Leupeptin (Sigma or ThermoFisher; 10 mg ml^{-1} stock in dimethyl sulfoxide).
- PMSF (Sigma; prepared fresh at 100 m*M* in 100% isopropanol).
- Sodium pyrophosphate (Sigma; 0.5 *M* stocks stored at −20 °C and thawed at 65 °C).
- NaF (Sigma; 0.5 *M* stocks stored at −20 °C).
- Okadaic acid (Axxora, suspected carcinogen, 100 μ*M* stocks in ethanol stored at −20 °C).
- Peptide *N*-glycosidase F (PNGase; New England Biolabs, Beverly, MA, 500 U μl^{-1}).

On the day of cell harvest, the cell lysis/RSB is first prepared and stored on ice. The 30-min serum starvation of cells is initiated and then ISO and ALP are prepared. Because there are many additives to the RSB, we make a 2× stock of the base buffer (40 m*M* HEPES, pH 7.4, 300 m*M* NaCl), add the DTT, protease inhibitors (PMSF, benzamidine, leupeptin) and phosphatase inhibitors (sodium fluoride, sodium pyrophosphate, okadaic acid), and bring the volume up with cold deionized water. DDM is added as a solid and dissolved on a nutator at 4 °C.

The media are removed from the transfected cells and replaced with 1 ml serum-free media. Approximately, 1 mg of ISO is dissolved in 1 m*M* ascorbic acid (an antioxidant to protect the vicinal hydroxyls on the catecholamine) to generate a 20 m*M* isoproterenol stock. This is further diluted to 1 m*M* in 1 m*M* ascorbic acid and stored on ice. The 20 m*M* ALP is diluted to 1 m*M* in ethanol.

Cells are treated for 5 min by staggering the addition of 10 μ*M* ISO or ALP to each well every 15 s to allow equal treatment. Following the 5-min incubation, the media are aspirated and the cells are washed two times with cold cell wash buffer. Buffer is removed, and each well is scraped in the presence of 200 μl RSB on ice. The lysate is transferred to a 1.5-ml microfuge tube and sonicated using a 90-cycle regimen of 1 s on/2 s off. The receptor is further solubilized by mixing 30 min on a nutator at 4 °C and then clarified by centrifugation at 14,000 × *g* at 4 °C. The soluble fraction, 3–5 mg ml^{-1} protein, may be stored at −20 °C.

3.7. Days 4 and 5: Peptide *N*-glycosidase F (PNGase) treatment and immunoblotting

Reagents and Equipment

- Electrophoresis and immunoblotting equipment.
- Nutator rocker/shaker at 4 °C.
- 4 × SDS receptor sample buffer/DTT (8% SDS, 250 m*M* Tris, pH 6.8, 20 m*M* EDTA, 40% glycerol, trace bromophenol blue, and 40 m*M* DTT).
- Precision Plus WesternC Standards for Chemiluminescent detection (Bio-Rad).
- Rocking platform at 4 °C
- β_2AR GRK phosphosite antibodies (Santa Cruz Biotechnology sc-16719-R).

- β_2AR carboxyl tail affinity purified polyclonal antibodies (Santa Cruz Biotechnology, sc-569).
- GRK2 polyclonal antibody (Santa Cruz Biotechnology, sc-562).
- Goat anti-rabbit horseradish peroxidase (HRP)-conjugated secondary antibody.
- StrepTactin HRP Conjugate (Bio-Rad).
- Tris buffered saline/Tween (TBS/Tween; 50 m*M* Tris, pH 7.5, 150 m*M* NaCl, 0.1 % (v/v) Tween-20).
- Blocking Buffer (TBS/Tween with 5% nonfat dried milk).
- Immunoblot stripping buffer (25 m*M* glycine, pH 2, 1% SDS).
- West Pico, Chemiluminescent Reagent (or West Femto if necessary; Thermo Scientific).

The stock PNGase is diluted fivefold to 100 U μl^{-1} in RSB (or 20 m*M* HEPES, pH 7.5, 150 m*M* NaCl if PNGase treatment occurs on a day other than the day of harvest). Thirty microliters of the soluble fraction are treated with 1 μl 100 U μl^{-1} PNGase for 2 h at 37 °C. Ten microliters of SDS sample buffer containing DTT at a final concentration of 40 m*M* is added, and the samples are incubated at 65 °C for 30 min with tap mixing every 5 min. Twenty microliters of the PNGase-treated samples are separated on 10% acrylamide SDS-PAGE alongside Precision Plus WesternC (Bio-Rad) molecular weight standards that can be visualized by chemiluminescent detection. Standard immunoblotting procedures are used employing 1:666 dilution of the primary antibody overnight with rocking at 4 °C. A 1:2000 dilution of the goat anti-mouse HRP secondary antibody is used in conjunction with a 1:20,000 dilution of the StrepTactin HRP Conjugate. We develop our blots with SuperSignal West Pico Chemiluminescent Substrate (Thermo Scientific) and image using the Bio-Rad ChemiDoc XRS System. The molecular weight standards are cut off and stored in TBS/Tween at 4 °C, whereas the remainder of the blot is stripped at room temperature for 5 min in stripping buffer. The blot is washed, reblocked, and probed with the β_2AR C-tail antibody diluted 1:1000, 1 h at room temperature, and then the secondary antibody and StrepTactin HRP as described above. The blot is developed with chemiluminescent reagent in the presence of the molecular weight standards, imaged, stripped, and reblocked. Finally, the blot is probed with the GRK2 antibody (1:1000, 1 h incubation at room temperature) followed by goat anti-mouse HRP secondary antibody (1:2000) and StrepTactin HRP (1:10,000) incubation as described above.

Densitometry is used to quantify *in vivo* phosphorylation of the β_2AR by transfected GRK2. The pSer and C-tail signals are normalized to the lane that contains transfected β_2AR alone treated with the inverse agonist, ALP. The ratio of pSer/C-tail is obtained and yields a measure of agonist-inducible receptor phosphorylation. To assess the effect of transfected G-proteins or mutant GRK, the pSer/C-tail ratio is normalized to the level of GRK in each GRK-transfected sample.

We found that the absence of okadaic acid in the RSB did not negatively impact the detection of pSer β_2AR (Fig. 19.3A), but that PNGase treatment is absolutely essential to allow detection of phosphorylated β_2AR at its monomer molecular weight (Fig. 19.3B). Exogenous GRK2 stimulated the ISO-dependent phosphorylation ~10-fold (Fig. 19.4). Cotransfection of GRK2 with G$\beta\gamma$ enhanced the GRK2-dependent pSer signal, but the ALP-treated samples also displayed elevated signal. Therefore, the strongest GRK2-dependent signal is observed with G$\beta\gamma$, but the strongest agonist-dependent signal is observed with GRK2 alone.

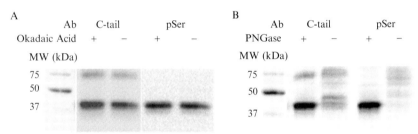

Figure 19.3 Test of the role of okadaic acid and PNGase F on agonist-induced β_2AR phosphorylation in intact cells using GRK phosphosite antibodies. COS-7 cells were grown in 6-well plates and transfected with cDNA encoding β_2AR, GRK2, Gβ_1, and Gγ_2 at a ratio of 0.7:0.7: 0.2: 0.2 μg. After 48 h, cells were treated with 10 μM isoproterenol for 5 min. (A) Cell lysates were prepared in the presence of reducing agent, protease inhibitors, phosphatase inhibitors sodium pyrophosphate, and sodium fluoride, but in the absence or presence of okadaic acid, and treated with PNGase F (PNGase). (B) Cell lysates were prepared in the presence of reducing agent, protease inhibitors, phosphatase inhibitors including okadaic acid and then treated in the absence or presence of PNGase F. Samples were separated by 10% SDS-PAGE and transferred to nitrocellulose. Agonist-induced phosphorylation of the β_2AR was detected with a phosphosite antibody that recognizes phosphorylated Ser^{355} and Ser^{356} (pSer), and total β_2AR was detected with an antibody that recognizes the carboxyl tail of the receptor independent of its phosphorylation status (C-tail).

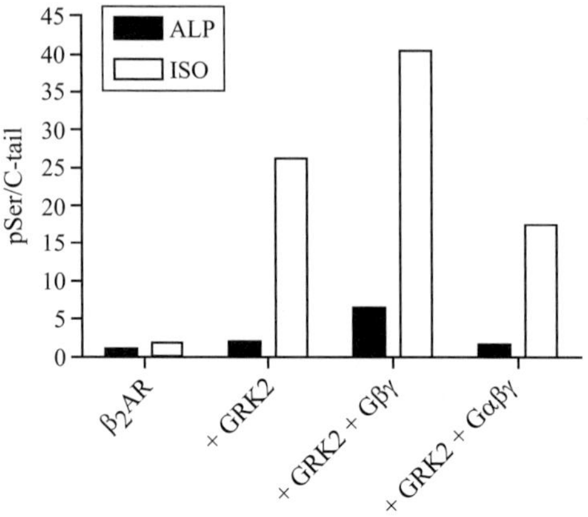

Figure 19.4 Role of exogenous GRK2 and heterotrimeric G-proteins on agonist-induced β_2AR phosphorylation in intact cells. COS-7 cells were grown in 6-well plates and transfected with 0.8 μg β_2AR cDNA in the absence or presence of cDNAs encoding GRK2 (0.3 μg), $G\alpha_s$ (0.3 μg), $G\beta_1$ (0.3 μg), and $G\gamma_2$ (0.3 μg) as indicated. Cells were treated for 5 min with 10 μ*M* alprenolol (ALP) or isoproterenol (ISO), and the detection of phosphorylated β_2AR was carried out as described in the legend to Fig. 19.3 and the text. The pSer and C-tail signals were normalized to that of the ALP-treated β_2AR lane and phosphorylation of the β_2AR was determined by taking the pSer/C-tail ratio. The levels of GRK2, when transfected, were with 20% of each other. The results of this experiment are representative of four similar experiments.

ACKNOWLEDGMENTS

This work was supported by National Science Foundation grants MCB0315888 and MCB0744739 (R. S. M.) and National Institute of Health grants HL071818, HL086865, and GM081655 (J. T.). We thank Dick Clark and Faiza Baameur (University of Texas Health Science Center, Houston) for advice and phosphorylated β_2AR controls, Valerie Tesmer (University of Michigan) for advice on insect cell culture, and Amber Cutter, Katelynn Mannix, Alex Leahey, Tim Clarke, and Devin McDonald (Siena College) for their contributions to these projects.

REFERENCES

Benovic, J. L., & Gomez, J. (1993). Molecular cloning and expression of GRK6. A new member of the G protein-coupled receptor kinase family. *The Journal of Biological Chemistry*, *268*, 19521–19527.

Benovic, J. L. (1991). Purification and characterization of beta-adrenergic receptor kinase. *Methods in Enzymology*, *200*, 351–362. PMID: 1659657

Benovic, J. L., Mayor, F., Jr., Staniszewski, C., Lefkowitz, R. J., & Caron, M. G. (1987). Purification and characterization of the β-adrenergic receptor kinase. *The Journal of Biological Chemistry*, *262*, 9026–9032.

Boguth, C. A., Singh, P., Huang, C. C., & Tesmer, J. J. (2010). Molecular basis for activation of G protein-coupled receptor kinases. *The EMBO Journal*, *29*, 3249–3259.

Cha, K., Bruel, C., Inglese, J., & Khorana, H. G. (1997). Rhodopsin kinase: expression in baculovirus-infected insect cells, and characterization of post-translational modifications. *Proceedings of the National Academy of Sciences of the United States of America, 94*, 10577–10582.

Chen, C. K., Zhang, K., Church-Kopish, J., Huang, W., Zhang, H., Chen, Y. J., et al. (2001). Characterization of human GRK7 as a potential cone opsin kinase. *Molecular Vision, 7*, 305–313.

Gan, X., Ma, Z., Deng, N., Wang, J., Ding, J., & Li, L. (2004). Involvement of the C-terminal proline-rich motif of G protein-coupled receptor kinases in recognition of activated rhodopsin. *The Journal of Biological Chemistry, 279*, 49741–49746.

Gan, X. Q., Wang, J. Y., Yang, Q. H., Li, Z., Liu, F., Pei, G., et al. (2000). Interaction between the conserved region in the C-terminal domain of GRK2 and rhodopsin is necessary for GRK2 to catalyze receptor phosphorylation. *The Journal of Biological Chemistry, 275*, 8469–8474.

Gurevich, E. V., Tesmer, J. J., Mushegian, A., & Gurevich, V. V. (2012). G protein-coupled receptor kinases: more than just kinases and not only for GPCRs. *Pharmacology & Therapeutics, 133*, 40–69.

Horner, T. J., Osawa, S., Schaller, M. D., & Weiss, E. R. (2005). Phosphorylation of GRK1 and GRK7 by cAMP-dependent protein kinase attenuates their enzymatic activities. *The Journal of Biological Chemistry, 280*, 28241–28250.

Huang, C. C., Orban, T., Jastrzebska, B., Palczewski, K., & Tesmer, J. J. (2011). Activation of G protein-coupled receptor kinase 1 involves interactions between its N-terminal region and its kinase domain. *Biochemistry, 50*, 1940–1949.

Huang, C. C., & Tesmer, J. J. (2011). Recognition in the face of diversity: interactions of heterotrimeric G proteins and G protein-coupled receptor (GPCR) kinases with activated GPCRs. *The Journal of Biological Chemistry, 286*, 7715–7721.

Huang, C. C., Yoshino-Koh, K., & Tesmer, J. J. (2009). A surface of the kinase domain critical for the allosteric activation of G protein-coupled receptor kinases. *The Journal of Biological Chemistry, 284*, 17206–17215.

Kannan, N., Haste, N., Taylor, S. S., & Neuwald, A. F. (2007). The hallmark of AGC kinase functional divergence is its C-terminal tail, a cis-acting regulatory module. *Proceedings of the National Academy of Sciences of the United States of America, 104*, 1272–1277.

Kim, C. M., Dion, S. B., Onorato, J. J., & Benovic, J. L. (1993). Expression and characterization of two β-adrenergic receptor kinase isoforms using the baculovirus expression system. *Receptor, 3*, 39–55.

Kitts, P. A., & Green, G. (1999). An immunological assay for determination of baculovirus titers in 48 hours. *Analytical Biochemistry, 268*, 173–178.

Kuhn, H., & Wilden, U. (1982). Assay of phosphorylation of rhodopsin in vitro and in vivo. *Methods in Enzymology, 81*, 489–496.

Kunapuli, P., & Benovic, J. L. (1993). Cloning and expression of GRK5: A member of the G protein-coupled receptor kinase family. *Proceedings of the National Academy of Sciences of the United States of America, 90*, 5588–5592.

Kunapuli, P., Onorato, J. J., Hosey, M. M., & Benovic, J. L. (1994). Expression, purification, and characterization of the G protein-coupled receptor kinase GRK5. *The Journal of Biological Chemistry, 269*, 1099–1105.

Lodowski, D. T., Barnhill, J. F., Pyskadlo, R. M., Ghirlando, R., Sterne-Marr, R., & Tesmer, J. J. (2005). The role of Gβγ and domain interfaces in the activation of G protein-coupled receptor kinase 2. *Biochemistry, 44*, 6958–6970.

Lodowski, D. T., Pitcher, J. A., Capel, W. D., Lefkowitz, R. J., & Tesmer, J. J. (2003). Keeping G proteins at bay: A complex between G protein-coupled receptor kinase 2 and Gβγ. *Science, 300*, 1256–1262.

Loudon, R. P., & Benovic, J. L. (1994). Expression, purification, and characterization of the G protein-coupled receptor kinase GRK6. *The Journal of Biological Chemistry*, *269*, 22691–22697.

Ohguro, H., Rudnicka-Nawrot, M., Buczylko, J., Zhao, X., Taylor, J. A., Walsh, K. A., et al. (1996). Structural and enzymatic aspects of rhodopsin phosphorylation. *The Journal of Biological Chemistry*, *271*, 5215–5224.

Palczewski, K., & Benovic, J. L. (1991). G-protein-coupled receptor kinases. *Trends in Biochemical Sciences*, *16*, 387–391.

Palczewski, K., Buczylko, J., Kaplan, M. W., Polans, A. S., & Crabb, J. W. (1991). Mechanism of rhodopsin kinase activation. *The Journal of Biological Chemistry*, *266*, 12949–12955.

Papermaster, D. S. (1982). Preparation of antibodies to rhodopsin and the large protein of rod outer segments. *Methods in Enzymology*, *81*, 240–246.

Pitcher, J. A., Tesmer, J. J., Freeman, J. L., Capel, W. D., Stone, W. C., & Lefkowitz, R. J. (1999). Feedback inhibition of G protein-coupled receptor kinase 2 (GRK2) activity by extracellular signal-regulated kinases. *The Journal of Biological Chemistry*, *274*, 34531–34534.

Pollok-Kopp, B., Schwarze, K., Baradari, V. K., & Oppermann, M. (2003). Analysis of ligand-stimulated CC chemokine receptor 5 (CCR5) phosphorylation in intact cells using phosphosite-specific antibodies. *The Journal of Biological Chemistry*, *278*, 2190–2198.

Premont, R. T., Macrae, A. D., Stoffel, R. H., Chung, N., Pitcher, J. A., Ambrose, C., et al. (1996). Characterization of the G protein-coupled receptor kinase GRK4. Identification of four splice variants. *The Journal of Biological Chemistry*, *271*, 6403–6410.

Pronin, A. N., Loudon, R. P., & Benovic, J. L. (2002). Characterization of G protein-coupled receptor kinases. *Methods in Enzymology*, *343*, 547–559.

Sallese, M., Mariggio, S., Collodel, G., Moretti, E., Piomboni, P., Baccetti, B., et al. (1997). G protein-coupled receptor kinase GRK4. Molecular analysis of the four isoforms and ultrastructural localization in spermatozoa and germinal cells. *The Journal of Biological Chemistry*, *272*, 10188–10195.

Seibold, A., Williams, B., Huang, Z. F., Friedman, J., Moore, R. H., Knoll, B. J., et al. (2000). Localization of the sites mediating desensitization of the β_2-adrenergic receptor by the GRK pathway. *Molecular Pharmacology*, *58*, 1162–1173.

Shenoy, S. K., Drake, M. T., Nelson, C. D., Houtz, D. A., Xiao, K., Madabushi, S., et al. (2006). β-arrestin-dependent, G protein-independent ERK1/2 activation by the β_2 adrenergic receptor. *The Journal of Biological Chemistry*, *281*, 1261–1273.

Singh, P., Wang, B., Maeda, T., Palczewski, K., & Tesmer, J. J. (2008). Structures of rhodopsin kinase in different ligand states reveal key elements involved in G protein-coupled receptor kinase activation. *The Journal of Biological Chemistry*, *283*, 14053–14062.

Sterne-Marr, R., Leahey, P. A., Bresee, J. E., Dickson, H. M., Ho, W., Ragusa, M. J., et al. (2009). GRK2 activation by receptors: Role of the kinase large lobe and carboxyl-terminal tail. *Biochemistry*, *48*, 4285–4293.

Studier, F. W., & Moffatt, B. A. (1986). Use of bacteriophage T7 RNA polymerase to direct selective high-level expression of cloned genes. *Journal of Molecular Biology*, *189*, 113–130.

Tesmer, J. J. (2009). Structure and function of regulator of G protein signaling homology domains. *Progress in Molecular Biology and Translational Science*, *86*, 75–113.

Tesmer, J. J. (2011). Activation of G protein-coupled receptor (GPCR) kinases by GPCRs. In J. Giraldo & J.-P. Pin (Eds.), *G protein-coupled receptors: From structure to function. RSC Drug Discovery Series 8*. London: The Royal Society of Chemistry, London 297–315.

Tran, T. M., Friedman, J., Qunaibi, E., Baameur, F., Moore, R. H., & Clark, R. B. (2004). Characterization of agonist stimulation of cAMP-dependent protein kinase and G protein-coupled receptor kinase phosphorylation of the β_2-adrenergic receptor using phosphoserine-specific antibodies. *Molecular Pharmacology*, *65*, 196–206.

CHAPTER TWENTY

Modern Methods to Investigate the Oligomerization of Glycoprotein Hormone Receptors (TSHR, LHR, FSHR)

Marco Bonomi*, **Luca Persani***,†,1

*Lab of Experimental Endocrinology and Metabolism, and Division of Endocrine and Metabolic Diseases, Ospedale San Luca, IRCCS Istituto Auxologico Italiano, Milan, Italy

†Department of Medical Sciences, University of Milan, Milan, Italy

[1]Corresponding author: e-mail address: luca.persani@unimi.it

Contents

1. Introduction 368
2. Resonance Energy Transfer Techniques 369
   2.1 FRET technique 370
   2.2 BRET technique 372
   2.3 HTRF-RET technique 374
3. Experimental Procedures 376
   3.1 Construct generation and transfection 376
   3.2 FRET experiment 376
   3.3 BRET experiment 377
   3.4 HTRF-RET experiment 379
Acknowledgments 381
References 381

Abstract

As for other GPCRs, the oligomerization of glycoprotein hormone receptors (GPHRs) appears as critical event for receptor function. By means of modern techniques based on the BRET or FRET principle, GPHR oligomerization has been reported to explain several physiological and pathological conditions. In particular, the presence of oligomers was demonstrated not only in *in vitro* heterologous systems but also in *in vivo* tissues, and GPHR homodimerization appears associated with strong negative cooperativity, thus suggesting that one hormone molecule may be sufficient for receptor dimer stimulation. In addition, oligomerization has been reported to occur early during the posttranslational maturation process and to be involved in the dominant negative effect exerted by loss-of-function TSH receptor (TSHR) mutants, that are prevalently retained inside the

Methods in Enzymology, Volume 521
ISSN 0076-6879
http://dx.doi.org/10.1016/B978-0-12-391862-8.00020-X

cell, on the surface expression of wild-type receptors. This molecular mechanism thus explains the dominant inheritance of certain forms of TSH resistance. Here, we provide the description of the methods used in the original BRET, FRET, and HTRF-RET experiments.

1. INTRODUCTION

Thyroid-stimulating hormone receptors (TSHRs) (Persani et al., 2010), luteinizing hormone receptors (LHRs) (Puett, Angelova, da Costa, Warrenfeltz, & Fanelli, 2010), and follicle-stimulating hormone receptors (FSHRs) (Meduri et al., 2008) constitute the glycoprotein hormone receptors (GPHRs) (Vassart, Pardo, & Costagliola, 2004) which belong to the subfamily A of the G-protein-coupled receptors (GPCRs). They are dimeric 30 kDa proteins with important roles in the control of metabolism and reproduction and are characterized by an ectodomain containing leucine-rich repeat (LRR) motifs, in addition to the canonical heptahelical serpentine domain typical of GPCRs. Indeed, they are also called leucine repeat-containing receptors (LGRs) (Van Loy et al., 2008). The bipartite structure of GPHRs and LGRs is accompanied by a functional dichotomy: their LRR-containing ectodomain is responsible for the specificity of binding of their respective agonists, which translates into activation of the rhodopsin-like serpentine domain, itself responsible for transducing the signal within the cell, mainly via activation of the G-protein Gs. GPCRs were traditionally depicted as monomeric entities that activate a single heterotrimeric G-protein upon ligand binding. In recent years, this classic GPCR stoichiometry (one ligand binding to one GPCR, binding to one G-protein, binding to one effector) has been challenged by the results consistent with the existence of GPCRs as dimers or higher-order oligomers (Milligan, 2004; Park, Filipek, Wells, & Palczewski, 2004; Terrillon & Bouvier, 2004). Original evidence for high-molecular-weight receptor complexes was obtained 20–30 years ago by means of classical molecular biology methods (Park et al., 2004). Among these, coimmunoprecipitation represents the most widely used technique for GPCR oligomerization studies (Hanyaloglu, Seeber, Kohout, Lefkowitz, & Eidne, 2002). This technology is relatively simple and low-cost and has the advantage over other methods that it can evaluate interactions between proteins that are synthesized in their natural subcellular context. Coimmunoprecipitation is, however, affected by several potential drawbacks (Eidne, Kroeger, & Hanyaloglu, 2002; Milligan & Bouvier, 2005); therefore, in recent years, an effort has been made to setup modern techniques for protein–protein interaction studies with improved characteristics.

Also, the receptors belonging to the GPHRs family have been reported to dimerize firstly by the use of coimmunoprecipitation technique, then followed by the application of novel resonance energy transfer techniques (Calebiro et al., 2005; Calebiro, 2011; Fan & Hendrickson, 2005; Horvat, Roess, Nelson, Barisas, & Clay, 2001; Ji et al., 2004; Latif, Graves, & Davies, 2002; Osuga, Kudo, Kaipia, Kobilka, & Hsueh, 1997; Tao, Johnson, & Segaloff, 2004; Urizar et al., 2005). Apart from the demonstration of the GPHRs oligomerization, the importance of these studies resides in the effort to explain the role of the TSHR, LHR, and FSHR homo-oligomers in the thyroid and gonads: physiological and pathological states. Using these techniques, the presence of oligomers was demonstrated not only in *in vitro* heterologous systems but also in *in vivo* tissues (Urizar et al., 2005). Moreover, it was demonstrated that homodimerization is associated with strong negative cooperativity (an allosteric mechanism where ligand binding to one site reduces the binding affinity to another site on a molecule or dimer), as demonstrated by receptor binding and desorption experiments. These experimental data suggest that one hormone molecule may be sufficient for receptor dimer stimulation. Negative cooperativity might be the way for the endocrine system to respond over a wider range of agonist concentrations, with maximal sensitivity displayed in the lower concentration range. Further, a study from our group demonstrated the intracellular entrapment of wild-type TSHR by oligomerization with mutants linked to dominant TSH resistance and unable to undergo a full posttranslational glycosylation and reach the thyroid cell surface (Calebiro et al., 2005; Persani, Calebiro, & Bonomi, 2007; Persani et al., 2010). Our findings indicate that TSHR oligomerization occurs already in the endoplasmic reticulum, and it may be a required event for posttranslational maturation and targeting to the cell surface.

To comply with the needs of investigators in the field, we hereby provide detailed methods for fluorescence resonance energy transfer (FRET), bioluminescence resonance energy transfer (BRET), and homogenous time-resolved fluorescence-resonance energy transfer (HTRF-RET) techniques.

2. RESONANCE ENERGY TRANSFER TECHNIQUES

Resonance energy transfer takes place when part of the energy of an excited donor is transferred to an acceptor fluorophore, which reemits light at another wavelength, only if the emission spectrum of the donor molecule and the absorption spectrum of the acceptor molecule overlap sufficiently.

This physical phenomenon was first described by Förster (1948) and is governed by Förster's equation: $KET = 1/\tau_d\ (R_0/r)^6$, where $R_0 = 8.785 \times 10^{-25}\ cm^3\ mol^{-1}\ \kappa^2\ \varphi_d J/n^4$. Accordingly, the amount of energy transferred per unit of time (KET) depends on the lifetime of donor in the absence of acceptor (τ_d), the distance between donor and acceptor (r), and the distance at which FRET efficiency is 50% (R_0). This value is in turn determined by the following parameters: κ^2, a dimensionless factor accounting for the angular relationship between donor and acceptor; φ_d, the quantum yield of the donor fluorophore; J, a measure of the spectral overlap between donor emission and acceptor excitation; and n, the refraction index of the medium.

2.1. FRET technique

FRET is the nonradiative transfer of energy from an excited donor to an acceptor fluorophore. As a result, donor emission is reduced and emission from the acceptor fluorophore is observed. Since the efficiency of RET depends on the inverse sixth power of the distance between donor and acceptor, the detection of FRET is a very sensitive measure of the proximity of two fluorophores. Some conditions must, however, be satisfied in order for FRET to occur (1) the two fluorophores must be in close proximity; (2) a favorable orientation between donor and acceptor must exist; and (3) a certain overlap between the emission spectrum of the donor and the excitation spectrum of the acceptor is required (Milligan & Bouvier, 2005). Several pairs of FRET donor and acceptor have been utilized, depending on the experimental design. In studies of protein–protein (e.g., GPCR) interactions, a frequently devised strategy is to generate chimeric constructs of the molecules of interest fused to fluorescent proteins. Because of their favorable spectral overlap, enhanced cyan fluorescent protein (ECFP) and enhanced yellow fluorescent protein (EYFP)—two mutants of the jellyfish *Aequorea victoria* green fluorescent protein (GFP)—are frequently utilized for this purpose. For this fluorophore pair, FRET occurs within a radial distance of 10 nm (100 Å). Fusion proteins (i.e., proteins produced by genetic engineering of the protein of interest, so that it is fused with a fluorescent protein) can then be cotransfected into cells to study their interactions *in vivo* (Fig. 20.1). Several methods have been developed to detect FRET. The most obvious approach is to excite the donor and measure emission intensity from the acceptor fluorophore. This can be done either with a fluorometer or a fluorescence microscope equipped with appropriate excitation and emission filters. A major drawback of this technique is represented by the

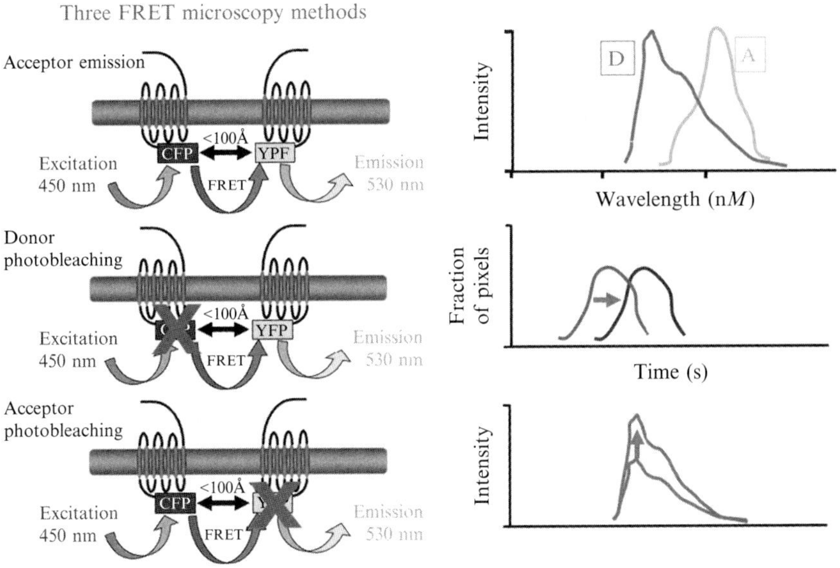

Figure 20.1 The principles of three methods for the evaluation FRET. (A) Acceptor emission: if FRET occurs between the donor CFP and the acceptor YFP, it may be evaluated by detecting the light emitted by the YFP acceptor, after excitation of the donor CFP by a light of specific wave length; (B) Donor photobleaching: if FRET occurs, the time required to bleach the donor increases because part of the energy is transferred to the acceptor; and (C) Acceptor photobleaching: if FRET occurs, the intensity of donor CFP emission should increase after the photobleaching of the YFP acceptor. (For color version of this figure, the reader is referred to the online version of this chapter.)

possible direct excitation of the acceptor fluorophore at the wavelength used for donor excitation; this can, nevertheless, be determined and compensated for (Eidne et al., 2002; Milligan & Bouvier, 2005). To avoid such problems, researchers have developed alternative methods that do not require direct measurement of the energy transferred to the acceptor. One of these, acceptor photobleaching, exploits the increase in donor emission that follows the exhaustion of the acceptor fluorophore. In this procedure, a baseline measurement of donor emission is performed. The specimen is then irradiated at the wavelength of acceptor excitation until exhaustion (bleaching), followed by a new measurement of donor emission. If FRET occurred between donor and acceptor, the donor is now free to emit all its energy by irradiation because of disappearance of energy transfer to the bleached acceptor, and an increase in the intensity of donor emission can be measured at the appropriate specific wavelength. Another recent development is donor photobleaching (Patel, Lange, & Patel, 2002). In this case, emission from the donor is recorded during constant illumination until it is exhausted and the time constant of donor photobleaching (τbl) is calculated by fitting intensity values to an exponential decay. In the presence of FRET, a delay in donor photobleaching is observed because of some energy being transferred to the acceptor. The efficiency of FRET can then be calculated by comparing τbl values in the presence or absence of the acceptor fluorophore (Patel et al., 2002). The major advantages of donor photobleaching compared with other methods are the independence of FRET efficiency on the expression levels of the donor fluorophore and the relatively limited cost of the instrumentation required (Eidne et al., 2002; Milligan & Bouvier, 2005; Patel et al., 2002). In addition to donor photobleaching, other time-resolved (TR) applications have been devised for measuring FRET. In the simplest version—fluorescence lifetime imaging—the specimen is excited with a short optical pulse, and the decay of donor fluorescence intensity is observed by confocal scanning or a high-speed camera. When present, FRET reduces the fluorescence lifetime of the donor. This method can be very sensitive but requires expensive equipment and often rather complex calculations (Suhling, French, & Phillips, 2005). Recently, a new application of TR-FRET allowing the evaluation of protein–protein interactions at the cell surface of intact cells has been developed. In this case, two antibodies—one directed against the extracellular portion of one of the proteins of interest, the other directed against the second protein of interest—are conjugated, respectively, with a long-lived donor fluorophore (e.g., europium) and a suitable fluorescent acceptor. The emission from donor and acceptor is then measured after a 50-μs delay to allow the decay of short-lived fluorescence from endogenous fluorophores. TR-FRET has been recently applied to the study of GPCR oligomerization (Maurel et al., 2004).

2.2. BRET technique

Bioluminescence is the principal source of light in dark environments, such as the deep sea, and BRET—the nonradiative transfer of energy from a bioluminescent donor to an acceptor fluorophore—is an innate biophysical phenomenon occurring in many marine organisms, for example, the sea pansy *Renilla reniformis*. BRET has, however, been brought only recently to the laboratory bench (Xu, Piston, & Johnson, 1999). In the BRET methodology, the first interaction partner is fused to *Renilla luciferase* (RLuc), which is spectrally similar to ECFP, whereas the second interaction partner is fused to a fluorescent protein (e.g., EYFP). RLuc oxydates its substrate,

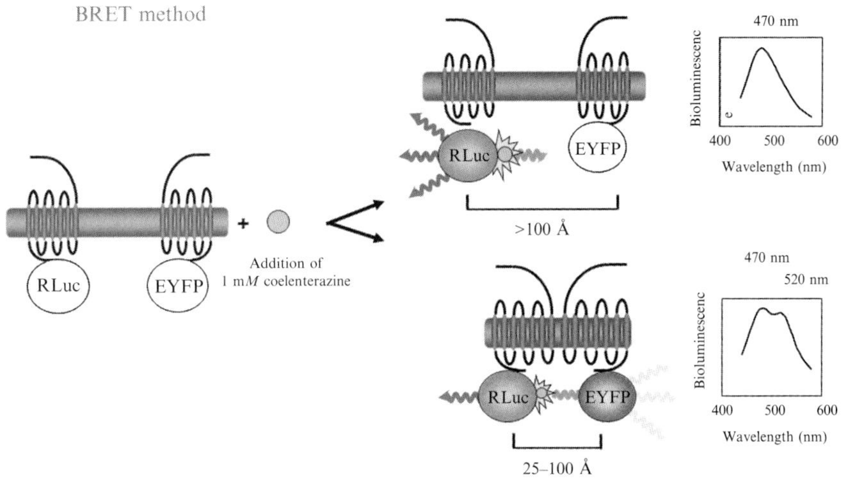

Figure 20.2 The principles of the BRET method. One of the partners of interest is fused to RLuc and the other to the EYFP. If the two partners do not interact, only one signal, emitted by the luciferase, can be detected after addition of its substrate, coelenterazine. If the two partners interact, resonance energy transfer occurs between the luciferase and the EYFP, and an additional signal, emitted by the EYFP, can be detected. (For color version of this figure, the reader is referred to the online version of this chapter.)

coelenterazine, and it produces photons with a wavelength of 480 nm. In the case, the two partners are not in the sufficiently close proximity to each other, the signal emitted by the luciferase will be detected (around 480 nm). If the two proteins are closer than 100 Å, RET will occur and a second emission will be registered, corresponding to the emission of the excited EYFP (approximately at 535 nm) (Fig. 20.2) (Eidne et al., 2002; Milligan & Bouvier, 2005; Xu et al., 1999). BRET detection is in general performed with a light detector equipped with adequate filters or a wavelength scanner. The BRET index is then determined by comparing the emission spectra of cells expressing RLuc alone and those producing both RLuc and EYFP. The overlap between the spectrum generated via RLuc enzymatic activity and the EYFP emission subsequent to energy transfer is substantial, resulting in a relatively high background signal. In order to improve this, the BRET2 method has been subsequently proposed. For this, the RLuc substrate coelenterazine is substituted by DeepBlueC, with the emission of light that is substantially blueshifted. In this case, a better spectral overlap for the RET phenomenon is provided by the use of modified fluorescent protein GFP2 as acceptor (Milligan & Bouvier, 2005). The characteristics of current BRET detection methods make this technique more suitable for screening

Table 20.1 Advantages and disadvantages of modern methods of analyzing protein–protein interactions for the GPHRs family

Method	Advantages	Disadvantages
FRET	Performed in native membranes and in the natural protein context Can be performed in living cells Possibility to execute time-course analysis Very sensitive, can be performed in single cells Permits visualization in subcellular compartments	Requires tagging of proteins Requires a light source Potential problems with photobleaching or direct excitation of acceptor fluorophore Possible dimerization of fluorescent proteins Complex and expensive instrumentation Most of the fluorescence resonance energy transfer methods are not suitable for screening
BRET	Performed in native membranes and in the natural protein context No artifact resulting from direct excitation of acceptor fluorophore No dimerization of tags Can be performed in living cells Possibility to execute time-course analysis Suitable for high-throughput screening	Requires tagging of proteins Relatively large tags Expensive instrumentation Less sensitive, not amenable to single cells Does not permit visualization in subcellular compartments
HTRF	Same as FRET, but: Applicable to native untagged receptors in their natural protein context	Same as FRET, but: Requires tagging of specific antibodies Requires the selection of suitable antibodies recognizing adequate epitopes on native proteins

Modified from Persani et al. (2007).

applications than FRET is. The advantages and drawbacks of current FRET and BRET technologies are illustrated in Table 20.1.

2.3. HTRF-RET technique

Recently, a new approach, based on FRET, was introduced to study GPCR (Maurel et al., 2004) and then GPHR (Urizar et al., 2005) oligomerization, a variant of the homogenous time-resolved fluorescence (HTRF/HTRF-RET) technology (see Fig. 20.3). The rare earth element

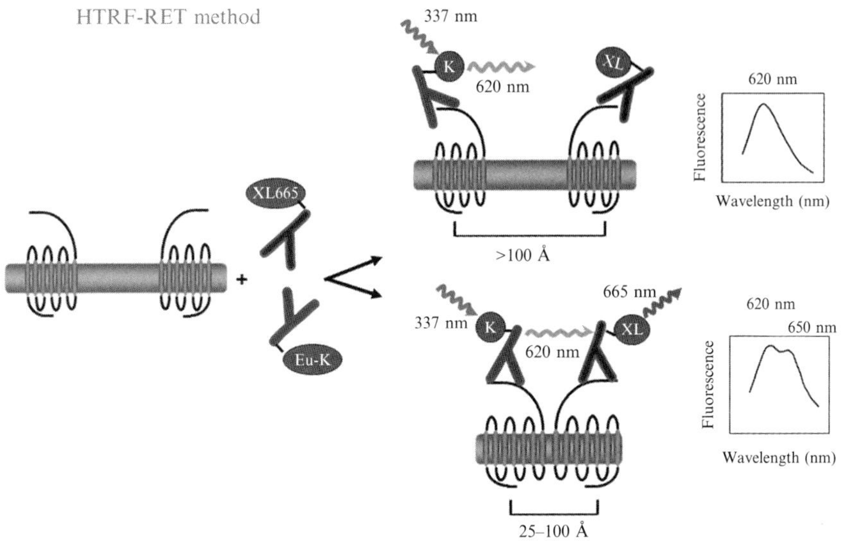

Figure 20.3 The principles of the HTRF-RET method. One antibody is fused to europium and the other one to XL-665. If the two partners do not interact, XL-665 emits light at 665 quite efficiently but is a short-lived if not in proximity to europium cryptate. If these two fluorophores are brought together by a biomolecular interaction, the long-lived europium cryptate donor induces a long-lived emission from XL665 which can be clearly distinguished from the short-lived X-665 signal using time-resolved detection. (For color version of this figure, the reader is referred to the online version of this chapter.)

lanthanide or europium is protected in the macrocyclic earth complex, called cryptate, and is used as the donor, while the modified allophycocyanin XL-665 is used as the acceptor. Both of the labels are stable, resist fluorescence quenching, and are easily conjugated to biomolecules of interest, such as specific antibodies able to recognize membrane protein. The importance of this methodology is testified by the possibility of specifically monitoring the interaction that occurs in the membrane in a TR manner and in an easier way (discriminating the fluorescence coming from serum or membrane components). On the other hand, since the energy transfer is happening between two labeled antibodies that recognize the putative cell surface interacting proteins, it is sufficient to label the antibodies without transfecting chimeric constructs. Further, since the antibodies are unable to cross the membrane, the antibody-based HTRF measures only the interactions occurring at the cell surface. Moreover, when available antibodies directed against the wild-type receptor, they can be used to determine the physical interaction *in vivo* in the native systems ruling out the effect of protein over expression.

3. EXPERIMENTAL PROCEDURES

3.1. Construct generation and transfection

To perform BRET and FRET studies, it is necessary to have plasmid construct harboring the GPHRs of interest. As mentioned above, receptor construct must be then fused in frame at its C-terminal either to the humanized RLuc or the EYFP variant of the GFP for the BRET and to the ECFP or EYFP reporters for the FRET study. These constructs will be then transfected in the cell system that will integrate the transfected DNA and will produce the mature protein (Bonomi, Busnelli, Persani, Vassart, & Costagliola, 2006).

Required materials

- DNA of interest: DNA of high quality (endotoxin-free plasmid DNA) is recommended, although lower-grade DNA would also allow reasonable yields of transfection.
- Growth medium: The growth medium for the cells contains DMEM supplemented with 10% FCS, Glutamax, and antibiotics (ampicillin/streptomycin mixture) could be used during cell growth.
- Transfection reagents: Lipofectamine (Life Technology, Italy) or calcium–phosphate precipitation methods could be used.

Other reagents

- Trypsin–EDTA solution (trypsin 0.25%, EDTA 0.05%).

Disposable

- 1.5 ml (Eppendorf) and 10 ml tubes.
- 10 cm tissue culture plates.

3.2. FRET experiment (from Calebiro et al., 2005)

Day 1: cells seeding: Cells are seeded at a density of 40,000 cells/well on sterile coverslips placed in 35 mm Petri dishes.

Day 2: cells transfection: 24 h after plating, expression vectors encoding fluorescent chimeras are cotransfected at 1:1 (1 + 1 μg DNA) or 1:4 (1 + 4 μg DNA) donor (ECFP) to acceptor (EYFP) ratios.

Day 3: FRET assay: 48 h after transfection, cells must be rinsed with PBS, fixed with paraformaldehyde 3.7% in PBS at 37 °C for 10 min, and washed twice with PBS. Coverslips are then removed, mounted with glycerol on microscope slides, and sealed. FRET is then measured as follows: the specimen is irradiated at the wavelength of 436 ± 20 nm and a time-lapse series of images of donor fluorescence is recorded at the wavelength of 480 ± 40 nm during continuous illumination. From the first image of the series, a binary mask is prepared to select pixels in the region of interest (ROI). The time series data for each pixel within an ROI are fit to an exponential decay function to determine decay constants of photobleaching. When FRET occurs between donor and acceptor fluorophores, the time constant for donor photobleaching increases (Jovin & Arndt-Jovin, 1989). Thus, the efficiency (*E*) of FRET is calculated as the percentage of change in the average time constant of donor photobleaching measured in specimens cotransfected with donor and acceptor fluorophore ($\tau_{donor+acceptor}$), with respect to that measured in specimens transfected with donor alone (τ_{donor}), via the following equation:

$$E = \left[1 - \left(\tau_{donor}/\tau_{donor+acceptor}\right)\right] \times 100.$$

One of the advantages of this method for measuring FRET is that the measurements do not end on absolute values of fluorescence. The photobleaching time constants were found to have skewed distributions, which became normal after logarithmic transformation. Therefore, data were analyzed using the natural logarithms of the photobleaching time constants, and efficiencies and statistics were derived by retransformation of the pertinent values.

Required materials

- Cells of interest: COS-7 can be obtained from the American Type Culture Collection (ATCC).
- Growth medium: The growth medium for the cells contains DMEM supplemented with 10% FCS, Glutamax, and antibiotics (ampicillin/streptomycin mixture).

Other reagents

- PBS 1×.
- Glycerol.

Disposable

- 35 mm tissue culture plates.
- Coverslips and microscope slides.

3.3. BRET experiment (from Urizar et al., 2005)

Day 1: cells seeding: Cells are seeded at a density of 2×10^6 cells/plate on sterile 100-mm culture plate.

Day 2: cells transfection: 24 h after plating, using the calcium–phosphate precipitation method, expression vectors encoding receptor-RLuc and receptor-EYFP are cotransfected, and empty vector is always added in order to transfect constant amount of plasmid DNA.

Day 3: cells splitting: 24 h after transfection, cells were detached with PBS 5 m*M* EDTA, centrifuged, and resuspended in the culture medium without phenol red. Approximately, 3×10^4 cells were seeded in tissue culture, treated, sterile 96-well plates (VWR International PBI, Italy). The next day, cells were used for BRET.

Day 4: BRET assay: At 48 h posttransfection, the culture medium should be replaced by PBS 0.1% glucose at room temperature (RT), and cells could be kept attached to the plastic surface during the BRET assay. Readings (except for the kinetic studies; see below) might be collected using a Mithras LB 940 Multireader (Berthold Technologies, Italy) and a MicroWin2000 software that allows the sequential integration of the signals detected in the 440–500 and 510–590 nm windows using filters with the appropriate band pass. The BRET signal is calculated by the ratio of the light emitted by EYFP (510–590 nm) over that emitted by the RLuc (440–500 nm). The measurements should be performed just after the addition of the substrate and 5, 10, and 15 min after. The net BRET values are derived by subtracting the background signal detected when RLuc-tagged construct was expressed alone. Total fluorescence and luminescence direct measurement allow the determination of the expression level of each protein. The EYFP total fluorescence is measured by using an excitation filter at 485 nm, an emission at 530 nm, and the following parameters: lamp energy 22,000; reading time 1.0 s. After fluorescence measurement, 5 μ*M* coelenterazine H must be added in the same cells and then the total luminescence of cells measured (Mithras LB 940) during 1 s, and the reading repeated 5, 10, and 15 min after the addition of the substrate in order to always monitor the total luminescence in the equilibrium. To avoid variations in the BRET signal that could result from fluctuation in the relative expression levels of the energy donor and acceptor, transfection conditions should be designed (except in the full titration curve experiments, see below) so to maintain constant EYFP/RLuc expression ratio in each experiment set. For BRET titration curves, cells must be cotransfected with a fixed amount of the RLuc-tagged receptor and increasing amounts of the EYFP-tagged receptor. Net BRET signal is plotted as a function of the EYFP over RLuc fusion expression (EYFP/RLuc). The curves must be fitted using a nonlinear regression equation assuming a single binding site. In the kinetic analysis of receptor interactions upon ligand addition, cH should be added prior to the injection of the ligand using the Mithras LB 940. Readings is then collected at 1 s intervals. The BRET signals are determined for each time by calculating the ratio of the light emitted by EYFP over that emitted by the RLuc. For agonists stimulations longer than 1 min, cells must be first treated with the drug (i.e., endogenous ligand), incubated in different reaction conditions during 10–60 min, and cH added immediately before the readings. BRET, total fluorescence, and total luminescence readings are then performed as explained before.

Required materials

- Cells of interest: HEK293 can be obtained from the ATCC.
- Growth medium: The growth medium for the cells contains DMEM supplemented with 10% FCS, Glutamax, and antibiotics (ampicillin/streptomycin mixture).

Other reagents

- Trypsin-EDTA.
- PBS 1×.
- Coelenterazine H (Molecular Probes, Life Technology).

Disposable

- 100 mm tissue culture plates.
- Tissue culture sterile 96-well plate (VWR International PBI).

3.4. HTRF-RET experiment (from Urizar et al., 2005)

3.4.1 Antibodies labeling

The first step is the antibodies activation with 4 equiv. of *N*-succinimidyl 3-[2-pyridyldithio] propionate already dissolved in ethanol. This step is followed by a 30 min incubation at RT. A 10 m*M* dithiothreitol is then added in the activated antibody solution, and the reduction step is performed 15 min at RT. The activated antibody is then purified on a Sephadex-G25SF HR10/10 column (Amersham Pharmacia Biotech) that must be previously equilibrated with a 0.1 *M* phosphate, 5 m*M* EDTA buffer, pH 6.9. The europium cryptate labeling of the antibody should be done in the presence of a 10-fold excess of the cryptate per antibody. Coupling reactions are incubated for 2 h at RT. Europium cryptate conjugates are then purified on a G25SF HR10/30 column (Amersham Pharmacia Biotech) conveniently pre-equilibrated with a 0.1 *M* phosphate buffer, pH 7.0. The number of europium cryptate per antibody (final molar ratio) could be determined on a spectrophotometer by measuring their absorbance at 280 and 305 nm and inserting the measured values into the equation:

$$\text{Molar ratio} = \frac{\left(\frac{\text{OD317}}{\varepsilon\,\text{cryptate(diMP)}}\right)}{\left(\frac{\text{OD280}-\left(=\frac{\text{D305}}{\text{A}}\right)}{\varepsilon\,\text{antibody}}\right)}.$$

The molar extinction coefficient (ε) of the europium cryptate is taken as 30,000 M^{-1} cm^{-1} at 305 nm, and the molar extinction coefficient of the antibodies is taken as 150,000 M^{-1} cm^{-1} at 280 nm. The factor A expressed the ratio (OD 305 nm/OD 280 nm) for europium cryptate and is determined to be 2.8. The acceptor *N*-hydroxysuccinimide ester derivative of XL665 could be used to label the different antibodies previously dissolved in a 0.1 *M* carbonate buffer, pH 9. An excess of 15 XL665 per antibody should be used in the labeling reaction during 1 h at RT in the presence of β-cyclodextrin. Final conjugates must be purified using a Superdex 75 HR10/30 column (Amersham Pharmacia Biotech.). The final number of dyes per antibody should be determined on a spectrophotometer as described above for the europium cryptate conjugates. OD at 305 nm must be replaced by the OD at 650 nm, which is the maximum of absorption for XL665. The molar extinction coefficient of the derivative of XL665 at 650 nm is taken as 250,000 M^{-1} cm^{-1}, and the factor A expressed the ratio (OD 650 nm/OD 280 nm) for XL665 is determined to be 20.

3.4.2 Homogenous time-resolved fluorescence assay

Day 1: cells seeding: Cells are seeded at a density of 2×10^6 cells/plate on sterile 100-mm culture plate.

Day 2: cells transfection: 24 h after plating, using the calcium–phosphate precipitation method, expression vectors encoding the wild-type receptor or the different receptors of interest are transfected or cotransfected in the cells.

Day 3: HTRF measurement: 24 h after transfection, cells are harvested and resuspended in PBS. Approximately 100,000 cells in 50 μl are distributed in black 96-well reading plate, where 0.5 μg/ml europium cryptate-labeled donor antibody and 5 μg/ml acceptor antibody must be previously added, in a dilution buffer with 0.4 *M* KF that prevent any potential fluorescence quenching effect from serum media (Maurel et al., 2004); fluorescence emissions are then monitored both at 620 nm and at 665 nm after 1 h incubation at 37 °C. A 400 μs integration time is used after a 50 μs delay to remove the short-lived fluorescence background from the specific signal. Both emissions must be measured on a dual-wavelength (665/620 nm) time-resolved fluorimeter. The ratio (fluorescence 665 nm/fluorescence 620 nm) is then computed. The specific signal over background called Delta *F* (ΔF) should be calculated using the following formula: $\Delta F = (\text{Ratio rec} - \text{Ratio neg})/(\text{Ratio neg})$. "Ratio neg" corresponded to the ratio for the negative energy transfer control, whereas "Ratio rec" corresponded to the ratio for the positive energy transfer control. The negative control is represented by HEK 293T cells expressing the empty vector or the receptor lacking the epitopic region for the antibody.

Required materials

- Cells of interest: HEK293 can be obtained from the ATCC.
- Growth medium: The growth medium for the cells contains DMEM supplemented with 10% FCS, Glutamax, and antibiotics (ampicillin/streptomycin mixture).

Other reagents

- PBS 1×.
- *N*-succinimidyl 3-[2-pyridyldithio] propionate (Pierce, Italy).
- Dithiothreitol (Sigma).

Disposable

- 100 mm tissue culture plates.
- Black 96-well reading plate (VWR International PBI).

ACKNOWLEDGMENTS

Dr. Marco Bonomi is presently supported by Funds from the Italian Minister of Health for Young Investigator (GR-2008-1137632).

REFERENCES

Bonomi, M., Busnelli, M., Persani, L., Vassart, G., & Costagliola, S. (2006). Structural differences in the hinge region of the glycoprotein hormone receptors: Evidence from the sulfated tyrosine residues. *Molecular Endocrinology, 20*(12), 3351–3363.

Calebiro, D. (2011). Thyroid-stimulating hormone receptor activity after internalization. *Annales d'endocrinologie, 72*(2), 64–67.

Calebiro, D., de Filippis, T., Lucchi, S., Covino, C., Panigone, S., Beck-Peccoz, P., et al. (2005). Intracellular entrapment of wild-type TSH receptor by oligomerization with mutants linked to dominant TSH resistance. *Human Molecular Genetics, 14*(20), 2991–3002.

Eidne, K. A., Kroeger, K. M., & Hanyaloglu, A. C. (2002). Applications of novel resonance energy transfer techniques to study dynamic hormone receptor interactions in living cells. *Trends in Endocrinology and Metabolism, 13*(10), 415–421.

Fan, Q. R., & Hendrickson, W. A. (2005). Structure of human follicle-stimulating hormone in complex with its receptor. *Nature, 433*(7023), 269–277.

Förster, T. (1948). Intermolecular energy migration and fluorescence. *Annals of Physics (Leipzig), 2*, 55–75.

Hanyaloglu, A. C., Seeber, R. M., Kohout, T. A., Lefkowitz, R. J., & Eidne, K. A. (2002). Homo- and hetero-oligomerization of thyrotropin-releasing hormone (TRH) receptor subtypes. Differential regulation of beta-arrestins 1 and 2. *Journal of Biological Chemistry, 277*(52), 50422–50430.

Horvat, R. D., Roess, D. A., Nelson, S. E., Barisas, B. G., & Clay, C. M. (2001). Binding of agonist but not antagonist leads to fluorescence resonance energy transfer between intrinsically fluorescent gonadotropin-releasing hormone receptors. *Molecular Endocrinology, 15* (5), 695–703.

Ji, I., Lee, C., Jeoung, M., Koo, Y., Sievert, G. A., & Ji, T. H. (2004). Trans-activation of mutant follicle-stimulating hormone receptors selectively generates only one of two hormone signals. *Molecular Endocrinology, 18*(4), 968–978.

Jovin, T. M., & Arndt-Jovin, D. J. (1989). Luminescence digital imaging microscopy. *Annual Review of Biophysics and Biophysical Chemistry, 18*, 271–308.

Latif, R., Graves, P., & Davies, T. F. (2002). Ligand-dependent inhibition of oligomerization at the human thyrotropin receptor. *Journal of Biological Chemistry, 277*(47), 45059–45067.

Maurel, D., Kniazeff, J., Mathis, G., Trinquet, E., Pin, J. P., & Ansanay, H. (2004). Cell surface detection of membrane protein interaction with homogeneous time-resolved fluorescence resonance energy transfer technology. *Analytical Biochemistry, 329*(2), 253–262.

Meduri, G., Bachelot, A., Cocca, M. P., Vasseur, C., Rodien, P., Kuttenn, F., et al. (2008). Molecular pathology of the FSH receptor: New insights into FSH physiology. *Molecular and Cellular Endocrinology, 282*(1–2), 130–142.

Milligan, G. (2004). G-protein coupled receptors dimerization: Function and ligand pharmacology. *Molecular Pharmacology, 66*, 1–7.

Milligan, G., & Bouvier, M. (2005). Methods to monitor the quaternary structure of G protein-coupled receptors. *FEBS Journal, 272*(12), 2914–2925.

Osuga, Y., Kudo, M., Kaipia, A., Kobilka, B., & Hsueh, A. J. (1997). Derivation of functional antagonists using N-terminal extracellular domain of gonadotropin and thyrotropin receptors. *Molecular Endocrinology, 11*(11), 1659–1668.

Park, P. S., Filipek, S., Wells, J. W., & Palczewski, K. (2004). Oligomerization of G protein-coupled receptors: Past, present, and future. *Biochemistry, 43*(50), 15643–15656.

Patel, R. C., Lange, D. C., & Patel, Y. C. (2002). Photobleaching fluorescence resonance energy transfer reveals ligand-induced oligomer formation of human somatostatin receptor subtypes. *Methods, 27*(4), 340–348.

Persani, L., Calebiro, D., & Bonomi, M. (2007). Modern methods to monitor protein-protein interactions: Insights on TSH receptor oligomerization. *Nature Clinical Practice. Endocrinology & Metabolism, 3*, 180–190.

Persani, L., Calebiro, D., Cordella, D., Weber, G., Gelmini, G., Libri, D., et al. (2010). Genetics and phenomics of hypothyroidism due to TSH resistance. *Molecular and Cellular Endocrinology, 322*(1–2), 72–82.

Puett, D., Angelova, K., da Costa, M. R., Warrenfeltz, S. W., & Fanelli, F. (2010). The luteinizing hormone receptor: Insights into structure-function relationships and hormone-receptor-mediated changes in gene expression in ovarian cancer cells. *Molecular and Cellular Endocrinology, 329*(1–2), 47–55.

Suhling, K., French, P. M., & Phillips, D. (2005). Time-resolved fluorescence microscopy. *Photochemical and Photobiological Sciences, 4*(1), 13–22.

Tao, Y. X., Johnson, N. B., & Segaloff, D. L. (2004). Constitutive and agonist-dependent self-association of the cell surface human lutropin receptor. *Journal of Biological Chemistry, 279*(7), 5904–5914.

Terrillon, S., & Bouvier, M. (2004). Roles of G-protein-coupled receptor dimerization. *EMBO Reports, 5*(1), 30–34.

Urizar, E., Montanelli, L., Loy, T., Bonomi, M., Swillens, S., Gales, C., et al. (2005). Glycoprotein hormone receptors: Link between receptor homodimerization and negative cooperativity. *EMBO Journal, 24*(11), 1954–1964.

Van Loy, T., Vandersmissen, H. P., Van Hiel, M. B., Poels, J., Verlinden, H., Badisco, L., et al. (2008). Comparative genomics of leucine-rich repeats containing G protein-coupled receptors and their ligands. *General and Comparative Endocrinology, 155*(1), 14–21.

Vassart, G., Pardo, L., & Costagliola, S. (2004). A molecular dissection of the glycoprotein hormone receptors. *Trends in Biochemical Sciences, 29*(3), 119–126.

Xu, Y., Piston, D. W., & Johnson, C. H. (1999). A bioluminescence resonance energy transfer (BRET) system: Application to interacting circadian clock proteins. *Proceedings of the National Academy of Sciences of the United States of America, 96*(1), 151–156.

AUTHOR INDEX

Note: Page numbers followed by "*f*" indicate figures, and "*t*" indicate tables, and "*np*" indicates footnotes.

A

AbdAlla, S., 220–221, 306
Abdel-Baset, A., 220–221
Abramowitz, J., 132
Achour, L. O., 132–133
Adams, D. R., 82, 84
Adams, P. D., 181–183
Adessi, C., 4
Aebi, M., 21
Agnati, L. F., 282
Ahmad, M., 162–164
Ahn, S., 158–159
Aittomaki, K., 23–24
Alayash, A. I., 153
Albizu, L., 260–261
Alevizaki, M., 23–24
Alexander, R. W., 48
Algeri, S., 282
Alken, M., 133
Allalou, A., 289–290
Allen, J. A., 48
Alm, C., 133–134
Almen, M. S., 330
Alonso, G., 343
Alt, C., 152–153
Altman, M. D., 87–88
Altschuler, Y., 190
Alvarez-Curto, E., 72–73, 74–75, 82, 88, 302–303
Amaral, M. D., 4
Ambrose, C., 350, 355–356
Ambrosino, C., 145
Ambrosio, M., 319–320
Amherdt, M., 190
An, Y., 190–191
Anand, J. P., 70
Andersen, T. T., 36
Andrieu, E. U., 82
Angelotti, T., 173–175, 176*f*, 177–179, 179*f*, 181, 181*f*, 185–186, 185*f*
Angelova, K., 368–369
Angers, S., 313, 337
Angulo, E., 241–242
Anisman, H., 220–221
Annibale, P., 296
Ansanay, H., 370–372, 374–375, 380–381
Anselmo, A., 153
Antic, D., 141
Appelbe, S., 71, 88
April, H., 53–54
Aragao, D., 255
Arana, Q., 51
Arancibia-Carcamo, I. L., 153–154
Areces, L., 324
Arens, C., 314–316
Arlow, D. H., 255
Armstrong, D., 297–298
Armstrong-Gold, C., 110–111
Arndt-Jovin, D. J., 376–377
Arthus, M. F., 7, 12–13, 209
Ascoli, M., 22, 33
Ashby, B., 53–54
Ashby, M. C., 112
Ashton, W. T., 5
Assini, A., 4
Aucouturier, P., 4
Audet, M., 101–102
Auzan, C., 209
Ayad, N. G., 8
Ayala-Yanez, R., 146
Ayers, D. F., 221–222
Ayoub, M. A., 75, 325–326
Aziz, A. S., 146

B

Baameur, F., 359
Baba, T., 40–41
Babak, T., 133–134
Baccetti, B., 355–356
Bachelot, A., 368–369
Baddeley, S. M., 110–111, 116, 119, 127*f*
Badisco, L., 368–369

Baggiolini, M., 152
Bahram, F., 289–290, 291
Bailey, C. P., 70
Bailey, T. J., 53–54
Baillie, G. S., 82, 84
Baker, J. G., 205, 255
Balass, M., 110–111
Balboni, G., 220–221
Balch, W. E., 190, 192–193
Baldassa, S., 145
Baldys, A., 134, 137, 138, 139–141, 140*f*, 142*f*, 144*f*, 145–146
Balenga, N. A., 220
Baler, R., 282
Ballet, S., 82
Banerji, S., 153
Banks, P., 8–9
Bannykh, S., 190–191
Baradari, V. K., 359
Baragli, A., 305
Barak, L. S., 101–103, 104*t*, 132
Barisas, B. G., 369
Barisas, G., 30–31
Bariteau, J. T., 25, 33, 36, 38–39
Barlowe, C., 190
Barnhill, J. F., 356
Barret, C., 29–30
Barth, F., 220
Barthet, G., 302–303
Baserga, R., 337
Bazzini, C., 313
Beau, I., 23–24
Beauchamp, J., 18–19
Beaulieu, J. M., 101–102
Becamel, C., 343
Beck, P., 241
Beck-Peccoz, P., 369, 376–377
Behnke, C. A., 205
Beilharz, T. H., 190–191
Benedek, G. B., 4
Benfenati, F., 282
Benke, D., 110–111
Benovic, J. L., 22, 92–93, 94, 96–97, 97*f*, 100–103, 104*t*, 132, 174–175, 348–349, 355–356
Berg, K. A., 220–221
Berlot, C. H., 266, 275–276
Bermak, J. C., 191, 195–196
Bernier, V., 4, 5–6, 7, 12–13, 204, 209
Berrendero, F., 220
Bertani, I., 4
Bertrand, D., 119
Bertrand, S., 119
Bettler, B., 110
Bhalla, S., 146, 172, 209
Bhari, N., 82, 84
Bhaskaran, R. S., 22, 33
Bhushan, R. G., 220–221
Bi, J., 145–146
Bichet, D. G., 4, 5–6, 29–30, 204
Bickford, L. C., 190–191
Biechele, T. L., 337
Birdsall, N. J. M., 240, 308
Birnbaumer, L., 132
Blab, G. A., 61
Blackburn, P. E., 153
Blacklock, B. J., 50
Blair, E., 153
Blake, A. D., 132
Blake-Palmer, K., 260–261
Blaser, H., 152–153
Blobel, G., 132–133
Bobadilla-Lugo, R. A., 220–221
Bockaert, J., 205, 343
Bodduluri, H., 152–153
Boeddrich, A., 4
Bogardus, A. M., 53–54
Bogdanov, Y., 110–111
Bogerd, J., 18–19
Boguth, C. A., 348–349, 350–351, 356, 357
Bohn, L. M., 70
Bokoch, M. P., 70
Bolanowski, M. A., 132, 134
Boldajipour, B., 152–153
Boldt, K., 337
Bonecchi, R., 152–153, 158–159, 161–162
Bonner, T. I., 220
Bonomi, M., 369, 374–375, 374*t*, 376, 377–381
Borleis, J. A., 53
Borowsky, B., 241
Borroni, E. M., 152–153, 158–159
Borroto-Escuela, D. O., 282, 289–290, 291
Borta, H., 23–24
Bot, G., 220–221
Botta, G., 313

Bottomley, S. P., 4
Boulay, F., 34
Boulo, T., 19
Bourne, H. R., 304
Bourrier, E., 75
Bouvier, M., 4, 5–6, 70–71, 82, 87–88, 101–102, 146, 172, 195–196, 204, 209, 244–246, 260–261, 282, 301, 312–313, 368–369, 370–374
Boykins, R. A., 153
Brady, A. E., 70
Brand, L., 321
Brann, M. R., 29–30
Breit, A., 313, 325–326
Brenner, C., 92
Bresee, J. E., 355–356
Breton, B., 101–102
Briddon, S. J., 262, 275–276
Brock, C., 75
Brothers, S. P., 5, 146, 172
Broussard, C., 330, 340, 342–343
Brown, A. M., 132
Bruel, C., 350
Brum, P. C., 173
Brumberg, H. A., 24, 25–27
Buczylko, J., 348–349, 350, 355–356, 358
Buehler, P. W., 153
Bullock, C., 191, 195–196
Bulteau-Pignoux, L., 4
Bunemann, M., 132
Burgueño, J., 241–242
Burnstock, G., 241, 255
Burtey, A., 92, 100–101, 104*t*
Bush, C. F., 241–242
Bush, E., 5–6, 12–13
Bushlin, I., 220, 221, 224–226, 228–229, 230–233
Busillo, J. M., 92–93, 96–97, 97*f*, 102
Busnelli, M., 376
By, K., 4

C

Cabral, G., 220
Calebiro, D., 368–369, 374*t*, 376–377
Cameroni, E., 153
Campbell, P. T., 132
Campbell, R. E., 270*np*
Canals, M., 70–71, 84–86, 87–88, 220–221, 241–242
Candelore, M. R., 132
Canela, E. I., 241–242, 298–299
Cantrell, D. A., 164–166
Cao, Y. Q., 101–102
Capel, W. D., 349, 354–355
Capra, V., 319–320
Caron, M. G., 70, 92, 100–103, 104*t*, 132, 282, 355–356
Carpenter, A. E., 289
Casadó, V., 241–242, 298–299
Casas-Gonzalez, P., 20–21, 24, 29–30, 33
Casellas, P., 220
Cassier, E., 302–303
Cassoni, P., 324
Castrillon, J., 283, 291
Cerovina, T., 133–134
Cha, K., 350
Chai, Y., 20
Chakir, K., 301
Chan, A. Y., 164–166
Chan, F. K., 30–31, 32
Chang, K., 204–205
Changeux, J. P., 110–111, 119
Charest, P. G., 101–102
Charo, I. F., 152
Chase, P., 8
Chattopadhyay, A., 48
Chauhan, S. S., 162–164
Chavrier, P., 121
Chazot, P. L., 72
Chelsky, D., 313
Chemel, B. R., 240, 260–261, 263, 264, 265–266, 267–268, 270–271, 272, 274, 275–276
Chen, C. K., 220–221, 350
Chen, J., 4, 263, 264, 265–266, 267–268, 270–271, 275–276
Chen, K., 134, 137–138
Chen, L., 4
Chen, Y. J., 350
Cherezov, V., 205
Cheung, C. C., 8–9
Chien, E. Y., 205, 255
Chikaraishi, D. M., 265–266
Chini, B., 324

Chinol, M., 324
Chooback, L., 51
Choy, E. W., 35
Chung, H. T., 29–30
Chung, N., 350, 355–356
Chung, T. D., 7
Church-Kopish, J., 350
Cichewicz, D. L., 220
Cimermancic, P., 100–101
Cimmino, M., 282
Ciruela, F., 241–242, 282, 291, 298–299
Clark, J. I., 4
Clark, R. B., 359
Clarke, C., 289
Clary, G., 330, 340, 342–343
Clauser, E., 191, 209
Clausson, C. M., 289–290
Clay, C. M., 369
Claycomb, W. C., 192–193
Cocca, M. P., 368–369
Cognet, L., 61
Cohen, B. D., 18–19, 21, 25, 33, 36, 38–39
Cohen, R. S., 143–144, 145–146
Colapietro, A. M., 102–103, 104*t*
Coleman, R. A., 53–54
Collins, A., 190
Collins, S., 132
Collodel, G., 355–356
Colzi, A., 305
Comerford, I., 152–153, 161–162
Comps-Agrar, L., 75
Conklin, B. R., 304
Conn, P. M., 4–6, 7, 9–11, 12–13, 20–21, 24, 29–30, 70, 132–133, 146–147, 172, 204, 209, 306–307
Conner, D. A., 92, 97, 102
Connolly, C. N., 114
Conti, M., 29–30
Corbin, J. D., 8–9
Corchero, J., 220
Cordella, D., 368–369
Cornea, A., 5
Corringer, P. J., 110–111
Corsi, M. M., 152–153
Corsini, G. U., 302–303
Costagliola, S., 18–19, 301–302, 368–369, 376
Cotecchia, S., 302–303
Covino, C., 369, 376–377
Crabb, J. W., 348–349, 358
Craig, D. A., 241
Craig, S., 158–159
Crepieux, P., 19–20, 22, 33, 34–35
Crooke, S. T., 134
Cubitt, A. B., 112*t*
Cui, Z. Q., 263
Cunningham, D. D., 36
Curatolo, L. M., 36
Cvejic, S., 235
Czaplinski, K., 133–134, 141

D

da Costa, M. R., 368–369
Dahan, S., 190
Dai, M., 241
Dai, W. W., 164–166
Dal Sacco, Z., 51
D'Ambrosi, N., 241–242
Damke, H., 40–41
Daniel, E., 260–261
Daniels, D. J., 220–221
Dascher, C., 190
Daulat, A. M., 330, 340, 342–344
Daunt, D. A., 173–175, 176*f*, 177–179, 179*f*, 181, 181*f*, 185–186, 185*f*
Daut, J., 190
Davies, D. R., 112*t*
Davies, T. F., 369
Davis, D. P., 20, 21
Davis, M., 190
De Blasi, A., 22
de Filippis, T., 369, 376–377
De Keijzer, S., 48, 49–50, 51–53, 56–57, 59, 60*f*, 61, 62
De Matteis, M. A., 184–185
De Weerd, W. F. C., 100
Decaillot, F. M., 230–233
Decourtye, J., 19–20
Defino, M. C., 337
Dehe, M., 172
Dehghani-Tafti, E., 110–111, 116, 119, 127*f*
Dejgaard, K., 190
Demey, E., 343
Deng, N., 355–356
Deng, X., 263, 266, 270–271, 270*np*

Dennis, M., 313
Derand, R., 4
Desai, S., 53–54
DesGroseillers, L., 145
Devane, W. A., 220
Devi, L. A., 220–222, 228–229, 230–233, 235, 260–261, 282, 299–300, 301–302, 304–305, 306
Devillers-Thiery, A., 119
Devost, D., 325–326
Devreotes, P. N., 48, 50, 51–53, 60*f*, 61
Di Marzio, P., 164–166
Di Marzo, V., 220
Dias, J. A., 18–19, 20–21, 23–24, 25–27, 26*f*, 28–31, 32, 33, 36, 38–39, 40–41
Dickson, H. M., 355–356
Dictenberg, J., 146–147
Didriksen, M., 101–102
Dietis, N., 220
Ding, J., 355–356
Dion, S. B., 349
Diwan, M., 220–221
Dixon, R. A., 132, 134
Dobberstein, B., 132–133
Dohlman, H. G., 132, 134
Dolphin, A. C., 110–111
Dominguez, M., 190
Donello, J. E., 53–54
Dong, C., 172, 190, 191, 192–193, 194, 195–196, 197–199, 198*f*, 199*f*
Doni, A., 153
Donnellan, P. D., 191
Dormer, R. L., 4
Downey, W. E. III., 102–103, 104*t*
Drakas, R., 337
Drake, M. T., 359
Dunham, J. H., 146, 147, 204
Dunlop, J., 9–11
Durand, G., 19–20
Durand, S., 4
Duren, H. M., 264
Durkin, M. M., 241
Durroux, T., 282, 325–326
Duthey, B., 172
Duvernay, M. T., 20–21, 132–133, 172, 191, 192–193, 194
Dziedzicka-Wasylewska, M., 220–221

E

Ecker, D. J., 134
Eda, M., 4
Edidin, M., 30–31
Edwards, P. C., 205, 255
Edwards, S. W., 184–185
Eidne, K. A., 244, 368–369, 370–374
Eisele, J. L., 119
Ejendal, K. F. K., 260–261, 262, 275–276
el Massiery, A., 220–221
Elalouf, J. M., 22
El-Asmar, L., 82
Ellgaard, L., 172
Ellis, J., 71
Ellis, R. E., 264
Elson, E. L., 64
Emerson, J. D., 12
Ersoy, B., 8–9
Escriche, M., 241–242
Ey, P. L., 226–227

F

Fagni, L., 343
Fairfax, B. P., 153–154
Falck, J. R., 92, 102–103, 104*t*
Fan, G. H., 155–157
Fan, J., 209
Fan, J. Q., 4
Fan, J. Y., 263
Fan, Q. R., 21, 29–30, 369
Fan, T., 29–30, 220–221, 301
Fan, X. C., 141
Fanelli, F., 319–320, 368–369
Farfel, Z., 304
Farhan, H., 190–191
Farnebo, M., 289–290
Faron-Gorecka, A., 220–221
Favre, H., 8–9
Fazel, A., 190
Feldhammer, M., 4
Feller, D. B., 337
Fenalti, G., 255
Feng, F. Y., 174–175, 177, 179–181
Feng, X., 29–30
Ferguson, S. S. G., 92, 100–103, 104*t*, 268

Fernandez-Ruiz, J. J., 220
Ferre, S., 220–221, 241–242, 282
Fessart, D., 101–102
Fiedler, K., 190
Field, M. E., 35
Filipeanu, C. M., 132–133, 172, 191, 192–193
Filipek, S., 209, 368–369
Filipovska, J., 220–221, 232–233
Filizola, M., 260–261, 275–276
Filliol, D., 101–102
Firsov, D., 22
Fletcher, P. J., 220–221
Foerster, B. R., 153
Fogliarino, S., 4
Ford, M. G. J., 92, 100–101, 104*t*
Forloni, G., 4
Förster, T., 369–370
Fox, B. A., 205
Fra, A. M., 153
Fraga, S., 4
Franchin, G., 164–166
Francis, D. J., 92
Franco, R., 298–299
Frankowska, M., 282
Fredriksson, R., 330
Freedman, N. J., 132
Freeman, J. L., 354–355
Freissmuth, M., 5–6, 190–191
French, P. M., 370–372
Fridkin, M., 110–111
Friedman, J., 359
Frielle, T., 132, 134
Friman, O., 289
Frings, S., 263
Froehlich, W. M., 50
Froestl, W., 241
Froment, C., 330
Frossard, M. J., 4
Fuentes, J. A., 220
Fugetta, E. K., 192–193
Fujino, H., 53–54
Fujiwara, T. K., 296, 307–308
Fukui, K., 240
Fukui, Y., 51–52
Fullekrug, J., 190
Furic, L., 145
Furst, J., 313
Futcher, B., 337
Fuxe, K., 220–221, 282

G

Gagne, A., 8–9
Gagnidze, K., 220–221, 232–233
Gagnon, A. W., 92, 97, 100–101, 102–103, 104*t*
Gaidarov, I., 92, 102–103, 104*t*
Gainetdinov, R. R., 70, 101–102
Gales, C., 369, 374–375, 377–381
Galet, C., 22, 33
Galietta, L. J., 4
Gallant, M. A., 316
Galliera, E., 152–153, 161–162
Galloway, J., 48, 51–53, 60*f*, 61
Gallwitz, D., 190–191
Galzi, J. L., 119
Gan, X. Q., 350, 355–356
Gao, Y., 209
Gao, Z. G., 255
Garcia, G. L., 49–50
Garcia-Negredo, G., 282
Garriga, P., 282
Gassmann, M., 110
Gaus, K., 63–64
Gauthier, C., 19–20, 22, 33, 34–35
Gaven, F., 302–303
Gavis, E. R., 145
Geleotti, N., 343
Gelmini, G., 368–369
George, S. R., 29–30, 220–221, 301
Gerard, N. P., 158–159
Gerst, J. E., 145–146
Gesty-Palmer, D., 101–102
Gharahdaghi, F., 336
Ghirlando, R., 356
Ghisi, V., 101–102
Giepmans, B. N., 270*np*
Giguere, P., 316
Giguere, V., 316
Gimenez, L. E., 92
Ginés, S., 241–242
Giorgi, F., 302–303
Glass, M., 260–261
Gloeckner, C. J., 337
Godinez-Hernandez, D., 220–221
Goldberg, J., 190–191

Goldenberg, S. J., 337
Gomes, I., 220–221, 224–226, 228–229, 230–233, 301–302, 306
Gomez, J., 349
Gonzalez-Hernandez Mde, L., 220–221
Gonzalez-Maeso, J., 260–261
Goodman, O. B. Jr., 92, 97, 100–101, 104*t*
Gorecki, A., 220–221
Gorvel, J. P., 121
Goueli, S. A., 8–9
Goulet, M. T., 5–6, 12–13, 87–88
Graham, G. J., 161–162
Grampp, T., 110–111
Granada, B., 270*np*
Graves, P., 369
Green, G., 352
Greer, J., 5–6, 12–13
Greschniok, A., 23–24
Griffith, M. T., 205
Grinde, E., 268
Gromoll, J., 18–19, 23–24
Gross, L. A., 112*t*
Gruber, C. W., 5–6
Grunewald, B., 145
Gualandi, L., 289–290, 291
Guan, J. S., 221
Guan, R., 29–30
Guan, X.-M., 34, 133, 146
Guan, Z., 92
Gudermann, T., 18–19, 23–24
Guidolin, D., 282, 289–290, 291
Guillaume, J. L., 330
Guillou, F., 22, 33, 34–35
Guldbrandsen, S., 190
Gullberg, M., 282–283, 287, 288
Gunther, R., 18–19
Guo, J., 191, 195–196, 197–199, 198*f*, 199*f*
Guo, W., 275–276
Gupta, A., 220–221, 224–226, 228–229, 230–233, 301–302, 306
Gupta, M., 153
Gurevich, E. V., 348
Gurevich, V. V., 92–93, 94, 100–101, 348
Gurkan, C., 190
Guss, J. M., 110–111
Gustafsdottir, S., 282–283
Gutierrez, M. A., 241–242
Gutierrez-Sagal, R., 21–22, 25
Guzzi, F., 319–320

H

Hadcock, J., 132
Haddadin, M. J., 4
Haga, T., 241–242, 244
Hagen, G. M., 30–31
Hahn, K., 190
Hall, A., 164–166
Hall, R. A., 146, 147, 204, 241–242
Halverson-Tamboli, R. A., 48
Hamamoto, S., 190–191
Hamano, F., 204–205
Hamdan, F. F., 101–102
Hampson, R. K., 221–222
Han, G. W., 255
Han, Y., 304
Hannan, S., 110–111, 116, 119, 127*f*
Hanninen, P., 8–9
Hanson, M. A., 205, 255
Hanson, S. M., 92
Hanyaloglu, A. C., 368–369, 370–374
Harden, T. K., 253–254
Harel, M., 110–111
Harma, H., 8–9
Harms, G. S., 48, 51–53, 60*f*, 61
Hasbi, A., 220–221
Hashidate, T., 205, 208–209, 210, 212, 213–214
Haskell, K. M., 134
Haste, N., 348
Hauri, H. P., 190
Hausdorff, W. P., 132
Hawrot, E., 110–111
Hayashi, M., 29–30
Haydon, P. G., 110–111
He, S. Q., 221
Head, B. P., 48
Hebert, C. A., 152
Hebert, D., 20
Hebert, T. E., 29–30, 191, 192–193, 194
Heid, J., 241
Hein, L., 177
Heinis, M., 162–164
Heiser, V., 4
Heitzler, D., 19
Helenius, A., 20, 21, 172

Henderson, G., 70
Henderson, R., 205
Hendrickson, W. A., 21, 29–30, 369
Henley, J. M., 112
Henstridge, C. M., 220
Hereld, D., 220–221
Hermosilla, R., 172
Herrick-Davis, K., 268
Hiatt, S. M., 264
Highfield Nichols, H., 70
Hilairet, S., 313
Hilenski, L. L., 48
Hill, S. J., 275–276
Hillion, J., 241–242
Hipkin, R. W., 22
Hirakawa, T., 22, 33
Hirasawa, N., 241–242, 246
Hirota, N., 205, 208–209, 210, 212, 213–214
Hislop, J. N., 100–101
Hnatowich, M., 132
Ho, W., 355–356
Hoeffel, G., 162–164
Hogue, M., 146, 172, 195–196, 209
Hokfelt, T., 282
Holl, R., 255
Holliday, N. D., 262, 275–276
Hollt, V., 220–221
Holmes, K. L., 30–31, 32
Holt, J. A., 92, 100–103, 104*t*
Holtgreve-Grez, H., 18–19
Homann, U., 190
Honda, A., 53–54
Honda, Z., 204–205
Honey, S., 337
Hoogenraad, N. J., 226
Hope, B. T., 241–242
Hori, T., 205, 208–209, 210, 213–214
Horner, T. J., 350
Horuk, R., 152
Horvat, R. D., 369
Hosey, M. M., 92, 132, 349
Hosmalin, A., 162–164
Hosobuchi, M., 190
Hosoda, R., 240
Hotellier, F., 330, 340, 342–343
Houslay, T. M., 82, 84
Houtz, D. A., 359
Howlett, A. C., 220
Hsiao, K., 8–9
Hsueh, A. J., 29–30, 369
Hu, C. D., 240, 260–261, 262, 263, 264, 265–266, 267–268, 270–271, 270*np*, 272, 274, 275–276
Huang, B., 70
Huang, C. C., 348–349, 350–351, 356, 357
Huang, W., 191, 195–196, 197–199, 198*f*, 199*f*, 350
Huang, Z. F., 359
Hubbard, S. R., 5
Hubbell, W. L., 92
Huganir, R. L., 110–111, 117
Huhtaniemi, I. T., 21, 23–24
Hurt, C. M., 173–175, 176*f*, 177–181, 179*f*, 181*f*, 185–186, 185*f*
Huttelmaier, S., 146–147
Hynes, T. R., 266, 275–276

I

Iacovelli, L., 22
Iafrate, M., 241–242
Ibaraki, K., 112
Ibata, K., 270*np*
Iglesias, P. A., 48, 51–53, 60*f*, 61
Ijzerman, A. P., 220–221
Imai, B. S., 336
Imanishi, Y., 209
Inahata, K., 263
Inglese, J., 350
Innamorati, G., 302
Insel, P. A., 48
Invernizzi, R., 4
Irie, A., 53–54
Irudayaraj, J. M., 263, 264, 265–266, 267–268, 270–271, 275–276
Isaacs, N. W., 18–19
Ishii, S., 204–205
Ito, J., 4–5, 7, 132–133, 146–147, 204, 306–307
Izumi, T., 204–205

J

Jaakola, V. P., 205
Jach, G., 263
Jackson, D. G., 153

Jacobson, K. A., 255
Jaeschke, H., 18–19
Jager, S., 100–101
Jala, V. R., 161–162
Jan, L. Y., 191
Jan, Y. N., 190, 191
Janovick, J. A., 4–6, 7, 9–11, 12–13, 87–88, 132–133, 146–147, 172, 204, 209, 306–307
Jardon-Valadez, E., 21–22, 25
Jarvius, J., 282–283, 287
Jarvius, M., 282–283, 287, 288, 289–290, 291
Jastrzebska, B., 357
Javitch, J. A., 260–261, 304
Jenkin, C. R., 226–227
Jeoung, M., 306–307, 369
Jeromin, A., 110–111
Ji, I., 18–19, 306–307, 369
Ji, T. H., 18–19, 306–307, 369
Jin, J., 253–254
Jin, T., 220–221
Jockers, R., 330, 343–344
Johnson, C. H., 372–374
Johnson, N. B., 369
Joly, E., 313
Jones, K. A., 241
Jones, S. V., 241–242
Jones, T. R., 289
Jordan, B. A., 220–221, 232–233, 299–300, 301–302, 306
Joseph, K., 134–138, 136*f*, 139–141, 140*f*, 142*f*, 144*f*, 145–146
Jouin, P., 343
Jovin, T. M., 376–377
Julius, D., 304

K

Kaberich, K., 190
Kabli, N., 220–221
Kaelin, W. G. Jr., 181–183
Kaipia, A., 369
Kallal, L., 101–102, 174–175
Kallio, J., 177
Kallio, K., 112*t*
Kamali-Moghaddam, M., 282–283
Kamiya, T., 241–242, 244
Kandasamy, K., 134–138, 136*f*, 139–141, 140*f*, 142*f*, 144*f*, 145–146
Kaneko, H., 241–242, 244
Kang, D. S., 92–93, 94, 100–101
Kang, I. H., 289
Kannan, N., 348
Kapa, I., 133, 146
Kaplan, M. W., 348–349, 358
Kappeler, F., 190
Kara, E., 22, 33, 34–35, 298–299
Karcz-Kubicha, M., 241–242
Kardash, E., 152–153
Kargl, J., 220
Karjalainen, R., 18–19
Kasai, R. S., 296, 307–308
Kascsak, R. J., 4
Kasher, R., 110–111
Kasila, P., 8–9
Katritch, V., 255
Kaupmann, K., 110, 241
Kaushal, S., 4
Kearn, C. S., 260–261
Keefer, J. R., 173
Keen, J. H., 92, 97, 100–103, 104*t*
Keene, J. D., 141
Keene, J. K., 143–144
Keller, P., 132, 134
Kellett, E., 321
Kelly, E., 70
Kenworthy, A. K., 30–31
Kern, R. C., 92–93, 94, 100–101
Kerppola, T. K., 261, 262, 263, 264, 270–271
Khiroug, L., 110–111
Khorana, H. G., 350
Kihara, Y., 204–205
Kilpatrick, L. E., 275–276
Kim, C. M., 349
Kim, J. Y., 53, 152–153, 158–159
Kim, Y. K., 145
Kim, Y.-M., 92, 97, 102–103, 104*t*
Kimbembe, C. C., 191
King, M. L., 145–146
Kinsella, B. T., 191
Kirscht, S., 220–221
Kishi, H., 22, 33
Kittler, J. T., 153–154
Kitts, P. A., 352

Kiuchi, T., 263
Klovins, J., 133, 146
Kluetzman, K. S., 38–39, 40–41
Klug, C. S., 92
Kniazeff, J., 370–372, 374–375, 380–381
Knight, G. E., 255
Knoll, B. J., 359
Knollman, P. E., 146
Kobayashi, H., 70–71, 87–88
Kobilka, B. K., 29–30, 34, 70, 132, 133, 134, 146, 173–175, 176*f*, 177–181, 179*f*, 181*f*, 185–186, 185*f*, 330, 369
Kobilka, T. S., 34, 133, 146, 205
Koch, S., 289–290, 291
Koch, T., 220–221
Kochl, R., 133
Kodama, Y., 263
Kohler, G., 224–226
Kohout, T. A., 92, 97, 102, 368–369
Konig, I. R., 314–316
Konig, P., 314–316
Konijn, T. M., 51
Koo, Y., 369
Korkhov, V. M., 190–191
Kouyama, T., 205
Kowal, D. M., 9–11
Koyama-Honda, I., 296, 307–308
Kralikova, M., 275–276
Krasteva, G., 314–316
Krause, G., 133, 172
Krebs, M. P., 4
Krishek, B. J., 114
Krishnamurthy, H., 22, 33
Kroeger, K. M., 368–369, 370–374
Kroening, S., 145
Krogan, N. J., 100–101
Krupnick, J. G., 92, 97, 100–101, 102–103, 104*t*
Kubota, M., 270*np*
Kudo, M., 29–30, 369
Kuge, O., 190
Kuhn, H., 357–358
Kuksa, V., 209
Kumar, M., 8–9
Kumar, R., 162–164
Kumasaka, T., 205
Kummer, W., 314–316
Kunapuli, P., 349
Kunapuli, S. P., 253–254
Kuroda, Y., 240
Kusemider, M., 220–221
Kuszak, A. J., 70
Kuttenn, F., 368–369
Kwatra, M. M., 132

L

Labbe-Jullie, C., 132–133
Laemmli, U. K., 27
Lagace, M., 4, 101–102, 204
Lager, P. J., 143–144
Lagerstrom, M. C., 205
Lam, C. M., 158–159
Lamb, M. E., 100
Lambert, D. G., 220
Lambert, N. A., 191, 195–196, 197–199, 198*f*, 199*f*
Lamers, G. E. M., 48, 49–50, 51, 56–57, 59, 61
Lamprecht, M. R., 289
Landegren, U., 289–290, 291
Lane, J. R., 205
Lange, D. C., 370–372
Langemeijer, E., 72
Laperriere, A., 7, 12–13, 146, 172, 195–196, 209
Lapointe, P., 190
Laporte, S. A., 92, 100–103, 104*t*
Laroche, G., 316
Larsson, L. G., 288
Larsson, R., 291
Latif, R., 369
Lauffer, B., 100–101
Laugsch, M., 220–221
Laustriat, D., 101–102
Lavoie, C., 301
Lazari, M. F., 22
Lazova, M. D., 49–50
Le Crom, S., 241–242
Le Minh, P. N., 301–302
Le Novere, N., 110–111
Leahey, P. A., 355–356
Leanos-Miranda, A., 4, 5, 7, 9–11, 12–13
Lecureuil, C., 19
Lecuyer, E., 133–134
Lederer, M., 146–147
Lee, C., 306–307, 369

Lee, K. B., 132
Lee, M. C., 190–191
Lee, S. P., 29–30
Leeb-Lundberg, L. M., 100
Lee-Ramos, D., 220–221, 232–233
Lefkowitz, R. J., 19–20, 22, 33, 70, 92, 97, 101–102, 103–105, 132, 161–162, 282, 312, 349, 354–356, 368–369
Lei, Y., 30–31
Lemay, J., 162–164
Lenzi, P., 282
Lepine, M. C., 316
Leslie, A. G., 255
Leuchowius, K. J., 282–283, 286–287, 288, 291
Leung, S. W., 4
Levac, B. A., 220–221
Li, C., 4
Li, J., 220–221
Li, L., 355–356
Li, M., 191, 195–196
Li, S., 220–221
Li, T., 337
Li, X., 110–111, 117, 119, 289–290, 291
Li, Y., 263
Li, Z., 350
Libri, D., 368–369
Lichtarge, O., 70–71, 87–88
Lim, M. P., 220–221, 232–233
Limbird, L. E., 70, 173, 184–185, 282
Lin, F. S., 92, 97, 102
Lin, H., 297
Lin, J., 263
Lin, R., 305
Lin, Y. F., 190
Lindau-Shepard, B., 18–19, 21, 24, 25–27
Lindstrom, J., 110–111
Lingappa, V. R., 132–133
Liu, F., 350
Liu, H. R., 221, 263, 266, 270–271, 270*np*
Liu, J., 30–31, 110–111
Liu, T., 51
Liu, W., 255
Liu, X., 18–19, 21, 22
Liu, Z., 263
Liu-Chen, L. Y., 220–221
Lledo, P. M., 145–146
Lluis, C., 298–299
Locati, M., 152–153, 158–159
Lodowski, D. T., 349, 356
Loh, H. H., 145–146
Lohmann, T., 172
Lohse, M. J., 132, 302–303
Loisel, T. P., 29–30
Lommerse, P. H. M., 48, 49–50, 51, 56–57, 59, 61
Lonergan, M., 7, 12–13, 209
Loosfelt, H., 24
Lopez, P., 181–183
Lopez-Gimenez, J. F., 70–71, 74–75, 84–86, 87–88
Lopez-Sanchez, P., 220–221
Lorenz, M., 146–147
Lother, H., 220–221, 306
Loudon, R. P., 349
Love, R. A., 112*t*
Loy, T., 369, 374–375, 377–381
Lu, H. Y., 145–146
Lucchi, S., 369, 376–377
Luini, A., 184–185
Luján, R., 241–242
Luker, G. D., 153
Luker, K. E., 153
Luo, J., 92–93, 96–97, 97*f*, 102
Lurz, R., 4
Lustig, K. D., 304
Luttrell, L. M., 35, 134, 137, 138, 139–141, 140*f*, 142*f*, 144*f*, 145–146
Lyle, A. N., 241–242
Lyons, J. A., 255
Lyssand, J. S., 337

M

Ma, D., 190
Ma, Z., 355–356
Macchi, P., 145
MacCoss, M. J., 337
Machamer, C. E., 190
Mackie, K., 260–261
Macrae, A. D., 350, 355–356
MacWilliams, H., 51
Madabushi, S, 359
Madoux, F., 8
Maeda, T., 350–351, 354–355, 356
Magenis, L. M., 25, 33, 36, 38–39
Maggio, R., 220–221, 302–303, 306–307

Magliery, T. J., 261, 262
Mahabaleshwar, H., 152–153
Mahmoudi, S., 289–290
Maillet, E. L., 220–221
Major, F., 145
Maki, R. A., 209
Malbon, C. C., 132, 134
Maldonado, R., 220
Malenka, R. C., 145–146
Malhotra, I. J., 221–222
Malhotra, R., 4
Malitschek, B., 241
Malkus, P. N., 190–191
Mallol, J., 298–299
Mancias, J. D., 190–191
Mann, M., 331
Manna, P. R., 23–24
Mannoury la Cour, C., 73, 74, 76–77
Mantovani, A., 152–153, 158–159
Manzanares, J., 220
Maquat, L. E., 145
Maranon, C., 162–164
Marcellino, D., 220–221, 282
Marciniak, S. J., 5
Margeta-Mitrovic, M., 191
Mariggio, S., 355–356
Marin, P., 343
Marion, S., 22, 34
Markovic, B., 110–111
Marquardt, D. W., 61
Martenson, C., 101–102
Martikkala, E., 8–9
Martin, N., 220–221
Martinat, N., 22, 33, 34–35
Martinez-Munoz, L., 220–221
Marullo, S., 92, 100–101, 104*t*, 132–133, 244–246, 312–313
Masago, K., 204–205
Masri, B., 101–102
Massardi, M. L., 153
Mathis, G., 370–372, 374–375, 380–381
Matifas, A., 101–102
Matsushima, K., 152
Matteson, J., 190–191
Maurel, D., 75, 370–372, 374–375, 380–381
Maurel, M. C., 19–20
Mauri, M., 319–320
Maurice, P., 330, 340, 342–344
Maya-Nunez, G., 20–21, 24, 29–30, 209
Mayor, F. Jr., 355–356
Mazurkiewicz, J. E., 26*f*, 28, 29–31, 32, 268
McBride, H., 121, 155–157
McCulloch, C. V., 161–162
McDonald, B. J., 114
McDowell, J. H., 4
McKinney, M., 296
McMillian, M., 265–266
McNeilly, C. M., 4
McVey, M., 321
Meagher, D. A., 336
Mechoulam, R., 220
Meduri, G., 24, 368–369
Ménard, L., 102–103, 104*t*
Meng, X., 146–147
Mercier, J. F., 313
Meresse, S., 121
Mervine, S. M., 266, 275–276
Messitt, T. J., 145–146
Metaye, T., 4
Mettey, Y., 4
Meyer, E. L., 145
Meyer, G., 313
Meyts, P. D., 282
Mi, D., 8–9
Michal, A., 92–93, 94
Michaud, D. E., 101–102
Michels, G., 110–111
Miki, I., 204–205
Mikosch, M., 190
Mikoshiba, K., 270*np*
Milano, S. K., 92, 97
Milasta, S., 71, 152–153, 161–162
Milgrom, E., 24
Millan, M. J., 73, 74, 76–77, 220–221
Miller, E. A., 190–191
Miller, W. E., 35
Miller, W. L., 8–9
Milligan, G., 70–71, 72–73, 74–75, 76–78, 80–82, 84–86, 87–88, 152–153, 161–162, 260–261, 300*f*, 303, 312, 321, 368–369, 370–374
Mills, I. G., 92, 104*t*
Milstein, C., 224–226
Minami, M., 204–205
Minina, S., 152–153

Minneman, K. P., 241–242
Mir, R. A., 162–164
Mirolo, M., 153
Mirschberger, C., 51
Mische, S. M., 336
Mishra, J. K., 8
Misrahi, M., 23–24
Miyawaki, A., 270*np*
Mizrachi, D., 18–19, 20
Mizuno, K., 263
Mizuno, N., 241–242, 246
Mobarec, J. C., 275–276
Moffatt, B. A., 350
Moffett, S., 29–30
Moise, L., 110–111
Mol, C. D., 255
Monsarrat, B., 330, 331, 334, 343
Montanelli, L., 369, 374–375, 377–381
Moon, R. T., 337
Moore, C. A., 92, 97
Moore, R. H., 359
Moreira, I. S., 304
Morello, J. P., 4, 7, 12–13, 29–30, 146, 172, 204, 209
Moreno, J. L., 260–261
Moretti, E., 355–356
Moriya, T., 253–254
Morrow, V., 152–153, 161–162
Mortier, A., 152–153
Mosbacher, J., 110, 172
Mosberg, H. I., 70
Mosley, R., 87–88
Moss, S. J., 114, 153–154
Mossessova, E., 190–191
Motoshima, H., 205
Mouillac, B., 325–326
Moukhametzianov, R., 205, 255
Mowry, K. L., 145–146
Moyer, B., 190–191
Mozo, J., 330, 340, 342–343
Muchowski, P. J., 4
Muda, M., 26*f*, 28, 29–31, 32
Mudo, G., 291
Mueller, S., 18–19
Mulder, J., 221, 224–226, 228–229, 230–233
Mundell, S. J., 92
Murakami, M., 205
Murphy, D. B., 50
Murphy, P. M., 152
Murray, F., 48
Mushegian, A., 348
Musnier, A., 19–20
Mutoh, H., 204–205
Muttenthaler, M., 5–6

N

Nagai, T., 270*np*
Nagamune, T., 205, 208–209, 210, 212, 213–214
Nagy, V., 220–221, 232–233, 301–302, 306
Nakada, C., 296, 307–308
Nakada, M. T., 134
Nakae, S., 152–153
Nakagawa, C., 263
Nakahata, N., 241–242, 246, 253–254
Nakamura, K., 22
Nakamura, M., 204–205
Nakanishi, H., 204–205
Nakata, H., 240, 241–242, 244, 246, 248, 250, 252*f*, 253–254
Nakatani, Y., 337
Namba, K., 241–242, 244, 246, 248, 250, 252*f*, 253–254
Namba, T., 53–54
Namkung, Y., 337
Narumiya, S., 53–54
Nath, N., 162–164
Naumann, U., 153
Nawoschik, S. P., 9–11
Nechamen, C. A., 18–19, 21, 23–24, 25–27, 26*f*, 28–31, 32, 38–39, 40–41
Neel, N. F., 155–157
Negishi, M., 53–54
Nehmé, R., 255
Nelson, C. D., 359
Nelson, S. E., 369
Nemoto, W., 240
Neubig, R., 260–261
Neuwald, A. F., 348
Ng, G. Y., 29–30
Nguyen, T., 29–30, 220–221
Ni, J., 153
Nibbs, R., 152–153
Nicchitta, C. V., 143–144
Nicholas, R. A., 253–254

Nichols, C. D., 172, 191, 192–193, 194, 195–196, 197–199, 198*f*, 199*f*
Nicolas, A., 110–111
Nicoll, R. A., 145–146
Niedzinski, E. J., 4
Nieschlag, E., 18–19
Niesman, I. R., 48
Nilsson, I., 289–290, 291
Nishimura, N., 190
Nishimura, S., 263
Nivarthi, R., 220–221
Nobles, K. N., 92
Noguchi, K., 204–205
Noorwez, S. M., 4, 209
Nordhoff, E., 4
Nordstrom, K. J., 330
Novi, F., 302–303, 305
Nufer, O., 190
Nwaneshiudu, C., 53–54

O

Oakley, R. H., 92, 100–103, 104*t*
Oas, T. G., 92
Obara, Y., 253–254
O'Dowd, B. F., 29–30, 132, 220–221
Ogawa, K., 70–71, 87–88
Ogryzko, V., 337
O'Hara, M., 153
Ohashi, K., 263
Ohguro, H., 350, 355–356
Ohyama, T., 152–153
Oksche, A., 133, 172
Okuno, T., 205, 208–209, 210, 213–214
Oldenburg, K. R., 7
Ong, E., 221, 224–226, 228–229, 230–233
Onorato, J. J., 349
Oosterom, J., 158–159
Oppermann, M., 359
Orban, T., 357
Orci, L., 190–191
Ormo, M., 112*t*
Orsini, M. J., 92, 102–103, 104*t*
Ortiz-Elizondo, C., 20–21
Osawa, S., 350
Ostrom, R. S., 48
Ostrowski, J., 132
Osuga, Y., 29–30, 369
Otero, K., 153
Otte, S., 190
Owen, D. M., 63–64
Oyanagi, K., 241–242, 253–254

P

Paccaud, J. P., 190
Padlan, E. A., 112*t*
Pagano, A., 172
Pakarinen, P., 23–24
Palacios, I. M., 145–146
Palczewski, K., 205, 209, 348–349, 350–351, 354–355, 356, 357, 358, 368–369
Palmer, A. E., 270*np*
Palmer, R. K., 241
Palmer, S., 26*f*, 28, 29–31, 32
Pals-Rylaarsdam, R., 132
Pande, J., 4
Pangalos, M., 110–111
Panigone, S., 369, 376–377
Pantel, J., 8–9
Papermaster, D. S., 357–358
Papi, S., 324
Pardo, L., 18–19, 368–369
Parent, C. A., 49–50
Parenti, M., 302
Parenty, 82
Parisien, M., 145
Park, B. S., 6
Park, E. S., 270*np*
Park, P. S., 368–369
Parmentier, M., 82, 301–302
Parola, A. L., 134
Parolaro, D., 220
Parrillas, V., 220–221
Parthasarathy, N., 133–134
Paruch, S., 162–164
Pasapera, A. M., 33
Paschke, R., 18–19
Patel, H. H., 48
Patel, R. C., 370–372
Patel, Y. C., 370–372
Patny, A., 87–88
Pausch, M. H., 9–11
Peak-Chew, S. Y., 92, 104*t*
Pediani, J. D., 71, 73, 74–75, 76–78, 80–82, 84–86, 87–88, 302–303
Pei, G., 350

Pei, L., 220–221
Pekel, E., 18–19
Pellissier, L. P., 302–303
Pello, O. M., 220–221
Peltoketo, H., 18–19
Peng, Y., 4
Penn, R. B., 92, 100–101
Pepperl, D. J., 53–54
Perez-Alea, M., 291
Perez-Solis, M. A., 20–22, 24, 25, 29–30
Peri, R., 9–11
Permanne, B., 4
Perreault, M. L., 220–221
Perroy, J., 343
Perry, S. J., 92, 97, 102, 209
Persani, L., 368–369, 374*t*, 376
Pesch, M., 263
Petaja-Repo, U. E., 8–9, 146, 172, 195–196, 204, 209
Peterson, A. J., 18–19, 21, 24, 25–27
Petit, P. X., 191
Petrovska, R., 133, 146
Pfeiffer, M., 220–221
Pfleger, K. D., 244
Phillips, D., 370–372
Pichon, C., 24
Pierce, K. L., 35, 53–54
Piketty, V., 19–20, 22, 33, 34–35
Pin, J. P., 205, 260–261, 302–303, 370–372, 374–375, 380–381
Pineda, D. B., 20–21
Pineiro, A., 21–22, 25
Pintar, J. E., 220–221, 232–233
Piomboni, P., 355–356
Piscitelli, F., 220
Piskorowski, R. A., 283, 291
Piston, D. W., 270*np*, 372–374
Pitcher, J. A., 132, 349, 350, 354–356
Pitchiaya, S., 70
Pitt, A. M., 82, 84
Plutner, H., 190
Poels, J., 368–369
Polans, A. S., 348–349, 358
Pollok-Kopp, B., 359
Popoli, P., 241–242
Portoghese, P. S., 220–221
Pou, C., 73, 74, 76–77
Poupon, A., 19–20
Praefcke, G. J. K., 92, 104*t*
Preisig-Muller, R., 190
Prelli, F., 4
Premont, R. T., 70, 350, 355–356
Prevo, R., 153
Prihandoko, R., 302–303
Prisco, M., 337
Pronin, A. N., 349
Proost, P., 152–153
Prossnitz, E. R., 296, 307–308
Prowse, S. J., 226–227
Pruenster, M., 152–153
Pryazhnikov, E., 110–111
Prystay, L., 8–9
Przybyla, J. A., 260–261, 262, 263, 264, 265–266, 268, 270–271, 272, 275–276
Pshezhetsky, A. V., 4
Pucadyil, T. J., 48
Puett, D., 368–369
Puthenveedu, M. A., 92, 100–101
Pyhtila, B., 143–144
Pyskadlo, R. M., 356

Q

Qi, Y., 265–266
Qian, H., 64
Qin, Y., 72
Quitterer, U., 220–221, 306
Qunaibi, E., 359

R

Rabilloud, T., 335–336
Radenovic, A., 296
Ragusa, M. J., 355–356
Raimondi, F., 319–320
Rajagopal, S., 158–159
Rajaram, R. D., 29–30
Ralevic, V., 241
Ramos, J. A., 220
Ramsay, D., 321
Ranadive, S. A., 8–9
Rands, E., 132
Rannikko, A., 23–24
Ransohoff, R. M., 152
Rao, A. G., 221–222
Rapoport, T. A., 145–146
Rapport, J. Z., 92, 100–101, 104*t*
Rasenick, M. M., 48

Rasmussen, S. G., 70, 205, 330
Raymond, J. R., 134–138, 136*f*, 139–141, 140*f*, 142*f*, 144*f*, 145–146
Raz, E., 153
Reedy, M. C., 143–144
Rees, S., 321
Regan, J. W., 53–54
Regan, L., 261, 262
Register, R. B., 132
Reichert, L. E. Jr., 36
Reichman-Fried, M., 152–153
Reid, H. M., 191
Reiter, E., 19–20, 22, 33, 34
Reiterer, V., 190–191
Remington, S. J., 112*t*
Rentero, C., 63–64
Ressler, K. J., 241–242
Reversi, A., 324
Richmond, A., 155–157
Richter, K., 263
Rickardson, L., 291
Ridderstrale, K., 282–283, 287
Ried, T., 18–19
Rigaut, G., 331
Rinne, S., 190
Rios, C. D., 220–221, 232–233, 306
Ritter, M., 313
Rivera, A., 289–290, 291
Rives, M. L., 75, 283, 291
Rizzo, M. A., 270*np*
Robert, F., 22, 34
Robert, J., 191, 209
Robitaille, M., 191, 192–193, 194
Rochdi, M. D., 101–102
Rodien, P., 368–369
Rodighiero, S., 313
Rodrigues, R. J., 241–242
Rodriguez-Frade, J. M., 220–221
Roess, D. A., 30–31, 369
Rogers, T. J., 220–221
Romero, J., 220
Romero-Fernandez, W., 282, 289–290, 291
Ron, D., 5
Rosa, P., 241–242
Rose, R. H., 262
Roseberry, A. G., 132
Rosenbaum, D. M., 205, 255, 330
Rosenthal, S. M., 8–9
Rosenthal, W., 133
Rot, A., 152–153
Roth, D. M., 48
Rothman, J. E., 190
Rottman, F. M., 221–222
Roudabush, F. L., 35
Roush, W. R., 8
Rovelli, G., 172
Rowan, M. P., 220–221
Rowbotham, D. J., 220
Rowe, T., 190
Royere, D., 19–20
Rozell, T. G., 18–19, 20
Rozenfeld, R., 220–222, 224–226, 228–229, 230–233, 260–261, 304–305
Rozwandowicz-Jansen, A., 8–9
Rubino, T., 220
Rudnicka-Nawrot, M., 350, 355–356
Rush, T. S., 87–88
Rutz, B., 331
Rutz, C., 133

S

Saba, E., 241–242
Sabo, J. L., 266, 275–276
Saborio, G. P., 4
Sai, J., 155–157
Saitoh, O., 240, 241–242, 244
Sakanaka, C., 204–205
Salahpour, A., 7, 12–13, 101–102, 209, 313
Salama, N., 190
Salata, R. A., 221–222
Sallese, M., 355–356
Sampei, K., 263
Sanchez, T. A., 220–221
Santini, F., 92, 97, 100–101, 102–103, 104*t*
Sartania, N., 71, 88
Saunders, C., 173
Sauter, K., 110–111
Savino, B., 152–153, 158–159
Scarselli, M., 296, 302–303, 305
Schaap, P., 53
Schachter, J. B., 253–254
Schaer, C. A., 153
Schaer, D. J., 153
Schaller, M. D., 350
Schallmach, E., 305
Schallmeiner, E., 282–283

Scherrer, G., 101–102
Scherzinger, E., 4
Schieltz, D. M., 337
Schindler, H., 62
Schioth, H. B., 133, 146, 205, 330
Schlichthorl, G., 190
Schmid, E. M., 92, 100–101, 104*t*
Schmid, J. A., 190–191
Schmid, S. L., 40–41
Schmidt, A., 18–19, 21
Schmidt, T., 49–50, 61, 62
Schneider, B. L., 337
Schoedon, G., 153
Schramm, N., 173
Schroder, H., 220–221
Schubert, A., 190
Schugardt, N., 4
Schulein, R., 172
Schuler, V., 241
Schulz, A., 23–24
Schulz, S., 220–221
Schumacher, A., 337
Schutyser, E., 155–157
Schutz, G. J., 62
Schwarz, D. A., 209
Schwarze, K., 359
Scott, M. G. H., 92, 100–101, 104*t*, 132–133
Scott, R., 241–242
Seachrist, J. L., 268
Sealfon, S. C., 260–261
Sebag, J., 8–9
Sebti, S., 4
Seeber, R. M., 368–369
Seeman, P., 29–30
Segal, D. M., 112*t*
Segaloff, D. L., 18–19, 20, 21, 29–30, 369
Seibold, A., 359
Sekine-Aizawa, Y., 110–111, 117
Sellers, W. R., 181–183
Seraphin, B., 331
Serge, A., 48, 49–50, 51, 56–57, 59, 61
Serradeil-Le Gal, C., 8–9
Serrano, A., 220–221
Serrano-Vega, M. J., 205
Sevier, C. S., 190
Seyama, Y., 204–205
Shaner, N. C., 270*np*
Shankar, H., 92–93, 94
Sharma, S. K., 220–221
Shcherbakova, O. G., 173–175, 176*f*, 177–179, 179*f*, 181, 181*f*, 185–186, 185*f*
Sheetz, M. P., 64
Shenoy, S. K., 92, 101–102, 161–162, 359
Shenton, F. C., 72
Sherry, B., 164–166
Shevchenko, A., 331
Shi, L., 275–276
Shi, M., 22, 33
Shibata, Y., 145–146
Shimizu, T., 204–205
Shoji, K., 263
Shukla, A. K., 101–102, 103–105
Shyu, Y. J., 261, 262, 263, 264, 266, 270–271, 270*np*
Siegel, R. M., 30–31, 32
Sievert, G. A., 369
Sigal, I. S., 132
Signorelli, P., 153, 161–162
Silva, M., 220–221
Silversides, D. W., 101–102
Simanski, S., 8
Simons, J. F., 20
Simpson, C. V., 153
Singer, R. H., 133–134, 141
Singer, S. J., 133, 143
Singh, P., 348–349, 350–351, 354–355, 356, 357
Sironi, M., 153
Sitte, H. H., 190–191
Slabough, S., 190
Sleister, H. M., 221–222
Smart, T. G., 110–111, 114, 116, 117, 119, 127*f*
Smith, K. A., 4
Smith, N. J., 300*f*
Smith, S. M., 30–31
Smith, W. L., 53–54
Snaar-Jagalska, B. E., 49–50, 61, 62
Soderberg, O., 282–283, 286–287, 288
Soede, R. D., 53
Soloviev, M. M., 241–242
Song, Y. S., 18–19

Soto, C., 4
Spaink, H. P., 48, 49–50, 51, 56–57, 59, 61, 62
Spicer, E. K., 134, 137
Springael, J. Y., 82, 301–302
Springer, G. H., 270*np*
Springsteel, M. F., 4
Sromek, S. M., 253–254
St Johnston, D., 133–134, 145–146
Stadel, J. M., 134
Stagg, S. M., 190
Stamnes, M. A., 190
Stamp, G. W., 18–19
Stanasila, L., 302–303
Staniszewski, C., 355–356
Stauffer, D., 172
Steele, J. M., 153
Stefano, F., 92
Stein, J. V., 158–159
Steinbach, P. A., 270*np*
Steiner, L., 224–226
Steitz, J. A., 141
Sterne-Marr, R., 355–356
Stockton, S. D. Jr., 260–261
Stoddart, L. A., 73, 74, 76–77
Stoffel, R. H., 350, 355–356
Stone, W. C., 354–355
Strader, D. J., 132
Strange, P. G., 296, 297–298
Strauss, L., 18–19
Strenio, J., 12
Stroud, R. M., 112*t*
Studier, F. W., 350
Stumm, R., 220–221
Suarez, C. D., 262
Subramaniam, K., 134–138, 136*f*
Sudhof, T. C., 145–146
Sugimoto, K., 263
Sugimoto, Y., 53–54
Suhling, K., 370–372
Sunahara, R. K., 70
Suzuki, K. G., 296, 307–308
Suzuki, T., 241–242, 244, 246, 248, 250, 252*f*, 253–254
Swaney, J. S., 48
Swillens, S., 369, 374–375, 377–381
Swofford, R., 30–31, 32
Symons, M., 164–166
Szabo, I., 220–221
Szeto, H. H., 220–221, 232–233

T

Tag, C., 314–316
Tajima, Y., 204–205
Takuwa, Y., 204–205
Tam, V., 301
Tamm, J. A., 241
Tan, C. M., 70, 184–185
Tang, X. B., 337
Tao, Y. X., 18–19, 369
Tarakanov, A. O., 291
Taulan, M., 325–326
Tautermann, C. S., 302–303
Taylor, J. A., 350, 355–356
Taylor, S. S., 348
Teerds, K. J., 23–24
Temkin, P., 100–101
Terasmaa, A., 220–221, 241–242
Terreni, L., 4
Terrillon, S., 101–102, 312, 325–326, 368–369
Tesmer, J. J., 348–349, 350–351, 354–355, 356, 357
Tesseraud, S., 19
Thelen, M., 158–159
Themmen, A. P., 21
Theriault, C., 316
Thian, F. S., 205
Thibodeau, P. H., 4
Tholanikunnel, B. G., 134–138, 136*f*, 139–141, 140*f*, 142*f*, 144*f*, 145–146
Thomas, P. J., 4, 110–111, 116, 119, 127*f*
Thomas, R. M., 26*f*, 28, 29–31, 32, 38–39, 40–41
Thorner, J., 132
Thorpe, C. J., 337
Thurston, G. M., 4
Tian, S., 263
Tiberi, M., 132
Timossi, C., 20–21
Toh, H., 240
Tomich, C. S., 221–222
Toro, M. J., 220–221
Torre, Y. M., 152–153
Torvinen, M., 220–221, 241–242
Tran, T. M., 359

Tranchant, T., 19–20
Tran-Van-Minh, A., 110–111
Trapaidze, N., 220–221, 232–233, 301–302, 306
Trent, J. O., 161–162
Trifilieff, P., 283, 291
Trinquet, E., 302–303, 370–372, 374–375, 380–381
Tripathi, V., 162–164
Troispoux, C., 22, 34
Trombetta, E., 20
Tryoen-Toth, P., 101–102
Tsai, N. P., 145–146
Tse, R., 29–30, 301
Tsien, R. Y., 30–31, 32, 112*t*, 261, 270*np*
Tsuga, H., 242, 246, 248, 250, 252*f*
Tsukiji, S., 210

U

Ueffing, M., 337
Uhlen, S., 133, 146
Uhrig, J. F., 263
Ullmer, C., 343
Ulloa-Aguirre, A., 4–6, 7, 9–11, 19–21, 33, 70, 132–133, 146–147, 204, 306–307
Uribe, A., 21–22, 25
Urizar, E., 275–276, 283, 291, 301–302, 304, 369, 374–375, 377–381
Ushio-Fukai, M., 48

V

Vago, G., 152–153
Valkema, R., 53
Valverde, O., 220
Van Craenenbroeck, K., 289–290, 291
Van Damme, J., 152–153
Van Dorpe, J., 4
Van Elsas, A., 158–159
Van Haastert, P. J. M., 51, 53
van Hemert, F., 48, 49–50, 51, 56–57, 59, 61
Van Hiel, M. B., 368–369
Van Lith, L. H., 158–159
Van Loy, T., 368–369
van Marle, A., 72
van Rijn, R. M., 72
Vandersmissen, H. P., 368–369
Vannier, B., 24
Varghese, G., 29–30, 301
Vassart, G., 18–19, 82, 301–302, 368–369, 376
Vasseur, C., 368–369
Veit, M., 190
Ventura, M. A., 191, 209
Verlinden, H., 368–369
Vidi, P. A., 240, 260–261, 262, 263, 264, 265–266, 267–268, 270–271, 272, 274, 275–276
Vidugiriene, J., 8–9
Vigano, D., 220
Violin, J. D., 101–102
Vishnivetskiy, S. A., 92
Vishwasrao, H. D., 283, 291
Vivo, M., 297
Voeltz, G. K., 145–146
Vogel, Z., 306–307
Volonté, C., 241–242
von Zastrow, M., 92, 100–101
Votsmeier, C., 190–191
Vulcano, M., 153

W

Wacker, J. L., 4, 337
Wahlby, C., 289
Waldhoer, M., 220
Walker, P., 195–196
Walsh, K. A., 350, 355–356
Walter, N. G., 70
Wang, B., 350–351, 354–355, 356
Wang, H. B., 18–19, 221
Wang, J. K., 265–266
Wang, J. Y., 350, 355–356
Wang, M., 220–221
Wang, N., 263
Wang, Q., 70
Wang, X. T., 4, 190–191
Wang, Y. P., 263
Ward, R. J., 71, 72–73, 74–75, 76–78, 80–82, 84–86, 88
Warne, T., 205, 255
Warnock, D. E., 40–41
Warrenfeltz, S. W., 368–369
Wasylewski, Z., 220–221
Watabe, A., 53–54
Watanabe, T., 204–205

Watts, V. J., 240, 260–261, 262, 263, 264, 265–266, 267–268, 270–271, 272, 274, 275–276
Weaver, B. A., 268
Weaver, C. D., 8–9
Weber, G., 368–369
Weber, M., 153
Wehbi, V., 19–20
Wei, H. P., 263
Wei, N., 145–146
Weibrecht, I., 286–287, 288, 289–290
Weinberg, C. R., 336
Weiner, R. S., 25
Weinstein, H., 304
Weiss, E. R., 350
Weisz, O. A., 190
Wells, J. W., 296–297, 368–369
Wess, J., 306–307
Wettlaufer, D. G., 5–6, 12–13
Whalen, E. J., 101–102
Whorton, M. R., 70
Wickstrom, M., 291
Wiesner, B., 172
Wilden, U., 357–358
Wiley, H. S., 36
Wilkins, M. E., 110–111, 116, 117, 119, 127*f*
Wilkinson, G., 70–71, 87
Williams, B., 359
Williams, J. C., 92, 100–101
Williams, S. Y., 8–9
Williamson, D., 63–64
Wilm, M., 331
Wilson, C. G., 261, 262
Wilson, S., 70–71, 87
Wisniewski, T., 4
Woods, A. S., 289–290, 291
Wozniak, M., 173
Wraight, C. J., 226
Wreggett, K. A., 296–297
Wu, B., 255
Wu, D., 220–221
Wu, G., 20–21, 132–133, 172, 190, 191, 192–193, 194, 195–196, 197–199, 198*f*, 199*f*
Wu, H., 255
Wu, P., 321
Wu, Q. J., 190, 192–193
Wu, X., 29–30

X

Xia, B., 263
Xiao, K., 92, 103–105, 359
Xie, Z., 220–221, 301
Xu, F., 255
Xu, T. R., 73, 75, 80, 82, 84
Xu, Y., 372–374

Y

Yamagishi, R., 241–242, 244
Yamaguchi, S., 205, 208–209, 210, 212, 213–214
Yamamoto, T., 205, 208–209, 210, 212, 213–214
Yanagida, K., 204–205
Yang, D., 301
Yang, Q. H., 350
Yao, R., 70–71, 87–88
Yasuda, D., 205, 208–209, 210, 212, 213–214
Yates, J. R., 337
Ye, K., 220–221
Yeh, L. A., 8–9
Yeung, T., 190
Yi, X., 337
Yokomizo, T., 204–205, 208–209, 210, 213–214
Yoo, J. S., 190–191
Yoshida, H., 133–134
Yoshino-Koh, K., 348–349, 350–351
Yoshioka, K., 240, 241–242, 244
Yost, E. A., 266, 275–276
Yost, S. M., 275–276
Yu, M., 190
Yu, S. S., 132
Yue, H., 4
Yumura, S., 51–52
Yumura, T. K., 51–52

Z

Zabel, B. A., 152–153
Zacharias, D. A., 30–31, 32
Zaman, G. J., 158–159
Zare, R. N., 70
Zarinan, T., 20–22, 24, 25, 29–30, 33

Zenklusen, D., 146–147
Zerangue, N., 190
Zerial, M., 121, 155–157
Zerwes, H. G., 153
Zhang, C., 255
Zhang, H., 350
Zhang, J., 92, 100–103, 104*t*
Zhang, J. H., 7
Zhang, J. Y., 9–11
Zhang, K., 350
Zhang, M., 18–19, 29–30
Zhang, S., 301
Zhang, X. M., 4, 145–146, 172, 190, 191, 192–193, 194
Zhang, Z. N., 221
Zhang, Z. P., 263
Zhao, B., 221
Zhao, L., 263
Zhao, Q., 255
Zhao, X., 350, 355–356
Zhao, Y., 101–102
Zheng, N., 337
Zheng, T., 143–144
Zhou, C., 263
Zhou, F., 20–21, 172, 191, 192–193, 194
Zhou, J., 263
Zhou, Q. Y., 191, 195–196
Zhou, Y. F., 263
Zhu, L., 209
Zhu, W. Z., 301
Zingg, H. H., 325–326
Zini, I., 282
Zuniga, L., 152–153
Zuo, L., 48
Zuzarte, M., 190
Zwier, J. M., 302–303

SUBJECT INDEX

Note: Page numbers followed by "*f*" indicate figures, and "*t*" indicate tables.

A

ACRs. *See* Atypical chemokine receptors (ACRs)
Agonist-elicited intracellular Ca^{2+} response, 210–211
Agonist/inverse agonists
  cell lysis/RSB, 361
  ISO/ALP, 361
  serum-free media, 361
Amino acid residues, GPCRs
  lipid mediators, 204–205
  mutant GPCRs (*see* Mutant GPCRs)
  pharmacological chaperones/ pharmacoperones, 204
  structural and functional analysis, 204–205
  surface-trafficked mutant, living cells (*see* Surface-trafficked mutant GPCRs)
Antibody feeding and cytofluorimetric analysis
  cytofluorimetric analysis, D6, 153–154, 154*f*
  D6 constitutive internalization protocol, 155
  description, 153–154
  ligand-dependent internalization, 153–154
Arrestin. *See also* β-Arrestins
  GPCR trafficking measurement (*see* GPCR trafficking)
  overexpression
    cell culture, 95–96
    knockdown, 96
    wild-type, mutant/GFP-tagged β-arrestins, 96
β-Arrestins
  antibodies, 92–95
  cell lysate preparation, 93–94
  chemokine source, 158–159
  description, 92–93
  electrophoresis and immunoblotting, 94–95
  GPCR trafficking (*see* GPCR trafficking)
  immunoblotting and immunohistochemistry, 92–93
  recruitment and activation, 158–159
  recruitment, FSHR, 35–36
  secondary structure, 92, 93*f*
  siRNA technology, 161–162
Atypical chemokine receptors (ACRs)
  description, 152
  "homeostatic" and "inflammatory", 152–153
  nonconventional chemokine, 152–153
  signaling
    CXCR7 and C5L2, 158–159
    description, 158–159
    DRY motif, 158–159
    D6 signaling pathways, 158–159
    Gαi-protein, HTRF technology, 159–161
    immunoblotting analysis (*see* Erk1/2 phosphorylation)
    siRNA technology (*see* β-Arrestins)
  structure-based classification, 152
  trafficking
    antibody feeding and cytofluorimetric analysis, 153–155
    confocal microscopy analysis (*see* Rabs-D6 colocalization)
    CXCR7, 153
    description, 153
    dominant negative Rab proteins, 157–158

B

Baculovirus, GRKs
  Bac-to-Bac manual, 351
  cDNAs, 350–351
  description, 350–351

Baculovirus, GRKs (*Continued*)
GRK2-H6 expression (*see* GRK2-H6 expression, transfected cells)
Invitrogen's pFastBac Dual vector, 350–351, 351*f*
Sf9 cells, 350–351
VS_1 virus, 352–353
β_2-Adrenergic receptors (β_2-ARs)
cell-based detection, GPCR, 359
cleavable signal sequence, 133
COS-7 cells, 359
description, 132
functional characterization
cotransfection studies, 138
RNA interference-mediated knockdown, 137
TIAR and HuR, 138
3'-UTR, 137–138
human coding regions, 132
identification
TIA-1 and TIAR, 135–137
western blot and electrophoretic mobility shift assays, 135–137
localization, cells
coding, mRNAs, 143–144
cytoskeleton and actin, 145–146
DDT_1-MF2 cells, 144–145
defective trafficking, 141, 142*f*
Drosophila and *Xenopus*, 145–146
FISH analysis, 141
HuR knockdown, 141–142
integral membrane protein, 143
long-lasting synaptic plasticity, 145–146
mRNA translational silencing, 141
mRNP complex formation, 144–145, 144*f*
peripheral cytoplasmic regions, 142–143
RNA-affinity purification, 145
RNA-based localization pathways, 143–144
stabilization, mRNA, 141–142
posttranscriptional regulation, 134
protocol, 359
purification
DDT1-MF2 cells, 134–135
RNA-affinity method, 134–135
SDS-PAGE and silver stain analysis, 134–135, 136*f*
β2-ARs. *See* β_2-Adrenergic receptors (β_2-ARs)
BBS. *See* α-Bungarotoxin binding site (BBS)
BiFC. *See* Bimolecular fluorescence complementation (BiFC)
BiFC fluorescent signal detection
cells and transfection, 267, 274
fluorescence intensity measurements, 275
imaging, GPCR oligomerization, 267
preparation, cell suspensions, 274
Bimolecular fluorescence complementation (BiFC)
advantages and limitations, 262
dimeric complexes, 262
fluorescent approaches, 261
fluorometric detection, GPCR dimerization, 273–275
GPCR–BiFC fusion proteins (*see* GPCR-BiFC fusion proteins)
GPCR interactions
detection, fluorescence microscopy, 266–270
microscopic detection, mBiFC, 270–273
intracellular signaling machinery, 260–261
protein–protein interactions, 262
receptors fusion, fragments, 261, 261*f*
regulation and physiological implications, 260–261
signal and signal-to-noise ratio, 262
Bioluminescence resonance energy transfer (BRET)
advantages and disadvantages, 372–374, 374*t*
assay
protein expression levels, 324
sample preparation, 324
biophysical assays, 244
$BRET_{max}$, 325
by-stander BRET signal, 325
coexpression, mammalian cells, 323
cotransfected HEK293T cells, 244, 245*f*
dark environments, 372–374
defined, DeepBlueC, and GFP^2 variant, 321

description, 244
experiment
assay, 378–379
cells seeding, 377
cells splitting, 378
cells transfection, 377–378
total fluorescence and luminescence readings, 378–379
EYFP emission, 372–374
generation, fusion constructs, 323
HA-A_1R-Rluc and HA-A_1R-GFP^2 plasmids, 245*f*, 246
light detector, wavelength scanner, 372–374
marine organisms, 321
measurement, 247, 324–325
methods, 322
nonlinear regression equation, 325
orexin OX_1 receptor oligomerization, 82–83
OTR–Rluc coexpression, 325–326
overexpression, noninteracting proteins, 244–246
oxytocin receptor homodimerization, 325–326, 326*f*
plasmid construction and transient transfection, 246–247
principles, 372–374, 373*f*
Renilla reniformis, 372–374
RLuc and EYFP, 321, 372–374
RLuc substrate coelenterazine, 372–374
subcellular localization, probes, 321
Biotinylation
CB_1 and OX_1, 84–86
description, 84–86
protection assay, 86–87
protocols, 84–86
SDS-PAGE and western blotting, 84–86
steps, protection assay, 84–86, 85*f*
BRET. *See* Bioluminescence resonance energy transfer (BRET)
α-Bungarotoxin binding site (BBS)
cloning, GPCRs
crystal structure, 114
DNA sequencing, 114
hippocampal neurons, 114, 115*f*
mGluR2 receptor, 113–114
terminal signal sequences, 114
Tm sequence, 114
description, 111
experimental applications
intracellular trafficking, internalized receptors, 121
live cell imaging and receptor internalization, 121–122
photobleaching, 123
receptor insertion, cell surface, 120–121
IgG antibody, 112, 113*f*
physical properties, molecular imaging tags, 111–112, 112*t*
tagging method, 111–112
tag validation
functional neutrality determination, 118–119
$GABA_B$ receptors, 117–118
saturation binding, 119–120
α-Bungarotoxin-linked fluorophores
BBS-tagging technique (*see* α-Bungarotoxin binding site (BBS))
binding affinity of BTX, 111
biotinylation/antibodies, 110–111
calcium phosphate transfection, 116–117
confocal microscopy, 111
description, 110
$GABA_B$ receptors, 110
GABA, BTX and BBS, 110
GIRK, 110
GPCR heteromer internalization
dual labeling, 126–128
$GABA_B$ receptors, 126–128, 127*f*
MTSES, 126–128
$R1a^{BBS-CC}$ and $R2^{BBS-SS}$, 126–128
HEK-293 (*see* Human embryonic kidney-293 (HEK-293))
image analysis (*see* Image analysis, BBS)
nAChRs, 110–111
N-terminal antibodies, 110–111

C

Ca^{2+} assay
fura-2-acetoxymethyl ester (Fura-2-AM), 253
mobilization, 253–254, 254*f*
protocol, 254
purinergic receptor subtypes, 253–254

cAMP assay
cellular signaling transduction, 251–252
protocol, 252–253
A_1R and $P2Y_2R$ agonists and antagonists, 251–252, 252*f*
Cannabinoid (CB_1)
htrFRET measurement, 77–78, 81*f*
"protection" assay, 84–86
cAR1 labeling, Halo-TMR
cAR1 receptor, 52–53
description, 52–53
HaloTag protein, 52–53
SNAP/CLIP-tag, 52–53
TIRFM, 52–53
CB_1. *See* Cannabinoid (CB_1)
Cells/tissues lysis, 233–234
Coat protein complex II (COPII)
α_{2B}-AR, 197, 198*f*
cargo interaction, 197
description, 196–197
GST fusion, 197
peptide-conjugated agarose beads, 197
Coimmunoprecipitation
cotransfected HEK293T cells, 242
description, 242
dimerization, 32
FLAG immunoprecipitates, 32–33
HRP-conjugated FLAG M2 mAb, 32–33
immunoprecipitation procedure, 32
protocol, 243–244
A_1R/Myc-$P2Y_2R$ coexpression, 242, 243*f*
Confocal microscopy, 207–208
COPII. *See* Coat protein complex II (COPII)

D

Data analysis, single-molecule
localization, 61
mobility, single GPCRs
diffusion and diffusion coefficient, 63–64
tracing single molecules, 62–63
Degraded FSH
binding and internalization, 40
internalization, 40–41
radiolabeled FSH, 40
DNA transfection, 359–360
Domain swapping, receptor oligomerization
binding and functional characteristics, 307
description, 306–307
LH and FSH, 306–307
nonfunctional receptor mutants, 306–307
Downward trafficking, FSHR
β-arrestin-1 and -2, 33
cell surface residence, 33
internalization, 36–39
intracellular domains, PM, 33
phosphorylation and β-arrestin recruitment, 34–36
recycling and degradation, 39–41

E

ECD. *See* Extracellular domain (ECD)
Electron-multiplying CCD cameras (EMCCD), 57–58
ELISA. *See* Enzyme-linked immunosorbent assay (ELISA)
ELISA detection, receptor heteromers
chronic treatment, drugs, 229–230
description, 228–229
μ-δ mAb clones, 228–229
pathological techniques, 228–229
problems, 230
regional differences, antibody recognition, 230
Endoplasmic reticulum (ER)
cell plasma membrane, 70–71
golgi trafficking (*see* ER/golgi trafficking)
identification, trafficking motifs, 172–173
marker calreticulin, 176*f*
oligomerization-defective mutants, 70–71
QCS, 5–6
stress responses, 5
translocation of GPCRs, 70
trapping, pharmacological chaperones and synthetic ligands, 87–88
Endoplasmic reticulum (ER) export
motifs
α_{2B}-AR, 191
cargo transport and recognition sites, 190–191
CD8 glycoprotein transport, 191
characteristics, 190–191
COPII, 190

description, 190
diacidic and dihydrophobic motifs, 190
GPCRs (*see* G protein-coupled receptors (GPCRs))
molecular mechanisms, 191
Sec24 isoforms, 190–191
mutant GPCRs
pharmacological chaperones, 209
receptor-specific ligands, 209–210
and surface trafficking, 210
Enhanced yellow fluorescent protein (EYFP)
BRET, 376
and RLuc, 372–374
spectral overlap, 370–372
Enzyme-linked immunosorbent assay (ELISA). *See also* ELISA detection, receptor heteromers
description, 97–99
epitope-tagged receptor, 97–99
wild-type/mutant β-arrestins, 97–99
ER. *See* Endoplasmic reticulum (ER)
ER/golgi trafficking
glycosidic processing
α2C AR trafficking, 180–181, 181*f*
description, 180–181
GPCR, glycolytic analysis, 179–180
Erk1/2 phosphorylation, 162–164
Extracellular domain (ECD)
hinge region, 18–19
NH_2-terminal, 18–19
EYFP. *See* Enhanced yellow fluorescent protein (EYFP)

F

FACS. *See* Flow-activated cell sorting (FACS); Fluorescence activated cell sorting (FACS)
FISH. *See* Fluorescent *in situ* hybridization (FISH)
Flow-activated cell sorting (FACS)
analysis, 183–184
interpretation, 184, 185*f*
receptor expression, 181–183
Flow cytometry, 206
FLP-IN™ T-REX™ expression system
heteromers
description, 73–74
pcDNA3.1, 73–74
pharmacology and trafficking, 74
homomers
description, 72–73
HEK293 modification, 72–73
pFRT/lacZeo and pcDNA6/TR, 72–73
Fluorescence
fluorophore-conjugated antibodies, 100–101
GFP, 101–102
Fluorescence activated cell sorting (FACS), 99
Fluorescence recovery after photobleaching (FRAP)
description, 195–196
prebleach frames, 196
protocol, 195–196
Fluorescence resonance energy transfer (FRET)
acceptor and donor photobleaching, 370–372
acceptor fluorophore, 370–372
"acceptor photobleaching" method, 314–316
advantages, 313
Alexa 568 and Alexa 647 chromofluor pair, 31
assay, 318–319
biochemical techniques, 316
"bleed-through", 314
chromophore basal bleaching, 320
coexpression, in mammalian cells, 318
couples, chromophores, 314, 315*f*
"cross talk", 314
C-tail fusion proteins, 30–31
description, 30–31
donor and acceptor, 370–372
ECFP and EYFP, 370–372
efficiency calculation, 370–372
experiment
advantages, 377
assay, 376–377
calculated equation, 376–377
cells seeding, 376
cells transfection, 376
fluorometer/fluorescence microscope, 370–372

Fluorescence resonance energy transfer (FRET) (*Continued*)
fluorophores, 370–372
FSHR oligomers, 30–31
generation, fusion constructs, 317
GPCR homo-/heterodimerization, 316
HEK-293, 31–32
immunostaining, 318
LSM FRET tool, 32
materials, HEK 293 cell lines, 316–317
methods, evaluation, 370–372, 371*f*
monomeric TM1 variant, TPs, 316
oligomerization, OX_1 receptor, 83–84
principles, evaluation, 370–372, 371*f*
protein–protein interactions, 370–372
"sensitized emission", 314
"sentinel" ROIs, 319–320
setup, 314
signaling processes and cellular reactions, 314
TM1 TP receptors, 320, 320*f*
TR-FRET, 370–372
Fluorescent *in situ* hybridization (FISH)
DDT_1-MF2 cells, 141
HuR knockdown, 141–142
Fluorescent microscopy. *See* Orexin (OX_1)
Fluorometric detection, GPCR dimerization
description, 273–274
fluorescence intensity measurements, 275
microscopic BiFC, 273–274
preparation, cell suspensions, 274
receptor interactions and interpretation, 275
materials and equipment, 274
transfection, 274
Follicle-stimulating hormone receptor (FSHR)
agonist-induced phosphorylation, 22
β-arrestin, 22
BXXBB motif, 20–21
calnexin and calreticulin, 20
canonical Gαs/cAMP/PKA signaling pathway, 19
coding gene, 18–19
COOH-terminus, 21–22
COPII-coated vesicles, 20–21
description, 18–19
downward trafficking (*see* Downward trafficking, FSHR)
ECD, 18–19
ectodomain, 18–19
functional selectivity, 19–20
glycosylation, 21
gonadal function, 18–19
GRK, 22
hFSHR, 20–21
LHCGR and, 18–19
oligomerization (*see* Oligomerization)
outward trafficking (*see* Outward trafficking, FSHR)
PM, 23
posttranslational modification, GPCR, 21–22
Follitropin receptor. *See* Follicle-stimulating hormone receptor (FSHR)
FRAP. *See* Fluorescence recovery after photobleaching (FRAP)
FRET. *See* Fluorescence resonance energy transfer (FRET)
FSHR. *See* Follicle-stimulating hormone receptor (FSHR)

G

GABAB receptors
expression, BBS, 117–118
as heterodimers, 127*f*
Gαi-protein activation
cAMP cell-based assays, 159–161
description, 159–161
Glutathione S-transferase (GST) fusion
COPII vesicles, 197
GST–ICL3 fusion proteins, 198*f*
protein pull-down and coimmunoprecipitation assays, 197
Glycoprotein hormone receptors
bipartite structure, GPHRs and LGRs, 368–369
BRET experiment, 377–379
coimmunoprecipitation, 368–369
description, 368–369
FRET experiment, 376–377
generation and transfection, 376
GPCR stoichiometry, 368–369
HTRF-RET experiment, 379–381
in vitro heterologous systems, 369

resonance energy transfer techniques, 369–375
TSHR, LHR, and FSHR homo-oligomers, 368–369
GnRHR. *See* Gonadotropin releasing hormone receptor (GnRHR)
Gonadotropin releasing hormone receptor (GnRHR)
description, 9–11
indole, 9–11
LiCl, 9–11
TR-FRET signal, 9–11
GPCR-associated protein complexes
description, 330
groups, GAPCs, 330
purification (*see* Peptide affinity chromatography)
tag expression, 330
TAP (*see* Tandem affinity purification (TAP))
GPCR-BiFC fusion proteins
construction, vectors, 265
experimental design parameters, 262–263
expression and function, 265–266
generation planning, 263–265, 264*f*
materials, 263, 264*f*
molecular cloning techniques, 262–263
GPCR dimerization analysis
BRET (*see* Bioluminescence resonance energy transfer (BRET))
coimmunoprecipitation, 312
description, 312
FRET (*see* Fluorescence resonance energy transfer (FRET))
protein–protein interactions, 312–313
RET (*see* Resonance energy transfer (RET))
GPCR heterodimerization, brain. *See* Proximity ligation assay (PLA)
GPCR heteromerization
chimeric adenylyl cyclase ACV/VI, 305–306
coupling selectivity and G protein dilution, 305
description, 304–305
dopamine D2 and D3 receptors, 305
MOR association, 306
nonfunctional receptors, 306
possibilities, 306
protomers, 304–305
GPCR homomerization
dopamine receptor homomers, 304
energy transfer-based techniques, 303
RASSLs, 302–303
scaffolding proteins, 304
signaling efficacy, 303
wild-type and mutant serotonin type 4, 302–303
GPCR interactions
BiFC and fluorescence microscopy
cells and transfection, 267
complemented receptor signal, 267–268
imaging, oligomerization, 267
nonfluorescent VN and VC fragments, 269–270
quantitative image analysis, 268–269
receptors A and B, 269–270
required materials and equipment, 266
microscopic detection, mBiFC (*see* Multicolor BiFC (mBiFC))
GPCR kinases (GRKs)
catalytic core, 348
characteristics, 348–349
crystal structure, GRK6, 348–349
description, 348
expression and purification
baculovirus (*see* Baculovirus, GRKs)
cation exchange chromatography, 354–356
GRK2-H6 purification, 353–356
recombinant GRK expression systems (*see* Recombinant GRK expression systems)
functional assays
agonist/inverse agonist, 360–361
β_2AR, 359
DNA transfection, 359–360
^{329}G-Rho preparation and activation assay, 358–359
peptide C phosphorylation assay, 358
PNGase treatment and immunoblotting, 361–363
Rho* phosphorylation assay, 357–358
GRK2, GRK5 and GRK6, 22
GRK1-GRK4 subfamily, 348

GPCR kinases (GRKs) (*Continued*)
phoshorylation, 22
GPCR oligomerization
biotinylation studies, 84–87
description, 70
ER trapping, pharmacological chaperones and synthetic ligands
β_2-adrenoceptor, 88
CXCR1 chemokine, 87
description, 87
FRET and BRET studies, 87
muscarinic receptor, 88
TMD, 87–88
TMDI–IV mutant receptor, 87–88
treatment, 87–88
FLP-IN™ T-REX™ (*see* FLP-IN™ T-REX™ expression system)
$GABA_B$ receptor, 70–71
homomers and heteromers, 71
OX_1 (*see* Orexin (OX_1))
protein–protein interactions, 70–71
resonance energy transfer techniques
BRET assays (*see* Bioluminescence resonance energy transfer (BRET))
description, 82
FRET (*see* Fluorescence resonance energy transfer (FRET))
rhodopsin, 71
SNAP–CLIP tagging, 75–82
translocation, 70
GPCRs. *See* G-protein coupled receptors (GPCRs)
GPCR trafficking
arrestin knockdown/knockout strategies
agonist-promoted changes, 102
description, 102
siRNAs, 102
arrestin mutants
agonist-promoted β_2AR sequestration, 102–103, 103*f*
description, 102–103
V54D in β-arrestin2, 102–103
FACS analysis (*see* Flow-activated cell sorting (FACS))
measurement
ELISA, 97–99
FACS, 99
fluorescence (*see* Fluorescence)
ligand binding, 100
G-protein-coupled heteromers regulation
development of tolerance, 221
dopamine D1–D2 receptors, 220–221
ELISA (*see* ELISA detection, receptor heteromers)
immunofluorescence (*see* Immunofluorescence, receptor heteromers)
immunoprecipitation and western blotting, 232–235
mAbs (*see* Heteromer-selective mAbs)
opioid receptor subtypes, 220–221
opioids and cannabinoids, 220
physiological response, 220
receptor-trafficking properties, 220–221
subtractive immunization strategy, 221
G-protein coupled purinergic receptors
adenosine and nucleotides, 241
altering ligand specificity, 241
BRET (*see* Bioluminescence resonance energy transfer (BRET))
Ca^{2+} assay, 253–254
cAMP assay, 251–253
coimmunoprecipitation (*see* Coimmunoprecipitation)
diversification, signal transduction, 240
homo- and hetero-oligomerization, 241–242
immunoelectron microscopy (*see* Immunoelectron microscopy)
ligand binding assay, 250–251
membrane trafficking, 241
physiological significance, oligomerization *in vivo*, 240–241
posttranslational processing, 240
protein–protein interactions, 242
G-protein coupled receptors (GPCRs)
COPII-coated vesicles (*see* Coat protein complex II (COPII))
C-terminus and intracellular loops, 191–192
deleted mutant, 192

ER export motifs transportation
CD8-3A, 199
CD8 glycoprotein, 197–199
cell surface expression, 197–199
PCR, 197–199
3R motif of α_{2B}-AR, 197–199, 199*f*
measurement, receptor expression
α_1-AR, β-AR and angiotensin II receptor, 192–193
cell surface expression, 192–193
constitutive receptor internalization, 194
fluorescence proteins, 193
HEK293 cells, 193
ligand binding assay, 192–193
ligand binding, membrane preparations, 193
receptor–ligand binding affinity, 193
traffic motif, 194
oligomerization (*see* GPCR oligomerization)
pharmacoperone drug (*see* Misfolded/mistrafficked mutants)
receptor export analysis
cell surface expression and subcellular distribution, 195
colocalization, 194
COS7 cells, 196
cover slips, 194–195
FRAP, 195–196
HEK293 cells, 194–195
mutated receptor, 194
PBS, 194–195
photobleaching, 196
receptor mutant, 192
trafficking (*see* GPCR trafficking)
Green fluorescence protein (GFP)
β-barrel-shaped protein, 101–102
fluorophore-labeled protein, 101–102
tagged β-arrestins, 96
Green fluorescent protein (GFP), 74–75
^{329}G-Rho preparation and activation assay, 358–359
GRK2-H6 expression, transfected cells
cation exchange chromatography
Q column, 355–356
S columns, 355–356
crystallization, 356
Ni-NTA agarose chromatography, 353–354

H

HEK-293. *See* Human embryonic kidney-293 (HEK-293)
Heteromer-selective mAbs
advantages, 221–222
antigen preparation, 222–223
"heteromer-specific" epitope, 221–222
hybridoma-secreting clones, 224–226
purification, hybridoma supernatant, 226–227
serum preparation, 227
subtractive immunization, 223–224
tolerization, 221–222
trouble-shooting and precautions, 228
Homogenous time-resolved fluorescence (HTRF). *See* Gαi-protein activation
Homogenous time-resolved fluorescence-resonance energy transfer (HTRF-RET)
allophycocyanin XL-665, 374–375
antibodies labeling
description, 379–380
equation, measured values, 379–380
G25SF HR10/30 column, 379–380
molar extinction coefficient, 380
N-hydroxysuccinimide ester acceptor, 380
Sephadex-G25SF HR10/10 column, 379–380
assay
cells seeding, 380
cells transfection, 380
measurement, 380–381
description, 374–375
GPCR heteromers
description, 78–80
FRET donor and acceptor, 78–80
GPCR homomers, 80
labeled antibodies, 374–375
principles, 374–375, 375*f*
wild-type receptor, 374–375

HTRF-RET. *See* Homogenous time-resolved fluorescence-resonance energy transfer (HTRF-RET)
Human embryonic kidney-293 (HEK-293)
α_{2B}-AR–CD8 chimeric proteins, 197–199
α_{2B}-AR plasmids, 193
cell surface expression, 199*f*
Flp-In™ T-Rex™, 75
FSHR detection, see
and hippocampal cell culture
cDNA expression, 115
description, 115
E18 Sprague-Dawley rat embryos, 116
pFRT/lacZeo and pcDNA6/TR, 72–73
Renilla reniformis, 82
RRA (*see* Radioreceptor assay)
wild type/mutated α_{2B}-AR tag, 194–195
HuR, receptor trafficking
DDT_1-MF2 cells, 139
^{125}I-CYP, 139
immunofluorescence staining, 139–141
knockdown, β2-AR, 139, 140*f*
pharmacological and functional properties, 139
Hybridoma-secreting clones, 224–226

I

Image analysis, BBS
confocal images, 123–124
internalization and insertion, membrane fluorescence, 124–125
metamorph, 123–124
protocol, fluorescence quantification, 123–124
Immunoelectron microscopy
anti-HA and anti-Myc immunolocalization, 247, 248*f*
description, 247
experimental steps, 247
HA-A_1R and Myc-P2Y2R localizations, 247, 248*f*
postembedding method, transfected HEK293T cells and brain tissue, 249–250
preembedding method, transfected HEK293T cells, 248–249
receptor pharmacology, 250–254
visualization, A_1R and P2Y_2R, 247, 248*f*
Immunofluorescence, receptor heteromers
CB_1R–AT1R, 230–232
common problems, 232
protocols, 230–232
tissue and subcellular distribution, 230–232
Immunoprecipitation
FSHR
antibody-protein complexes, 28–29
description, 28–29
Igepal/DOC lysis buffer, 28–29
immunoblotting, 29
and western blotting
antigenic epitope masking, 235
applications, 232–233
artifactual receptor aggregation, 235
cells/tissues lysis, 233–234
description, 232–233
harsh solubilization procedures, 235
nonspecific bands, 235
In situ PLA, 283, 285*f*
Internalization and insertion, membrane fluorescence
BBS-tagged $GABA_B$ receptors, 124–125
description, 124–125
receptor insertion estimation, 125–126
recycling and internalization, 124–125
Internalization, FSHR
description, 36
equilibrium conditions, 38
^{125}I-FSH, 38
intracellular signaling pathways, downward trafficing, 36, 37*f*
measurement, 36
nonequilibrium conditions, 38–39
preincubation, 38
surface binding and radiolabeled ligand, 36
Intracellular Ca^{2+} signaling, 213

L

LH. *See* Luteinizing hormone (LH)
Ligand binding assay
cell surface receptor density, 100
[^{3}H]CCPA saturation binding, 250, 251*f*
hydrophilic radioligands, 100

protocol, 250–251
radioligand binding, 100
"total binding", 100
Ligand-dependent internalization, 153–154
Luteinizing hormone (LH), 306–307

M

mBiFC. *See* Multicolor BiFC (mBiFC)
Misfolded/mistrafficked mutants
antagonists, 5
assay automation
automated liquid handling station, 11, 11*f*
automated protocol, 11–12
description, 11
required instruments, 11
data analysis, 12
description, 4
endpoint measures
GnRHR assay, 9–11
high-throughput settings assays, 7–8
V2R assay, 8–9
GnRHR, V2R and rhodopsin, 4–5
hit follow-up experiments
description, 12–13
doxycycline, 12–13
V2R $L^{83}Q$ pharmacoperone, 13, 13*f*
misfolded mutant receptors, 5
pharmacoperone model systems, 5–7
protein misfolding, 4
Morphine tolerance
determination, ELISA, 228–229
μOR–δOR heteromer-selective antibodies, 221
Motif screening, GPCRs
epitope tagging, 173
identification, trafficking motifs, 173–174
total interpretation *vs.* surface immunofluorescent staining, 175, 176*f*
mRNA localization
β2-AR and GPCRs, 133
β2-ARs (*see* β_2-Adrenergic receptors (β_2-ARs))
cells, β2-AR, 141–146
description, 132
Drosophila embryogenesis, 133–134
endocytic pathway, 132–133
GPCR phosphorylation, 132
HuR, receptor trafficking (*see* HuR, receptor trafficking)
RNP, 133–134
SRP, 132–133
subcellular destinations, 132–133
translation and translocation, amino acids, 133
Multicolor BiFC (mBiFC)
construction, fusion proteins, 272
depiction, 270–271, 271*f*
GFP-derived fluorescent proteins, 270–271
image acquisition, 272
materials and equipment, 271
properties, fluorescent proteins and BiFC fragments, 270–271, 270*t*
quantitative analysis and results assessment, 272–273
transfection and drug treatment, 272
Mutant GPCRs
ER export (*see* Endoplasmic reticulum (ER) export, mutant GPCRs)
generation
HA-hPAFR and HA-hBLT2, 205
rhodopsin-type, TMs, 205
trafficking (*see* Trafficking, mutant GPCRs)

N

nAChRs. *See* Nicotinic acetylcholine receptors (nAChRs)
Nicotinic acetylcholine receptors (nAChRs), 110–111, 117
Ni-NTA agarose chromatography
baculovirus-infected cells, 354, 355*f*
cell pellets, 354
NCLB column, 354
Nonconventional chemokines, 152–153

O

Oligomerization
c-myc-tagged and FLAG-tagged hFSHR (*see* Coimmunoprecipitation)
description, 29–30
FRET (*see* Fluorescence resonance energy transfer (FRET))
FSHR-ECD/FSHR-ECD, 29–30
hFSHR monomers, 29–30
ligand-bound ectodomain, 29–30

Orexin (OX_1)
BRET assays, 82–83
"colocalization", 74–75
description, 74–75
FRET imaging, living cells, 83–84
GFP, 74–75
homomers and heteromers, 75
Outward trafficking, FSHR
AFNQT, 23–24
amino terminal folding, 23–24
description, 23–24
expression detection methods
anti-human FSHR antibody, 25–27
description, 25–27
hFSHR visualization, immunoblotting, 25–27, 26*f*
immunoprecipitation (*see* Immunoprecipitation)
SDS-PAGE (*see* Sodium dodecyl sulfate polyacrylamide gel electrophoresis (SDS-PAGE))
western immunoblotting (*see* Western immunoblotting)
HEK-293 (*see* Human embryonic kidney-293 (HEK-293))
hFSHR ectodomain, 23–24
LRR, 23–24
radioligand receptor-binding assays, 24
OX_1. *See* Orexin (OX_1)

P

PAF. *See* Platelet-activating factor (PAF)
PBS. *See* Phosphate buffer saline (PBS)
Peptide affinity chromatography
cell/tissue lysate preparation, 340
GPCR subdomains, 339–340
materials, 340
peptide columns preparation, 341–342
purification, 342–343
Peptide C phosphorylation assay, 358
Peptide N-glycosidase F (PNGase)
densitometry, 363
description, 362
exogenous GRK2 and heterotrimeric G-protein, 363, 364*f*
okadaic acid, RSB, 363, 363*f*
PGE2. *See* Prostaglandin E2 (PGE2)
Pharmacological chaperones
ER accumulation, 209
mutant GPCRs, 210
pharmacoperones, 204
Pharmacoperone drugs. *See* Misfolded/mistrafficked mutants
Pharmacoperone model systems
disease-causing proteins, 5–6
false positives identification, 6
GPCRs, 5–6
HeLa cells, 6
human gonadotropin, 7
QCS and ER, 5–6
stable cell lines, 7
Phosphate buffer saline (PBS)
ER marker calregulin/calnexin, 194–195
incubation, 194
Phosphorylation, FSHR, 34–35
PLA. *See* Proximity ligation assay (PLA)
Platelet-activating factor (PAF), 204–205
PNGase. *See* Peptide N-glycosidase F (PNGase)
Prostaglandin E2 (PGE2)
aHA-Qdot conjugate preparation, 54–55
cAMP levels, 53–54
cyclooxygenase-catalyzed metabolism, 53–54
description, 53–54
HA Epitope Tag, 53–54
HEK293, HA-EP2, 54
labeling, HA-EP2-expression, 55
receptor tracking and signaling regulation, 53–54
Proximity ligation assay (PLA)
β adrenergic receptors, 282
advantages and disadvantages, 289–290
agonist/antagonist regulation, 291–292
brain slices *ex vivo*, 282
brain tissue preparation
blocking agents, 286–287
description, 283–284
fixed/unfixed cryostat sections, 284–286
immunohistochemistry, 283–284
preservation, tissue structure, 284–286
cellular processes and signaling, 282

detection, GPCR heteromers, 282–283, 284*f*
GPCR–GPCR and GPCR–RTK heteromers, 291, 291*f*
high selectivity and sensitivity assay, 283
in situ PLA, 283
proteins and protein–protein interactions, 291
proximity probes (*see* Proximity probes)
quantitative PLA image analysis, 285*f*, 289
RCPs, 282–283
reactions, reagents and solutions, 287–289
receptor–receptor interactions, 282
Proximity probes
conjugation, oligonucleotides, 287
description, 287

Q

QCS. *See* Quality control system (QCS)
Quality control system (QCS)
disease-causing proteins, 5–6
ER, 5–6
misfolded mutants, 5
Quantitative image analysis
fluorescent intensities, 268–269
ratiometric analysis, mBiFC measurements, 269
software, 268–269
transfection efficiency, 268
Quantitative PLA image analysis, 285*f*, 289

R

Rabs-D6 colocalization
D6 constitutive internalization and recycling pathways, 155–157
description, 155–157
"rapid" and "slow" pathway, 155–157
Radioligand binding
description, 296
equilibrium and kinetic binding experiments
receptor heteromerization, 299–302
receptor homomerization, 296–299
saturation and competition studies, 296
Radioreceptor assay (RRA)
description, 24
^{125}I-FSH, 25
NSB calculation, 25
RASSLs. *See* Receptors activated solely by synthetic ligands (RASSLs)
RCP. *See* Rolling circle product (RCP)
Receptor functionality analysis, GPCRs
α2C AR mutations, MAP kinase, 179, 179*f*
ligand binding, 175–177
saturation ligand binding assays, 177
signaling, 177–178
Receptor heterocomplex, 282, 291
Receptor heteromerization
coexpression, kappa and delta opioid, 299–300
cooperative interactions, 301–302
curve profiles, 301
DOR-KOR and DOR-MOR, 301
eteromeric assembly, 299–300
GPCR oligomerization, 301
Receptor homomerization
adenosine A_1 receptors, 298–299
allosteric and orthosteric sites, 299, 300*f*
antagonist/inverse agonist (+)-butaclamol, 298–299
B_{max}, 297–298
cooperative interactions, 298
"exponential decay" curve, 298
[^{3}H]spiperone and [^{3}H]raclopride binding, 297, 297*f*
muscarinic receptors solubilization, 296–297
off-target and in-target allosterisms, 299
presence/absence, ions, 297–298
receptor occupancy, 299
Receptor oligomerization
description, 296
domain swapping (*see* Domain swapping, receptor oligomerization)
radioligand binding (*see* Radioligand binding)
signal transduction (*see* Signal transduction)
Receptor pharmacology, 250–254
Receptors activated solely by synthetic ligands (RASSLs), 302–303, 304
Receptor trafficking. *See* GPCR oligomerization
Recombinant GRK expression systems
Autographa californica, 349

Recombinant GRK expression systems (*Continued*)
GRK1 and GRK4, 350
GRK1 and GRK7, 350
GRK5 and GRK6, 349
GRK7, GRK7, 350
insect cells, 350
Recycled FSHR
cell-associated radioactivity, 39–40
description, 39
^{125}I-hFSH, 39–40
transfected cells, 39–40
Resonance energy transfer (RET)
donor and acceptor molecules, 313
emission and excitation spectrums, 313
FRET and BRET, 313
Resonance energy transfer techniques
BRET technique, 372–374
description, 369–370
FRET technique (*see* Fluorescence resonance energy transfer (FRET))
HTRF-RET technique, 374–375
RET. *See* Resonance energy transfer (RET)
Rho* phosphorylation assay, 357–358
Ribonucleoprotein (RNP), 133–134
RNP. *See* Ribonucleoprotein (RNP)
Rolling circle product (RCP), 282–283
RRA. *See* Radioreceptor assay (RRA)

S

SDS-PAGE. *See* Sodium dodecyl sulfate polyacrylamide gel electrophoresis (SDS-PAGE)
Serum preparation, 227
Signal recognition particle (SRP)
mRNA and recognition, 143
signal sequence and recognition, 132–133
Signal transduction
GPCR heteromerization, 304–306
GPCR homomerization, 302–304
qualitative and quantitative changes, 302
Single-molecule imaging technique
data analysis (*see* Data analysis, single-molecule)
description, 48, 56–57
Dictyostelium cAMP, 48
eYFP-labeled receptor, 56–57
image acquisition, 58–60
labeling, GPCRs (*see* Single-particle tracking (SPT))
signaling efficiency, 48
SPT setup (*see* Single-particle tracking (SPT))
TIRF microscopy, 56–57
wide-field technique, 56–57
Single-particle tracking (SPT)
cAR1 labeling, eYFP
culture and transformation, *Dictyostelium* cells, 50–51
description, 49–50
Dictyostelium cells and sample preparation, 51
Dictyostelium discoideum, 49–50
TIRF (*see* Ternal reflection fluorescence (TIRF))
cAR1 labeling, Halo-TMR (*see* cAR1 labeling, Halo-TMR)
description, 57–58
EMCCD, 57–58
EP2 receptor labeling, quantum dots (*see* Prostaglandin E2 (PGE2))
glass cleaning
description, 49
protocol, 49
labeling techniques, GPCRs, 48–49, 49*f*
refractive index differences, 57–58
types, GPCR labeling, 48–49
wide-field setup, TIRFM, 57–58, 58*t*
size exclusion chromatography, 356
Small G protein activation assay
description, 164–166
GST-PBD, 164–166
human, murine/rat Rac1-GTP, 164–166
SNAP–CLIP tagging
autofluorescent proteins
construction and labeling, 76–77, 76*f*
description, 76–77
htrFRET, 77
"suicide" enzymes, 76–77
Tag-lite® technology, 77
description, 75
htrFRET, 78–80
internalization, coexpressed GPCRs
CB_1 and OX_1, 80–82

N-terminally tagged GPCRs
measurement, 80–82, 81*f*
Tag-lite® htrFRET donor, 80–82
microscopy
description, 77–78
fluorescent species, 77–78
Sodium dodecyl sulfate polyacrylamide gel electrophoresis (SDS-PAGE)
gel solutions preparations, 27
protein extracts, 27
Sortase-A-mediated labeling, 211–213
SPT. *See* Single-particle tracking (SPT)
SRP. *See* Signal recognition particle (SRP)
Subtractive immunization, 223–224
Surface-trafficked mutant GPCRs
agonist-elicited intracellular Ca^{2+} response, 210–211
hPAFR/P247A accumulation, 213–214
intracellular Ca^{2+} signaling, 213
pharmacological chaperones, 210
recycling deficiency, hPAFR/P247A, 214
sortase-A-mediated labeling, 211–213
Systematic and quantitative analysis, GPCR
α2A and α2C ARs, 173
biochemical analysis, ER/golgi trafficking (*see* ER/golgi trafficking)
description, 172
ER/golgi export/retention, 172
GPCR ER export motif, 172–173
GPCR trafficking (*see* GPCR trafficking)
identification, ER trafficking motifs, 172–173
motif screening, immunofluorescent staining (*see* Motif screening, GPCRs)
receptor functionality analysis (*see* Receptor functionality analysis, GPCRs)

T

Tag-lite® technology, 77
Tandem affinity purification (TAP)
cells amplification, 334
constructs, stable cell line and cell amplification, 339
crude membranes and receptor solubilization, 334
description, 331
generation, TAP-tagged constructs, 333
IgG BD/CBP-tagged GPCRs and associated protein complexes, 334–335
methods, 331, 332*f*
MT_1 and MT_2 melatonin receptors, 331
reagents, 337–338
required materials, 331–333
sample preparation, MS analysis, 335–337
SBP/CBP-tagged GPCRs and associated protein complexes, 339
stable cell lines expression, 333–334
surface expression, SBP/CBP-tagged melatonin MT1 receptor, 337, 338*f*
tag combinations, 337
TAP. *See* Tandem affinity purification (TAP)
T-cell-restricted intracellular antigen-1 (TIA-1), 135–137
Ternal reflection fluorescence (TIRF)
basal plasma membrane, 51–52
description, 51–52
measurement, single molecules, 51–52
TIA-1. *See* T-cell-restricted intracellular antigen-1 (TIA-1)
Time-resolved fluorescence energy transfer (TR-FRET)
high-throughput applications, GPCRs, 9–11
IP-One™, 9–11
TIRF. *See* Ternal reflection fluorescence (TIRF)
TMD. *See* Transmembrane domain (TMD)
Trafficking, mutant GPCRs
description, 205
ER retention
glyco-modification analysis, 206–207
localization, confocal microscopy, 207–208
surface expression analysis, flow cytometry, 206
requirements, conserved residues and helix 8, 208–209
transfected cells, 351–352

Transient transfection, 157–158
Transmembrane domain (TMD), 87–88
TR-FRET. *See* Time-resolved fluorescence energy transfer (TR-FRET)

V

Vasopressin type 2 receptor (V2R)
  assay endpoint measure
    cAMP-Glo™ reagent, 8–9, 9*f*
    cAMP levels, 8–9
    description, 8–9
    SR121463, 8–9
  SR121463B, 13
  validated drug targets, 5–6
V54D in β-arrestin2. *See* V53D point mutant in β-arrestin1 (V54D in β-arrestin2)
V53D point mutant in β-arrestin1 (V54D in β-arrestin2), 102–103
V2R. *See* Vasopressin type 2 receptor (V2R)
VS_1 virus, 352–353

W

Western immunoblotting, 28

Y

YFP-labeled intracellular markers, 267–268
YS/3R mutants, 265

Z

Zeiss LSM 710 confocal laser scanning microscope, 317, 318

David C. Smithson *et al.*, Figure 1.2 A view of the automated liquid handling station used to perform the assays described in this report.

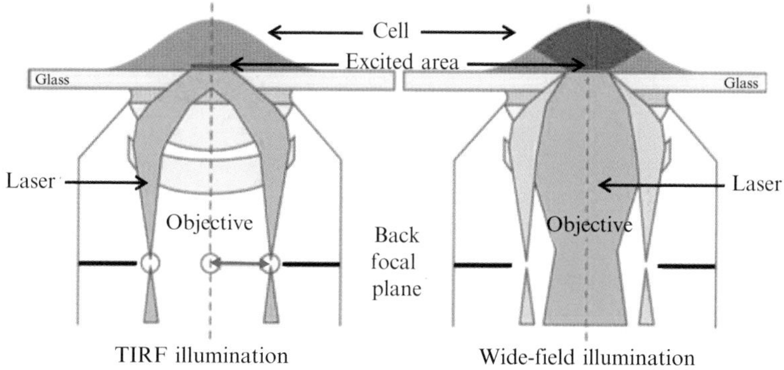

B.E. Snaar-Jagalska *et al.*, Figure 3.2 Switching from wide-field imaging to TIRF. The fluorophores in the cell are excited by laser-mediated excitation. In wide field, where the light comes from the center of the objective, most of the cells are excited. In TIRF, the position of the beam focus at the back focal plane of the objective is shifted from the center (wide field) to the edge of the objective changing the incident angle of the light beam. The evanescent wave (a very thin electromagnetic field) generated by the reflected light at the glass–cell surface selectively excites fluorophores in a restricted region of less than a 100 nm vertical distance to the cover slip–specimen interface.

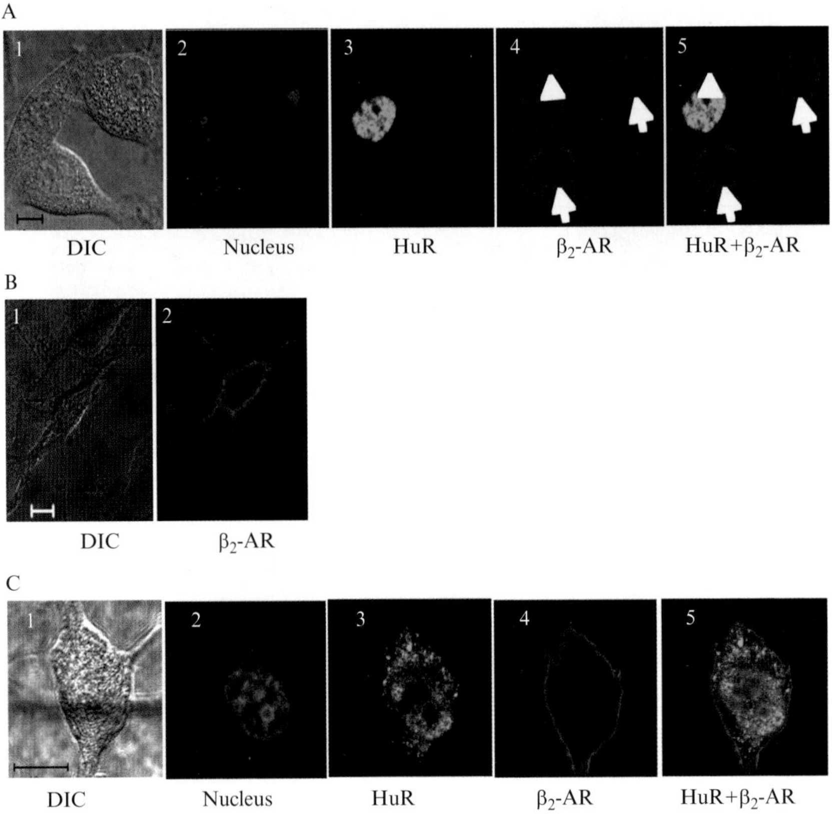

Kusumam Joseph *et al.*, Figure 7.2 Knockdown of the β_2-AR mRNA-binding protein HuR resulted in defective trafficking of receptors to the plasma membrane in DDT_1-MF2 cells. (A) Confocal microscopy images of DDT_1-MF2 cells show immunofluorescence staining of HuR (green) and β_2-AR (red) in HuR knockdown and control cells. The presence of cells with (arrowhead) and without HuR (arrow) was chosen within the same microscopic field to provide an internal control. The accumulation of receptors around the nucleus is seen only in HuR knockdown cells. (B) β_2-AR when overexpressed can traffic to the plasma membrane. Confocal microscopy images show immunofluorescence staining of β_2-AR (red) in DDT_1-MF2 cells transfected with full-length β_2-AR cDNA. (C) Magnified view of a single cell overexpressing β_2-AR cDNA. Confocal microscopy was performed using a Zeiss LSM510META laser-scanning microscope (Carl Zeiss, Inc.). *From Tholanikunnel et al. (2010).*

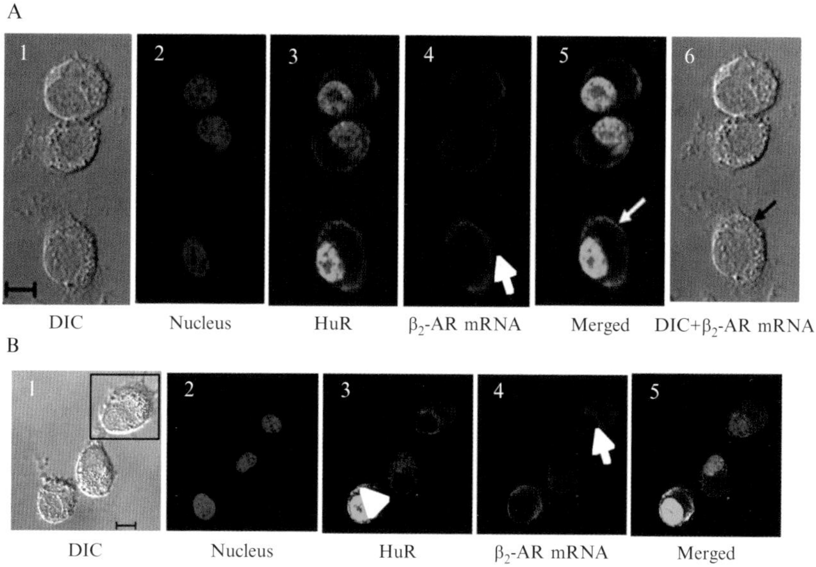

Kusumam Joseph *et al.*, Figure 7.3 β_2-AR mRNA and cytoplasmic HuR are colocalized to the cell periphery and knockdown of HuR results in defective trafficking of receptor mRNA. (A) Confocal images of DDT_1-MF2 cells with fluorescent *in situ* hybridization analysis using digoxigenin-labeled riboprobes directed against β_2-AR mRNA and immunofluorescence staining of HuR protein. Prehybridization and hybridization was performed as described in Tholanikunnel et al. (2010). The RNA-probe signal was amplified using biotinylated antidigoxin followed by streptavidin coupled to Cy3. (B) Knockdown of HuR resulted in decreased levels of β_2-AR mRNA that failed to traffic to plasma membrane and appeared around the nucleus. HuR knockdown (inset) and control cells are shown within the same microscopic field to provide an internal control. *From Tholanikunnel et al. (2010).*

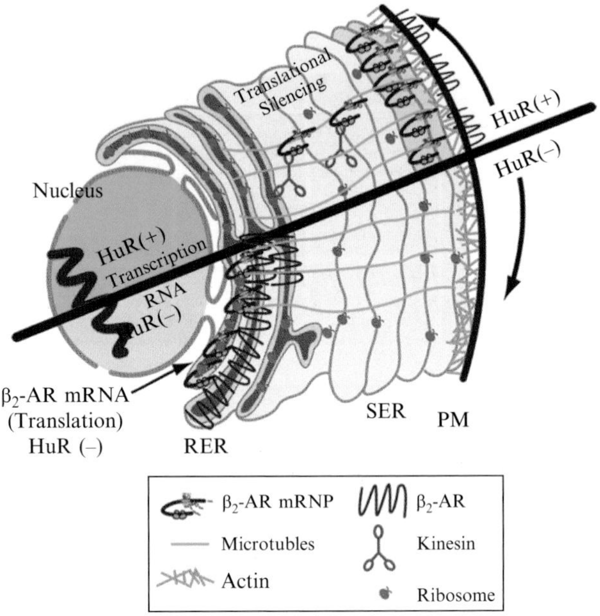

Kusumam Joseph *et al.*, Figure 7.4 Proposed model for β_2-AR mRNP complex formation, its transport and localization to the cell periphery. β_2-AR mRNA is recognized in the nucleus by the nucleocytoplasmic shuttling RNA-binding protein HuR. Upon export, the β_2-AR mRNA–HuR protein complex associates with additional RNA-binding proteins such as Staufen and cytoskeletal elements (actin, tubulin, and kinesin) and is transported to the cell periphery. Continued association of HuR protein silences translational initiation while chaperoning the mRNP complex to the plasma membrane (upper half). The mechanisms involved in β_2-AR mRNA dissociation from TIAR, HuR, and other RNA-binding proteins and translational activation are topics of current interest. When HuR expression is downregulated (lower half) β_2-AR mRNA translation is initiated in perinuclear ER leading to overproduction of receptors but defective trafficking to the plasma membrane. *From Tholanikunnel et al. (2010).*

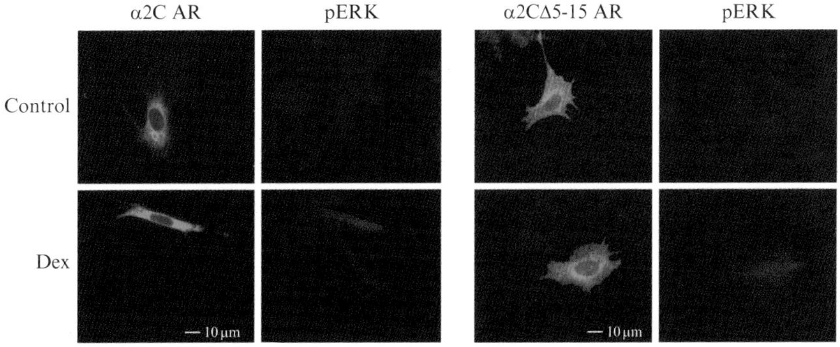

Carl M. Hurt *et al.*, Figure 9.2 Functional screening of α2C AR mutations by MAP kinase. An immunofluorescent assay for MAP kinase (pERK) in response to agonist (Dex = dexmedetomidine) allows for screening of GPCR chimeras and mutations to ensure pharmacological function prior to further trafficking analysis. Permeabilized cells were stained with monoclonal antibody 16B12 to examine total cellular HA-α2 AR expression. Note minimal pERK staining in the absence of agonist. Both α2CΔ5-15 and α2CA7D ARs can activate MAP kinase and bind RX-821002 (data not shown) suggesting that they are not misfolded and are functional. *Adapted from Angelotti et al. (2010).*

A

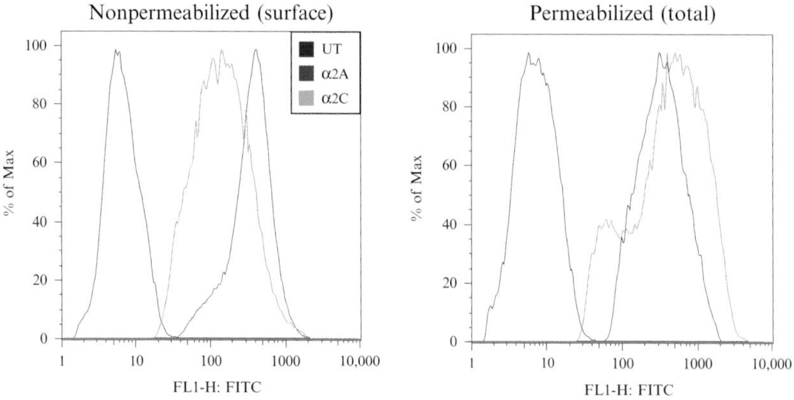

B

	Average surface median fluorescent intensity	Average total median fluorescent intensity	Average % surface expression	Average % intracellular expression
α2A AR	347.3 ± 50.2	376.3 ± 52.2	92 ± 3.9	8 ± 3.9
α2C AR	126 ± 16.6	495.9 ± 59.1	25 ± 1	75 ± 1
α2CΔ5-15 AR	305.8 ± 50	582.8 ± 115.8	52 ± 7.3	48 ± 7.3
α2CA7D AR	271 ± 60.4	558 ± 131.8	49 ± 5.8	51 ± 5.8

Fluorescent intensity represented as arbitrary units. Values given are the average ± SEM.

Carl M. Hurt *et al.*, Figure 9.4 Quantitative FACS analysis of plasma membrane trafficking. (A) Relative plasma membrane expression can be easily quantified by measuring GPCR levels in permeabilized (total) and nonpermeabilized (surface) cells using a one-step fluorescent labeling procedure (UT = untransfected). Following proper FACS gating to remove dead cells, measurement of single cell fluorescence allows for the determination of median fluorescent intensity for each GPCR under both conditions. Notice the shift in α2C AR median fluorescence intensity following permeabilization, reflecting the larger intracellular pool of receptor B. Calculation of relative surface and intracellular expression levels can be performed using median fluorescent measurements, demonstrating that both α2CΔ5-15 and α2CA7D ARs enhance plasma membrane expression. *Adapted from Angelotti et al. (2010).*

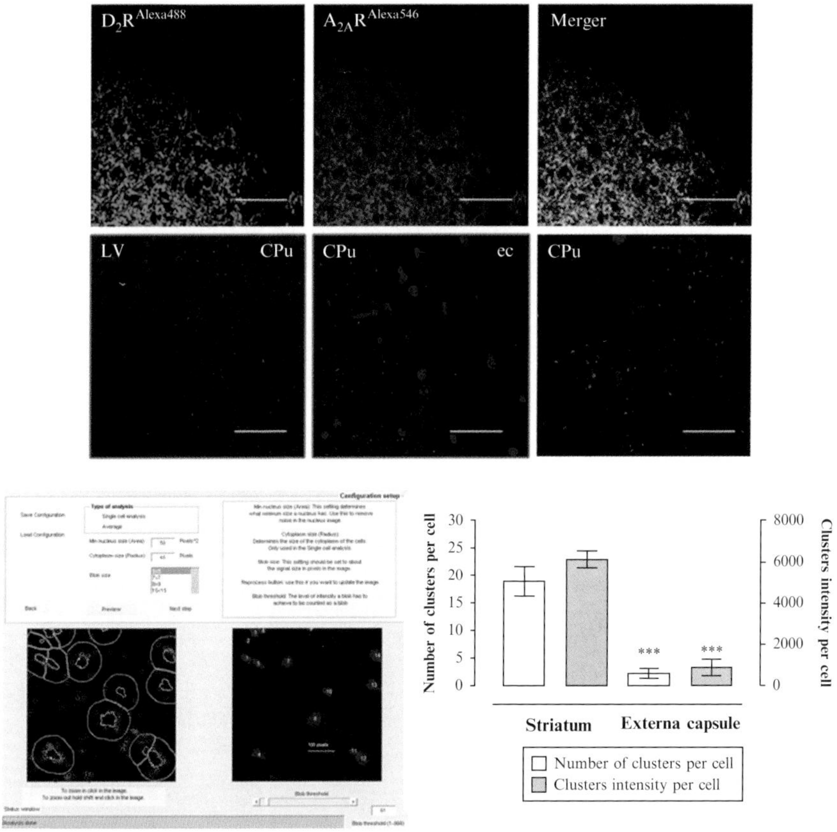

Dasiel O. Borroto-Escuela *et al.*, Figure 15.2 Upper-upper panel: Specific D_2R (green) and $A_{2A}R$ (red) immunoreactivities and colocalization (yellow) in striatal sections. D_2R immunoreactivity was high in the striatum of rat surrounded by the external capsule (ec) using fluorescence immunohistochemistry (left). $A_{2A}R$ immunoreactivity (middle) showed a high level of colocalization with D_2R (right) in the striatum of wild-type rat. Scale bars, 75 μm. Upper-down panel: PLA-positive $A_{2A}R$–D_2R heteromers in striatal sections adjacent to the sections with immunoreactivity $A_{2A}R$–D_2R heteromers were visualized as red clusters (blobs, dots) within the striatum which were almost absent within the lateral ventricle (LV, left) and the external capsule (ec, middle panel). Higher magnification image revealed a large number of PLA-positive red clusters within the caudate putamen (CPu, right). Each cluster represents a high concentration of fluorescence from the single-molecule amplification resulting from several hundred-fold replication of the DNA circle formed as a result of the probe proximity; the cluster/dot number can be quantified independently of the intensity. Nuclei are shown in blue (DAPI). Scale bars, 50 μm for left and middle panels; 20 μm for the right panels. Lower-left panel: Screendump from the corresponding BlobFinder analysis. The left pictures show how the software has identified the PLA signals, the nuclear limit, and the approximate limit of the cytoplasm based on a user-defined radius. The right picture shows the raw image based on 14 *Z*-planes with the nuclei enumerated. Lower-right panel: PLA-positive red clusters in striatum were quantified per cell using BlobFinder and the results are presented. Quantification of $A_{2A}R$–D_2R heteromers demonstrates highly significant differences in PLA clusters per cell between caudate putamen and external capsule (***$P < 0.001$ by Student's *t*-test).

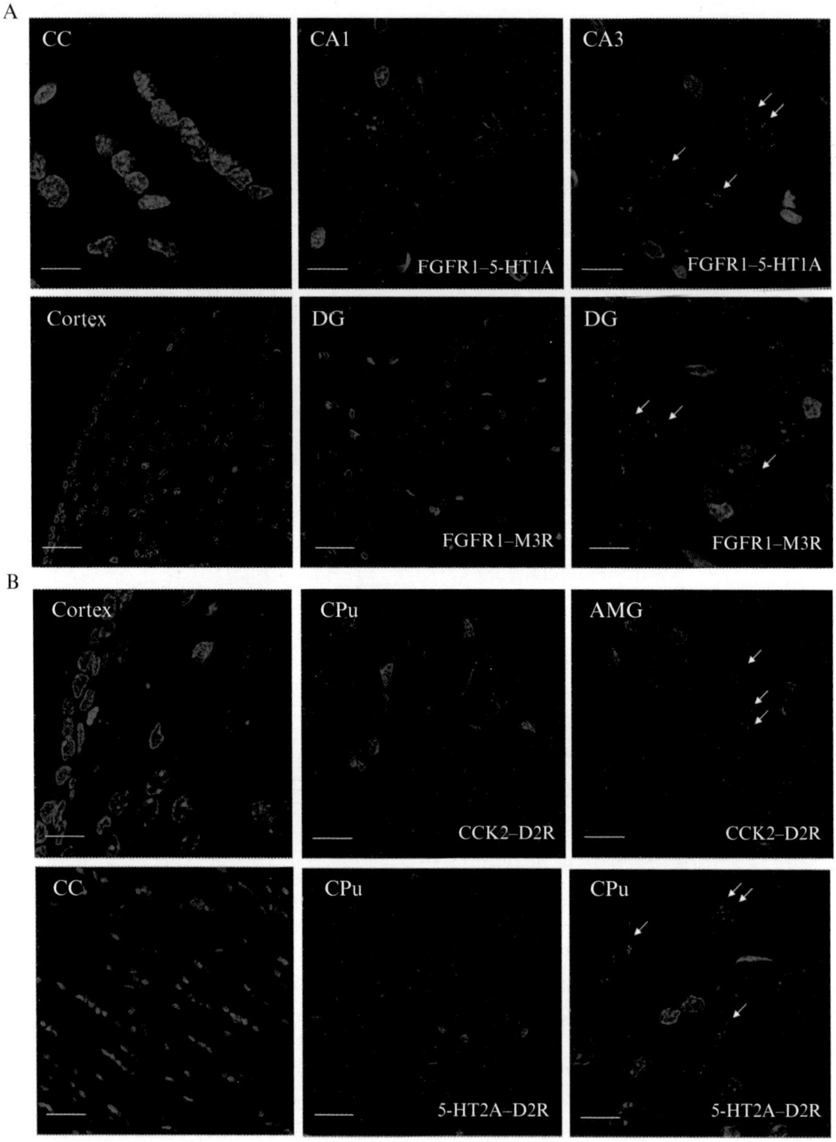

Dasiel O. Borroto-Escuela *et al.*, Figure 15.3 Detection of different GPCR–GPCR heteromers and GPCR–RTK heterocomplexes in dorsal rat hippocampal and striatal sections by *in situ* PLA. (A-upper panel) Constitutive FGFR1–5-HT1A heteroreceptor complexes are detected by *in situ* PLA (red clusters) in dorsal rat hippocampus (Ammon's horn 1 and 3 (CA1, CA3) but not, as an example, in the corpus callosum (cc). Scale bars, 20 μm. (A-lower panel) Constitutive FGFR1–M3R heteroreceptor complexes are detected by *in situ* PLA (red clusters) in dorsal rat hippocampus (granular layer of the dentate gyrus [DG]) but not in the cortex cerebri. Scale bars from the left to the right, 50, 50, and 20 μm. (B-upper panel) Constitutive CCK2–D_2R heteromers are detected by *in situ* PLA (red clusters) in striatal sections (caudate putamen: CPu; amygdaloid cortex: AMG) but not, for example, in the cortex cerebri. Scale bars, 20 μm. (B-upper panel) Constitutive 5-HT2A–D_2R heteromers are detected by *in situ* PLA (red clusters) in striatal sections (caudate putamen: CPu; amygdaloid cortex: AMG) but not, for example, in the corpus callosum (cc). Scale bars from the left to the right, 75, 50, and 20 μm. Nuclei appear as a blue color in all panels and the white arrows indicate the red cluster formation (PLA signal).

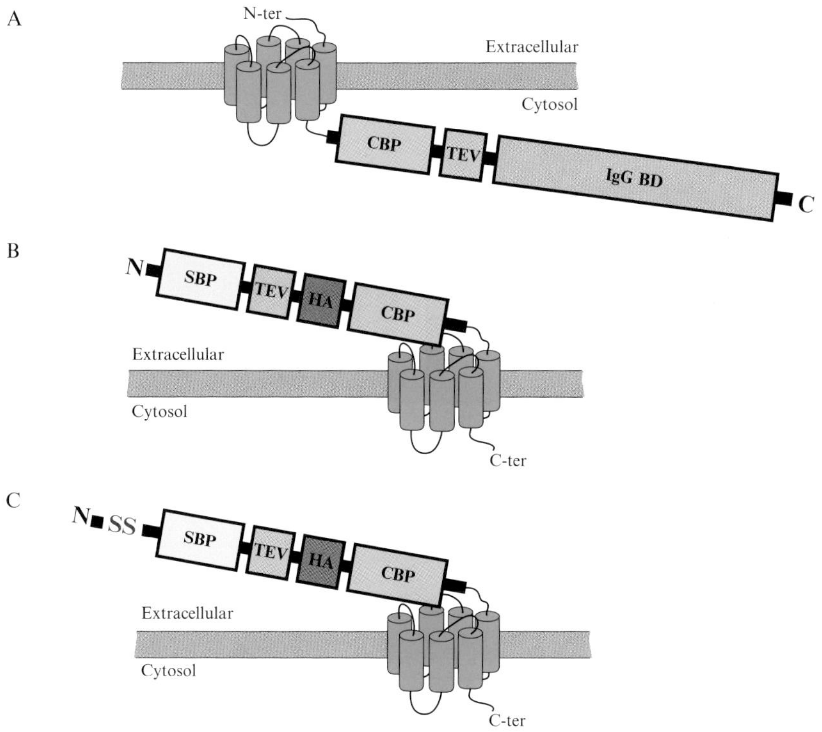

CBP tag = KRRWKKNFIAVSAANRFKKI (≈ 3.2 kDa)

SBP tag = AGELEQLRARLEHHPQGQRE (≈ 2.5 kDa)

IgG BD tag = KTAALAQHDEAVDNKFNKEQQNAFYEILHLPNLNEEQRNAFIQSLKDDPSQSANLLAE AKKLNDAQAPKVDNKFNKEQQNAFYEILHLPNLNEEQRNAFIQSLKDDPSQSANLLAEAKKLNGAQ APKVDANSAGKST (≈ 15.3 kDa)

SS = MRCSPGGVWLGLAASLLHVSLHGEGT

TEV site = ENLYFQG

HA epitope = YPYDVPDYA

Avais Daulat *et al.*, Figure 18.1 Different TAP methods. Schematic representation of TAP tags used to purify GPCR-associated protein complexes. (A) IgG BD/CBP tag fused to the receptor C-tail. (B) SBP/CBP tag fused to the N-ter of the receptor. (C) Same as (B) but with the addition of an SS sequence to facilitate cell surface expression. CBP, calmodulin binding domain; HA, hemagglutinin tag; IgG BD, IgG binding domain; SBP, streptavidin binding domain; SS, signal peptide sequence; TEV, tobacco etch virus protease cleavage site.

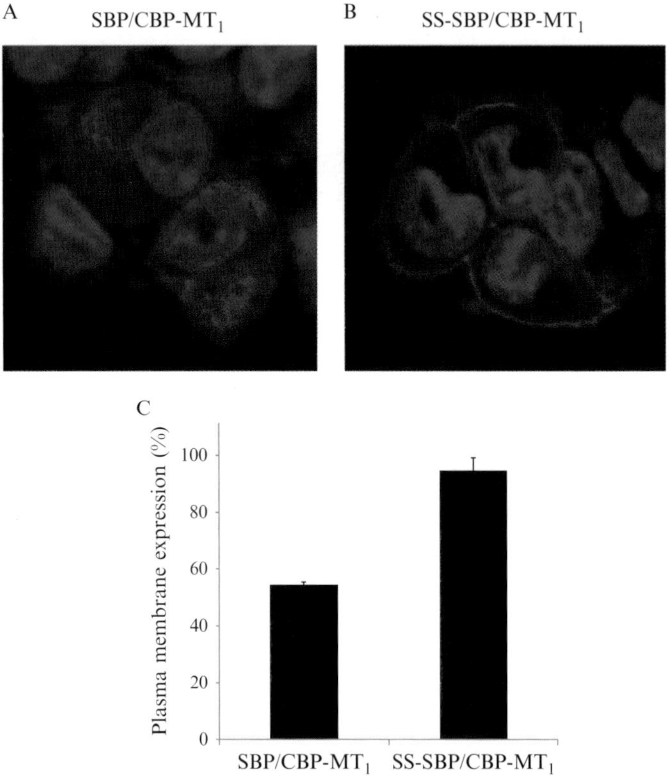

Avais Daulat *et al.*, Figure 18.2 Surface expression of SBP/CBP-tagged melatonin MT_1 receptor. Expression profile of SBP/CBP-tagged MT_1 receptor without (A) or with (B) a signal peptide sequence (SS) determined by immunofluorescence in permeabilized HEK293 cells. Receptors are labeled with anti-HA antibodies and cell nuclei with DAPI. Cell surface expression of these constructs is quantified by Western in-cell non-permeabilized cells using anti-HA antibodies. Total expression is determined in permeabilized cells and the % of surface expression calculated.